Discrete Optimization

This is a volume in
COMPUTER SCIENCE AND SCIENTIFIC COMPUTING

(formerly "Computer Science and Applied Mathematics")
Werner Rheinboldt and Daniel Siewiorek, editors

Discrete Optimization

R. Gary Parker

School of Industrial and Systems Engineering
Georgia Institute of Technology
Atlanta, Georgia

Ronald L. Rardin

School of Industrial Engineering
Purdue University
West Lafayette, Indiana

ACADEMIC PRESS, INC.
Harcourt Brace Jovanovich, Publishers

Boston Orlando San Diego
New York Austin London Sydney
Tokyo Toronto

ACADEMIC PRESS, INC.
1250 Sixth Avenue, San Diego, CA 92101

United Kingdom Edition published by
ACADEMIC PRESS, INC. (LONDON) LTD.
24-28 Oval Road, London NW1 7DX

Library of Congress Cataloging-in-Publication Data

Parker, R. Gary.
Discrete optimization.

(Computer science and scientific computing)
Includes index.
1. Mathematical optimization. I. Rardin, Ronald L.
II. Title. III. Series.
QA402.5.P39 1988 511′.6 87-1076
ISBN 0-12-545075-3 (alk. paper)

Printed in the United States of America
88 89 90 91 9 8 7 6 5 4 3 2 1

For their enduring patience and support, we dedicate this book to our wives Anne and Blanca and our children, Robbie, Robert, David, and Christopher.

Contents

Preface

From the outset, our aim in pursuing this project has been to produce a book that would span, in a single volume, the host of fundamental issues and algorithmic strategies that have emerged as the general-purpose core in the discipline of discrete optimization. With roots in mathematical programming, computer science, and combinatorial mathematics, this field now attracts students, researchers, and practitioners alike having a variety of backgrounds and interests. While this diversity and its concomitant effect on fundamental results makes the task of unification arduous, we firmly believe that it also has been a source of substantial richness and has contributed directly to the ultimate maturation of the discipline.

Following a brief introduction, we begin in Chapter 2 with complexity theory. Here, we provide an overview of results that have proven central to the development of the intellectual framework of discrete optimization. Computations and supporting theory of generic approaches relevant to various discrete model contexts follow. For problems falling into the polynomial-time solvable category, we treat both matroid (Chapter 3) and linear programming-based (Chapter 4) procedures. In Chapters 5 and 6, we take up approaches that are inherently exponential in character, including in-depth coverage of enumerative (branch and bound) as well as cutting/polyhedral methods. We conclude (Chapter 7) with a generic treatment of common nonexact approaches.

We also would like to call attention to the inclusion of three appendices covering basic results from the theory of convex sets and polytopes, graphs, and linear programming. Our principal purpose in adding this material is to alleviate some potential frustration felt by those having uneven backgrounds in the aforementioned areas. Still others may find the coverage useful as a refresher.

Throughout this undertaking, we have tried to remain steadfast in our goal of making accessible to students, instructors, and researchers the substantial array of elegant results that have emerged in this field in the past quarter century. Algorithms are presented in computational format and demonstrated throughout with examples. Still, we have made every effort to maintain what we believe to be a suitable and consistent treatment of underlying theory. We have compiled lists of exercises for each chapter that span the range from routine applications of algorithms to new research results, with the more challenging problems appropriately marked with a star. In short, we have attempted to produce a book possessing both pedagogical value as a graduate text and reference value in discrete optimization research.

The material in this volume has been used at Georgia Tech and Purdue in both Masters- and PhD-level courses carrying such titles as "Integer Programming," "Combinatorial Optimization," and "Discrete Deterministic Models in Operations Research." Accordingly, we have found the modular organizational structure within chapters to be nicely suited for tailoring courses to fit varying student backgrounds and topical objectives. It also facilitates the capability of "entering" the book at specific sections of interest.

As a final comment regarding style we should mention the convention we have adopted pertaining to references. Naturally, we have attempted to be accurate, providing what we believe to be the principal citing(s) relative to given concepts, algorithms, theorems, and the like. On the other hand, we have not sought to be comprehensive in the sense of a literature review, choosing, instead, to compile a fairly extensive bibliography. Even so, there will no doubt be some references we have overlooked. Our only defense in this regard is an apology to those authors of works omitted and a hope that what we have included sufficiently supports the coverage comprising this book.

A large group of individuals have contributed both directly and indirectly to this probject. Some have used various parts of the book (or at least some manuscript version) in their courses at other universities. Some have read the manuscript and offered valuable comments on important matters regarding style, content, and clarity. Still others graciously provided us the benefit of their substantial expertise in a host of technical issues that necessarily arise in a book of this scope. A very partial list of names would include J. Kennington, M. Karwan, R. Bulfin, V. Chandru, C. Tovey, J. Vande Vate, J. Jarvis, M. Bazaraa, and R. Jeroslow.

R. Gary Parker
Ronald L. Rardin

1

Introduction to Discrete Optimization

Anyone who has ever made responsible decisions of any kind has certainly encountered discrete decisions; decisions among a finite set of mutually exclusive alternatives. Stark choices between building and not building, turning left versus turning right, or visiting city A instead of city B, simply cannot be escaped. Of course, discreteness of the decision space offers the advantage of concreteness and indeed, elementary graphs or similar illustrations can often naturally and intuitively represent the meaning of a particular choice. Discreteness however, also brings forth a heavy burden of dimensionality. If more than a few choices are to be made, the decision-maker confronts an incomprehensible expanse of cases, combinations and possibilities requiring his or her evaluation.

Since the 18th century, this intriguing paradox of problems, possessing intuitive simplicity of presentation coupled with mind-boggling complexity of solution, has combined with the reality that discrete decisions abound in all areas of management and engineering to attract researchers to the study of discrete problems. Interest has multiplied with the revolutionary development of computing machines during the second half of the 20th century. For some problems, elegant solution procedures have been discovered. For most, though, a host of properties and algorithms have been developed, and considerable progress has attended, but no really complete resolution has yet appeared. In a few cases, problems with apparently simple form have yielded to almost no advances at all.

But why is this the case? Why, in fact, does a class of problems that in many cases appear to pose modest demands often create severe difficulties? Moreover, given the present state of affairs, what are the prospects for resolution and, in the interim, what are we to do regarding solutions to discrete models of practical, real-world problems? The response to these sorts of questions constitutes, in large measure, the substance of this book.

1.1 Discrete Optimization Defined

Our concern is *discrete optimization,* the analysis and solution of problems mathematically modeled as the minimization or maximization of a value measure over a feasible space involving mutually exclusive, logical constraints. Enforcement of such logical constraints can be viewed abstractly as the arrangement of given elements into sets. Thus, in their most abstract mathematical form, discrete optimization problems can be expressed as

$$\begin{array}{ll} \text{min (or max)} & \alpha(T) \\ \text{subject to} & T \in F \end{array}$$

where T is an arrangement, F is the collection of feasible arrangements, and $\alpha(T)$ measures the value of members of F.

The study of arrangements is at the heart of the definition of *combinatorics.* Consequently, the reader will observe that we wish to view discrete optimization as a branch of combinatorics. This stance is motivated by our desire to keep absolutely central to the notion of discrete optimization the element of unavoidable choice among mutually exclusive alternatives. Discreteness is at the core, not the perifery of the problems we shall discuss.

Certainly, discrete optimization does not encompass all of the vast field of combinatorics. One popular classification recognizes four forms of combinatorial problems, distinguished by whether the question is *existence* of specific arrangements, versus *exhibition* or *evaluation* of required arrangements, versus *enumeration* or *counting* of possible arrangements, versus *extremization* of some measure over arrangements. To this extent, we equate discrete optimization with the last, extremization, branch of combinatorics.

Much of what follows is concerned with defining and presenting known results about particular discrete optimization problems. We shall briefly introduce some here in order to illustrate the enormous diversity of model forms that have been studied. The informed reader will also note that we have included in this introductory list problems spanning the field in terms of tractability. Some are extremely well solved, while others continue to frustrate researchers after two centuries of attention.

Traveling Salesman Problem. Given a graph (directed or undirected) with specified weights on the edges, determine a closed traversal that includes every vertex of the graph exactly once and that has minimum total edge weight.

Postman's Problem. For a given graph (directed or undirected) with specified weights on the edges, determine a traversal that includes every edge in the graph at least once and that has minimum total weight.

Knapsack Problem. Determine a set of integer values x_i, $i = 1, 2, \ldots, n$ that minimizes $f(x_1, x_2, \ldots, x_n)$ subject to the restriction $g(x_1, x_2, \ldots, x_n) \geq b$ where b is a parameter.

Parallel Machine Scheduling. Given a set T of single operation tasks, each with processing time τ_j, $1 \leq j \leq |T|$, assign each task to exactly one of m machines so that the completion time of all tasks is minimized.

Vertex Coloring. Given an undirected graph, determine the minimum number of colors needed to color each vertex of the graph in order that no pair of adjacent vertices (vertices connected by an edge) share the same color.

Spanning Tree. Given an undirected graph with specified weights on the edges, determine a minimum total weight subset of edges that forms a connected, acyclic graph having at least one edge incident to every vertex.

Shortest Path. For a given graph (directed or undirected) with specified weights or lengths on the edges, find a minimum total length nonrepeating sequence of edges that connects two specified vertices and that conforms to any directions on edges.

Bin Packing. For a list of n weights, w_i, $1 \leq i \leq n$ and a set of bins, each with fixed capacity, say W, find a feasible assignment of weights to bins that minimizes the total number of bins used.

Matching. Given a list of items $i = 1, 2, \ldots, n$ and weights w_{ij} associated with pairing item i with item j, find a maximum total weight scheme for pairing items in the list so that each item is paired with one other at most.

Set Covering. Given a finite set S, a family of subsets $\{S_j \subseteq S: j \in J\}$ and costs, c_j, associated with the S_j, choose a minimum total cost collection of the subsets that includes every element of S at least once.

Maximum Flow. Given a graph (directed or undirected) and specified capacities on the edges, find a maximum edge flow between two specified vertices that conforms to capacities and has total flow into all other vertices equal to total flow out.

p-Median Problem. For a graph (directed or undirected) with specified weights on the edges, choose p vertices so that the sum of distances from all vertices to the closest of the chosen p is minimized.

Fixed Charge Problem. Given a feasible set S of nonnegative activity or traffic levels, $\mathbf{x} = (x_1, x_2, \ldots, x_n)$, unit costs v_j for employing x_j, and fixed costs f_j assessed whenever x_j is positive, choose a minimum total cost $\mathbf{x} \in S$.

1.2 Discrete Optimization and Integer Programming

All of the above discrete optimization problems, and every model we shall encounter, can be expressed in the form

$$
(IP) \qquad
\begin{array}{lll}
\text{min (or max)} & \displaystyle\sum_{j=1}^{n} c_j x_j & \\
\text{s.t.} & \displaystyle\sum_{j=1}^{n} a_{ij} x_j \geq b_i & \text{for } i = 1, 2, \ldots, m \\
 & x_j \geq 0 & \text{for } j = 1, 2, \ldots, n \\
 & x_j \text{ integer} & \text{for } j \in I
\end{array}
$$

Here the x_j are decision variables constrained by nonnegativity, m linear inequalities with coefficients a_{ij} and b_i, and the requirement that each x_j with $j \in I$ $(I \subseteq \{1, 2, \ldots, n\})$ be integral. We seek to minimize or maximize the sum of the x_j times given weights, c_j.

Problems in the form (*IP*) are known as linear integer programming problems, or more briefly, *integer programs.* If $I = \{1,2, \ldots ,n\}$ we call the problem a *pure integer program* and otherwise *mixed integer.* Aside from the obvious advantage of treating all problems in a single format, there can be considerable payoff in viewing discrete optimizations as integer programs. If the last, "x_j integer" constraints are dropped in (*IP*), we are left with a *linear program*—the best solved of all broad classes of optimization problems. General, and sometimes quite efficient, algorithms can be derived for (*IP*) by exploiting this association with linear programming.

In the development to follow we shall present the more current of these general algorithms and emphasize how many procedures for specific discrete optimization problems can be interpreted as adaptions of general techniques. At the same time, it would be seriously oversimplifying matters to suggest that discrete optimization begins and ends with the study of general techniques for integer programming. Many discrete models can be forced into the (*IP*) format only at the cost of introducing immense numbers of variables and constraints. In such cases, general integer programming methods have little or no value, if for no other reason than that the underlying linear program is far beyond the capabilities of even the best of current algorithms. Furthermore, many discrete problems, including most of those that can be regarded as truly well-solved, have been conquered precisely because they have exploitable combinatorial structure, which is not present in general problems and thus not addressed by general algorithms.

It is mainly to avoid such oversimplification that we have chosen to define discrete optimization as a branch of combinatorics rather than a division of mathematical programming, even though it is without doubt a part of both. There is also a more subtle reason. The integrality constraints of (*IP*) formulations, which are totally or partially relaxed in general algorithms, are exactly the combinatorial, disjunctive characteristics that make the problems discrete. If discrete optimization problems are too casually couched in the integer programming format, it is very easy to be lulled into believing that relaxed structure is unimportant. How much difference can it make that some variable takes on the value 7/10 instead of 0 or 1? In truly discrete problems, it makes all the difference. The phenomenon coded by a 0–1 variable may simply have no interpretation when the variable assumes a fractional value; rounding to either 0 or 1 is in effect guessing a solution that might very well have been obtained without any optimization at all. Even though the integer programming point of view is highly relevant to much of discrete optimization, we believe one is less likely to be mislead if he or she keeps primary focus on combinatorial aspects of the problems.

1.3 Why are Discrete Optimization Problems Difficult to Solve?

Anyone who has had even the most passing introduction to discrete optimization problems is at least subtly aware of their inherent difficulty. What is it about discrete problems that makes them difficult to solve?

It is turning out that trying to provide a satisfactory answer to this question is one of the most intriguing phases of theoretical investigation in discrete optimization. Considerable progress has been achieved, and we shall devote a whole chapter to it in this book. Even at this introductory point, however, it seems appropriate to provide some preliminary insight into the fundamental causes of the intransigence that surrounds many discrete problems.

Superficially, it is quite clear why discrete problems are difficult. Their feasible solution spaces are enormous in size, and they grow explosively with the number of discrete choices to be resolved. For example, a problem requiring a modest 200 independent, binary decisions has 2^{200}, or about 10^{60} solutions to consider. A case with 201 binary decisions has twice as many.

This immense and exponentially growing size of discrete solution spaces categorically rules out a complete enumeration of the solutions in all but the smallest of cases. Often, the enumeration of even a tiny fraction of the set of solutions is computationally untenable.

A cynic might observe that the simplest of continuous optimization problems has a truly infinite solution space, and some such problems have been well-solved since the era of Newton and Liebnitz. There must be more to problem complexity than mere cardinality of the solution space.

Size of the solution space is relevant only if the problem must be solved by enumerating all, or at least a significant fraction, of the solutions. Unfortunately, that is precisely the state of our present knowledge regarding most discrete optimization problems. Furthermore, there is strong reason to believe that quite fundamental limits of our mathematics and computing machines will leave most discrete problems permanently in this "enumeration required" category.

To escape such enumeration, one must be able to find short cuts. Many successful algorithms take advantage of proofs that optimal solutions occur within a very small subset of the feasible points, e.g., the extreme points, inflextion points, etc. Others depend heavily on formal characterizations of progress. They stop only when theory is available to show that, since no further progress of the designated kind can be achieved, the current solution is necessarily optimal.

Numerous simplifying results of these two types are available in discrete optimization. With the exception of the minority of problems that are well-solved, however, none of the available results appears even close to providing completely satisfactory algorithms. Although they may reduce the number of solutions that need to be considered, and/or provide a conclusive progress test, the problem or test left to be resolved is usually a difficult discrete optimization problem.

1.4 Progress in Discrete Optimization

There is a degree of gloom in the picture we have painted so far. As a consequence, even the most acclaimed of discrete optimizers is sometimes (at least briefly) overcome by pessimism about the prospects for the field.

We find the challenge of dealing, however partially, with such difficult but important problems, and of establishing the boundaries of tractability in discrete optimization more than a little exciting. Our principal motivation in preparing this book is to collect the immense knowledge that is available.

There have been significant success stories in discrete optimization. Among these, one would necessarily include the landmark work on network flows of Ford and Fulkerson, the early work on cutting planes by Gomory, and the elegant results of Edmonds on optimum matchings. The past 10 to 15 years alone have brought forth some major developments including the initial efforts of Cook and later of Karp, which gave impetus to many of the results in complexity theory that now saturate the literature. During this period, the four color conjecture was converted to theorem status, a fairly complete theory of cutting planes emerged, Lagrangean techniques were successfully adapted to discrete optimization, and the ellipsoid and related methods provided a polynomial time solution scheme for linear programs.

But recent years have not been marked by theoretical developments alone. Recognition that general methods were not on the horizon focused attention on a host of special cases and examples of moderate size. A great deal of practical progress occurred. Genuinely large knapsack problems can now be treated efficiently (empirically speaking); certain scheduling problems have been conquered; and even the notorious traveling salesman problem has been humbled somewhat. Problems of sizes unheard of even a decade ago can now be rather routinely solved.

1.5 Organization of the Book

This book presents general theory and algorithms relevant to all parts of discrete optimization. We begin (in Chapter 2) on what has come to be the common language of discrete optimization—complexity theory. Included are the ideas of computability, worst-case analysis, and the notions of *P*, *NP*, and *NP*-Complete.

The remainder of this book is composed of five chapters addressed to the algorithmic categories delineated by complexity theory. In Chapters 3 and 4, we address polynomial procedures. Primary topics include matroids, polynomial solution of linear programs, and their relation to combinatorial polyhedra. Chapter five covers the class of procedures commonly referred to as partial enumeration or branch and bound. Standard linear programming based methods are fully treated, along with newer Lagrangean dual ideas and Benders decomposition. In Chapter 6, polyhedral description or cutting methods are presented. All of the modern general theories of these approaches are discussed along with an introduction to the specialized facetial techniques that have proven useful on specific models. Finally, in Chapter 7, we turn to nonexact, heuristic procedures. After general discussions of how these algorithms may be evaluated, we define and investigate three broad classes: greedy, local search, and truncated exponential. We have also included three appendices in order to provide some background pertaining to fundamental results from linear programming, graph theory, convex sets and polytopes.

EXERCISES

1-1. Formulate each of the following problems defined in Section 1.1 in the (mixed-)integer format (*IP*) of Section 1.2. Then comment on the interpretation (if any) of a solution satisfying all constraints except the integrality requirements. Each formulation should employ a number of constraints and variables that grows as a low order polynomial (e.g., n^2, m^3) with the problem size.

(a) Traveling salesman problem (directed graph)
(b) Traveling salesman problem (undirected graph)
(c) Postman's problem (directed graph)
(d) Postman's problem (undirected graph)
(e) Knapsack problem ($f(x_1, x_2, \ldots, x_n) \triangleq \sum_{j=1}^{n} c_j x_j$, $g(x_1, x_2, \ldots, x_n) \triangleq \sum_{j=1}^{n} a_j x_j$ for integer c_j and a_j)
(f) Parallel machine scheduling
(g) Vertex coloring
(h) Spanning tree (directed graph, at most one inbound arc per vertex in the solution)
(i) Spanning tree (undirected graph)
(j) Shortest path (directed graph, weights nonnegative)
(k) Shortest path (undirected graph, weights nonnegative)
(l) Bin packing
(m) Matching
(n) Set covering

(o) Maximum flow (directed graph)
(p) Maximum flow (undirected graph)
(q) p-Median (directed graph)
(r) p-Median (undirected graph)
(s) Fixed charge problem ($S \triangleq \{\mathbf{x} \in \mathbb{R}^n: \mathbf{x} \geq \mathbf{0}, \mathbf{A}\mathbf{x} \geq \mathbf{b}\}$ for integer matrix $\mathbf{A}$ and vector $\mathbf{b}$)

1-2. Formulate each of the following problems in the (mixed-)integer format (*IP*) of Section 1.2. Then, comment on the interpretation (if any) of a solution satisfying all constraints except the integrality requirements. Each formulation should employ a number of constraints and variables that grows as a low order polynomial (e.g., n^2,m^3) with the problem size.

(a) (Uncapacitated Facilities Location). Given a set of candidate facility locations $i = 1, 2, \ldots, m$, a set of customer demand points $j = 1, 2, \ldots, n$, positive construction costs f_i for constructing facility i, and nonnegative transportation cost v_{ij} for supplying demand j from facility i, find a minimum total cost collection of facilities to build such that each demand is allocated to one open facility.

(b) (Capacitated Facilities Location). Same as above except each candidate facility has an associated capacity, s_i, and each customer has an associated demand, d_j. Total demand of customers (partially or fully) assigned to a facility cannot exceed its capacity.

(c) (Assignment). Given n objects, n locations, and weights w_{ij} of assigning object i to location j, find a minimum total weight assignment of objects to locations (each object and location assigned once).

(d) (Generalized Assignment). Given m objects and n locations, weights w_{ij} of assigning object i to location j, capacities u_j of locations $j = 1, 2, \ldots, n$, and positive integer sizes a_i of objects $i = 1, 2, \ldots, m$, find a minimum total weight assignment of objects to locations that conforms to capacities (each object assigned once, total size assigned to a location $\leq$ capacity).

(e) (Edge Covering). Given an undirected graph with nonnegative weights w_{ij} on edges $e = (i, j)$, find a minimum total weight collection of edges such that each vertex is incident to at least one edge of the collection.

(f) (Tardiness Machine Scheduling). Given a set of tasks $j = 1, \ldots, n$, associated positive integer times τ_j required by the tasks on a single machine, due dates d_j for tasks and a collection $<$ of precedence pairs (i, j) indicating that task i must complete before task j can begin, find a sequencing of tasks on the machine that minimizes total due date violation.

1-3. Formulate each of the following problems (treated further in Section 2.3.2) in the (mixed-)integer format (*IP*) of Section 1.2. Then comment on the interpretation (if any) of a solution satisfying all constraints except the integrality requirements. Each formulation should employ a number of constraints and variables that grows as a low order polynomial (e.g., n^2, m^3) with the problem size.

(a) (Max Clique). Given an undirected graph, find the maximum k such that the graph contains a clique (vertex induced complete subgraph) of size k.

(b) (Vertex Cover). Given an undirected graph, and vertex weights, w_i, find the minimum weight collection of vertices such that every edge is incident to at least one vertex in the collection.

(c) (Set Packing). Given a finite set S, a family of subsets $\{S_j \subseteq S: j \in J\}$ and values v_j associated with each subset S_j, find a maximum total value collection of the subsets that includes every element of S at most once.

(d) (Set Partitioning). Given a finite set S, a family of subsets $\{S_j \subseteq S: j \in J\}$ and weights w_j associated with each subset S_j, find a minimum total weight collection of the subsets that forms a partition of S (every element of S occurs exactly once).

(e) (Steiner Tree). Given an undirected graph, a subset of distinguished vertices, S, and nonnegative edge weights, w_e, find a minimum weight subset of edges that forms a tree (connected, cycle free subgraph) having edges incident to every vertex of S.

(f) (Partition). Given positive integers $a_1, a_2, \ldots, a_n$ and corresponding weights $w_1, w_2, \ldots, w_n$, find a partition of $\{1, 2, \ldots, n\}$ into two subsets of equal a_j sum, such that the first has maximum total w_j weight.

(g) (Minimum Cut). Given a directed graph, two distinguished vertices $s \neq t$, and positive weights w_{ij} on arcs, find a minimum total weight cut separating s and t (arc subset including at least one element of every s-to-t path).

2

Computational Complexity

Very early in our lives we begin to categorize problems in terms of their relative difficulty. From the point of view of primary school students, addition is routine, but long division is "higher math". At a more advanced level, college undergraduates experiencing them for the first time usually find linear programming manipulations nasty because of the tedia of simplex calculations, while rating attacks on discrete problems rather straightforward.

These student points of view are, of course, subjective. Still, even fairly sophisticated observers of discrete optimization must be perplexed by very great differences in difficulty among problems that appear quite similar. Consider, for example, a graph, $G(V, E)$, with vertex set V and edge set E (when a graph is directed, we shall denote it by $G(V, A)$ where A is the set of arcs) and let each edge in E have a weight and assume, further, that these are nonnegative integers. Now, if we desire a shortest or least weight (simple) path between two vertices, s and t, our problem is universally considered among the most elementary in optimization.

Now make one minor change: let the edge weights have arbitrary sign. Although it appears to be only cosmetic, this modification opens the door to cases with negative total length cycles and renders the problem immensely more difficult than its nonnegative analog.

When problems can change from very easy to very hard by the simple reversal of a few coefficient signs, it is certain that even the most seasoned observer will find it impossible to consistently guess whether problems will prove difficult. Can a systematic theory be developed that distinguishes objectively among problem difficulties? In this chapter we survey the available results.

2.1 Fundamental Concepts

The search for fundamental distinctions in the tractability of problems constitutes the area known as *complexity theory.* Mathematicians and computer scientists have studied complexity for many years; however, most persons interested in discrete optimization did not begin to consider its implications for their problems until approximately 1970. Initiated in large measure by the seminal papers of Cook (1971) and Karp (1972), the discrete optimization literature on complexity has profoundly affected all phases of that research. Put briefly, complexity theory seeks to classify problems in terms of the mathematical order of the computational resources required to solve the problems via digital computer algorithms.

In this chapter, we shall be concerned with only one resource—computer time. Naturally, however, anything as subtle as a general theory of problem tractability is bound to assign abstract, and sometimes elusive, meaning to the simple words of such a definition. So we can begin by introducing some basic notions.

2.1.1 Problems

In discrete optimization, the "Shortest Path Problem", the "Assignment Problem", or the "Knapsack Problem" represent a whole family of specific examples. The notion of a problem in complexity theory is parallel. A *problem* is a (usually infinite) collection of *instances* of like mathematical form but differing in size and the values of numerical constants in the problem form. Thus, for example, the problem of finding a shortest path between a specific pair of vertices in a given acyclic graph $G(V, A)$ where all arc weights are 1 would represent one instance of the general shortest path problem.

2.1.2 Computational Orders

Implicit in the above definition is the fact that a problem includes instances of very different size. Leaving for a moment the issue of how one defines size, and indeed how one defines computation, we would certainly expect that the magnitude of computation would increase with instance size.

Complexity theory categorizes problems in terms of the form of mathematical functions expressing amounts of computation as a function of instance size. If, for example, some algorithm requires a number of computations that can be expressed as a fourth order polynomial of the problem size, n, we say that algorithm is order n^4, or simply $O(n^4)$. Other algorithms may be $O(n^2)$, $O(n \log n)$, $O(2^n)$, etc.

If our goal is to define the fundamental frontiers separating problems and algorithms into truly distinct classes, then we must discuss computation in such broad, order groupings. There can be a dramatic difference in the value of a given fourth order polynomial, depending on the magnitudes of its coefficients; however, such differences by a constant are impacted by details of the computer, the algorithm, and the problem formulation. If we content ourselves with discussing only computational orders, such minor differences will not interfere with our ability to draw more fundamental distinctions.

Throughout most of the discussion to follow, we shall deal with an even cruder breakdown of computational order. Our interest will center on distinctions among (i) computation that grows logarithmically (or better) with instance size, (ii) computation that grows polynomially, and (iii) computation that increases exponentially (or at least worse than polynomially). While this subdivision of computational growth is broad, it will provide enough resolution for us to gain a great deal of insight regarding problem tractability. Moreover, it greatly simplifies our analysis, since a function $g(n)$ that is logarithmic (respectively polynomial or exponential) remains so when we replace n by any polynomial in n. Hence, we need not be concerned with differences in problem statement, algorithm design, or computer machinery that result in only a polynomial change in the function relating computation to size.

Lest the reader think that all real differences in computation time have been erased by so broad a treatment of computational orders, consider a quick numerical example. Suppose, for size $n = 10$, there is a logarithmically growing algorithm that requires a full hour, an $O(n^5)$ algorithm that requires only one minute, and an $O(2^n)$ scheme that needs just one second. For $n = 100$, the logarithmic algorithm will require two hours, the polynomial one consumes 69.4 days, and the exponential scheme needs more than 10^{17} centuries! As problem size becomes large, such dramatic differences clearly make logarithmic growth much preferred to polynomial, and polynomial much preferred to exponential.

2.1.3 Computation Time and Turing Machines

Thus far, we have viewed computation time as the number of ill-defined elementary steps needed by an algorithm to solve a specified problem instance. In order to characterize computation time more carefully, we have to be more precise regarding what is meant by algorithms and computing machines. These notions must be founded on a model of computation that possesses the dual qualities of generality and simplicity. That is, we need to exhibit a computing device that can perform the same functions as the most powerful and sophisticated computers we have, and at that same time, do so by making use of only the most primitive of operations. We would also like this primitive computing device to be able to perform its work nearly as fast as any other machine we have.

The Turing machine is such a device. A Turing machine is an abstract model of a computer named for its originator, the British logician, Alan Turing. Of course, one does not literally purchase a Turing machine, yet

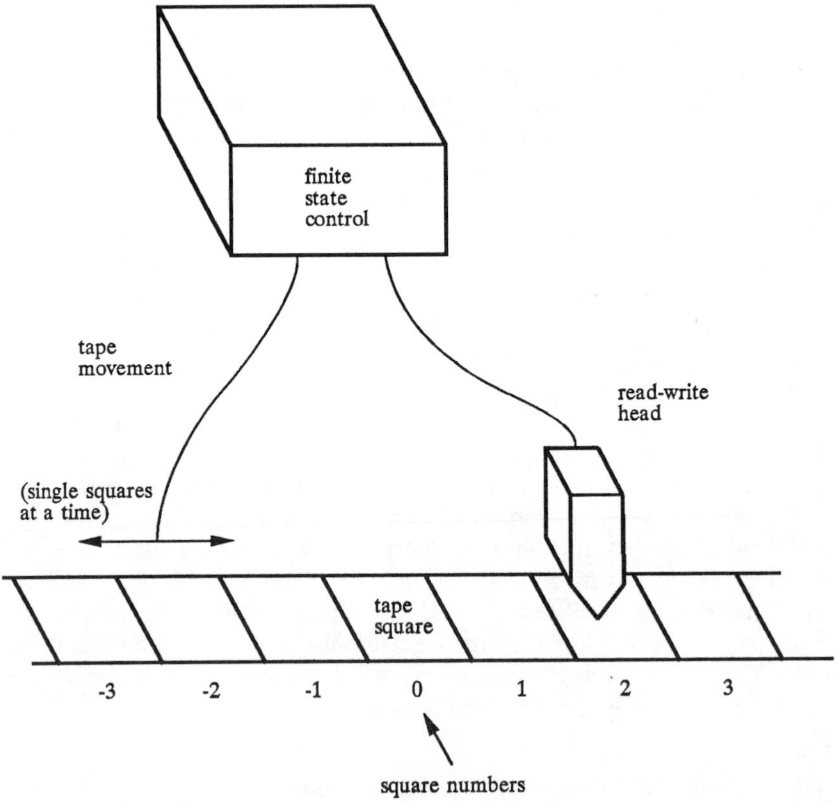

Fig. 2.1. Conceptualization of a Turing machine.

TABLE 2.1. States and Transition Function for Example 2.1

State	Symbol			
	0	1	x	ḇ
k_0 = initial	$(k_1, 0, +1)$	$(k_1, 1, +1)$	$(k_N, ḇ, -1)$	$(k_N, ḇ, -1)$
k_1 = forward symbol verifying	$(k_1, 0, +1)$	$(k_1, 1, +1)$	$(k_N, ḇ, -1)$	$(k_2, ḇ, -1)$
k_2 = rewind	$(k_2, 0, -1)$	$(k_2, 1, -1)$	$(k_2, x, -1)$	$(k_3, ḇ, +1)$
k_3 = even and searching	$(k_4, x, +1)$	$(k_5, x, +1)$	$(k_3, ḇ, +1)$	$(k_Y, ḇ, -1)$
k_4 = need a 1	$(k_4, 0, +1)$	$(k_2, x, -1)$	$(k_4, x, +1)$	$(k_N, ḇ, -1)$
k_5 = need a 0	$(k_2, x, -1)$	$(k_5, 1, +1)$	$(k_5, x, +1)$	$(k_N, ḇ, -1)$

States k_Y = accepting halt state
k_N = rejecting halt state

this imaginary computer can theoretically perform any operation that our current computers are capable of, although more slowly.

Turing machines conceive of all algorithms as schemes for manipulating a finite *alphabet, T,* of symbols on an infinite, lineal storage medium referred to as a *tape.* This tape is marked off into *squares* that are labeled . . . , −2, −1, 0, 1, 2, . . . (see Figure 2.1). Each square on the tape may contain any of the alphabet's finitely many symbols.

At each step, the machine finds itself in one of a finite set of internal *states, K.* The *program* of a given Turing machine is a function $\delta: K \times T \rightarrow K \times T \times \{-1, +1\}$. To begin. we take as *input* to our Turing machine some *string* of length s in tape squares 1 through s, allowing but one symbol per square. All other squares (those other than $1 - s$) contain the blank symbol, ḇ. The program begins its computation with the read–write head positioned over square 1 and in the initial state k_0.

Each processing step considers the current state k and the current symbol in the square under the read–write head (call it α). If $\delta(k, \alpha) = (\bar{k}, \bar{\alpha}, p)$ the machine places $\bar{\alpha}$ in the current square, moves either one square to the left or one to the right ($p = \pm 1$), and shifts to state $\bar{k}$. Should $\bar{k}$ be a halting state, computation terminates.

Example 2.1. Turing Machine Computations. To illustrate, let us construct a Turing machine to check whether an input string from the alphabet $T = \{0, 1, x, ḇ\}$ is a sequence of 0's and 1's with equal numbers of each. We shall use states $k \in K$ and transition function δ as shown in Table 2.1.

On the input string "0110", this Turing machine would first verify that all symbols in the string are 0 or 1 as follows (an arrow indicates the position of the read–write head):

k_0:	. . . $\not b$	↓ 0	1	1	0	$\not b$ . . .
k_1:	. . . $\not b$	0	↓ 1	1	0	$\not b$ . . .
k_1:	. . . $\not b$	0	1	↓ 1	0	$\not b$ . . .
k_1:	. . . $\not b$	0	1	1	↓ 0	$\not b$ . . .
k_1:	. . . $\not b$	0	1	1	0	↓ $\not b$ . . .

The search for a first 0–1 pair begins with a rewind.

k_2:	. . . $\not b$	0	1	1	↓ 0	$\not b$ . . .
k_2:	. . . $\not b$	0	1	↓ 1	0	$\not b$ . . .
k_2:	. . . $\not b$	0	↓ 1	1	0	$\not b$ . . .
k_2:	. . . $\not b$	↓ 0	1	1	0	$\not b$ . . .
k_2:	. . . ↓ $\not b$	0	1	1	0	$\not b$ . . .

The first encountered unmatched character is tagged with an x.

k_3:	. . . $\not b$	↓ 0	1	1	0	$\not b$ . . .
k_4:	. . . $\not b$	x	↓ 1	1	0	$\not b$ . . .

After a pair has been tagged with the x character, another rewind is executed.

k_2:	. . . $\not b$	↓ x	x	1	0	$\not b$ . . .
k_2:	. . . ↓ $\not b$	x	x	1	0	$\not b$ . . .
k_3:	. . . $\not b$	↓ x	x	1	0	$\not b$ . . .
k_3:	. . . $\not b$	$\not b$	↓ x	1	0	$\not b$ . . .
k_3:	. . . $\not b$	$\not b$	$\not b$	↓ 1	0	$\not b$ . . .
k_5:	. . . $\not b$	$\not b$	$\not b$	x	↓ 0	$\not b$ . . .

A last rewind precedes the final search, which finding no unmatched 0's or 1's, accepts the string.

$$
\begin{array}{lccccccc}
 & & & & \downarrow & & & \\
k_2: & \ldots\not{b} & \not{b} & \not{b} & x & x & \not{b}\ldots \\
 & & & \downarrow & & & & \\
k_2: & \ldots\not{b} & \not{b} & \not{b} & x & x & \not{b}\ldots \\
 & & & & \downarrow & & & \\
k_3: & \ldots\not{b} & \not{b} & \not{b} & x & x & \not{b}\ldots \\
 & & & & & \downarrow & & \\
k_3: & \ldots\not{b} & \not{b} & \not{b} & \not{b} & x & \not{b}\ldots \\
 & & & & & & \downarrow & \\
k_3: & \ldots\not{b} & \not{b} & \not{b} & \not{b} & \not{b} & \not{b}\ldots \\
 & & & & & \downarrow & & \\
k_Y: & \ldots\not{b} & \not{b} & \not{b} & \not{b} & \not{b} & \not{b}\ldots
\end{array}
$$

■

In order to gain a more pragmatic perspective, it is useful to consider how modern digital computer computations are modeled by this simple Turing device. The Turing tape represents storage. The alphabet corresponds roughly to the possible contents of a bit, byte, word, or similar storage unit manipulated by a computer. It is not important what unit we choose; we only wish to assume that each storage unit has finite capacity for storing distinct characters.

States of a Turing machine model the contents of various registers, accumulators, buffers, etc., in modern computers. Again, it is not important how many states there are, as long as there are only finitely many. We only wish to preclude those impossible machines that can multiply arbitrarily large numbers without saving intermediate results and those that can execute an infinite number of instructions in parallel.

Programs for Turing machines could be included as part of the input, as they are typically in general purpose computers. We usually think of the program for a particular algorithm, however, as being hard-wired by the deterministic mapping that tells the machine what to do for each combination of current square contents and current state.

The significance of Turing machines in complexity theory is that they provide a precise definition of *time,* i.e., the number of steps required for a Turing device to halt. Of course, there are other models (including several alternative Turing machine models) that could provide equally satisfactory bases for complexity evaluation. But here we can take advantage of the fact that we are interested only in broad computational orders. All known models are polynomially equivalent in the sense that an algorithm requires polynomially many steps of one model if and only if it is polynomial using the other.

As an example, suppose we include the notion of random access in our computer model. Specifically, we might assume that at any step the computing device could look at any single square, rather than the particular one under the read–write head, in deciding how to proceed. If an algorithm on such a machine halted in polynomial time, it could look at only a polynomial number of squares. Thus, random access could be accomplished by our sequential Turing machine through the simple expedient of a polynomial spacing out and spacing back to and from each needed square. If we only wish to know whether the number of steps is polynomial, it does not seem to matter whether we treat storage as sequential or random access.

2.1.4 Problem Size and Encoding

With a Turing machine model of computation in mind, we are now able to describe exactly what we mean by the size of an instance. Turing machines do not solve problems directly. Instead, they process an *encoding* of the problem in symbols drawn from the machine alphabet. By the *size* of an instance we mean the *length of this encoding,* i.e., the number of squares required to store it on the Turing machine tape.

Defining size as the length of a problem's encoding suggests that the same problem could potentially have quite different sizes depending on how we choose to encode it. For the most part, however, our analysis of complexity is not impacted by these coding differences because all reasonable encodings are equivalent in the sense that the length of one can be expressed as a polynomial function of another.

Unfortunately, there are encoding differences that are too dramatic to ignore totally. Familiar binary, octal, decimal, or similar encodings of problem constants employ a fixed-base number system. The exact base is inconseqential, but it is important to observe that the length of the encoding will grow only with the logarithm of coefficient magnitude. For example, $\lceil \log_2(x+1) \rceil$ binary digits are required to encode any positive integer x.

But what if we choose (as we often do in hand data tabulation) to encode integers by strokes? Letting the symbol for a stroke be "/", the number 4 would then be encoded as "////"; the number 1,000,000 would be encoded by one million strokes. Clearly, the length of such a stroke encoding grows polynomially with coefficient magnitude. Since polynomial growth is much faster than logarithmic, it is entirely conceivable that an algorithm could solve a problem in polynomial time with respect to stroke encoding (known formally as *unary encoding*), and yet experience exponential time growth under a binary encoding.

We shall see subsequently that such anomalies occur fairly often. For this reason it is customary to classify a problem as being solvable in polynomial time only when its solution requires a number of steps that grows polynomially in the logarithm of problem constants, i.e., when it is polynomial under the more compact fixed-base encoding. Time is *pseudo-polynomial* if it is polynomial only under unary encoding.

The requirement that a problem is polynomially solvable only if it remains so under all reasonable encodings does not seem to provide any difficulty in problem categorization. Still, this dependence of problem classification on encoding is slightly disquieting. It is not inconceivable that some problems we are now willing to consider polynomially-solved could cease to be so upon discovery of a new, super sparse encoding scheme that logarithmically shortened the required input.

2.1.5 Problem Reductions

It is common in actual computation to solve one problem by employing an efficient subroutine for another. For example, a procedure for inverting an nxn matrix can be used to solve a system of n equations in n unknowns. Such relationships between problems play a central role in complexity analysis.

We say a problem (P) *reduces* to another (P') if any algorithm that solves problem (P') can be converted to an algorithm for solving problem (P). Throughout the discussion of this book, we shall need only the more limited notion of polynomial reduction. Problem (P) *polynomially reduces* to problem (P') if a polynomial time algorithm for (P') would imply a polynomial time algorithm for (P).

Example 2.2. Polynomial Reduction. To illustrate, consider the following scheduling problem: a set of jobs are to be processed on two machines where no job requires in excess of three operations. A job may require, for example, processing on machine one first, followed by machine two, and finally back on machine one. Our objective is to minimize *makespan,* i.e., complete the set of jobs in minimum time. Let us refer to this problem as (P_J).

Now, take the one-row integer program or knapsack problem that we state in the equality form: given integers $a_1, a_2, \ldots, a_n$ and b, does there exist a subset $S \subseteq \{1, 2, \ldots, n\}$ such that $\Sigma_{j \in S} a_j = b$? Calling the latter problem (P_K), our objective is to show that (P_K) polynomially reduces to (P_J).

For a given (P_K) we construct an instance of (P_J) wherein the first n jobs require only one operation, this being on machine one. Each has processing time a_j for $j = 1, 2, \ldots, n$. Job $n + 1$ possesses three operations constrained in such a way that the first is on machine two, the second on machine one, and the last on machine two again. The first such operation has duration b, the second duration 1, and the third duration $\Sigma_{j=1}^{n} a_j - b$.

Clearly, one lower bound on the completion time of all jobs in this instance of (P_J) is the sum of processing times for job $n + 1$, i.e., $\Sigma_{j=1}^{n} a_j + 1$. Any feasible schedule for all jobs achieving this makespan value must be optimal. Suppose a subset S exists such that the knapsack problem is solvable. For (P_J) we can schedule jobs implied by S first on machine one, followed by the second operation of job $n + 1$, and complete with the remaining jobs (those not given by S). The first and last operations for job $n + 1$ (on machine two) finish at times b and $\Sigma_{j=1}^{n} a_j + 1$, respectively. Thus, the completion time of this schedule is $\Sigma_{j=1}^{n} a_j + 1$. The chart in Figure 2.2(a) illustrates the idea.

If, conversely, there is no subset $S \subseteq \{1, 2, \ldots, n\}$ with $\Sigma_{j \in S} a_j = b$ our scheduling instance would be forced to a solution like that of Figures 2.2(b) or (c). Either job $n + 1$ waits before it obtains the needed unit of time on machine one or some of jobs $1, 2, \ldots, n$ wait to keep job $n + 1$ progressing. Either way the last job will complete after time $\Sigma_{j=1}^{n} a_j + 1$.

We can conclude that the question of whether (P_K) has a solution can be reduced to asking whether the corresponding (P_J) has makespan no greater than $\Sigma_{j=1}^{n} a_j + 1$. Since (as is usually the case) the size of the required (P_J) instance is a simple polynomial (in fact linear) function of the size of (P_K), we have a polynomial reduction. Problem (P_K) indeed reduces polynomially to (P_J). ■

Polynomial reducibility of (P) to (P') is denoted $(P) \propto (P')$. In the above example $(P_K) \propto (P_J)$. Quite obviously the relation $\propto$ is reflexive in that $(P) \propto (P)$. Polynomial reducibility is also transitive; $(P) \propto (P')$ and $(P') \propto (P'')$ together imply $(P) \propto (P'')$. Thus, whenever we can show $\propto$ to be symmetric (i.e., $(P) \propto (P')$ implies $(P') \propto (P)$), the relation will provide a powerful analytic tool by grouping problems into equivalence classes.

Some polynomial reductions of a problem (P) to another, (P'), behave as in the above example in that the problem playing the role of (P) can be rewritten so that a single call to a subroutine for (P') is sufficient to solve (P). These reductions are sometimes referred to as *polynomial transformations* or *Karp reductions.*

The notion of polynomial reduction that we wish to employ is conceptually more general (although it is an open research question whether it is theoretically so). We shall include in the relationship $(P) \propto (P')$ the possibility that a polynomial number of calls to a subroutine for (P') are needed to solve (P). Of course, by polynomial number we mean polynomial in the length of the (P) encoding. This form is called a *Turing reduction.*

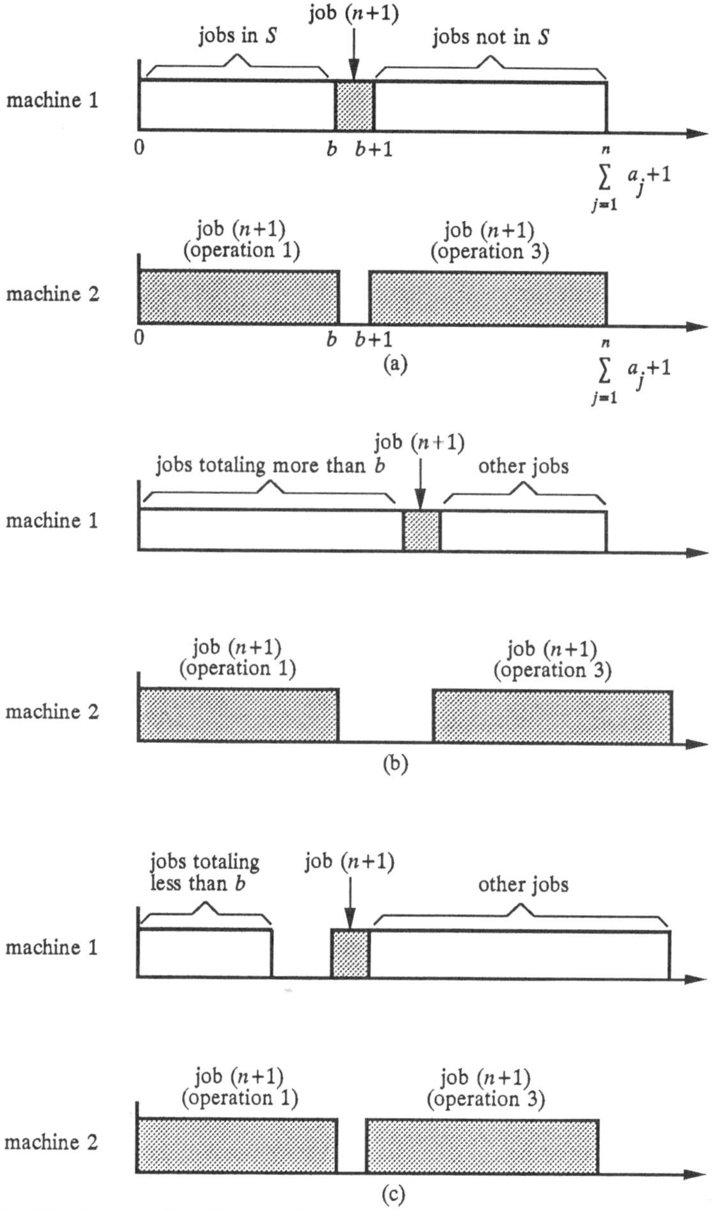

Fig. 2.2. Schedules employed in reduction example 2.2; (a) Knapsack is feasible, (b) Knapsack infeasible, job $(n + 1)$ waits, (c) Knapsack infeasible, small jobs wait.

2.1.6 *Worst Case Behavior*

To this point we have been intentionally vague about exactly what is required in order for us to classify a given problem as polynomial-time solvable, exponential-time solvable, etc. It is standard in complexity theory to make such definitions in a *worst case* sense. For example, a problem is considered polynomial-time solvable if there exists a Turing machine algorithm that solves every instance of the problem in a number of steps bounded by a polynomial function of encoding length. This approach is termed worst case because it requires that we be able to efficiently solve absolutely every instance, however unusual or contrived.

To be sure, this worst case concept of tractability provides a desirable degree of certainty about a problem. If every instance of a problem is polynomial-time solvable, we can be quite sure it should be classed among the relatively tractable cases.

The principal value of the worst case point of view in the theory, however, arises in connection with the problem reductions defined in the previous section. In order to group problems into equivalence classes, we need to be assured that whenever, say $(P) \propto (P')$, an efficient algorithm for (P') will provide an equally satisfactory one for (P). If (P) is in some sense a hard problem, it would not be surprising to find that the reduction process leads to the solution of particularly nasty instances of (P'). Thus, unless we know that every instance of (P') can be effectively solved, there is little we can infer about the solvability of (P).

In spite of this theoretical convenience, worst case concepts are often among the most criticized and controversial aspects of complexity theory, because they may conflict with typical experience. Many complexity theorists are willing to term an algorithm good only if it is polynomial-time bounded in the worst case, and to call a problem tractable only if there exists a good algorithm for it. As we have seen, such definitions do seem necessary to make the theory consistent, but it is sometimes the case that an algorithm is not good by this definition, yet performs very well in practice. The reason is that worse case instances of such problems are very very rare.

A classic example of this phenomenon can be found in the area of linear programming. The Soviet researcher, Khachian, proved in 1979 that linear programs are solvable by an algorithm of polynomial time complexity. (We present his results in Chapter 4). Shortly thereafter, and amid great commotion, rather compelling evidence was presented that suggested that void of much refinement, the Khachian ellipsoid procedure was not of great practical value. In all typical linear programs its computational performance is easily bested by the famous simplex algorithm (see Appendix

C) because the ellipsoid technique actually requires the high order polynomial number of steps in its theoretical bound while simplex computations on typical data grows as a low order polynomial in instance size. Still, Klee and Minty (1972) demonstrated that the simplex algorithm could require exponential time in the worst case. Curiously then, we have a circumstance where a so-called inefficient algorithm (the simplex procedure) out-performs one that is, technically speaking, efficient or good.

Of course, the Khachian result should not be diminished by this apparent anomaly. Indeed, the way out of such a dilemma is obvious when one fixes on the fact that in complexity theory, we seek to classify problems, not algorithms. Certainly the best algorithm for a given problem depends on the types of instances actually found in the situation where the problems arise. On the other hand, to fix a problem in the prestigious polynomial-time solvable class, we would like to know there is a polynomially bounded procedure for every instance, even if it is not the algorithm we would recommend for most instances.

Recent history has born out the potential value of the polynomial-time solvability standard for even such a well-solved class as linear programs. Karmarker (1984) has provided a new, formally efficient algorithm for linear programming that has clear roots in Khachian's insights. This new algorithm, however, may very well compete with the simplex on large typical examples.

2.2 Decision Problems

Having introduced some preliminary notions in complexity theory, we can now begin to define fundamental problem classifications. Persons interested in discrete optimization are naturally most concerned with problems that are solved by an algorithm that either exhibits an optimal solution or proves that none exists. Alternately, computer scientists are more accustomed to concentrating on a class of more succinctly defined cases called *decision problems* or *language recognition problems.* These problems are solved by a correct *yes* or *no* response to any well formed input, rather than by a full solution vector or an optimal value. Both Examples 2.1 and 2.2 had this *yes/no* character.

To see that the relationship between an optimization problem and one placed in a decision problem format is rather straightforward, consider the famous traveling salesman problem. In the optimization context, we seek a closed (simple) cycle (sometimes called a *tour*) through a given set of vertices in a graph that has minimum total edge weight. The analogous decision problem is:

Given a graph $G(V, E)$, weights w_{ij} for vertex pairs (i, j) and w, does there exist a closed cycle in G meeting every vertex of V having total edge weight no greater than w?

In general, we convert optimization problems to their decision problem counterparts by posing the question of whether there is a feasible solution to a given problem having objective function value equal or superior to a specified threshold. (w in the problem above.)

2.2.1 *The Class P*

Decision problems come in all degrees of difficulty—up to and including so-called *undecidable* ones that are provably not solvable by any algorithm. The best solved family of decision problems are those belonging to the Class P, the set of all problems admitting algorithms of polynomial time complexity. That is, P includes all those decision problems for which there is a Turing machine algorithm that always halts with the correct *yes/no* answer in a number of steps bounded by a polynomial in the length of the problem encoding.

Obviously, P stands for polynomial. It certainly deserves, however, some much more glorious appellation! Virtually all the decision problems that readers would agree are well solved have been shown to belong to P. Examples in the optimization context are the decision problem analogs of linear programming, matching, spanning trees, network flows, shortest path, and several one-machine scheduling models. On the other hand, none of the problems we traditionally view as hard is known to belong to P. Membership in P is prestigious indeed.

2.2.2 *Complements of Problems in P*

Complements of decision problems are obtained by reversing the roles of the *yes* and *no* responses.[1] For example, the complement of the traveling salesman decision problem stated may be taken as

For a given graph $G(V, E)$, specified weights w_{ij} and w, does there exist no closed cycle in G meeting every vertex of V and having total edge weight no greater than w?

Input is exactly as before; however, instances for which the response was previously *yes* are now answered *no* and vice versa.

[1] Actually, the complement of a problem's acceptable *yes* string set includes both *no* cases and strings that do not constitute well-formed inputs. The latter, however, is usually trivially, or at least polynomially, verified.

This switch from an existence problem to a nonexistence one is much more significant than it might first appear (as we shall shortly show). For problems in the Class *P*, however, the change truly is inconsequential. Complements of problems in *P* must also belong to *P*. This is obvious from our definition of *P*, since the implied Turing machine always stops in polynomial time with a correct answer. The same machine will answer the complementary problem if we merely read *yes* when the machine responds *no,* and vice versa. Even if we had conformed to more technical computer science convention and defined *P* to include problems for which a Turing machine is guaranteed to halt only if the answer is *yes,* there would have been no difficulty. Given a polynomial time bound, we could simply add a step counting clock to our Turing machine, and stop with a *no* response whenever the clock showed we had exceeded the bound without halting on *yes.*

2.2.3 The Class NP

In order to qualify for the Class *P*, a decision problem must actually be resolvable in polynomially-bounded time. But what if someone merely handed us a candidate solution? Often, it would be relatively easy to check whether this solution demonstrated that the answer to the original decision problem was *yes.* In the traveling salesman example, if someone proposed a particular cycle, we could very rapidly determine whether the cycle conformed to all feasibility requirements and the threshold value on total edge weight. If the supplied cycle proved feasible, the issue regarding the existence of such a cycle would be settled. If the particular cycle, however, was in violation of some constraint (including the threshold requirement), all we would know is that we needed to seek another candidate.

This notion of easy to verify, but not necessarily easy to solve decision problems is at the heart of the Class *NP.* Specifically, *NP* includes all those decision problems that could be polynomial-time answered if the right (polynomial-length) clue or guess was appended to the problem input string. The associated verifying Turing machine takes both the problem encoding and the clue string as input. Whenever the correct conclusion for an instance is *yes,* there must exist some clue that will cause this Turing machine to halt with the answer *yes.* Naturally, there must also be no clue that would cause the same Turing machine to halt with the answer *yes* when the correct response is *no.*

The letters *NP* denote *nondeterministic polynomial.* This terminology derives from the fact that problems in *NP* are nondeterministically polynomial-time solvable in the sense that a suitable random proof generator has some positive probability of guessing a polynomial-length proof of *yes* whenever the correct response is *yes.*

Persons accustomed to working in discrete optimization are likely to find it more natural to adopt the equivalent conceptualization that problems in *NP* are ones that can be solved by total enumeration of some polynomial-length decision vector. If we totally enumerate the possible clue strings and apply the Turing machine verification algorithm to each possibility, we must ultimately encounter a string that leads to a *yes* conclusion when one exists.

The Class *NP* certainly contains the Class *P*, the set of decision problems that are polynomial-time solvable without benefit of a clue. But *NP* also includes a great many problems not known to belong to *P*. In fact, *NP* contains the decision problem analogs of virtually all popular discrete optimization problems. Any optimization problems that can be (theoretically) solved by mixed-integer linear programming, or by polynomial-depth backtrack search, or by straightforward discrete dynamic programming have decision problem analogs in the Class *NP*.

2.2.4 Complements of Problems in NP—the Class CoNP

The class of complements of problems in *NP*, i.e., the class of decision problems for which a *yes* response corresponds to a *no* for some member of *NP*, is called *CoNP*. We have already observed that the polynomial-time algorithms available for decision problems in *P* can easily be adapted to solve their complementary *no* counterparts. That is, $P = CoP$.

There is a (verifying) Turing machine associated with every member of *NP*, and the Turing machine can be made to stop in polynomial time on every input. Unlike the *P* case, however, the clue or guessed solution is part of that input. Thus, if the machine stops with *yes,* we know the answer to the underlying problem is *yes.* But if the machine stops without proving *yes,* we can conclude very little. Perhaps the correct answer is *yes,* and we have used a wrong clue. Perhaps it is *no.* All we can be sure of is that the problem encoding plus the given clue did not lead to a *yes* result.

How can we be sure the proper answer is *no,* i.e., how can we solve problems in *CoNP*? It is entirely possible that the only viable approach is to enumerate all possible clues—concluding *no* if none of the clues leads to a verifiable *yes.* Since such enumeration is exponential in nature, it may not even be possible to verify *no* answers with a polynomial-length clue. That is, there are likely to be many members of *CoNP* that do not belong to *NP*.

We can display this idea in Figure 2.3. The entire family of decision problems is represented by the large diamond. Classes *NP* and *CoNP* form subsets of the family. The Class *P* lies in the intersection of *NP* and *CoNP*.

It is instructive to think for a moment about the practical consequences of a problem in *CoNP* also belonging to *NP*. Consider, for example, the

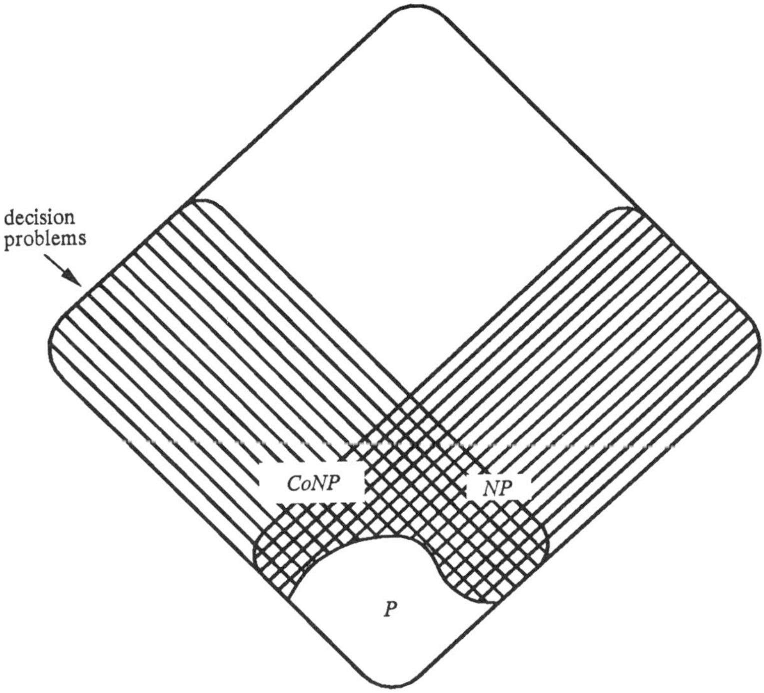

Fig. 2.3. The family of decision problems.

threshold version of (consistent) linear programming, which belongs to P and thus to $NP \cap CoNP$. The problem is

> Given vectors $\mathbf{b}$ and $\mathbf{c}$, a matrix $\mathbf{A}$, and a constant v such that $\mathbf{Ax}' \geq \mathbf{b}$ for at least one nonnegative $\mathbf{x}'$, does there exist a nonnegative vector $\mathbf{x}$ such that $\mathbf{Ax} \geq \mathbf{b}$ and $\mathbf{cx} \leq v$?

Its complementary, nonexistence problem is

> Given vectors $\mathbf{b}$ and $\mathbf{c}$, a matrix $\mathbf{A}$, and a constant v such that $\mathbf{Ax}' \geq \mathbf{b}$ for at least one nonnegative $\mathbf{x}'$, does there exist no nonnegative vector $\mathbf{x}$ such that $\mathbf{Ax} \geq \mathbf{b}$ and $\mathbf{cx} \leq v$?

At face value, the complement appears no easier to deal with than say a comparable complement for the traveling salesman problem. We know, however, from the optimal condition/duality theory of linear programming (see Appendix C) that there is a much more concise way to state the complement. It is equivalent to

> Given vectors **b** and **c**, a matrix **A**, and a constant v, does there exist a nonnegative vector **u** such that $\mathbf{uA} \leq \mathbf{c}$ and $\mathbf{ub} > v$?

When the answer to this version is *yes,* there is a dual feasible solution **u** with objective function value $\mathbf{ub} > v$, i.e., there can be no primal feasible solution **x** with primal cost $\mathbf{cx} \leq v$. The availability of this theory shows directly why the complementary problem belongs to *NP*. Enumeration (or guessing) over dual solutions (actually dual bases) will nondeterministically uncover short proofs of *no* cases in polynomially bounded time.

Loosely speaking, we can see that any problem of *CoNP,* which also belongs to *NP*, must admit some duality/optimality conditions of this sort. There must be a relatively concise clue that, if properly guessed, will demonstrate nonexistence of solutions without actually trying all the exponentially many candidates.

2.3 *NP*–Equivalent Problems

In the previous section, we explained that the Class *NP* includes the decision problem versions of virtually all the widely studied discrete optimization problems and that its subset *P* contained all those that have been conquered with well-bounded, constructive algorithms. In a very real sense, the entire history of discrete optimization can thus be viewed as a quest for schemes relegating members of *NP* to the Class *P*.

2.3.1 NP–Hardness and Cook's Theorem

Of course, if we set out on a process seeking to prove every member of *NP* in *P*, we would face a most tedious undertaking. There must be a better way, and, at least in principle, there is. We could take advantage of the notion of a polynomial reduction outlined earlier. Specifically, we could search for a problem to which every member of *NP* polynomially reduces. A polynomial–time algorithm for such a problem would then prove in one step that $P = NP$.

Problems to which all members of *NP* polynomially reduce are called *NP–Hard.* Although none is even close to being solved polynomially, it has turned out that the family of *NP*–Hard problems is amazingly rich. It includes, in particular, virtually all the problems that have frustrated discrete optimizers for decades. A very partial list would include traveling salesman, graph-coloring, precedence constrained scheduling, parallel-processor scheduling, knapsack, set covering/packing/partitioning, various fixed charge models, and $0-1$ integer/mixed-integer programming.

The key to the discovery of so many *NP*–Hard problems, and indeed the seminal development of all recent complexity theory was achieved by Stephen Cook (1971). Cook showed that a particular problem in logic, known as the *Satisfiability Problem,* was *NP*–Hard. This result started the wheel in motion, and Richard Karp's (1972) work provided real momentum by demonstrating the reducibility of satisfiability to various well-known problems in the discrete optimization literature.

As suggested, the satisfiability problem is one that is perhaps more nearly suited for the domain of the logician than say the operation researcher.

Define a set of Boolean variables $\{x_1, x_2, \ldots, x_n\}$ and let the complement of any of these variables x_i be denoted by $\bar{x}_i$. In the language of logic, these variables are referred to as *literals.* To each literal we assign a label of *true* or *false* such that x_i is *true* if and only if $\bar{x}_i$ if *false.*

Using standard notation where the symbol $\vee$ denotes *or* and $\wedge$ is the symbol for *and,* we can write any Boolean expression in what is referred to as *conjunctive normal form,* i.e., as a finite conjunction of disjunctions using each literal once at most. For example, with the set of variables $\{x_1, x_2, x_3, x_4\}$ we might encounter the following conjunctive normal form expression:

$$(x_1 \vee x_2 \vee x_4) \wedge (\bar{x}_1 \vee x_2 \vee \bar{x}_3) \wedge (\bar{x}_2 \vee \bar{x}_4)$$

Each disjunctive grouping in parentheses is referred to as a *clause.* The satisfiability problem is

> Given a set of literals and a conjunction of clauses defined over the literals, is there an assignment of values to the literals for which the Boolean expression is *true?*

If so, then the expression is said to be *satisfiable.*

A clause is satisfiable if at least one of its member literals is *true* and the entire expression is *true* (satisfiable) if all of its clauses are simultaneously satisfiable given a particular assignment to the variables. The Boolean expression above is satisfiable via the following assignment: $x_1 = x_2 = x_3 =$ *true* and $x_4 =$ *false.*

Suppose we denote the satisfiability problem by *(SAT)* and let (Q) be any member of *NP*. We can formally state Cook's Theorem as follows:

Theorem 2.1. (Cook (1971). *NP*-Hardness of *(SAT)*. *Every problem $(Q) \in NP$ polynomially reduces to (SAT).*

Proof. The one thing all $(Q) \in NP$ have in common is a Turing machine that will accept, in polynomial time, instances for which the correct answer is *yes* (perhaps with the aid of an appropriate clue). For any (Q), let that

Turing machine, say T_Q, be characterized (as outlined above) by an alphabet, T, a state set, K, and a mapping $\delta: K \times T \rightarrow K \times T \times \{-1, +1\}$. Given a ($Q$) input string, say $(d_1, d_2, \ldots, d_s)$, of length s, we shall show how a corresponding expression for *(SAT)* is constructed that has an assignment making it *true* if and only if T_Q would conclude *yes* for the string.

Let $p(s)$ be the polynomial which bounds T_Q computation, i.e., the one proving $(Q) \in NP$. The set X of literals for our *(SAT)* instance details the operation of T_Q.

- a_{ik}: implies the *ith* step of computation finds T_Q in state k; $0 \leq i \leq p(s)$, $k \in K$
- b_{ij}: implies the *ith* step of computation has the read–write head of T_Q scanning tape square j, $0 \leq i \leq p(s)$, $-p(s) \leq j \leq p(s)$
- c_{ijl}: implies the *ith* step of computation finds the contents of tape square j equal to symbol l; $0 \leq i \leq p(s)$, $-p(s) \leq j \leq p(s)$, $l \in T$

Note that, although the T_Q tape is infinite, we need be concerned with only the squares $-p(s)$ through $p(s)$ that might be examined in the $p(s)$ steps of computation.

We begin detailing operation of T_Q by clauses specifying the machine is in exactly one state at each step of computation:

$$\left(\bigvee_{k \in K} a_{ik}\right) \qquad \text{for all } 0 \leq i \leq p(s) \qquad (2\text{-}1)$$

$$(\bar{a}_{ik} \vee \bar{a}_{ik'}) \qquad \text{for all } 0 \leq i \leq p(s);\ k, k' \in K;\ k \neq k' \qquad (2\text{-}2)$$

System (2-1) demands at least one a_{ik} be *true,* and (2-2) precludes more than one from holding.

In a similar way, we add clauses assuring T_Q is always scanning exactly one tape square and that all tape squares contain exactly one element of the alphabet T.

$$(b_{i,-p(s)} \vee b_{i,-p(s)+1} \vee \cdots \vee b_{i,p(s)}) \qquad \text{for all } 0 \leq i \leq p(s) \qquad (2\text{-}3)$$

$$(\bar{b}_{ij} \vee \bar{b}_{ij'}) \qquad \text{for all } 0 \leq i \leq p(s);\ -p(s) \leq j < j' \leq p(s) \qquad (2\text{-}4)$$

$$\left(\bigvee_{l \in T} c_{ijl}\right) \qquad \text{for all } 0 \leq i \leq p(s);\ -p(s) \leq j \leq p(s) \qquad (2\text{-}5)$$

$$(\bar{c}_{ijl} \vee \bar{c}_{ijl'}) \qquad \text{for all } 0 \leq i \leq p(s);\ -p(s) \leq j \leq p(s);\ l, l' \in T;\ l \neq l' \qquad (2\text{-}6)$$

At time zero we want T_Q to be in initial state k_0 of K, with squares $1, 2, \ldots, s$ occupied by our input string and the read–write head positioned over square 1. These properties are enforced by clauses

$$(a_{0k_0}) \tag{2-7}$$

$$(c_{0jd_j}) \qquad \text{for all } 1 \le j \le s \tag{2-8}$$

$$(b_{01}) \tag{2-9}$$

Similarly, we wish the machine to stop in acceptance state k_Y within $p(s)$ steps:

$$(a_{1k_Y} \vee a_{2k_Y} \vee \cdots \vee a_{p(s)k_Y}) \tag{2-10}$$

The remaining clauses enforce the program function δ: $K \times T \to K \times T \times \{-1, 1\}$. The step after i must match in state, square revision, and tape position:

$$(\bar{a}_{ik} \vee \bar{b}_{ij} \vee \bar{c}_{ijl} \vee a_{(i+1)k'}) \qquad \begin{array}{l}\text{for all } 0 \le i \le p(s) - 1,\\ -p(s) \le j \le p(s),\ l \in T,\\ k \in K,\ \delta(k, l) \triangleq (k', l', j' - j)\end{array} \tag{2-11}$$

$$(\bar{a}_{ik} \vee \bar{b}_{ij} \vee \bar{c}_{ijl} \vee c_{(i+1)jl'}) \qquad \begin{array}{l}\text{for all } 0 \le i \le p(s) - 1,\\ -p(s) \le j \le p(s),\ l \in T,\\ k \in K,\ \delta(k, l) \triangleq (k', l', j' - j)\end{array} \tag{2-12}$$

$$(\bar{a}_{ik} \vee \bar{b}_{ij} \vee \bar{c}_{ijl} \vee b_{(i+1)j'}) \qquad \begin{array}{l}\text{for all } 0 \le i \le p(s) - 1,\\ -p(s) \le j \le p(s),\ l \in T,\\ k \in K,\ \delta(k, l) \triangleq (k', l', j' - j)\end{array} \tag{2-13}$$

where k', l', and j' are implicitly defined by δ. Also, no tape square may change except the one being scanned,

$$(b_{ij} \vee \bar{c}_{ijl} \vee c_{(i+1)jl}) \qquad \begin{array}{l}\text{for all } 0 \le i \le p(s) - 1,\\ -p(s) \le j \le p(s),\ l \in T\end{array} \tag{2-14}$$

The expression formed by clause systems (2-1) through (2-14) is satisfiable exactly when T_Q stops in state k_Y. Thus, we have shown (Q) reduces to (SAT).

We want, of course, to also be assured that the reduction is polynomial in the length of (Q) input. Examination of the details on clauses (2-1) to (2-14), however, will show that the number of required clauses of each type is no more than $(p(s))^3(|K|)^2(|T|)^2$. For any specific problem (Q), the machine T_Q and thus its state and alphabet dimensions $|K|$ and $|T|$, are constants. It follows that the reduction creates an instance of (SAT) having an input of length polynomial in s and the proof is complete. ■

2.3.2 NP-Completeness

Cook's Theorem 2.1 shows (SAT) is as difficult to polynomially solve as any problem of NP, but we already know (SAT) belongs to NP. Given a guessed truth value for the literals of X, it is trivial to verify that a specified set of clauses is satisfiable when it is.

Problems that are both *NP*–Hard and members of *NP* are called *NP-Complete.* Any two *NP*-Complete problems, say (Q) and (R), have the property that $(Q) \propto (R)$ and $(R) \propto (Q)$. Thus, since we have already noted the operation $\propto$ is transitive and reflexive, we see that *NP*-Complete problems form an equivalence class. If any member of the class admits a polynomial–time algorithm, then all members do.

The satisfiability problem holds an honored place in contemporary complexity theory, if for no other reason than that it was the first *NP*-Complete problem. We would like to classify, however, as many problems as we can in *NP*-Complete. Knowing a problem is *NP*-Complete is something of a consolation prize for not being able to fix its precise status. By its *NP*-Completeness, we know at least that if the status of any member of the class is resolved, the problem of interest will also be.

Beyond this, there is something of a self-perpetuating phenomenon in that the greater the set of *NP*-Complete problems, the greater the opportunity to discover additional members of the class. Cook's theorem allows us to skip the difficult task of treating an arbitrary member of *NP*. To establish the *NP*-Completeness of a problem, $(\overline{Q})$, we need only to find a candidate *NP*-Complete problem $(\hat{Q})$ and produce a suitable reduction, $(\hat{Q}) \propto (\overline{Q})$. This and a proof that $(\overline{Q}) \in NP$ establishes *NP*-Completeness of a new problem. It in turn could be used in the reduction process to show that others are *NP*-Complete.

We now know of many *NP*-Complete problems; the list having grown rapidly since Cook's fundamental result. In Garey and Johnson's (1979) excellent book, over 300 *NP*-Complete entries are presented. In the beginning, however, there was only the satisfiability problem, and this, as we suggested earlier, was not exactly a household name in discrete optimization.

The link to more familiar problems came with Richard Karp's (1972) proofs that (SAT) reduced to 20 problems in discrete optimization. Karp's reductions can best be shown by a tree such as that given in Figure 2.4. A branch from some parent node in that tree to a descendant implies the reduction of the former problem to the latter. Since the tree is rooted on the satisfiability problem, all listed problems are *NP*-Hard. *NP*-Completeness follows by checking that each belongs to *NP*.

One of the early reductions (Cook, in fact, showed this) involved a restricted version of the satisfiability problem. Referred to as *3-Satisfiability,* we seek a satisfying truth assignment to the literals defined over a variable set X, where every clause has exactly three literals. We have

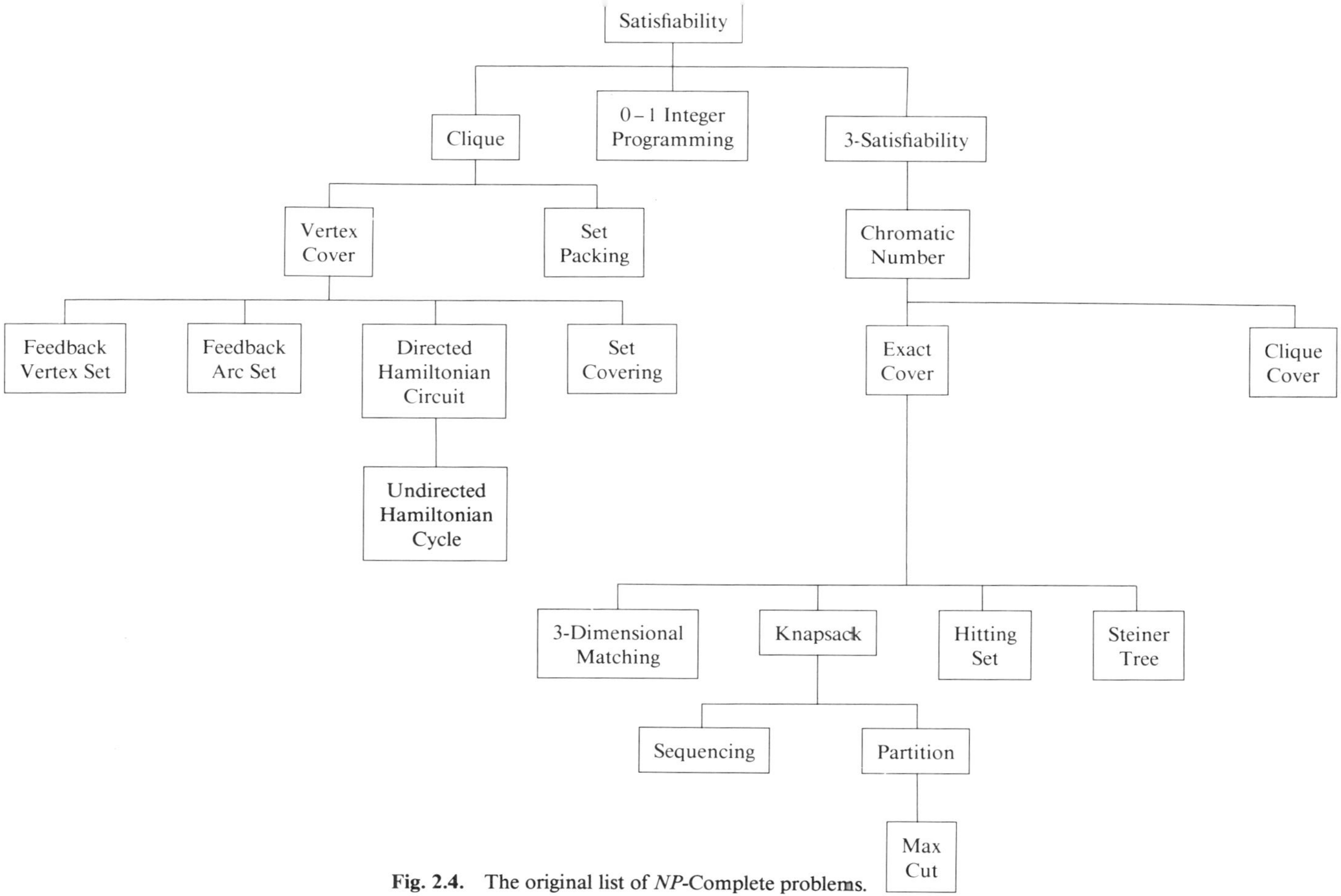

Fig. 2.4. The original list of *NP*-Complete problems.

Theorem 2.2. Complexity of 3-Satisfiability. *The 3-Satisfiability problem (3-SAT) is NP-Complete.*

Proof. Clearly, (3-*SAT*) belongs to *NP* for the same reasons as *(SAT)*. To show that (3-*SAT*) is *NP*-Hard, we will demonstrate that *(SAT)* $\propto$ (3-*SAT*).

Let the (*SAT*) variable set be $X = \{x_1, x_2, \ldots, x_n\}$ and denote the set of clauses by $C = \{c_1, c_2, \ldots, c_m\}$. We will show that an associated variable set $\bar{X} \triangleq X \cup (\cup_{i=1}^{m} \bar{X}_i)$ can be constructed as well as a new set of clauses $\bar{C} \triangleq \cup_{i=1}^{m} \bar{C}_i$ such that $\bar{C}$ is satisfiable if and only if C is. Of course, we require that no clause in $\bar{C}$ have more than three literals, i.e., $\bar{C}$ constitutes our instance of (3-*SAT*).

Select any clause $c_i \in C$ with say $c_i = \{w_1 \lor w_2 \lor \cdots \lor w_k\}$, where each literal w_j in c_i is either x_j or $\bar{x}_j$. If $k = 1$, 2 or 3 we construct variable and clause extensions $\bar{X}_i$ and $\bar{C}_i$ as follows:

$$k = 1: \quad \bar{X}_i \longleftarrow \{y_i^1, y_i^2\}$$

$$\bar{C}_i \longleftarrow \{(w_1 \lor y_i^1 \lor y_i^2), (w_1 \lor y_i^1 \lor \bar{y}_i^2), (w_1 \lor \bar{y}_i^1 \lor y_i^2), (w_1 \lor \bar{y}_i^1 \lor \bar{y}_i^2)\}$$

$$k = 2: \quad \bar{X}_i \longleftarrow \{y_i^1\}$$

$$\bar{C}_i \longleftarrow \{(w_1 \lor w_2 \lor y_i^1), (w_1 \lor w_2 \lor \bar{y}_i^1)\}$$

$$k = 3: \quad \bar{X}_i \longleftarrow \varnothing$$

$$\bar{C}_i \longleftarrow \{c_i\}$$

Clearly, the clauses in $\bar{C}_i$ are satisfiable if and only if c_i is.

The more interesting case where $k \geq 4$ is handled analogously:

$$\bar{X}_i \longleftarrow \{y_i^t\colon 1 \leq t \leq k-3\}$$

$$\bar{C}_i \longleftarrow \{(w_1 \lor w_2 \lor y_i^1)\} \cup \{(w_{t+2} \lor \bar{y}_i^t \lor y_i^{t+1})\colon 1 \leq t \leq k-4\} \cup \{(w_{k-1} \lor w_k \lor \bar{y}_i^{k-3})\}.$$

If c_i is *true*, then at least one w_j in c_i is also *true*. We make each clause of $\bar{C}_i$ *true* by assigning *true* to all y_i^l with $l \leq j-2$ and *false* to y_i^l having $l > j-2$.

Alternately, assume some assignment to the y_i's in $\bar{X}_i$ creates a satisfying truth assignment for the given $\bar{C}_i$. Now, if y_i^1 is *false*, then either w_1 or w_2 must be *true*. Also, if y_i^{k-3} is *true*, then either w_{k-1} or w_k must be *true*. If y_i^1 is *true* and y_i^{k-3} is *false*, then for some l, $1 \leq l \leq k-4$, y_i^l is *true* and y_i^{l+1} is *false* and thus, w_{l+2} must be *true*. That is, at least some w_j in original clause c_i must be *true*. It follows that $\bar{C}$ is satisfiable if and only if C is.

Concluding, we easily observe that the length of each replacement expression $\bar{C}_i$ is only bounded by a constant multiple of the length of the respective clause that is replaced, c_i. Clearly, we have the desired polynomial transformation and the proof is complete. ■

Another reduction that appears early in the original tree of reductions given in Figure 2.4 is that regarding the *Clique* problem. For a graph $G(V, E)$ consisting of p vertices, a *clique* is any complete subgraph. If the complete subgraph has k vertices, we say the clique is of size k or simply that it is a *k-clique.* Problem Clique asks:

> For some $G(V, E)$ where $|V| = p$ and for a positive integer $k \le p$, does there exist a subgraph of G that is complete and is of order no less than k?

Theorem 2.3. *Clique is NP-Complete*

Proof. That Clique is in *NP* is obvious. Any nondeterministic algorithm can guess a subset of vertices and check in polynomial time whether the size of the subset is correct and if every pair of vertices in the subset are connected by an edge. Following, we shall show that *(SAT)* $\propto$ Clique.

Let an instance of *(SAT)* be given by variable set $X = \{x_1, x_2, \ldots, x_n\}$ and clause set $C = \{c_1, c_2, \ldots, c_k\}$. Now, we can construct a graph $G(V, E)$ where $V = \{[\alpha, i]: \alpha$ is a literal occurring in clause $c_i\}$ and $E = \{([\alpha, i], [\beta, j]): i \ne j$ and $\alpha \ne \bar{\beta}\}$. Clearly, from a given C, the corresponding graph, G can be constructed in time bounded by a polynomial in the length of C since V has cardinality less than the length of C and the cardinality of E is $O(|V|^2)$.

Assume that C is satisfiable. Then each clause in C is *true* and in each, there must be at least one literal having value *true.* For clause c_i, let a *true* literal be $w^i_{j[i]}$. Then the vertex set $\{[w^i_{j[i]}, i]: 1 \le i \le k\} \subseteq V$ forms a k-clique in $G(V, E)$, since otherwise there must be a vertex pair $([w^i_{j[i]}, i], [w^m_{j[m]}, m])$, $i \ne m$, which are not adjacent in G. However, this would mean that $w^i_{j[i]} = \overline{w}^m_{j[m]}$ by definition of E, which is not possible since $w^i_{j[i]}$ and $w^m_{j[m]}$ both have value *true.*

Conversely, suppose $G(V, E)$ has a k-clique. For any edge in this clique, the incident vertices represent different clauses in C; otherwise, there would be no edge. Now, if the vertices in the clique are given as before by $w^i_{j[i]}$, let us construct two disjoint sets U_1 and U_2 where

$$U_1 = \{x \in X: w^i_{j[i]} = x, \text{ for some } 1 \le i \le k\}$$

$$U_2 = \{x \in X: \overline{w}^i_{j[i]} = \bar{x}, \text{ for some } 1 \le i \le k\}.$$

Obviously, $U_1 \cap U_2 = \emptyset$, for otherwise we would have an invalid edge $([\alpha, i], [\beta, j])$ where $\alpha = \bar{\beta}$. By assigning literals, defined for our variables in U_1, the value *true* and those in U_2, the value *false,* each clause $c_i \in C$ is satisfied and hence C is satisfied. We have thus shown that $G(V, E)$ has a k-clique if and only if C is satisfiable. ■

Example 2.3. Reduction of (SAT) to Clique. Let an instance of *(SAT)* be given by $X = \{x_1, x_2, x_3\}$ and $C = \{\{x_1, x_2\}, \{x_2, \bar{x}_3\}, \{\bar{x}_1, \bar{x}_3\}\}$. From the construction developed in the proof of Theorem 2.3, we can construct from C, the graph $G(V, E)$ below:

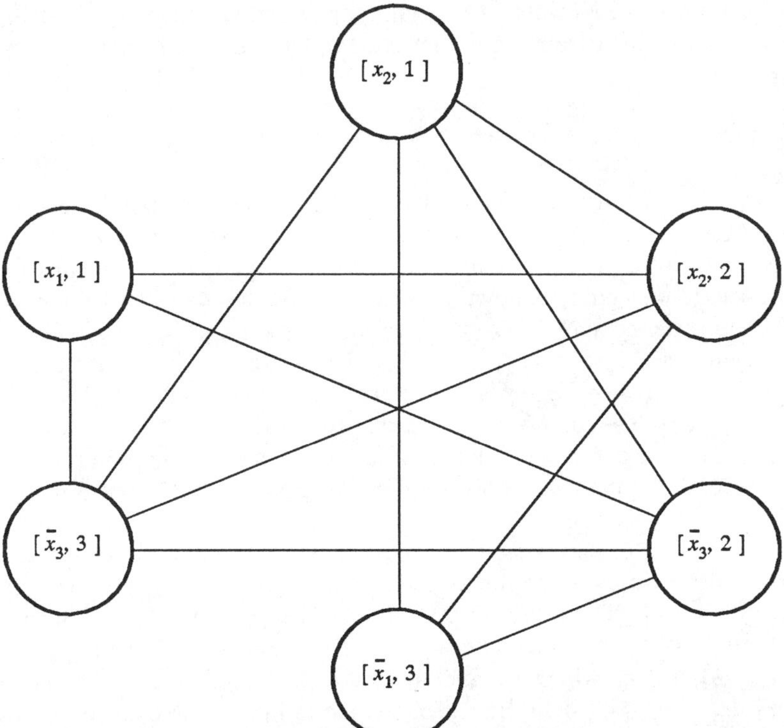

Since there are three clauses in C, we seek a 3-clique in $G(V, E)$. It is easy to see that there are several such cliques. One of these is given by the vertex set $\{[x_2, 1], [x_2, 2], [\bar{x}_3, 3]\}$. Thus, $U_1 = \{x_2\}$ and $U_2 = \{x_3\}$ and letting x_2 take value *true* and x_3 value *false,* the literals x_2 and $\bar{x}_3$ are *true.* Clearly, this truth assignment (regardless of the assignment for x_1) renders C satisfiable. ■

Turn now in the tree of Figure 2.4 to the reduction of Clique to *Vertex Cover.* For a graph $G(V, E)$, a vertex cover is given by $V' \subseteq V$ such that every edge in E is incident to at least one vertex in V'. In the decision context, we seek a subset V' having cardinality no greater than $l \leq |V|$; that is, does $G(V, E)$ possess a l-vertex cover?

Now, let us examine another question regarding a graph $G(V, E)$. A subset of vertices $\bar{V} \subseteq V$ is called an *independent set* if no two vertices in $\bar{V}$ are adjacent in G. That is $\bar{V}$ is an independent set if for all pairs of vertices i and j in $\bar{V}$ edge $(i, j) \notin E$. For some positive integer $q \leq |V|$, we seek an independent set $\bar{V}$ such that $|\bar{V}| \geq q$. It is trivial to observe that if $\bar{V}$ is an independent set for some graph $G(V, E)$, then $\bar{V}$ is a clique in the complement of G say G^c. (For any graph G, the complement, G^c, is the graph having none of the edges in G and every edge not in G.) Clearly, if the clique problem is *NP*-Complete, then so is the independent set problem.

Let us return to Vertex Cover. In showing that Clique $\propto$ Vertex Cover, we find that this would follow immediately once we invoke a particular, well-known property of graphs. Specifically, $\bar{V} \subseteq V$ is an independent set in $G(V, E)$ if and only if $V \backslash \bar{V} = V'$ is a vertex covering in $G(V, E)$. So, for an instance of Clique with given $G(V, E)$ and positive integer $k \leq |V|$, we can create the corresponding instance of Vertex Cover by creating $G^c(V, E^c)$ and simply setting $l = |V| - k$.

We shall not show any additional, explicit reductions displayed in Figure 2.4 but rather conclude the present section by showing, for a given pair $(Q, \bar{Q})$, where $Q \propto \bar{Q}$, the construction of the instance of $\bar{Q}$ from that for Q. Our presentation parallels that in Karp (1972).

(i) *0-1 Integer Programming:* Given an integer matrix **A** and an integer vector **b**, does there exist a 0-1 vector **x** such that $\mathbf{Ax} = \mathbf{b}$?
Satisfiability $\propto$ *0-1 Integer Programming*

$$a_{ij} = \begin{cases} 1 & \text{if } x_j \text{ is literal in clause } c_i \\ -1 & \text{if } \bar{x}_j \text{ is a literal in clause } c_i \\ 0 & \text{otherwise} \end{cases}$$

for $1 \leq i \leq m$ and $1 \leq j \leq n$.

$$b_i = 1 - |U_i|$$

where U_i is the set of complemented variables in c_i, $1 \leq i \leq m$.

(ii) *Set Packing:* Given a family of sets $S = \{S_j\}$ and a positive integer q, does S contain q mutually disjoint sets?
Clique $\propto$ *Set Packing*
Let $G(V, E)$ be defined for *Clique*. Elements in sets S_j, $1 \leq j \leq n$ are two-element subsets of V $\{i, j\} \notin E$. Set $q \leftarrow k$, where k is defined for *Clique*.

(iii) *Set Covering:* Given a finite family of sets $S = \{S_j\}$ and a positive integer k, does there exist a subfamily $W \subseteq S$ containing no more than k sets such that $\cup W_i = \cup S_j$?

Vertex Cover $\propto$ *Set Covering*
For given $G(V, E)$ let S_j be the set of edges incident to vertex j. Set $k \leftarrow l$ where l is defined for the vertex covering problem.

(iv) *Feedback Vertex Set:* Given a directed graph $G(V', A)$ and a positive integer k, does there exist a subset $R \subseteq V'$ with $|R| \leq k$ such that every directed cycle (circuit) of G contains a vertex in R?
Vertex Cover $\propto$ *Feedback Vertex Set*
Let $V' = V$ and $A = \{\langle i, j\rangle, \langle j, i\rangle: (i, j) \in E\}$. Set $k \leftarrow l$ where l, V and E are defined for the vertex covering problem. Here $\langle i, j\rangle$ denotes an arc directed from i to j.

(v) *Feedback Arc Set:* Given a directed graph $G(V', A)$ and a positive integer k, does there exist a subset $S \subseteq A$ with $|S| \leq k$ such that every directed cycle of G contains an arc from S?
Vertex Cover $\propto$ *Feedback Arc Set*
Let $V' = V \times \{0, 1\}$ and $A = \{\langle [i, 0], [i, 1]\rangle: i \in V\} \cup \{\langle [i, 1], [j, 0]\rangle: (i, j) \in E\}$. Set $k \leftarrow l$ where l, V and E are defined for the vertex covering problem.

(vi) *Directed Hamiltonian Circuit:* Given a directed graph $G(V', A')$, does G possess a directed cycle that includes every vertex in V' exactly once?
Vertex Cover $\propto$ *Directed Hamiltonian Circuit*
Let $E = \{0, 1, 2, \ldots, m-1\}$ and fix l as defined in the vertex covering problem. Let V' be defined such that $V' = \{a_1, a_2, \ldots, a_l\} \cup \{[u, i, \alpha]: u \in V$ and u is incident to $i \in E$, $\alpha \in \{0, 1\}\}$.

$A' = \{\langle [u, i, 0], [u, i, 1]\rangle: [u, i, 0] \in V'\} \cup$

$\{\langle [u, i, \alpha], [v, i, \alpha]\rangle: i \in E$, u and v are incident to i, $\alpha \in \{0, 1\}\} \cup$

$\{\langle [u, i, 1], [u, j, 0]\rangle: u$ is incident to i and j and there exists no h, $i < h < j$ such that u is incident to $h\} \cup$

$\{\langle [u, i, 1], a_f\rangle: 1 \leq f \leq l$ and there exists no $h > i$ where u is incident to $h\} \cup$

$\{\langle a_f, [u, i, 0]\rangle: 1 \leq f \leq l$ and there exists no $h < i$ where u is incident to $h\}$.

(vii) *Undirected Hamiltonian Cycle:* Given a graph $G(V', E')$ does G possess a cycle which includes every vertex in V' exactly once?
Directed Hamiltonian Circuit $\propto$ *Undirected Hamiltonian Cycle*
Let $V' = V \times \{0, 1, 2\}$ and

$$E' = \{([u, 0], [u, 1]), ([u, 1], [u, 2]): u \in V\} \cup \{([u, 2], [v, 0]): [u, v] \in A\}$$

where V and A are defined for *Directed Hamiltonian Circuit.*

(viii) *Chromatic Number:* Given a graph $G(V, E)$ and a positive integer k, can the vertices in V be colored with k or fewer colors such that no pair of adjacent vertices have the same color?
3-SAT $\propto$ *Chromatic Number*
Let $m \geq 4$ and construct $G(V, E)$ where

$$V = \{x_1, x_2, \ldots, x_m\} \cup \{\bar{x}_1, \bar{x}_2, \ldots, \bar{x}_m\} \cup \{v_1, v_2, \ldots, v_m\} \cup \{c_1, c_2, \ldots, c_r\} \text{ and}$$

$$E = \{(x_i, \bar{x}_i): i = 1, 2, \ldots, n\} \cup \{(v_i, v_j): i \neq j\} \cup \{(v_i, x_j): i \neq j\} \cup \{(v_i, \bar{x}_j): i \neq j\} \cup \{(x_i, c_f): x_i \notin c_f\} \cup \{(\bar{x}_i, c_f): \bar{x}_j \in c_f\}.$$

Set $k \leftarrow r + 1$ where r is the number of clauses in *3-SAT.*

(ix) *Clique Cover:* For a graph $G(V, E)$ and a positive integer l, is V the union of l or fewer cliques?
Chromatic Number $\propto$ *Clique Cover*
For G and k as given for the chromatic number problem, set $G' = G^c$ and $l \leftarrow k$.

(x) *Exact Cover:* Given a family $S = \{S_j\}$ where $S_j \subseteq \{u_i, 1 \leq i \leq t\}$, is there a subfamily $W \subseteq S$ such that the sets W_i of W are disjoint and $\cup W_i = \cup S_j = \{u_i: 1 \leq i \leq t\}$?
Chromatic Number $\propto$ *Exact Cover*
Let the set of elements be given by

$$V \cup E \cup \{[u, e, f]: \text{vertex } u \text{ is incident to edge } e \text{ and } 1 \leq f \leq k\}$$

where V, E and k are defined for *Chromatic Number.* The sets S_j are: for each f, $1 \leq f \leq k$ and each $u \in V$, $\{u\} \cup \{[u, e, f]$: e is incident with $u\}$; for each edge e in E and each pair f_1, f_2 where $1 \leq f_1 \leq k$ and $1 \leq f_2 \leq k$ and $f_1 \neq f_2$, $\{e\} \cup \{[u, e, f]: f \neq f_1\} \cup \{[v, e, g]: g \neq f_2\}$. Here u and v are the vertices incident to edge e.

(xi) *Hitting Set:* Given a family $U = \{U_i\}$ where $U_i \subseteq \{s_j: 1 \leq j \leq r\}$, is there a set $W \subseteq \{s_j: 1 \leq j \leq r\}$ such that for each i, $|W \cap U_i| = 1$?
Exact Cover $\propto$ *Hitting Set*
The hitting set problem has sets U_i and elements s_j where $s_j \in U_i$ if and only if $u_i \in S_j$ where u_i and S_j are elements and sets of the exact cover problem.

(xii) *Steiner Tree:* Given a graph $G(V, E)$, a subset $V' \subseteq V$, a weighting function w that maps E to the set of positive integers and a positive integer k, does G possess a tree having weight no greater than k and that contains V'?

Exact Cover $\propto$ *Steiner Tree*

Let

$$V = \{n_0\} \cup \{S_j\} \cup \{u_i\}$$
$$V' = \{n_0\} \cup \{u_i\}$$
$$E = \{(n_0, S_j)\} \cup \{(S_j, u_i): u_i \in S_j\}$$
$$w_{n_0, S_j} = |S_j|$$
$$w_{S_j, u_i} = 0$$
$$k = |\{u_i\}|.$$

(xiii) *3-Dimensional Matching:* Given a set $U \subseteq T \times T \times T$ where T is a finite set, is there a subset $W \subseteq U$ such that $|W| = |T|$ and no two elements of W are the same in any coordinate?

Exact Cover $\propto$ *3-Dimensional Matching*

We can assume that $|S_j| \geq 2$ for each j. Let $T = \{[i, j]: u_i \in S_j\}$. Let α be an arbitrary one-to-one function mapping $\{u_i\}$ into T. Define π: $T \to T$ to be a permutation such that for each fixed j, $\{[i, j]: u_i \in S_j\}$ is a cycle of π. Construct U such that

$$U = \{[\alpha(u_i), [i, j], [i, j]]: [i, j] \in T\} \cup$$
$$\{[\beta, \sigma, \pi(\sigma)]: \text{for all } i, \beta \neq \alpha(u_i), \sigma \in T\}.$$

(xiv) *Knapsack:* Given integers $(a_1, a_2, \ldots, a_r)$ and b, does there exist a 0-1 vector $\mathbf{x}$ such that $\Sigma a_j x_j = b$?

Exact Cover $\propto$ *Knapsack*

Let

$$d = |S| + 1, \; a_{ij} = \begin{cases} 1 & \text{if } u_i \in S_j \\ 0 & \text{otherwise} \end{cases}$$

$$r = |S|, \qquad a_j = \sum_i a_{ij} d^{i-1} \text{ and } b = \sum_i d^{i-1}.$$

(xv) *Job Sequencing:* Given a set of jobs $\{1, 2, \ldots, t\}$ with integer processing times $(\tau_1, \tau_2, \ldots, \tau_t)$, an integer due-date vector $(d_1, d_2, \ldots, d_t)$, an integer penalty vector $(P_1, P_2, \ldots, P_t)$ and positive integer k, does there exist an ordering of the jobs $\pi = (\pi(1), \pi(2), \ldots, \pi(t))$ such that the total tardiness penalty is no greater than k? Note that for job j, the tardiness penalty is $P_{\pi(j)}$ if $\tau_{\pi(1)} + \tau_{\pi(2)} + \cdots + \tau_{\pi(j)} > d_{\pi(j)}$ and 0 otherwise.

Knapsack $\propto$ *Job Sequencing*
Let

$$t = r,\ \tau_i = P_i = a_i \text{ for all } i, \text{ and } d_i = b.$$

(xvi) *Partition:* Given a vector of positive integers $(i_1, i_2, \ldots, i_s)$, is there a set $I \subset \{1, 2, \ldots, s\}$ such that $\Sigma_{l \in I}\, i_l = \Sigma_{l \notin I}\, i_l$?
Knapsack $\propto$ *Partition*
Let

$$s = r + 2$$

$$i_j = a_j,\ 1 \le j \le r$$

$$i_{r+1} = b + 1$$

$$i_{r+2} = \left(\sum_{i=1}^{r} a_i\right) + 1 - b.$$

(xvii) *Max Cut:* Given a graph $G(V, A)$, and a function that maps arcs in A to the set of positive integers, and a positive integer f, does there exist a subset of $S \subseteq V$ such that

$$\sum_{\substack{(i,j) \in A \\ i \in S \\ j \notin S}} w_{ij} \ge f?$$

Partition $\propto$ *Max Cut*
Let

$$V = \{1, 2, \ldots, s\}$$

$$A = \{(i, j): i \in V, j \in V, i \ne j\}$$

$$w_{jk} = i_j \cdot i_k$$

$$f = \left\lceil \frac{1}{4} \sum i_j^2 \right\rceil.$$

2.3.3 NP-Equivalent Optimization Problems

Strictly speaking, the class *NP*-Complete contains no optimization problems. Every member is a decision problem that is answerable by a *yes* or a *no*. Since many important discrete optimization problems have an *NP*-Complete threshold decision analog, however, such optimization problems are *NP*-Hard. For example, an algorithm for minimizing the tour length in a traveling salesman problem would resolve the issue of existence of solutions within a specified tour length threshold for every possible threshold.

Since such discrete optimization problems are *NP*-Hard, a polynomial time algorithm for any of them would provide one for all the decision problems in *NP*. That is, every member of *NP* reduces polynomially to the specified optimization problem. It is interesting to investigate the converse: Do optimization problems polynomially reduce to members of *NP*?

We term *NP-Easy* problems that polynomially reduce to a member of *NP*. To see that most familiar discrete optimization problems are *NP*-Easy, we must be a little more precise about what it means to solve an optimization problem. The least demanding definition of a solution is a determination of the optimal solution *value* (or a demonstration that no optimum exists). More desirable still is an algorithm that actually exhibits a *solution* achieving the value.

Under quite general conditions the value version of the optimization problem reduces to the threshold analog. Suppose, for example, we can be assured that the optimal solution value is integral and lies in the interval $[0, 2^{p(n)}]$, where $p(n)$ is a polynomial function of the length, n, of the problem encoding. Then by a process of binary search we can, in $p(n)$ calls to a subroutine for the threshold problem, identify the optimal solution value or show that none exists.

To illustrate, consider the simple (inequality form) knapsack problem

$$
(KP) \qquad \begin{aligned} \max \quad & \sum_{j=1}^{n} c_j x_j \\ \text{s.t.} \quad & \sum_{j=1}^{n} a_j x_j \leq b \\ & x_j = 0 \text{ or } 1 \qquad \text{for } j = 1, 2, \ldots, n \end{aligned}
$$

with all coefficients integer. Assuming some of the values c_j are positive, a trivial upper bound on (*KP*)'s solution value is the sum of the positive c_j. Also, of course, 0 is a lower bound.

For a specific instance, the upper bound might be 32768. In one call to a subroutine answering the decision problem

> Does there exist a feasible solution, $\mathbf{x}$, to (*KP*) with $\sum_{j=1}^{n} c_j x_j$ at least α?

we could determine if the optimal solution value for this instance was at least $\alpha = 32768/2 = 16384$. If the subroutine response is *yes*, our next call would use $\alpha = 16384 + (16384/2) = 24576$. On a *no*, we would try $16384/2 = 8192$. Continuing this process we would isolate the optimal (integer) solution value in at most $\log_2(32768) = 15$ calls. The optimal value version of (*KP*) has been polynomially reduced to the threshold decision form because a polynomial number of calls to the latter will solve the first.

Note that correctness of this reduction for the value version depends on being able to deal with both *yes* and *no* threshold instances. Such reductions thus really only tell us something about the difficulty of the value version if threshold problems are in *P* (or at least in $NP \cap CoNP$).

Let us turn now to the matter of exhibiting optimal solutions. In our knapsack case, we wish to know a choice of the x_j that achieves the optimal solution value.

Suppose we have a subroutine for computing (polynomially) the value of an optimal solution. We can sometimes show that the problem of exhibiting optimal solutions reduces to that of computing optimal solution values by successively enumerating solutions, using the solution value routine to direct the enumeration.

Figure 2.5 illustrates this for a knapsack example. We enumerate by trying each x_j in turn at value 0 and then at value 1. The optimal value computing routine is applied to both cases. For example, values in Figure 2.5 show the best knapsack solution to a particular instance with $x_1 = 0$ has value 76892. If $x_1 = 1$ in the same problem, the knapsack can achieve only 44519. Thus, we choose $x_1 = 0$ and proceed to $(x_1 = 0, x_2 = 0)$ and $(x_1 = 0, x_2 = 1)$. Continuing in this way, we isolate a solution achieving the optimal 76892 value with at most $2n$ calls to the optimal value subroutine. The optimal solution version of our knapsack problem has been polynomially reduced to the optimal value case.

There is one "catch" in this last type of reduction. The optimal value subroutine we have available must be capable of solving a given problem of specified size and form. The instances, solved at each node in the Figure 2.5 tree are of slightly differing form: certain variables are fixed in value. In this knapsack case, we merely eliminate variables fixed at 0 and use right-hand-side $b - \Sigma_{\{j: x_j \text{ is fixed at } 1\}} a_j$ to obtain a new knapsack of the form to which our value routine is adapted. But, for example, in a problem defined on complete graphs, fixing variables may result in restricted problems on partial (not all pairs of vertices connected by an edge) graphs. Such difficulties are usually easily surmounted, but some adjustment in the optimal value computing routine may be required.

Summarizing, we have seen that in the great majority of discrete optimization cases, optimization problems are as easy (in the sense of polynomial time solution) as their decision analogs in the Class *NP*. Such optimization problems that are also *NP*-Hard may thus be viewed as *NP-Equivalent.* A polynomial time algorithm for any *NP*-Complete problem, or indeed for any *NP*-Hard problem, would provide a polynomially-bounded algorithm for all such discrete optimization models.

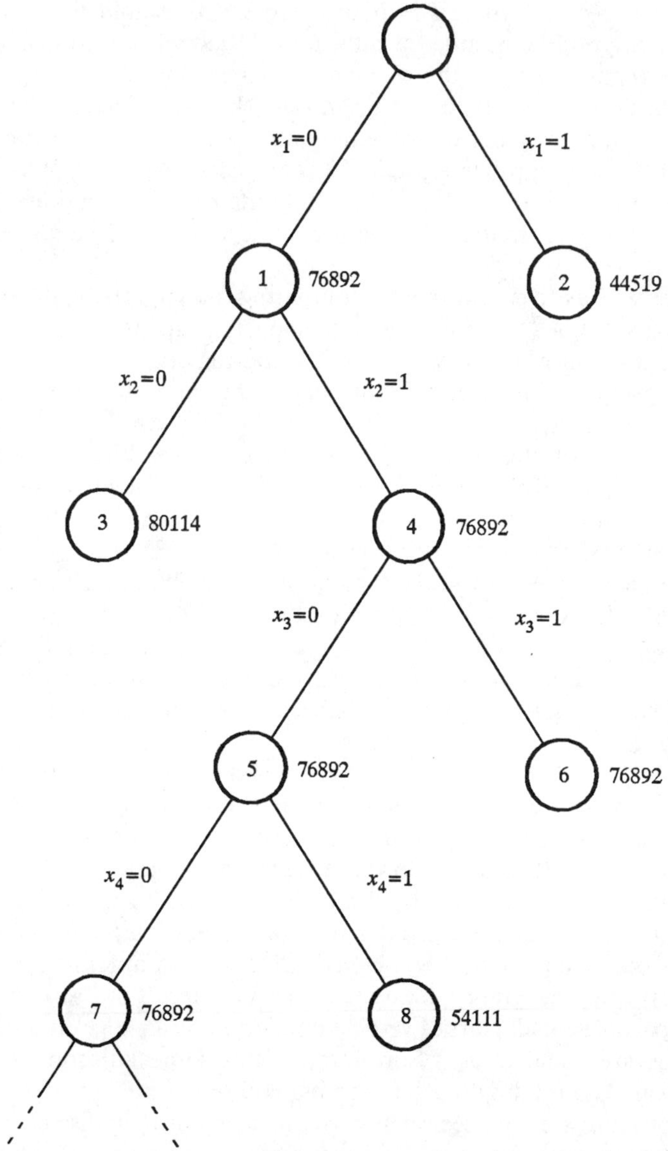

Fig. 2.5. Enumerating to exhibit an optimal solution.

2.4 The $P \neq NP$ Conjecture

We introduced the subject of complexity theory by asking whether there were fundamental boundaries in problem tractability. For some discrete problems, well-bounded constructive solution algorithms are known, but very small changes in the problem form, e.g., allowing arbitrary edge weights in shortest path problems, can relegate problems to an apparently quite different class. Will well-bounded, constructive algorithms ever be discovered for the latter cases? The conventional wisdom, for some time, has been that the answer is *no.* We are now in a position to see how complexity theory greatly strengthens such a conjecture and defines where the frontier lies.

Figure 2.6 updates our classification or problems. The set of decision problems now shows a *P* subset and an *NP*-Complete subset. Most optimization problems fall in the larger Polynomial and *NP*-Equivalent sets.

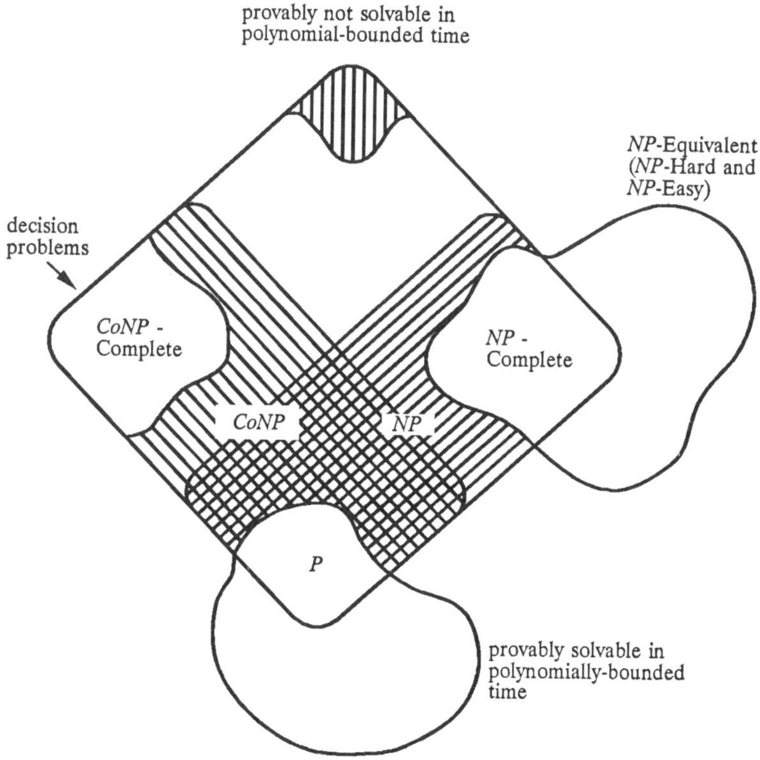

Fig. 2.6. Problem classes as they are conjectured to be.

The figure reflects the conjecture that P is a proper subset of NP. It is widely believed that $P \neq NP$ because enormous consequences would follow from a contrary finding. First, we know that if $P = NP$, all NP-Equivalent problems, including a great many that have perplexed researchers for a century or more, must admit polynomially bounded algorithms. No such algorithm is known. Furthermore, since complements of problems in P also belong to NP, it would follow that $NP = CoNP$. Earlier, we showed how that would imply that there are concise, easily checked optimality condition/duality theories for all NP-Equivalent problems.

Put briefly, a finding that $P = NP$ would in one step resolve most of the open problems in discrete optimization. The frontier between problems amenable to well-bounded, constructive algorithms and those requiring less elegant methods certainly seems to pass between the Polynomial and NP-Equivalent problem classes.

It is important to note, however, that a conjecture, even a widely held conjecture, is not a proof. Problems always appear difficult until someone finds an answer—usually someone not well enough educated to know the problem is hard. There are problems, however, that are provably not solvable in polynomial time. For example, Jeroslow (1973a) showed that quadratically constrained integer programs are not solvable at all by Turing machines. At this writing, however, no one is even close to showing that any common discrete optimization problems exhibit such behavior. It may be many years before we have a conclusive resolution of the $P \neq NP$ question.

2.5 Dealing with *NP*-Hard Problems

Very often in discrete optimization research, it is fairly easy to prove that problems being studied are NP-Hard or NP-Equivalent. Such a finding probably should discourage a search for the sort of well-bounded, constructive procedure that results in a polynomial-time algorithm. In attempting to find such a scheme, one is in a very direct sense simultaneously addressing all NP-Equivalent problems.

Some would jump to the gloomy notion that any further work on an NP-Hard problem is unwarranted. We find this a serious misrepresentation of complexity results.

NP-Equivalent problems are equivalent only in the sense that no algorithm is likely to be discovered that resolves every instance in polynomially bounded time. This does not preclude doing well on smaller instances with algorithms of exponential worst-case behavior or even finding polynomial-time algorithms for subsets of instances. Furthermore, there appear to be at least informal sub-classes within the NP-Equivalent group distinguished by their susceptibility to weaker algorithmic standards than polynomial-time solvability.

2.5.1 *Special Cases*

Perhaps the most fruitful opportunity for success with *NP*-Hard problems is the consideration of special cases. To illustrate, suppose we are interested in finding a spanning tree in a complete graph on p vertices. This problem is among the best solved in all discrete optimization. We can easily find minimum total weight, maximum total weight, and various other forms of optimal spanning trees.

If we now add the extra constraint that our trees cannot include more than k edges incident to any vertex, we obtain the *NP*-Equivalent *degree-constrained spanning tree problem.* Notice that the fact that the more general degree-constrained problem is *NP*-Equivalent does not preclude the result that the special case with a non-binding degree limit ($k = p - 1$ will serve) can be polynomially solvable.

More generally, suppose it were known that most degree-constrained spanning tree problems admit only the sort of exact optimization procedure that requires exponential time in the worst case. It may still be true that many important examples that arise in practice are among the easier instances, with exponential-time algorithms giving very satisfactory actual experience. Often the problem instances that arise in *NP*-Completeness results on worst case behavior involve very large coefficients, dense graphs, etc., even though such phenomena are comparatively rare in actual data.

2.5.2 *Pseudo-Polynomial Algorithms and Strong Sense NP-Hardness*

A second notion of variations within the *NP*-Equivalent class revolves around the unary versus binary encoding issue introduced in Section 2.1.4. A problem is said to be *pseudo-polynomial-time solvable* if there exists an algorithm that obtains a solution in a number of steps bounded by a polynomial in the length of its stroke encoding. That is, pseudo-polynomial algorithms are bounded by a polynomial that may include the magnitudes of problem constants. On the other hand, a problem is said to be *NP-Hard in the strong sense* if it remains *NP*-Hard even when it is unary encoded.

The knapsack problem

$$(KP) \qquad \begin{aligned} \max \quad & \sum_{j=1}^{n} c_j x_j \\ \text{s.t.} \quad & \sum_{j=1}^{n} a_j x_j \leq b \\ & x_j = 0 \text{ or } 1 \qquad j = 1, 2, \ldots, n \end{aligned}$$

is a classic example of an *NP*-Hard problem that admits a pseudo-polynomial-time algorithm. The straightforward, discrete dynamic programming approach to this problem that is found in any introductory text in operations research, requires $O(nb)$ computations. In the worst case, as b grows exponentially in the number of digits needed to express it, such an algorithm does not solve the knapsack in polynomial time. Still, the existence of a pseudo-polynomial procedure suggests that instances with moderate right-hand-sides, may be solved easily. In fact, current algorithms exist (e.g. Zoltners (1978), Balas and Zemel (1980)) for solving examples having reasonable coefficient sizes that involved many thousands of variables. Only the purest of theorists would term such performance intractable.

At the other extreme are problems that are *NP*-Hard in the strong sense. To know a problem has this undesirable property, we need only prove that it remains *NP*-Hard if constants in the problems are limited by some fixed bound. For example, the traveling salesman problem is *NP*-Hard if cost coefficients (edge weights) are limited to the integers 0 and 1. Thus, any pseudo-polynomial-time algorithm for the traveling salesman problem could be used as a polynomial-time algorithm for all *NP*-Equivalent problems. Unless $P = NP$, no such algorithm can exist.

2.5.3 Solving Over Relaxations

In simplest terms, a *relaxation* of an optimization problem (see Section 5.1) is an easier optimization problem with a feasible set containing that of the original problem. If an optimal solution to such a relaxation happens to be feasible in the original problem, it is optimal there.

Much more often than one might expect, such a lucky eventuality actually occurs for discrete problem instances derived from real data. Thus, it is often quite a practical idea to start dealing with an instance of a problem that is known to be hard in the worst case by forming and solving relaxations.

Rardin et al. (1987) have shown we can use this notion to obtain a rigorous subclassification of some *NP*-Hard problems. Focus on the value (not threshold) version of the optimization problem. Those instances for which the optimal value can be established by polynomial-time solution of a relaxation belong to P (if phrased in the *yes/no* fashion of whether the optimal value for the instance is a specified input α).

The technology we shall present in Chapter 6 shows that virtually every discrete optimization problem of interest in this book can be equivalently expressed as a linear programming problem. Thus there is a linear programming relaxation for almost every discrete optimization that can, in principle, establish the optimal value of the discrete problem. However, the constraints of that linear programming relaxation may require more than polynomial effort to derive. In fact, it can be shown that the complete list of needed constraints for most *NP*-Hard discrete optimization problems cannot be polynomially derived unless $NP = CoNP$ (see Section 6.3).

Therein lies the mechanism for identifying an "easier" subset of instances of an *NP*-Hard discrete optimization problem. When the optimal value question can be resolved over a polynomially derivable linear programming relaxation, the *yes/no* value question is in *NP*; the correctness of the given candidate value can be verified by guessing a solution that achieves it together with the derivation of needed linear programming constraints. But for essentially the same reasons that we cannot derive all the needed linear constraints for arbitrary instances in polynomial time unless $NP = CoNP$, the full set of value instances for most *NP*-Hard optimization problems cannot belong to *NP* unless $NP = CoNP$. Thus the subset that does belong to *NP* very probably characterizes an easier class.

2.5.4 Polynomial-Time Nonexact Algorithms

Another approach to dealing with hard optimization problems is to be less ambitious and settle for suboptimal solutions. Out of necessity we turn to heuristic or nonexact procedures when exact approaches seem beyond reach.

After one elects for a heuristic procedure, a perfectly legitimate question is "How close to optimal can one come in polynomial time?" Of course, to deal with such a question, we must be more explicit about what it means for an algorithm to come close to optimality. If $v^* > 0$ is the value of an optimal solution to a (minimize) problem and v_H is the value produced by some nonexact algorithm, it is conventional to measure accuracy in terms of the ratio $\rho = v_H/v^*$. We always have $\rho \geq 1$, and if $\rho = 1$, the heuristic yields an optimal solution.

Even after deciding to focus on the performance ratio ρ, there remains the issue of what standards to enforce. We detail several standards in Chapter 7. Here, we will merely treat one—performance guarantees.

A *performance guarantee* is an upper bound or worst case limit of the value of ρ for any problem instance. To demonstrate, we can consider an example pertaining to the traveling salesman problem.

Example 2.4. Twice-Around Heuristic for Traveling Salesman Problems. Consider a traveling salesman problem defined on a complete graph with p vertices, where all edge weights, w_{ij}, satisfy the triangle inequality (i.e., $w_{ij} + w_{jk} \geq w_{ik}$). One heuristic operates in the following manner. Determine a minimum weight spanning tree in the graph. Let this spanning tree be T and duplicate every edge of it. The resultant multigraph, say $\overline{T}$, will be such that it must be possible to find (starting at any vertex) a closed path that uses each edge in $\overline{T}$ exactly once. Taking this path to be given by an ordered list of vertices, we begin with an initial vertex and proceed through the list jumping over vertices that have been previously encountered in the list. The result is a traveling salesman tour.

The simple procedure described above is illustrated in Figure 2.7. Clearly, the process of jumping over duplicated vertices in the list described is nothing more than a scheme of short cutting from one vertex to another on the spanning tree, T. The question of interest, however, regards the quality of the proposed heuristic.

As it turns out, constructing this bound on ρ for the stated twice around heuristic is very simple. Let the total weight of T be $l(T)$ and of $\overline{T}$, $l(\overline{T})$. Clearly, $l(\overline{T}) = 2l(T)$. However, if l^* is the length of an optimal tour,

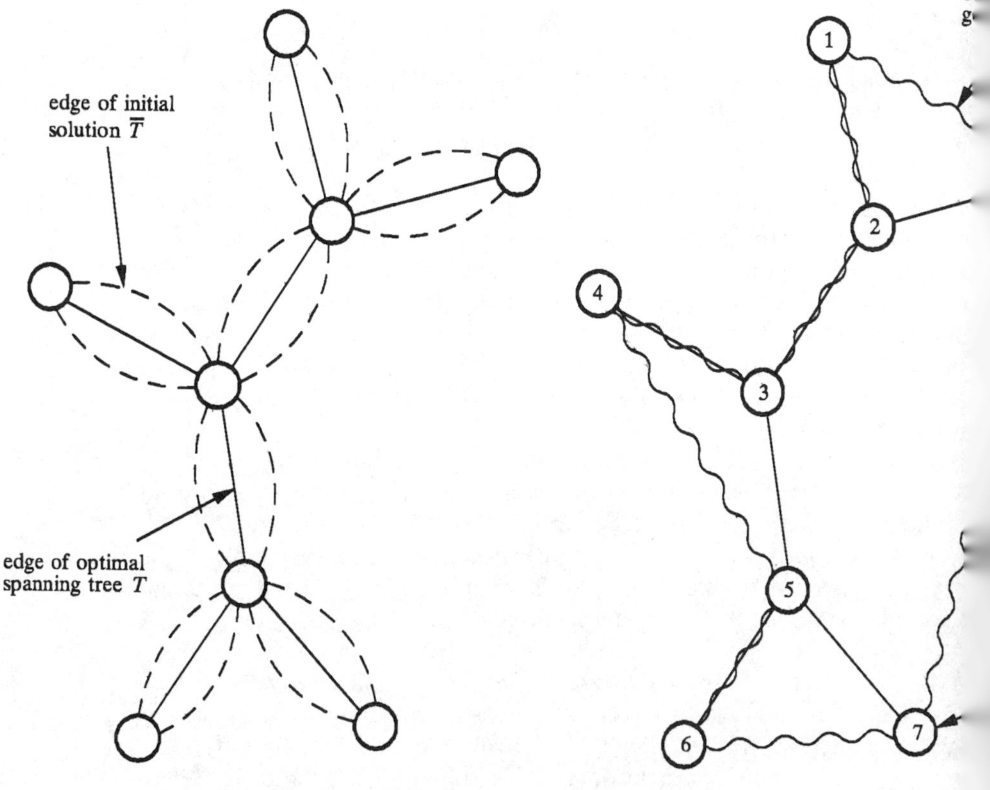

Fig. 2.7. Illustration of the twice around heuristic for the traveling salesman problem.

$l(T) \le l^*$ or $2l(T) \le 2l^*$ because no tour can cost less than the minimum weight spanning tree. Hence, $l(\overline{T}) \le 2l^*$. But the shortcutting process in the heuristic cannot add length due to the triangle inequality. That is, the heuristic solution $l_h \le l(\overline{T})$, and so $l_h \le 2l^*$. We have a performance guarantee of *2* for the stated heuristic. ■

We return to performance guarantees in Chapter 7, but it should be noted in the present discussion of complexity classes that there seem to be problem groupings distinguished by the sort of performance guarantees that can be achieved in polynomial time. In a few cases, nonexact algorithms are available that, in a certain sense, come as close as desired to an optimum in polynomial time. Computation time must be bounded by a polynomial in the length of the problem encoding and an allowed error. Time will increase with the required accuracy (sometimes exponentially), but for fixed accuracy, a polynomial time scheme is available.

For most of the common discrete optimization problems, no similar scheme for achieving any desired accuracy is known. It is often relatively easy to find, however, some nonexact algorithm with a uniform, constant performance guarantee. The bound of 2 derived above provides an example. It would not seem unfair to classify problems with such constant performance guarantees as the moderates of the *NP*-Equivalent class.

This third, least tractable group of *NP*-Equivalent problems has the rather negative property that unless *P* and *NP* are equivalent, no constant (finite) upper bound on the ratio ρ can be achieved for any polynomial time heuristic. Notable along these lines is the result of Sahni and Gonzales (1976) that states that unless $P = NP$, no polynomial time heuristic for an arbitrarily weighted traveling salesman problem can exist that possesses a finite worst case bound. If such a heuristic did exist, we could contrive edge lengths on missing edges of partial graphs and use the procedure to decide whether an arbitrary graph admits a Hamiltonian cycle. The latter problem is well-known to be *NP*-Complete.

EXERCISES

2-1. An undirected graph $G(V, E)$ can be encoded either (i) by the number of vertices $|V|$, the number of edges $|E|$, and a list of $|E|$ vertex pairs, or (ii) by the number of vertices $|V|$ and a $|V|$ by $|V|$ adjacency matrix $\mathbf{A}$ having $a_{ij} = 1$ when vertex i is adjacent to vertex j in the graph. Show that these two encodings are polynomially equivalent, i.e., an algorithm is polynomial for one if and only if it is polynomial in the other.

2-2. Follow the pattern of Example 2.1 to define Turing machines to perform each of the following tasks.

(a) Determine whether one given binary number is the sum of two others given.

(b) Determine whether the first of two given binary numbers is the largest in magnitude.

2-3. A k-tape Turing machine is defined exactly as the 1-tape variety of Section 2.1.3 except that storage is arrayed over a fixed number k tapes, and each step of computation can use/change the contents of the square under the read/write head for each tape, and/or move each tape ± 1 square. Show that computation is polynomially bounded for k-tape Turing machines if and only if it is polynomially bounded for 1-tape Turing machines.

2-4. A synchronous parallel Turing machine can be defined as a complex of finitely many, k, Turing machines operating on a problem in parallel. The machines share a k by $k-1$ common memory of squares μ_{ij}, where processor i can send one alphabet character of information to processor j. All processors start together. At each step each machine, i, can do everything that a single Turing machine can do, but, in addition, it can use input messages $\{\mu_{ki}: k \neq i\}$ and change output messages $\{\mu_{ik}: k \neq i\}$. Show that such a parallel Turing machine stops in polynomially-bounded time if and only if any ordinary serial Turing machine does.

2-5. The bounded integer programming problem is to min{$\mathbf{cx}$: $\mathbf{Ax} \geq \mathbf{b}$, $\mathbf{u} \geq \mathbf{x} \geq \mathbf{0}$, $\mathbf{x}$ integer}, where $\mathbf{A}$ is a given integer matrix, and $\mathbf{b}$, $\mathbf{c}$ and $\mathbf{u}$ are given integer vectors. The corresponding $0-1$ optimization is identical except $\mathbf{u} = \mathbf{1}$. Show that the two problems are polynomially equivalent, i.e., one can be solved in polynomial time if and only if the other can.

2-6. Roman numerals encode numbers by strings of separate symbols for each power of 10 (I, X, C, M, . . .) and half each power (V, L, D, . . .). Consider encoding a problem with instances having arbitrarily large integer constants.

(a) Show that assumptions of the Turing model of computation would be violated by an encoding in Roman numerals if new symbols continue to be used for ever higher powers (and half powers) of 10.

(b) One solution to the difficulty of (a) is to end the symbol sequence at some fixed power k, using more and more copies of the symbols for large values. Show that an algorithm applied to such an encoding would be polynomial (respectively pseudopolynomial) if and only if stroke (i.e., unary) encoding is.

★(c) Attempt to devise another solution to the issue of (a) that is similarly equivalent to binary encoding. (Hint: Consider symbol sequencing.)

2-7. Show that each of the following problems defined in Section 2.3 belongs to NP.

(a) Satisfiability
(b) Clique
(c) Exact Cover
(d) 3-Satisfiability
(e) Vertex Cover
(f) Set Partition
(g) Chromatic Number
(h) Clique Cover
(i) Feedback Arc Set
(j) Knapsack
(k) 0–1 Integer Programming
(m) Feedback Vertex Set
(n) Directed Hamiltonian Circuit
(o) Undirected Hamiltonian Cycle
(p) 3-Dimensional Matching
(q) Hitting Set
(r) Steiner Tree
(s) Job Sequencing
(t) Partition
(u) Max Cut

2-8. Show in detail how each of the following reductions sketched in Section 2.3.2 is a polynomial transformation, i.e., (i) that the answer to the second problem is *yes* exactly when that of the first is *yes*, and (ii) that the instance of the second produced is bounded in length by a polynomial in the length of the first.

(a) Satisfiability $\propto$ 0–1 Integer Programming
(b) Clique $\propto$ Set Packing
(c) Vertex Cover $\propto$ Set Covering
(d) Vertex Cover $\propto$ Feedback Vertex Set
(e) Vertex Cover $\propto$ Feedback Arc Set
(f) Vertex Cover $\propto$ Directed Hamiltonian Circuit
(g) Directed Hamiltonian Circuit $\propto$ Undirected Hamiltonian Cycle
(h) 3-SAT $\propto$ Chromatic Number
(i) Chromatic Number $\propto$ Clique Cover
(j) Chromatic Number $\propto$ Exact Cover
(k) Exact Cover $\propto$ Hitting Set
(l) Exact Cover $\propto$ Steiner Tree
(m) Exact Cover $\propto$ 3-Dimensional Matching
(n) Exact Cover $\propto$ Knapsack
(o) Knapsack $\propto$ Job Sequencing
(p) Knapsack $\propto$ Partition
(q) Partition $\propto$ Max Cut

2-9. For each of the following discrete optimization problems defined in Section 1.1, state the threshold existence (i.e., decision) version and prove it is *NP*-Complete:

(a) Traveling Salesman Problem (undirected graph)
(b) Traveling Salesman Problem (directed graph)
(c) Parallel Machine Scheduling Problem
(d) Bin Packing Problem
(e) *p*-Median Problem (directed graph)

(f) p-Median Problem (undirected graph)
(g) Fixed Charge Problem ($S \triangleq \{\mathbf{x} \in \mathbb{R}^n\colon \mathbf{x} \geq \mathbf{0}, \mathbf{Ax} \geq \mathbf{b}\}$ for integer matrix $\mathbf{A}$ and vector $\mathbf{b}$)

2-10. Consider the threshold linear programming problem "Given integer matrix $\mathbf{A}$, integer vectors $\mathbf{b}$ and $\mathbf{c}$, and an integer value v, does there exist $\mathbf{x} \geq \mathbf{0}$ such that $\mathbf{Ax} = \mathbf{b}$, $\mathbf{cx} \leq v$?" Let $L \triangleq$ the number of bits in the binary encodings of such instances.

(a) Show the determinant of any square submatrix of $(\mathbf{A}, \mathbf{b})$ has absolute value bounded by 2^L.
(b) Use the result of (a), together with Cramer's rule and the theory of basic solutions (see Appendices A and C) to show that such binary encodings of threshold linear problems belong to *NP*.

2-11. For each of the following discrete optimization problems defined in Exercise **1-2**, state the threshold existence (i.e., decision) version and prove it is *NP*-Complete:

(a) Uncapacitated Facilities Location
(b) Capacitated Facilities Location
(c) Generalized Assignment

2-12. Reductions of Figure 2.4 are done in only one direction, i.e., we show $(P_1) \propto (P_2)$ if a polynomial algorithm for (P_2) would provide one for (P_1).

(a) Apply Cook's Theorem to show that if (P_1) is *NP*-Complete and $(P_2) \in NP$, then $(P_2) \propto (P_1)$.
(b) Define explicit polynomial reductions (not using Cook's Theorem) that show how $(P_2) \propto (P_1)$ for as many as possible of the $(P_1) \propto (P_2)$ relationships shared in Figure 2.4.

★2-13. Consider the following problem: Given a graph G and a tree H, does G possess a subgraph isomorphic to H? Show that even if G is restricted to *series-parallel graphs* (graphs having no subgraph homeomorphic to K_4), this problem is *NP*-Complete.

2-14. Let $G(V, E)$ be a simple graph having nonnegative integer edge weights. Then for some integer $0 < k \leq |V|$, show that the (decision version) problem of producing a minimum weight spanning tree of G having no vertex degree in excess of k is *NP*-Complete.

★2-15. Given an arbitrary graph $G(V, E)$, show that simply deciding if G possesses a spanning subgraph that is Eulerian is *NP*-Complete.

2-16. The class *CoNP*-Complete can be defined as either (i) the set of complements of *NP*-Complete problems, or (ii) the set of members of *CoNP* to which all other members polynomially reduce. Show that these two definitions are equivalent.

2-17. Show that if the complement of any *NP*-Complete problem belongs to *NP* then *NP* = *CoNP*.

2-18. Show that every decision problem with finitely many instances belongs to *P*.

2-19. Show that if *P* = *NP*, every decision problem with finitely many instances is *NP*-Complete.

2-20. Papadimitriou and Yannakakis (1984) have shown the significance of a new complexity class D^p of problems defined by the intersection of a member of *NP* and a member of *CoNP*. That is, instances of problems in D^p must simultaneously be instances of both a defining problem in *NP* and a defining problem in *CoNP*.

(a) Explain how D^p is qualitatively different than $NP \cap CoNP$

(b) Let $C^{\leq}$ be a threshold problem of the form "Does there exist $\mathbf{x} \in F$, such that $\mathbf{cx} \leq v$?" where F, $\mathbf{c}$ and v are the feasible set of an instance, the objective function, and a threshold respectively. A corresponding exact value form $C^{=}$ is "Is $v = \inf\{\mathbf{cx} : \mathbf{x} \in F\}$?" Show that if $\mathbf{cx}$ is always integer valued, $C^{=} \in D^p$ whenever $C^{\leq} \in NP$.

2-21. One form of the knapsack problem is

$$\max\left\{\sum_{j=1}^{n} c_j x_j : \sum_{j=1}^{n} a_j x_j \leq b;\ x_i \geq 0 \text{ and integer for } j = 1, 2, \ldots, n\right\},$$

where b, $\{c_j\}$ and $\{a_j\}$ are given positive integers, and $a_j \leq b$ for all j. One heuristic procedure for approximate solution is the set $\tilde{x}_{\tilde{j}} \leftarrow \lfloor b/a_{\tilde{j}} \rfloor$, $\tilde{x}_j \leftarrow 0$ for $j \neq \tilde{j}$, where $c_{\tilde{j}}/a_{\tilde{j}} = \max\{c_j/a_j : j = 1, 2, \ldots, n\}$. Show that the ratio of the optimal solution value to this approximate solution is never worse than $\rho = 2$.

2-22. Consider the knapsack problem of Exercise **2-21.** Show that the problem can be solved in $O(nb)$ time by recursive application for $k \leftarrow 1, \ldots, n$ and $y = 0, 1, \ldots, b$ of the following construction

$$v_1(y) \longleftarrow c_1 \lfloor y/a_1 \rfloor$$

$$v_k(y) \longleftarrow \max_{x_k = 0, 1, \ldots, \lfloor y/a_k \rfloor} \{c_k x_k + v_{k-1}(y - a_k x_k)\}$$

where $v_j(y) \triangleq$ the optimal value of a knapsack problem in the first j variables with right-hand-side limit y.

2-23. Return to the knapsack problem of Exercises **2-21** and **2-22,** and compare to the knapsack decision problem of Section 2.3.2.

(a) Adapt the algorithm of **2-22** to produce a polynomial algorithm for Knapsack when input is stroke (unary) encoded.

(b) Show that this algorithm is not polynomially bounded when Knapsack is binary encoded.

(c) Show that the reduction of Section 2.3.2 from Exact Cover is not a polynomial transformation if Knapsack is stroke (unary) encoded.

2-24. Given a directed graph $G(V, A)$ and nonempty vertex subsets V_1, $V_2 \subset V$ with $V_1 \cap V_2 = \varnothing$,

(a) Develop a polynomial time algorithm for determining whether the graph contains a directed path from any element of V_1 to any element of V_2.

*(b) Show that the algorithm of (a) can be used to solve 2-*SAT* (the version of Satisfiability where every clause has at most two literals) in polynomial time.

***2-25.** Show that 3-*SAT* restricted to instances where no variable appears in more than three clauses is in class *P*. (Hint: See Hall's theorem on systems of distinct representatives in, for example, Berge (1971)).

2-26. Form a list of combinatorial problems where deciding the *existence* of a solution can be easily accomplished but where an effective (efficient) procedure for *producing* such a solution is not known.

***2-27.** All planar graphs are known to be (vertex) colorable (i.e., colored so that adjacent vertices have distinct colors) using no more than four colors. Consider the following "4-color completion" problem: Given a planar graph, where some subset of vertices are feasibly colored (using no more than four colors), can the remaining vertices be colored so that overall, no more than four colors are required? Prove this problem to be *NP*-Complete.

2-28. For a tree, *T*, let a subset of the vertices be 2-colored. Give a polynomial-time algorithm for deciding if the remaining vertices can be colored yielding a feasible 2-coloring overall. (Note that the biparticity of trees trivially establishes that they are 2-colorable.)

***2-29.** The path-length distribution of a tree, *T*, is the *t*-tuple $(x_1, x_2, \ldots, x_t)$, where x_k is the number of *k*-edge paths in *T*. Give a polynomial algorithm for computing the path-length distribution. Is the path-length distribution unique?

2-30. Give a problem (decision version), which is *NP*-Hard but not in *NP*.

2-31. Section 2.5.3 asserted that value versions of optimization problems solvable by linear programming over a polynomial-time derivable system of linear constraints belong to *NP*. Establish that fact rigorously.

3

Polynomial Algorithms—Matroids

The survey of complexity theory in Chapter 2 highlighted the fundamental distinction between algorithms for polynomial-time solvable problems and ones probably requiring procedures that are exponential-time in the worst case. In this chapter, we begin our treatment of algorithmic results by exploring the first of two major subcategories of polynomial-time algorithms for discrete optimization—ones over matroids. Chapter 4 treats the other main class of polynomial-time procedures, linear programming. Nonpolynomial procedures are addressed in Chapters 5 and 6.

Hassler Whitney introduced the concept of a matroid in his seminal work (Whitney (1935)) dealing with abstract properties of linear dependence. Initially, Whitney employed notions yielding an abstract generalization of a matrix (thus the term *matroid*); however, the language common in present-day matroid theory stems as much from graph theory as from linear algebra.

Matroid theory has generated an array of useful results and insights that have become central to the field of combinatorial optimization. Still, the need for even a dose of basic matroid theory often is met with skepticism. Is such a background as crucial for survival in the field of discrete optimization as say knowledge of branch and bound procedures, the simplex algorithm, or even computational complexity? In most (unbiased) quarters, the answer would likely be *no.* Indeed, the abstract combinatorial structure of matroids renders the necessity of their in-depth study difficult to justify for solely problem solving.

Why, then, do we consider them at all? We can justify the study of matroids on straightforward pedagogical grounds. That is, in attaining even a basic understanding of the theory of matroids, one's discrete optimization insight and understanding must certainly be enhanced. Matroid theory provides unification and generalization in a field frought with isolated results. Moreover, we shall see that it yields elegant proofs of phenomena leading to efficient solutions to numerous problems and, at least as important, draws generic boundaries for the set of problems likely to admit polynomial-time schemes.

Our treatment of matroids will be restricted to the essentials needed for this pedagogical goal. In this regard, the reader wishing more details and proofs of theorems that we only state is referred to the many excellent surveys of matroid theory. We particularly recommend Welsh (1976) and Bixby (1981). Readers unfamiliar with graph theory are also referred to the brief review in Appendix B for concepts used throughout the chapter.

3.1 Independence Systems and Matroids

In order to establish a natural setting for our discussion of matroids and to provide some essential background for later developments (e.g., Chapter 7), we begin by examining a class of structures more general than matroids.

Suppose we have a finite ground set G and a collection, $\mathscr{I}$, of subsets of G. Then the pair $(G, \mathscr{I})$, is said to be an *independence system* if the empty set is in $\mathscr{I}$ and $\mathscr{I}$ is closed under inclusion; that is

$$\text{(i) } \varnothing \in \mathscr{I} \tag{3-1}$$

$$\text{(ii) } H_1 \subseteq H_2 \in \mathscr{I} \Rightarrow H_1 \in \mathscr{I} \tag{3-2}$$

Elements in $\mathscr{I}$ are called *independent sets* and, naturally enough, subsets of G not in $\mathscr{I}$ are said to be *dependent.* A simple and easily recognized independence system can be illustrated by the following example.

Example 3.1. An Independence System. Let G be the edge set of an undirected graph and define $\mathscr{I}_1$ to be the collection of subsets of edges in G having the property that no two edges in a given subset are incident to the same vertex. For the simple graph shown below, we have $G = \{e_1, e_2, e_3, e_4\}$ and $\mathscr{I}_1 = \{\varnothing, \{e_1\}, \{e_2\}, \{e_3\}, \{e_4\}, \{e_2, e_3\}\}$.

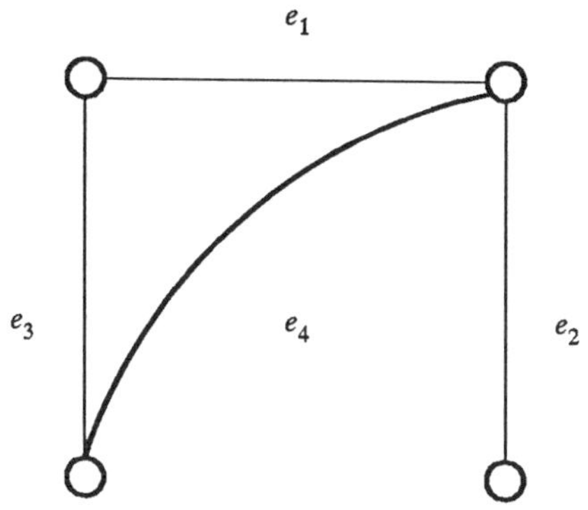

Clearly, members in $\mathscr{I}_1$ satisfy (3-1) and (3-2) and thus, the specific pair $(G, \mathscr{I}_1)$ is an independence system. ■

We say that a subset $Z \subseteq H$ is *maximal* with respect to some property if there exists no other subset $Z' \subseteq H$ that has the property and properly contains Z. Relativizing, for some $H \subseteq G$, a subset $Z \subseteq H$ is *maximal independent* if it is independent and augmenting Z with any element of H destroys its independence. In Example 3.1, $\{e_1\}$, $\{e_4\}$, and $\{e_2, e_3\}$ are maximal independent subsets in $\mathscr{I}_1$.

Now suppose we consider an alternative independence system over set G defined as in Example 3.1, with family $\mathscr{I}_2$ containing subsets having the property that none induces a cycle. In that case

$$\begin{aligned}\mathscr{I}_2 = \{&\varnothing, \{e_1\}, \{e_2\}, \{e_3\}, \{e_4\}, \{e_1, e_2\}, \{e_1, e_3\}, \{e_1, e_4\},\\ &\{e_2, e_3\}, \{e_2, e_4\}, \{e_3, e_4\}, \{e_1, e_2, e_3\}, \{e_1, e_2, e_4\},\\ &\{e_2, e_3, e_4\}\},\end{aligned}$$

with the maximal independent sets of G being the last three shown. Maximal independent subsets of $A \triangleq \{e_1, e_3, e_4\}$ are $\{e_1, e_3\}$, $\{e_1, e_4\}$ and $\{e_3, e_4\}$.

Note that here all of the maximal independent sets have the same cardinality (but the $\mathscr{I}_1$ ones do not). When the cardinality of all maximal independent sets is the same, the notions of maximal and maximum are equivalent. It is this additional structure (over all subsets) that distinguishes matroids among independence systems. More succinctly, if $(G, \mathscr{I})$ is an independence system such that

$$\text{(iii) for every } H \subseteq G, \text{ if } H_1, H_2 \subseteq H \text{ are maximal over subsets of } H \text{ in } \mathscr{I}, \text{ then } |H_1| = |H_2|, \tag{3-3}$$

we call the pair $(G, \mathscr{I})$ a *matroid*. The $\mathscr{I}_1$ independence system of Example

3.1 is not a matroid, while the one just described is, since in the latter, condition (3-3) is satisfied. While the $\mathscr{I}_2$ case is a matroid, it is important to note that further proof is required. We have seen that all maximal independent subsets of G have the same cardinality, but to be a matroid, we must also have this equal cardinality property for maximal independent subsets of all $H \subset G$.

There are numerous equivalent ways of defining a matroid from axioms like (3-1) to (3-3), but we shall not need them in our coverage. The interested reader is directed to one of the full treatises on matroid theory.

Let $M = (G, \mathscr{I})$ be a matroid. Then, a maximal independent subset of G is called a *base* of M, and for any subset $H \subseteq G$, the cardinality of a maximal independent subset of H, is referred to as the *rank* of H, denoted $r(H)$. The rank of a matroid is simply $r(G)$.

A *circuit* of M is a minimal dependent subset of G. Here we have our first taste of the graph-theoretic influence in matroid theory; for a certain class of matroids, these circuits correspond exactly to cycles in a graph. This is obvious from our illustration above where $\mathscr{I}_2$ contained all cycle-free subsets of edges in the graph employed in Example 3.1. The *span* of a subset $H \subseteq G$ is a maximal superset of H for which the rank is the same as H. The span of H is denoted by $sp(H)$. If $B \subseteq G$ is a base then $sp(B) = G$.

3.2 Examples of Matroids

Matroids can arise in a number of interesting and useful ways.

3.2.1 Uniform Matroids

Denoted by $U_{k,n}$, this independence system is particularly simple. It is defined on n elements in which every subset of cardinality no greater than k is independent. To see that $U_{k,n}$ is a matroid, note that subsets of independent sets are obviously independent. Also, given any subset H of elements, if $|H| \leq k$, then trivially, H is independent and maximal in itself; if $|H| > k$, then every maximal independent subset of H has cardinality equal to k.

3.2.2 Partition Matroids

Here we have a partition P of G (disjoint and exhaustive division into subsets) such that $P = \{G_1, G_2, \ldots, G_k\}$ and nonnegative integers $i_1, i_2, \ldots, i_k$. M_p is the independence system $(G, \mathscr{I})$, where $H \subseteq G$ is independent if and only if $|G_t \cap H| \leq i_t$, $1 \leq t \leq k$. That is, a subset of G is independent if and only if it has, for any t, at most i_t elements in common with the respective G_t. Again, it is easy to see (3-1) and (3-2) hold. To show (3-3) is satisfied, let H be any subset of G. Every maximal independent subset of H is formed by simply selecting as many elements as possible (but no more than i_t) from each subset G_t having nonempty intersection with H. The cardinality of each will be $\Sigma_{t=1}^{k} \min \{i_t, |G_t \cap H|\}$.

3.2.3 *Transversal Matroids*

Let $G(V, E)$ be an undirected graph having no odd cycles (cycles with an odd number of edges). Such a graph is *bipartite,* i.e., we can partition V into two subsets S and T such that every edge in E is incident to a vertex in S and one in T. Letting $G \triangleq S$, we may define $X \subseteq S$ as independent if there is a subset of edges in E such that exactly one member of the subset meets each vertex of X, and edges of the subset meet distinct members of T. That is, the members of X form a *system of distinct representatives* of the subsets of S adjacent to particular members of T. Denote this independence system by M_T.

To show that M_T satisfies (3-3), choose $X \subseteq S$ and let H_1 and H_2 be a pair of independent subsets of X. Further, let E_1 and E_2 be the subsets of E such that E_i meets S at exactly H_i, $i = 1, 2$. Assuming, then, that $|H_1| < |H_2|$, one can show that there exists an element $v \in X \backslash H_1$ such that $H_1 \cup \{v\}$ is an independent set, i.e., H_1 is not maximal. This is sufficient to establish that M_T is a matroid.

Transversal matroids play a central role in the theory of matching, particularly, bipartite matching. Also of interest is the fact that all uniform and partition matroids are transversal.

3.2.4 *Graphic Matroids*

If E is the edge set of a graph, and if $\mathscr{I}$ contains those subsets of edges forming cycle-free subgraphs (forests) in the graph, then $(E, \mathscr{I})$ is a matroid. Specifically, $(E, \mathscr{I})$ is called the *polygon* or cycle matroid of the graph denoted by M_G. The illustration given earlier provides an example. Any matroid occurring as the cycle (polygon)-matroid of a graph is said to be *graphic.*

3.2.5 *Representable, Regular and Binary Matroids*

Let $\mathbf{A}$ be a matrix over a field Q (i.e., field of real numbers, binary field, etc.) and let G correspond to the columns of $\mathbf{A}$. Then, if $\mathscr{I}$ is a collection of linearly independent subsets of columns of $\mathbf{A}$, $(G, \mathscr{I})$ is a matroid. Matroid axioms follow from the well-known property that all bases of vector spaces have the same cardinality. $(G, \mathscr{I})$ derived from matrices in this way are called *representable* matroids.

A representable matroid is *binary* if it is representable over the field of integers modulo 2 (sometimes given as $GF(2)$) and is *regular,* if it is representable over $\mathbb{R}^m$ by a matrix with each square submatrix having determinant 0, +1 or −1. The diagram in Figure 3.1 illustrates known relationships between these various classes of matroids. All indicated containments can be strict.

3.3 Matroid Duality

Like many other elegant mathematical structures, matroids admit a notion of duality. The matroid concept of duality also can be seen to subsume many dualities for special cases. A formal definition is contained in Whitney's (1935) classic theorem:

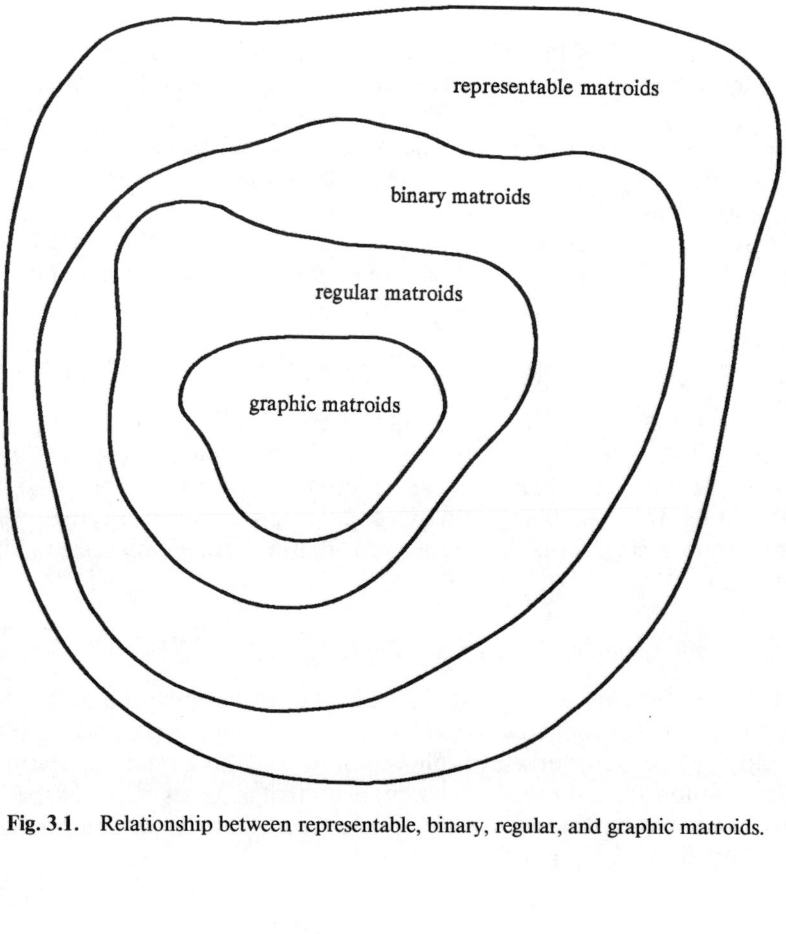

Fig. 3.1. Relationship between representable, binary, regular, and graphic matroids.

Theorem 3.1. Dual Matroid. *If G is a set of elements and* $\{B_k : k \in K\}$, *a set of bases of a matroid, M on G, then* $\{G \backslash B_k : k \in K\}$ *is the set of bases of another matroid, M* on G.*

M^* is referred to as the *dual matroid* of M and as one might suspect, the dual of M^* restores M, i.e., $(M^*)^* = M$.

Many properties of the dual matroid follow directly from the definition. For example, a subset $H \subseteq G$ is independent in M^* if and only if $G \backslash H$ contains a base of M. In addition, for given M and M^* on G, the rank of M^* is $|G| - r(M)$. More generally, we have:

Theorem 3.2. Rank Functions of *M* and *M.** *The rank functions of M and M*, say r and r* respectively, are related by*

$$r^*(G \backslash H) = |G| - r(G) - (|H| - r(H)), \qquad \text{for all } H \subseteq G \qquad (3\text{-}4)$$

For matroids M and M^*, the rank function r^* is called the *corank* function of M. A base of M^* is a *cobase* of M and a circuit of M^* is a *cocircuit* of M. Some of these duality relationships can be easily demonstrated by the following illustration.

Example 3.2. Matroid Duality. Consider the graphic matroid, M_G defined on the graph, $G(V, E)$, below:

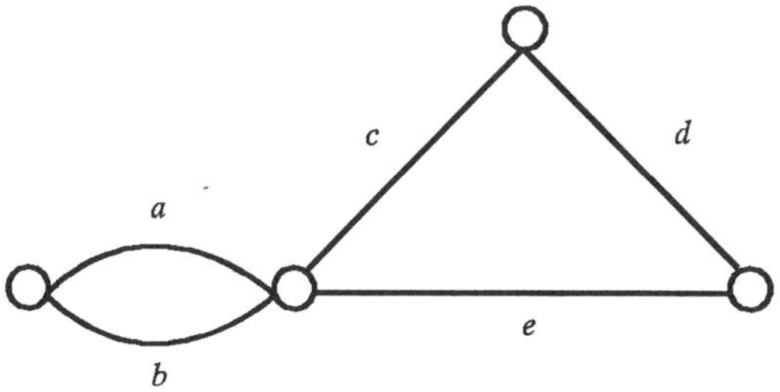

We have: circuits of M_G: $\{\{a, b\}, \{c, d, e\}\}$

bases of M_G: $\{\{a, c, e\}, \{a, c, d\}, \{a, d, e\}, \{b, c, e\}, \{b, c, d\}, \{b, d, e\}\}$

bases of M_G^*: $\{\{b, d\}, \{b, e\}, \{b, c\}, \{a, d\}, \{a, e\}, \{a, c\}\}$

circuits of M_G^*: $\{\{a, b\}, \{c, d\}, \{c, e\}, \{d, e\}\}$

Observe that the rank of $M = r(E) = 3$. That of M^*, $r^*(E) = 2 = |E| - r(E)$. For a nonbase set $H \triangleq \{a, b, c, d\}$ in M,

$$|E| - r(E) - (|H| - r(H)) = 5 - 3 - (4 - 3) = 1 = r^*(\{e\}) \triangleq r^*(E \backslash H) \blacksquare$$

Examine again the graph used in Example 3.2 and, in particular, let us locate those sets of edges whose removal disconnects the graph and that are minimal in this regard. We call such a subset of edges a *cocycle*. Accordingly, the cocycles of the graph shown are $\{a, b\}$, $\{c, d\}$, $\{c, e\}$, and $\{d, e\}$. More importantly, however, we note that these are just the circuits of the dual of the graphic matroid defined on the graph in question. Furthermore, this outcome is not at all coincidental. The next theorem demonstrates that.

Theorem 3.3. Cocycles and Circuits. *If C^* is the cocycle set of a graph, then C^* is the set of circuits of the graphic matroid's dual matroid M_G^*.*

M_G^* is referred to as the *cocycle matroid.*

A *planar graph* is a graph that can be drawn in the plane without edge crossings. Such a drawing is called the *plane imbedding* of the graph. For a given planar graph G, denote a plane imbedding by G_p.

Now, for any planar graph G and, in particular, for any plane imbedding G_p, we can define the geometric dual $G_p{}^*$ as follows: vertices in G_p^* correspond to faces in G_p and pairs of vertices in G_p^* are connected by an edge, if and only if the respective faces in G_p have an edge in common. Edges in G_p that are on the boundary of only one face (sometimes called the outer or exterior face) result in a self-loop in G_p^*. We illustrate the construction of a geometric dual in Figure 3.2.

It is easy to see that the geometric dual of any G_p is also planar. It is also easy to see that $(G_p^*)^*$ is isomorphic to G_p. The relevance of geometric duals here is that matroid duality can be seen to subsume it.

Theorem 3.4. Planar Graph Duality and Matroid Duality. *If G is a planar graph and G_p^* is any geometric dual of an imbedding of G, then the cycle matroids of G and G_p^* are dual matroids.*

The result of Theorem 3.4 is easy to demonstrate. Consider again the graph used in Example 3.2. If we let the graph as presented be G_p, then G_p^* appears below:

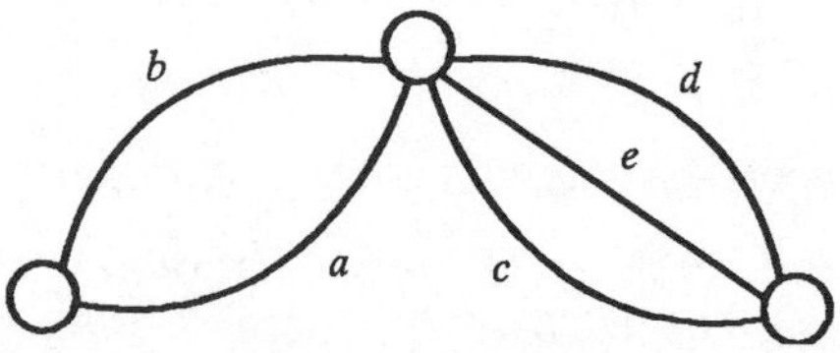

Observe that we have labeled the edges in G_p^* consistent with the labeling in G_p. This naming makes it easy to see that the cycle matroid of G_p^* has bases

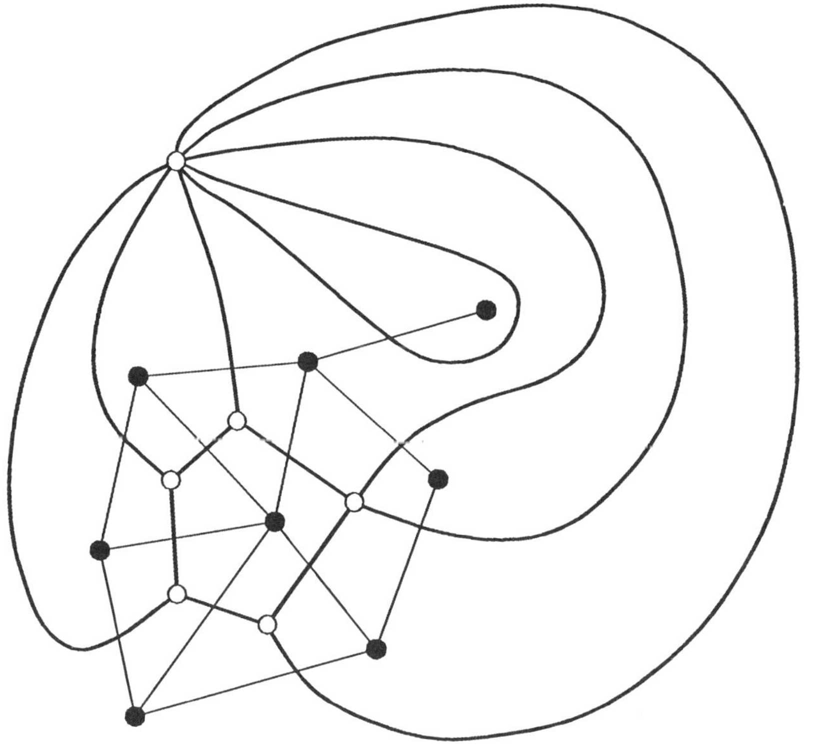

Fig. 3.2. G_p and G_p^* (bold lines).

given by $\{a, c\}$, $\{a, e\}$, $\{a, d\}$, $\{b, c\}$, $\{b, d\}$, and $\{b, e\}$ (the spanning trees in G_p^*). These sets, though, are exactly the bases of M_G^* in Example 3.2, i.e., the dual matroid of M_G, the cycle matroid of G_p.

Some readers may recall the fundamental result of Euler that relates the number of edges, vertices, and faces of a planar graph. Letting the set of faces defined by some connected G_p be $D(G_p)$, Euler showed that $|V| + |D(G_p)| - |E| = 2$. Using what we know regarding matroids, we can now see that this result is a straightforward consequence of property (3-4) developed earlier. We know that $r(M) + r(M^*) = |G|$. Thus, if we let M_G be the cycle matroid of G_p and, M_G^* the cycle matroid of G_p^*, then $r(M_G) = |V| - 1$, $r(M_G^*) = |D(G_p)| - 1$, and Euler's formula follows.

Of course, there are numerous other results that demonstrate the power and beauty of matroids as a generalization in graph theory. Classic among these is Tutte's work characterizing planar and graphic matroids (Tutte (1958)). Unfortunately, we do not have space in this book for coverage of these deep results.

3.4 Optimization and Independence Systems

Given an independence system, $(G, \mathscr{I})$, let us define nonnegative weights $\{w_e : e \in G\}$. Then the *maximum weight independent set problem* can be stated as

$$(IS) \qquad \max\left\{\sum_{e\in H} w_e : H \in \mathscr{I}\right\} \qquad (3\text{-}5)$$

3.4.1 The Greedy Algorithm

Suppose we adopt the following naive approach to problem (3-5). Simply arrange the elements in G in nonincreasing order of their weights and choose (in this order) elements, rejecting a given choice only if its selection would destroy independence. More formally, we have

Algorithm 3A. The Greedy Algorithm

STEP 0: *Initialization.* Arrange the elements in G as $e_{[1]}, e_{[2]}, \ldots, e_{[|G|]}$ in nonincreasing order of their weights. Set solution set $I \leftarrow \varnothing$ and $k \leftarrow 1$.

STEP 1: *Augmentation.* If $k > |G|$, stop. Otherwise, if $I \cup \{e_{[k]}\} \in \mathscr{I}$, accept $I \leftarrow I \cup \{e_{[k]}\}$. Set $k \leftarrow k + 1$ and repeat this step. ■

We can demonstrate Algorithm 3A with an example using an independence system introduced earlier.

Example 3.3. The Greedy Algorithm. Consider the independence system, $(G, \mathscr{I}_1)$, defined explicitly in Example 3.1 and let the edge weights be given such that $w_1 = 8$, $w_2 = w_3 = 6$ and $w_4 = 7$. Applying the algorithm, we see that $e_{[1]} = e_1$ and thus, $I = \{e_1\}$. Continuing, we have $e_{[2]} \triangleq e_4$; however, $\{e_1, e_4\} \notin \mathscr{I}_1$ so e_4 is not added to I. Similarly, we see that neither $\{e_1, e_2\}$ nor $\{e_1, e_3\}$ are in $\mathscr{I}_1$ and we stop with $I = \{e_1\}$ having weight of 8. ■

Of course, a hasty examination of the graph used for this example reveals that the edge subset $\{e_2, e_3\}$ having weight 12 is optimum over all subsets in $\mathscr{I}_1$. Clearly, the greedy algorithm can fail to produce an optimal solution. Are there interesting cases where such a naive scheme will always produce an optimum solution? Perhaps surprisingly, the answer is *yes*.

Theorem 3.5. Optimality of the Greedy Algorithm (Sufficiency). *Let $\{w_e : e \in G\}$ be nonnegative weights of the elements in a finite set G. Then if the independence system, $(G, \mathscr{I})$, is a matroid, Greedy Algorithm 3A will*

yield an optimum solution for the maximum weight independent set problem (IS).

Proof. Since $w_e \geq 0$ for all $e \in G$, there will clearly be an optimum for (IS) that is a maximal independent set (base) of $(G, \mathscr{I})$. Let I be the base produced by Algorithm 3A and B an optimal base in (IS). Also assume without loss of generality, that $|B \cap I|$ is maximum among the optimal bases.

If there is a first element $e \in G$ selected by Algorithm 3A and not belonging to B, since the latter is a maximal independent set, then $B \cup e$ contains a circuit U with $e \in U$. Also, there is at least one $\bar{e} \in U \backslash I$ because I can completely contain no dependent set.

Now, from an alternative axiomatic characterization of matroids, equivalent to (3-3), we know $B' = B \backslash \{\bar{e}\} \cup \{e\}$ is another base. (This is known as the *base axiom* and can be found in standard treatises on matroids including Welsh (1979)). But then, if B' is not to improve on the optimal B, $w_{\bar{e}} \geq w_e$. Since Algorithm 3A selected e before $\bar{e}$, we also know $w_e \geq w_{\bar{e}}$. We may conclude that $w_e = w_{\bar{e}}$ and B' is an alternative optimum base with $|B' \cap I| > |B \cap I|$. This is a contradiction. Thus, there must be no element $e \in I \backslash B$ and the theorem follows. ■

Theorem 3.5 demonstrates that Algorithm 3A always produces a maximum weight independent set in a matroid. The reader may wish to verify this result for the graphic matroid $(G, \mathscr{I}_2)$ discussed after Example 3.1 using the weights in Example 3.3 that failed for a nonmatroid independence system. In fact, we can show in general that the greedy algorithm will always fail for suitable weights on such independence systems.

Theorem 3.6. Optimality of the Greedy Algorithm (Necessity). *Let $(G, \mathscr{I})$ be an independence system. Then the greedy algorithm will produce an optimal solution to (IS) for all nonnegative weights $\{w_e : e \in G\}$ only if $(G, \mathscr{I})$ is a matroid.*

Proof. Assume that $(G, \mathscr{I})$ is not a matroid. Then we must show that there exists suitable weights $\{w_e : e \in G\}$ for which the greedy algorithm fails. Now, by hypothesis, there must exist a subset $H \subseteq G$ and two maximal subsets of H, say H_1 and H_2, such that $|H_1| < |H_2|$. Define

$$w_e \triangleq \begin{cases} 1 & \text{if } e \in H_1 \\ 1 - \epsilon & \text{if } e \in H_2 \backslash H_1 \\ 0 & \text{otherwise} \end{cases}$$

where $0 < \epsilon < (|H_2| - |H_1|)/|H_2|$. Clearly, the greedy algorithm will select $I \supset H_1$ with weight $|H_1|$. However, $\Sigma_{e \in H_2} w_e \geq |H_2|(1 - \epsilon) > |H_2|(|H_1|/|H_2|) = |H_1|$, which shows that the greedy algorithm produces a suboptimal set. ∎

Combining these two theorems leads to the following characterization.

Theorem 3.7. Matroid Characterization in Terms of the Greedy Algorithm. *Let G be a finite set of elements and $\mathcal{I}$ a collection of subsets of G. Then (G, $\mathcal{I}$) is a matroid if and only if*

(i) *(G, $\mathcal{I}$) is an independence system,*
(ii) *for all nonnegative weights $\{w_e : e \in G\}$, Greedy Algorithm 3A solves the maximum weight independent set problem (IS).*

What about the computational performance of Greedy Algorithm 3A? Up to this point, we have avoided this issue. It is clear, however, that the efficiency of the generic form in which Algorithm 3A is stated depends on the need to check for independence at Step 1.

It should come as no surprise that the computation needed to check independence varies greatly with the nature of the underlying matroid. For representable matroids, independence is easily checked in polynomial time by standard Gaussian elimination. This technique is also available in all the special cases shown in Figure 3.1, but much faster procedures are often known.

For nonrepresentable matroids checking independence can be *NP*-Hard. That is, Algorithm 3A is polynomial-time for all matroids only if $P = NP$. Fortunately, such examples rarely occur in matroids we encounter in discrete optimization. Still, it is important to note that Algorithm 3A is formally efficient only when we can check independence in polynomial time.

3.4.2 Polyhedral Interpretation

The treatment has proceeded so far on purely combinatorial grounds, disregarding the mixed-integer programming format to which all discrete optimization problems can be reduced. How can we obtain the latter representation?

For a matroid $M \triangleq (G, \mathcal{I})$, a *closed* set $S \subseteq G$ is a set for which $sp(S) = S$, i.e., no element $e \in G$ can be added to S without increasing $r(S)$. In terms of such sets, we can show that the maximum weight independent set problem (*IS*) can be formulated as

$$\max \quad \mathbf{wx} \tag{3-6}$$

$$\text{s.t.} \quad \mathbf{Ax} \leq \mathbf{r} \tag{3-7}$$

$$(W1) \qquad \mathbf{x} \geq \mathbf{0} \tag{3-8}$$

$$\mathbf{x} \text{ binary} \tag{3-9}$$

where $\mathbf{w}$ is the vector of nonnegative weights w_e, 0-1 decision vector $\mathbf{x}$ selects the optimal independent set, $\mathbf{A}$ is a matrix with rows equal to the incidence vectors of closed sets S_i of M and $\mathbf{r}$ is the corresponding right-hand-side vector $r(S_i)$.

The elegant fact demonstrated in the next theorem by Edmonds is that an optimal solution $\mathbf{x}$ can always be found in $(W1)$ without enforcing the binary requirement (3-9). That is, $(W1)$ can, in principle, be solved by linear programming.

Theorem 3.8. Polyhedral Characterization of Maximum Independent Sets. *For any matroid $M \triangleq (G, \mathcal{I})$, the vertices of the corresponding polyhedral system (3-7), (3-8) are the (binary) incidence vectors of the independent sets of $\mathcal{I}$.*

Proof. We shall proceed by demonstrating that for any weight vector $\mathbf{w}$ (not necessarily nonnegative), the linear program (3-6)–(3-8) solves at the incidence vector of an independent set I. Since weights can be chosen to make any element of $\mathcal{I}$ optimal in $(W1)$, this will establish the theorem.

Pick a weight vector $\mathbf{w}$ and construct an independent set $I \in \mathcal{I}$ by modifying a Greedy Algorithm 3A to stop when either $k > |G|$ or $w_{[k]} \leq 0$. We add independent elements to I in nonincreasing order of weight until none remains that can produce an independent set of greater weight.

Clearly, a set I constructed in this way will be independent and have incidence vector $\mathbf{x}^I$ that violates no constraint of (3-7)–(3-8). In particular, if $e_{j[1]}, e_{j[2]}, \ldots, e_{j[l]}$ is the sequence of elements placed in I, sets

$$S_1 \triangleq sp(\{e_{j[1]}\})$$
$$S_2 \triangleq sp(\{e_{j[1]}, e_{j[2]}\})$$
$$\vdots$$
$$S_l \triangleq sp(\{e_{j[1]}, e_{j[2]}, \ldots, e_{j[l]}\})$$

will correspond to rows of (3-7) satisfied as equalities.

Let these rows be the first l of (3-7) and consider the linear programming dual of (3-6)–(3-8)

$$\begin{aligned} \min \quad & \mathbf{ur} \\ \text{s.t.} \quad & \mathbf{uA} \geq \mathbf{w} \\ & \mathbf{u} \geq \mathbf{0} \end{aligned}$$

Vector $\mathbf{x}^I$ will be optimal in (3-6)–(3-8) if we can find a corresponding $\mathbf{u}^I$ feasible in the dual with the same objective function value. Construct

$$u_i^I \longleftarrow \begin{cases} w_{j[i]} - w_{j[i+1]} & \text{if } 1 \leq i < l \\ w_{j[l]} & \text{if } i = l \\ 0 & \text{otherwise} \end{cases}$$

Noting the $r_i \triangleq r(S_i) \triangleq i$, the dual objective value of this solution will be

$$\sum_{i=1}^{l} r_i u_i^I = \sum_{i=1}^{l-1} i(w_{j[i]} - w_{j[i+1]}) + l\, w_{j[l]} = \sum_{i=1}^{l} w_{j[i]}$$

This is exactly the value of $\mathbf{wx}^I$. The solution $\mathbf{u}^I$ is also nonnegative because we know from the nature of the Greedy Algorithm that $w_{j[i]} \geq w_{j[i+1]}$.

To see that $\mathbf{u}^I$ also satisfies $\mathbf{u}^I\mathbf{A} \geq \mathbf{w}$, first observe that constraints $(\mathbf{u}^I\mathbf{A})_{j[i]} \geq w_{j[i]}$ for $1 \leq i \leq l$ are satisfied as equalities by construction. Sum $(\mathbf{u}^I\mathbf{A})_{j[i]} = \Sigma_{k=i}^{l} u_k^I = w_{j[i]}$. For e_j not selected by the Greedy Algorithm, either $w_j \leq 0$ or $e_j \in S_i$ for all $t \leq i \leq l$, because e_j was skipped as dependent in conjunction with $e_{j[1]}, e_{j[2]}, \ldots, e_{j[t]}$.

In the first case, nonnegativity of $\mathbf{u}^I$ and $\mathbf{A}$ establishes $(\mathbf{u}^I\mathbf{A})_j \geq 0 \geq w_j$. In the second, $(\mathbf{u}^I\mathbf{A})_j = \Sigma_{k=t}^{l} u_k^I = w_{j[t]}$ and $w_{j[t]} \geq w_j$, because the Greedy Algorithm selected $e_{j[t]}$ first. This completes the proof. ■

Example 3.4. Matroid Polyhedron. To illustrate Theorem 3.8, let us return to the graphic ($\mathscr{I}_2$) matroid on the graph of Example 3.1 using weight vector $\mathbf{w} = (10, 5, 6, 7)$. The closed set incidence matrix $\mathbf{A}$ and rank right-hand-sides for that matroid are

$$\mathbf{A} = \begin{array}{c} \begin{matrix} e_1 & e_2 & e_3 & e_4 \end{matrix} \\ \begin{bmatrix} 1 & 0 & 0 & 0 \\ 1 & 0 & 1 & 1 \\ 1 & 1 & 1 & 1 \\ 0 & 0 & 0 & 1 \\ 1 & 1 & 0 & 0 \\ 0 & 1 & 0 & 0 \\ 0 & 1 & 1 & 0 \\ 0 & 1 & 0 & 1 \\ 0 & 0 & 1 & 0 \end{bmatrix} \end{array} \qquad \mathbf{r} = \begin{bmatrix} 1 \\ 2 \\ 3 \\ 1 \\ 2 \\ 1 \\ 2 \\ 2 \\ 1 \end{bmatrix}$$

For this problem, the Greedy Algorithm will construct optimal set $I = \{e_1, e_4, e_2\}$. To check that the corresponding incidence solution $\mathbf{x}^I = (1, 1, 0, 1)$ is optimal, we select rows of $\mathbf{A}$ and $\mathbf{r}$ corresponding to $sp\{e_1\}$, $sp\{e_1, e_4\}$, and $sp\{e_1, e_4, e_2\}$. These are already the first three rows in the above list. Dual values $u_1^I \leftarrow w_1 - w_4 = 3$, $u_2^I \leftarrow w_4 - w_2 = 2$, $u_3^I \leftarrow w_2 = 5$, $u_4^I \leftarrow u_5^I \leftarrow \cdots \leftarrow u_9^I \leftarrow 0$ give a solution proving the optimality of $\mathbf{x}^I$. ■

3.5 Matroid Intersection

We observed in the previous section that the independence system defined by the edge set of an undirected graph and the collection of cycle-free subsets of the edges was a matroid. Suppose we consider a somewhat related independence system where our graph is now directed, say $G(V, A)$. Specifically, create the independence system $(A, \mathcal{I})$, where $\mathcal{I}$ contains those arc subsets of A that form an acyclic subgraph of $G(V, A)$ with the additional property that no vertex has more than one arc directed into it. Is this independence system a matroid? We can answer this question easily by examining the graph in Figure 3.3. Clearly, the arc set $\{e_2, e_3, e_8, e_9\}$ is independent and is also maximal. Unfortunately, the arc set $\{e_2, e_3, e_4, e_5, e_7\}$ is also maximal independent in $\mathcal{I}$ and has greater cardinality. Thus, the stated system is not a matroid and the greedy algorithm will not generally be optimal.

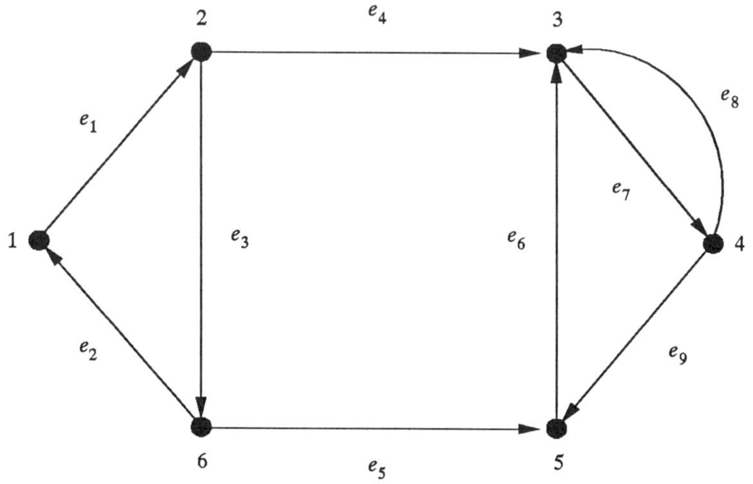

Fig. 3.3. Graph of Example 3.5.

Still there surely are matroids connected with the conditions we have stated for independence. We know the cycle free sets of edges in a graph form the independent sets of the graphic matroid, and since each arc of a directed graph is inbound at only one vertex, the arc sets with at most one inbound member at a vertex are the independent sets of a partition matroid.

Suppose then that we restate our problem as one of finding a maximum cardinality subset of arcs in A that is independent in both a graphic matroid and a partition matroid. That is, if our independence system is $(A, \mathscr{I})$, $\mathscr{I}$ can be expressed as $\mathscr{I}_1 \cap \mathscr{I}_2$ where $(A, \mathscr{I}_1)$ and $(A, \mathscr{I}_2)$ are graphic and partition matroids respectively. More simply, $(A, \mathscr{I})$ is the *intersection* of a pair of matroids. Finding a maximum cardinality subset in such $\mathscr{I}$ is referred to as the *matroid intersection problem.*

Example 3.5. Matroid Intersection. For the graph shown in Figure 3.3, let M_G be the graphic matroid defined on the edge set of the underlying undirected graph. Also, let M_P be the partition matroid defined by the arc set of the true graph and the arcs directed into each vertex. Elements of this partition P are $\{\{e_2\}, \{e_1\}, \{e_4, e_6, e_8\}, \{e_7\}, \{e_5, e_9\}, \{e_3\}\}$. It is easy to verify that all maximal sets independent in both matroids are of one of the forms $\{e_i, e_j, e_4, e_5, e_7\}$, $\{e_i, e_j, e_4, e_7, e_9\}$, $\{e_i, e_j, e_5, e_6, e_7\}$, $\{e_i, e_j, e_5, e_8\}$, and $\{e_i, e_j, e_6, e_9\}$, where $\{e_i, e_j\}$ is either $\{e_1, e_2\}$, $\{e_2, e_3\}$ or $\{e_1, e_3\}$. Any of the five element independent sets is an optimum for the *maximum* cardinality matroid intersection problem on this example. ■

3.5.1 Matroid Partition

Later in this section, we present an algorithm for the problem of determining a maximum cardinality intersection of two matroids. Our approach will follow Welsh (1976) in that we first develop a procedure for solving a more general problem—*matroid partitioning.* From this, we can show that the cardinality intersection problem is resolved as a corollary. For more direct approaches, interested readers should see, as an example, Lawler (1975).

Suppose we have a finite set of elements, G, and matroids M_1, $M_2, \ldots, M_k$ defined on G. The matroid partitioning problem seeks a partition of G into k disjoint subsets G_i such that G_i is independent in M_i for $1 \le i \le k$, i.e., $G_i \in \mathscr{I}_i$ where $\mathscr{I}_i$ is the family of independent sets in M_i.

Another way to think of the problem is that of assigning the elements in G to k color classes, so that all elements with color j constitute an independent set in matroid M_j, $1 \le j \le k$. Following, we present a procedure where a given partial (possibly empty) coloration of G is successively augmented until such time that either a complete, admissible coloring (partition) results or it is shown that one does not exist.

Suppose we have a partial coloring of G and denote by G_j the subset of elements currently assigned color j where $G_j \in \mathscr{I}_j$. Elements in G that are unassigned (uncolored) are specified by set $G_0 = G \backslash (\cup_j G_j)$. Clearly, for some $e \notin G_j$, we can color e with color j only if $\{e\} \cup G_j \in \mathscr{I}_j$. Alternately, if $\{e\} \cup G_j \notin \mathscr{I}_j$, then there must be a unique circuit $C \subseteq \{e\} \cup G_j$ such that e can be assigned color j if any $\hat{e} \in C \cap G_j$ is assigned another color $i \neq j$.

Imagine constructing a directed graph, $G(V, A)$, where $V = G \cup \{o_1, o_2, \ldots, o_k\}$ and A is defined as follows:

(i) for $e \in G_j$ and $e' \notin G_j$, $(e', e) \in A$, if and only if $G_j \cup \{e'\} \backslash \{e\}$ is independent in M_j. (3-10)

(ii) for $e \notin G_l$, $(e, o_l) \in A$, if and only if $\{e\} \cup G_l$ is independent in M_l. (3-11)

Then, the following lemma, which we state without proof (see Welsh (1976)), will play a central role in establishing the ensuing algorithm.

Lemma 3.9. Coloring Augmentation. *For the graph $G(V, A)$ defined above, let $(e_0, e_1, \ldots, e_t)$ be a directed path with e_t some vertex, say o_u and having the property that for any e_i in the path, there exists no nonpath arc $(e_i, e_l) \in A$, $l > i + 1$. If e_i is assigned the color of e_{i+1} for $0 \leq i \leq t$, then the elements having color j are independent in M_j for $1 \leq j \leq k$.*

Essentially, the lemma states that for a given e_0 to $e_t = o_u$ path, without (direct arc) shortcuts in the graph $G(V, A)$, we can alter the color assignment by changing the color of each element to that of its immediate successor in the path and preserve independence of the implied (partial) partition. This is an important property because it exposes a simple (yet elegant) algorithmic strategy. For each uncolored $e_0 \in G_0$ initiating such a path, we can use the result and color e_0 with e_1's color, e_1 with e_2's and so on, culminating with a new coloring that has an increased number of elements colored.

We are now ready to specify an algorithm. Actually, we solve the more general problem: for a given finite set, G, matroids $M_i = (G, \mathscr{I}_i)$, $1 \leq i \leq k$, and integers $m_1, m_2, \ldots, m_k$, does there exist a family of disjoint subsets $G_i \subset G$ such that $|G_i| = m_i$ and $G_i \in \mathscr{I}_i$ for $1 \leq i \leq k$?

Algorithm 3B. Matroid Partition

STEP 0: *Initialization.* Let $G_1, G_2, \ldots, G_k$ be disjoint (possibly empty) subsets of G where $G_i \in \mathscr{I}_i$ for $1 \leq i \leq k$.

STEP 1: *Graph Construction.* If $|G_i| = m_i$ for $1 \leq i \leq k$, stop; the present G_i provide the required partition. Otherwise, construct a directed graph on vertex set $V = G \cup \{o_1, o_2, \ldots, o_k\}$ as defined by $G_1, G_2, \ldots, G_k$ (3-10) and (3-11).

STEP 2: *Augmentation.* Find the first t such that $|G_1| = m_1$, $|G_2| = m_2, \ldots, |G_{t-1}| = m_{t-1}$, $|G_t| < m_t$ and try to find a path in $G(V, A)$ from some element $e_0 \in G \backslash \cup_{1 \leq i \leq k} G_i$ to o_t satisfying Lemma 3.9. If such a path exists, change the coloring of the elements on the path as indicated in the Lemma, update the sets G_i, $1 \leq i \leq k$ accordingly, and return to Step 1. If no such e_0 to e_t path can be found, stop and conclude that a suitable partition of G does not exist. ■

Prior to establishing the correctness of Algorithm 3B, we can demonstrate its application.

Example 3.6. Matroid Partition. Consider the complete graph on four vertices shown below and let $M_1 = M_2 = M_G$, where M_G is the graphic matroid. Suppose we seek a partition of the edge set, E into E_1 and E_2 each with cardinality $m_1 = m_2 = 3$ and each independent in their respective matroids.

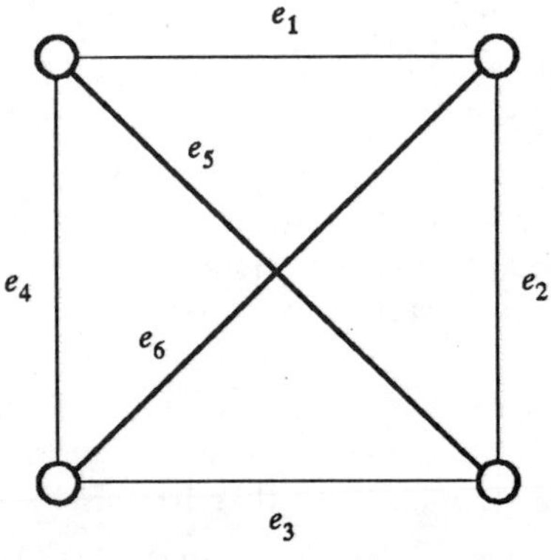

Proceeding in step-by-step fashion, we have:

STEP 0: Let us start with $E_1 = \{e_1, e_4, e_5\}$ and $E_2 = \{e_2, e_3\}$ where E_1 elements are colored *red* and E_2 elements, *blue*.

STEP 1: With the current E_1 and E_2 we construct $G(V, A)$ as shown in Figure 3.4. Observe that vertex o_1 is isolated since E_1 is maximal in M_1.

STEP 2: We have $|E_1| = 3 = m_1$ and $|E_2| = 2 < m_2$ and so $t = 2$. Since $E \backslash \{E_1 \cup E_2\}$ consists of only e_6, we seek a path in $G(V, A)$ of Figure 3.4

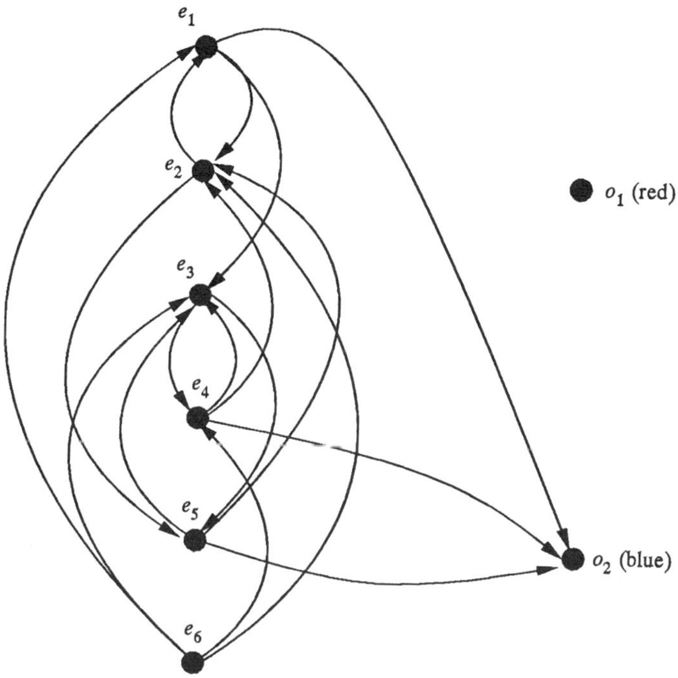

Fig. 3.4. Graph for initial coloring in Example 3.6.

from e_6 to o_2 having no shortcuts. Such a path is $e_6 - e_4 - o_2$. Thus, we color e_6 *red* and e_4 *blue*. The new subsets are $E_1 = \{e_1, e_5, e_6\}$ and $E_2 = \{e_2, e_3, e_4\}$. Clearly, these are independent in M_1 and M_2 respectively.

At this point, we would return to Step 1 with the new E_1 and E_2. It is obvious, however, that the corresponding $G(V, A)$ would have vertices o_1 and o_2 isolated since E_1 and E_2 are maximal in their respective matroids. Hence, no path from any element in E could exist to either o_1 or o_2. It is also the case that the maximality of E_1 and E_2 precludes any further augmentation in any event and we can stop the procedure with the desired partition. ■

The correctness of Algorithm 3B can be established with the aid of the following additional result:

Lemma 3.10. Matroid Union. *Let $M_1, M_2, \ldots, M_k$ be matroids on the set G where $\mathcal{I}_i$ is associated with M_i, $1 \le i \le k$. Let $\mathcal{I} = \{H : H =$*

$\cup_{1\le i\le k} H_i$, $H_i \in \mathscr{I}_i$, $1 \le i \le k\}$. Then $\mathscr{I}$ is the collection of independent sets of a matroid on G with rank function given by

$$r(T) = \min_{H\subseteq T}\left(\sum_{1\le i\le k} r_i(H) + |T\backslash H|\right) \tag{3-12}$$

where $r_i(\cdot)$ is the rank function identified with M_i, $1 \le i \le k$.

Some authors call this matroid, $(G, \mathscr{I})$, the *union* of $M_1, M_2, \ldots, M_k$ and denote it by $\vee_{1\le i\le k} M_i$.

Theorem 3.11. Correctness of Algorithm 3B. *Given matroids M_1, $M_2, \ldots, M_k$ on G, Algorithm 3B stops with the desired partition G_1, $G_2, \ldots, G_k$ having $|G_i| = m_i$ and $G_i \in \mathscr{I}_i$ for $1 \le i \le k$, or the correct conclusion that no such partition exists.*

Proof. Certainly the algorithm stops because the number of augmentations is bounded by $|G|$. If it stops at Step 1, the needed partition is at hand.

Assume instead that the procedure stops at Step 2 with sets G_1, $G_2, \ldots, G_k$, where $|G_i| = m_i$, $1 \le i \le t-1$ and $|G_t| < m_t$. In addition, we can assume without loss of generality, that $r(M_i) = m_i$ for $1 \le i \le k$. This follows since $r(M_i) < m_i$ assures an infeasible problem and if $r(M_i) > m_i$ we need only truncate M_i by deleting independent sets of more than m_i elements.

Now, if $G(V, A)$ is the directed graph constructed relative to G_1, $G_2, \ldots, G_k$, then for any vertex $e \in G_0 \triangleq G\backslash\cup_{1\le i\le k} G_i$, there exists no path from e to o_t since otherwise, the algorithm would not have stopped. Define sets U_i and V_i, $1 \le i \le t$ where

$$V_i \triangleq \{e \in G_i\text{: there exists a path in } G(V, A) \text{ from } e \text{ to } o_t\}$$

and

$$U_i \triangleq G_i\backslash V_i$$

with $U \triangleq U_1 \cup U_2 \cup \cdots U_t \cup G_o$. By $U_1 \subseteq G_1$, we have $r_1(U) \ge |U_1|$. If $r_1(U)$, however, is strictly greater than $|U_1|$, there must exist an element e in $U_2 \cup U_3 \cup \cdots \cup U_t \cup G_o$ independent of U_1 in M_1. Let us assume $e \in U_2$. But $|G_1| = m_1 = r(M_1)$ and so there is an arc $(e, \hat{e})$ in G for some $\hat{e} \in V_1$. But then since $V_1 \neq \varnothing$ this means that there is a path from $\hat{e}$ to o_t in G and thus one from e to o_t implying $e \in V_2$ or $e \notin U_2$. The same argument establishes a similar result when e is assumed to be a member of U_i for $3 \le i \le t-1$. Also, if $e \in G_o$ there obviously could be no $e - o_t$ path. Thus, we must have $r_i(U) = |U_i|$ for $1 \le i \le t-1$.

Also, repeating this argument for $e \in U_1 \cup U_2 \cup \cdots \cup U_{t-1} \cup G_o$ will reveal e dependent on U_t in M_t yielding $r_t(U) = |U_t|$. Combining this with the above equalities gives

$$\begin{aligned}\sum_{1 \leq i \leq t} r_i(U) + |G \backslash U| &= \sum_{1 \leq i \leq t} |U_i| + \sum_{1 \leq i \leq t} |V_i| \\ &= \sum_{1 \leq i \leq t} |G_i| \qquad (3\text{-}13) \\ &< \sum_{1 \leq i \leq t} m_i\end{aligned}$$

In order for a suitable partition to even exist, however, it must be that $\bigvee_{1 \leq i \leq t} M_i$ has rank $\Sigma_{1 \leq i \leq t}\, m_i$ and from Lemma 3.10, a necessary and sufficient condition for this to hold is that for all $H \subseteq G$

$$\sum_{1 \leq i \leq t} r_i(H) + |G \backslash H| \geq \sum_{1 \leq i \leq t} m_i, \qquad (3\text{-}14)$$

which is not true when we set $H = U$. Hence, if Algorithm 3B stops without the desired partition (coloring), then no such partition exists. This completes the proof. ■

For a partition problem defined by k matroids on G, the computational requirement of Algorithm 3B is $|G|^3 + |G|^2 k$ and thus $O(|G|^3)$ steps. As in all matroid algorithms, this bound is polynomial only if independence testing required to construct graphs at Step 1 is polynomial.

3.5.2 Cardinality Intersection

What about the matroid intersection problem? Recall that this requires a subset $H \subseteq G$ of maximum cardinality (or relativizing, of cardinality k), which is independent in the pair of matroids M_1 and M_2 both defined on G.

For given matroids M_1 and M_2 on G, and integer k, we may assume $k \leq r(M_1)$ and $k \leq r(M_2)$ or the needed intersection clearly cannot exist. When convenient, we could also construct truncated versions M_1^k, M_2^k of M_1 and M_2 with $r(M_1^k) = r(M_2^k) = k$. It is only necessary to delete from the list of independent sets in the two matroids all members with cardinality more than k. But over such truncated matroids, matroid intersection and matroid partition are closely linked.

Theorem 3.12. Matroid Intersection from the Partition Problem. *Let M_1^k and M_2^k be truncations of M_1 and M_2 on G. Then there exists a subset $H \subseteq G$, with cardinality k, independent in M_1 and M_2, if and only if G can* *be partitioned into two sets G_1 and G_2 independent in M_1^k and $(M_2^k)^*$ respectively.*

Proof. An intersection of size k will exist exactly when M_1^k and M_2^k have a common base. That is, an intersection exists if and only if there is an M_1^k base B such that $G \backslash B$ is a base in the dual matroid $(M_2^k)^*$. ■

Example 3.7. Connection Between Matroid Partition and Matroid Intersection. Consider the graph below with G the set of edges in the graph. Letting vertices 1 and 2 comprise vertex set V_1 and 3, 4 and 5 vertex set V_2, define matroids M_1 and M_2 on G where $\mathscr{I}_i$ contains subsets of edges no two of which share the same vertex in $V_i (1 \leq i \leq 2)$. The bases of M_1 are

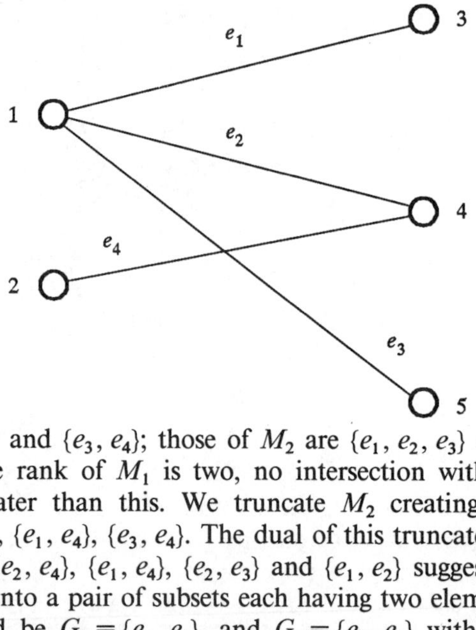

$\{e_1, e_4\}$, $\{e_2, e_4\}$ and $\{e_3, e_4\}$; those of M_2 are $\{e_1, e_2, e_3\}$ and $\{e_1, e_3, e_4\}$. Now, since the rank of M_1 is two, no intersection with M_2 can have cardinality greater than this. We truncate M_2 creating bases $\{e_1, e_2\}$, $\{e_1, e_3\}$, $\{e_2, e_3\}$, $\{e_1, e_4\}$, $\{e_3, e_4\}$. The dual of this truncated matroid, has bases $\{e_3, e_4\}$, $\{e_2, e_4\}$, $\{e_1, e_4\}$, $\{e_2, e_3\}$ and $\{e_1, e_2\}$ suggesting we seek a partition of G into a pair of subsets each having two elements. One such partition would be $G_1 = \{e_1, e_4\}$, and $G_2 = \{e_2, e_3\}$ with another, $G_1 = \{e_3, e_4\}$ and $G_2 = \{e_1, e_2\}$. Clearly, either choice of G_1 forms a desired intersection of M_1 and M_2. ■

3.5.3 Weighted Matroid Intersection

Rather than maximum cardinality intersections between pairs of matroids, let us now generalize and seek those having maximum total weight. That is, given matroids $M_1 = (G, \mathscr{I}_1)$ and $M_2 = (G, \mathscr{I}_2)$ and weights $\{w_e : e \in G\}$, find a set $I \in \mathscr{I}_1 \cap \mathscr{I}_2$ such that $\Sigma_{e \in I} w_e$ is maximum. Clearly, when all weights are equal, the cardinality case results.

We proceed as in Section 3.4.2, i.e., by working with a mixed-integer programming statement of the matroid intersection problem. Specifically, consider the form

$$(W2)\qquad \begin{array}{lll} \max & \mathbf{wx} & (3\text{-}15) \\ \text{s.t.} & \mathbf{1x} \le K & (3\text{-}16) \\ & \mathbf{Ax} \le \mathbf{r} & (3\text{-}17) \\ & \mathbf{Bx} \le \mathbf{s} & (3\text{-}18) \\ & \mathbf{x} \ge \mathbf{0} & (3\text{-}19) \\ & \mathbf{x} \text{ binary} & (3\text{-}20) \end{array}$$

where **A** and **B** are closed set incidence matrices for matroids M_1 and M_2, respectively, with corresponding rank right-hand-side limits given by **r** and **s**. K is a given cardinality limit and $x_j = 1$ for e_j in a solution.

Theorem 3.8 of Section 3.4.2 showed that maximum weight sets independent in a single matroid can be obtained by solving a linear program of the form (3-15)–(3-17), (3-19). We now proceed to an algorithm for a maximum weighted set independent in two matroids based on a corresponding result also due to Edmonds. (See Section 3.5.4 for comments establishing that further extension to 3-matroid intersection is highly unlikely.)

Theorem 3.13. Matroid Polyhedral Intersection. *For any two matroids M_1 and M_2, vertices of the polyhedron defined by (3-16)–(3-19) are the (binary) incidence vectors of sets I independent in both M_1 and M_2.*

Prior to stating the algorithm, it is instructive to consider an illustration of the structure the new theorem captures.

Example 3.8. Polyhedral Structure of the Intersection Problem. Let us again consider the bipartite graph of Example 3.7 and, specifically, the matroids M_1 and M_2 that are defined there. Taking M_1 first, the explicit constraints (3-17) are

$$x_4 \le 1$$

$$x_1 + x_2 + x_3 \le 1$$

$$x_1 + x_2 + x_3 + x_4 \le 2$$

With M_2 corresponding constraints (3-18) are

$$x_1 \le 1$$

$$x_2 + x_4 \leq 1$$
$$x_3 \leq 1$$
$$x_1 + x_2 + x_3 + x_4 \leq 3$$
$$x_1 + x_2 + x_4 \leq 2$$
$$x_1 + x_3 \leq 2$$
$$x_2 + x_3 + x_4 \leq 2$$

The reader can easily verify the following to be complete lists of integer feasible vectors for the respective systems:

M_1	M_2	
(0000)	(0000)	(0011)*
(1000)	(1000)	(1110)
(0100)	(0100)	(1011)
(0010)	(0001)	
(0001)	(1100)	
(1001)*	(1001)*	
(0101)	(1010)	
(0011)*	(0110)	

The *-marked vectors are feasible in both systems. For equal weights (cardinality) and K large, any of these marked vectors can be shown to be an optimum for the linear program of ($W2$). The algorithm below will constructively establish that the same is true for any weights. ■

The dual of the linear program of ($W2$) is

$$\min \quad Kt + \mathbf{ru} + \mathbf{sv} \tag{3-21}$$
$$\text{s.t.} \quad (\mathbf{uA} + \mathbf{vB})_j + t - w_j \geq 0 \quad \text{for all } j \tag{3-22}$$
$$\mathbf{u} \geq \mathbf{0},\ \mathbf{v} \geq \mathbf{0},\ t \geq 0 \tag{3-23}$$

Complementary slackness conditions necessary and sufficient for optimality of two feasible primal and dual solutions are

$$x_j > 0 \text{ implies } \quad \overline{w}_j \triangleq (\mathbf{uA} + \mathbf{vB})_j + t - w_j = 0 \tag{3-24}$$
$$u_i > 0 \text{ implies } (\mathbf{Ax})_i = r_i \tag{3-25}$$
$$v_j > 0 \text{ implies } (\mathbf{Bx})_j = s_j \tag{3-26}$$
$$t > 0 \text{ implies } \sum_j x_j = K \tag{3-27}$$

Our algorithmic approach will be a primal-dual one. That is, we begin with a primal and dual feasible solution satisfying all complementarity conditions with the possible exception of (3-27). A cardinality intersection algorithm is then applied to a restricted problem induced by $\overline{G} \triangleq \{e_j : \overline{w}_j = 0\}$. This either increases the cardinality of the current primal solution (maintaining complementarity on primal constraints) or allows an update to the dual solution. The dual is changed accordingly and either optimality is concluded or the cardinality algorithm is reapplied to a new restricted problem.

There are essentially four key elements that provide a summary of the primal-dual scheme. These are:

(a) A cardinality procedure must augment the cardinality of a current intersection when this is possible and stop when it is not.
(b) Whenever augmentation (primal change) occurs, it must maintain as equalities all primal constraints having positive dual variable values (i.e., (3-25) and (3-26)).
(c) The dual adjustment algorithm must either reduce the degree of violation on (3-27) or add at least one usable element to the set defining the restricted problem.
(d) The dual adjustment must not delete from the restricted problem any element that is in the current primal solution (that is, it must keep (3-24)).

Central among the four aspects above is (a), the cardinality improving scheme. Of course, we could incorporate the indirect strategy given earlier based upon matroid partition ideas. Here we shall utilize, however, the more conventional, labeling routine.

Assume we have some (nonempty) set $I \in \mathscr{I}_1 \cap \mathscr{I}_2$. Then our labeling routine seeks an *augmenting path, S,* that begins with some $e_i \notin I$ to add to the solution, follows with $e_j \in I$ to delete, and continues in this alternating manner, concluding with some $e_k \notin I$. Obviously, there is one more element to be added than deleted in this chain, and thus our intersection set cardinality can be increased. If we find such a path of elements, S, then augmentation yields $I \leftarrow I \oplus S \triangleq I \cup S \backslash (I \cap S)$.

Let us denote by $C_i^{(1)}$, the unique M_1 circuit in $I \cup \{e_i\}$ for any $e_i \in sp_1(I) \backslash I$ and $\{e_i\}$, if $e_i \in I$. Similarly, let $C_i^{(2)}$ be the unique M_2 circuit in $I \cup \{e_i\}$ for any $e_i \in sp_2(I) \backslash I$ and $\{e_i\}$ if $e_i \in I$.

We first label any $e_i \notin I$ such that $I \cup \{e_i\} \in \mathscr{I}_1$. If $I \cup \{e_i\} \in \mathscr{I}_2$, we stop and augment I (as $I \cup \{e_i\}$). Otherwise, we would label, for deletion, any $e_j \in C_i^{(2)} \backslash \{e_i\}$. This, in turn, might liberate some element, e_k, dependent in M_1 when e_j is present, but not after e_j is removed. Specifically, we seek e_k with $e_j \in C_k^{(1)}$. This element plays the role of a new e_i and the process continues. Since we are merely trading elements of circuits we must have

$|I| = |I \oplus S|$ for any path, S, created in this manner until an $e_i \in \mathscr{I}_1 \cap \mathscr{I}_2$ is located at which point we have $|I| + 1 = |I \oplus S|$. Observe that $I \oplus S \in \mathscr{I}_1 \cap \mathscr{I}_2$ since added elements are independent in both matroids.

The computational procedure below closely parallels Lawler (1976). We begin with some definitions. Let I_L and I_U denote the elements in I that are labeled and unlabeled respectively, when the cardinality algorithm stops. Define $U_0, U_1, \ldots, U_p$ and $V_0, V_1, \ldots, V_q$ to be sets such that

$$U_0 = \varnothing,\ U_{i-1} \subset U_i \text{ for } 1 \le i \le p;\ U_p = I \tag{3-28}$$

$$V_0 = \varnothing,\ V_{j-1} \subset V_j \text{ for } 1 \le j \le q;\ V_q = I \tag{3-29}$$

Also, let $D_i^{(1)} \triangleq C_i^{(1)} \backslash U_{k(i)-1}$, for all $e_i \in sp_1(I)$ and $D_i^{(2)} \triangleq C_i^{(2)} \backslash V_{l(i)-1}$, for all $e_i \in sp_2(I)$. Here, $U_{k(i)}$ is the smallest U_k with $e_i \in sp_1(U_k)$. $V_{l(i)}$ is defined similarly relative to $sp_2(V_l)$. Because of the manner in which successive U_i and V_j are defined in (3-28) and (3-29), we call $D_i^{(1)}$ and $D_i^{(2)}$ *incremental circuits.*

Algorithm 3C. Weighted Matroid Intersection

STEP 0: *Initialization.* Select a primal and dual feasible set $I \in \mathscr{I}_1 \cap \mathscr{I}_2$ and sets $\{U_i\}$ and $\{V_i\}$ satisfying (3-28) and (3-29). If $I = \varnothing$, set $t \leftarrow \max\{w_j\}$, $\mathbf{u} \leftarrow \mathbf{0}$, $\mathbf{v} \leftarrow \mathbf{0}$, $\mathbf{x} \leftarrow \mathbf{0}$, $p \leftarrow 0$, $q \leftarrow 0$. Eliminate all labels, if any, and for each $e_j \in G$ compute $\overline{w}_j$. Finally, construct $\overline{G} \triangleq \{e_j : \overline{w}_j = 0\}$.

STEP 1: *Labeling.*

1.0 Assign a label of 0+ to all elements in $\overline{G} \backslash sp_1(I)$.

1.1 Select the labeled element e_i possessing the oldest unscanned label. If no unscanned labels exist, go to Step 3. Otherwise, go to 1.2, if e_i is + labeled and to 1.3 if it is − labeled.

1.2 Scan the + label on e_i. If $I \cup \{e_i\} \in \mathscr{I}_2$, go to Step 2 and augment the current intersection. Otherwise, label by $i-$ any unlabeled elements in $D_i^{(2)}$ and go to 1.1.

1.3 Scan the − label on e_i. Label by $i+$ any unlabeled e_j with $e_i \in D_j^{(1)}$ and go to 1.1.

STEP 2: *Augmentation.*

2.0 Identify the augmenting sequence $S = \{e_{[1]}, e_{[2]}, \ldots, e_{[|S|]}\}$ by backtracking on the labels. (Here, $e_{[i]}$ denotes the ith element in S). Set $I \leftarrow I \oplus S$.

2.1 Replace each departing $e_{[i]}$ in all sets U_i, $1 \le i \le p$ to which it belongs by $e_{[i+1]}$.

2.2 Replace each departing $e_{[i]}$ in all sets V_j, $1 \le j \le q$ to which it belongs by $e_{[i-1]}$.

2.3 If this is the first time 2.3 is reached in the current application of the labeling procedure, set $p \leftarrow p+1$ and $q \leftarrow q+1$. In any case, set $U_p \leftarrow I$ and $V_q \leftarrow I$, erase all labels and return to Step 1.

STEP 3: *Dual Update.*

3.1 Create new sets U_k as necessary so that $U_k \backslash U_{k-1}$ either consists entirely of labeled or unlabeled elements. This can be done by creating $U = U_k \backslash U_{k-1} \cap I_U$ and forming the insertion, satisfying (3-28), between U_{k-1} and U_k. Increase p and subscripts on U_k, $U_{k+1}, \ldots, U_p$ with the new set now called U_k and having dual variable equal to zero.

3.2 Create new sets V_l as necessary so that $V_l \backslash V_{l-1}$ either consists entirely of labeled or unlabeled elements. This can be done by creating $V = V_l \backslash V_{l-1} \cap I_L$ and forming the insertion, satisfying (3-29), between V_{l-1} and V_l. Increase q and subscripts on V_l, $V_{l+1}, \ldots, V_q$ with the new set now called V_l and having dual variable equal to zero.

3.3 Update the dual solution as follows (where $u_k \triangleq u_{sp_1(U_k)}$ and $v_l \triangleq v_{sp_2(V_l)}$).

$$u_k \longleftarrow u_k - \delta \quad \text{if } U_k \backslash U_{k-1} \text{ is entirely labeled, } U_{k+1} \backslash U_k \text{ is entirely unlabeled and } k < p. \tag{3-30}$$

$$u_k \longleftarrow u_k + \delta \quad \text{if } U_k \backslash U_{k-1} \text{ is entirely unlabeled and either } k = p \text{ or } U_{k+1} \backslash U_k \text{ is entirely labeled} \tag{3-31}$$

$$v_l \longleftarrow v_l - \delta \quad \text{if } V_l \backslash V_{l-1} \text{ is entirely unlabeled, } V_{l+1} \backslash V_l \text{ is entirely labeled and } l < q \tag{3-32}$$

$$v_l \longleftarrow v_l + \delta \quad \text{if } V_l \backslash V_{l-1} \text{ is entirely labeled and either } l = q \text{ or } V_{l+1} \backslash V_l \text{ is entirely unlabeled} \tag{3-33}$$

$$t \longleftarrow t - \delta$$

where $\delta = \min\{t, \delta_u, \delta_v, \delta_w\}$ with

$\delta_u = \min \{u_k$ decreasing as stated at(3-30)$\}$ or ∞ if there are none.

$\delta_v = \min\{v_l$ decreasing as stated at (3-32)$\}$ or ∞ if there are none.

$$\delta_w = \min \left\{ \begin{array}{l} \bar{w}_j\text{: neither } D_j^{(1)} \text{ completely} \\ \quad\text{unlabeled nor } D_j^{(2)} \text{ is} \\ \quad\text{completely labeled} \end{array} \right\}$$

or ∞ if there are none.

3.4 If $t = 0$, stop, the present intersection, I, is optimum. If not, eliminate all elements (other than U_p and V_q) in the sets $\{U_k\}$ and $\{V_l\}$ having dual variable values equal to zero. If δ_u and δ_v determined δ at 3.3, go to 3.1. Otherwise, update $\bar{G}$ and return to Step 1. ■

Example 3.9. Weighted Matroid Intersection. Let the edges in the bipartite graph of Example 3.7 have weights $w_1 = 6$, $w_2 = 8$, $w_3 = 4$, and $w_4 = 3$. Then proceeding in step-by-step fashion, Algorithm 3C progresses as follows:

STEP 0: We shall initialize with $I = U_0 = V_0 = \varnothing$. Further, $t = w_2 = 8$, $p = q = 0$ and all primal and dual variables are set at zero. Finally, we have $\bar{\mathbf{w}} = (2, 0, 4, 5)$ and so, $\bar{G} = \{e_2\}$.

STEPS 1, 2: $\bar{G} \backslash sp_1(\varnothing) = \{e_2\}$ and e_2 takes on label at 0+. Upon scanning, we find $I \cup \{e_2\} \in \mathscr{I}_2$ and augmentation occurs. We set $U_1 = V_1 = \{e_2\}$ and reapply the labeling routine.

STEP 1: We now have $sp_1(I) = \{e_2\}$ with $\bar{G} \backslash \{e_2\} = \varnothing$ and, hence, a dual revision must be made.

STEP 3: All duals are presently zero and so $\delta = \min(t, \delta_w)$. The reader may verify that $\delta_w = \bar{w}_4 = 5$ and hence $\delta = 5$. We set $u_1 \triangleq u_{sp_1(U_1)}$, where $sp_1(U_1) = \{e_1, e_2, e_3\}$ and update t and $\bar{\mathbf{w}}$ as 3 and (2, 0, 4, 0) respectively the latter yielding $\bar{G} = \{e_2, e_4\}$.

STEP 1: We now have $\bar{G} \backslash sp_1(I) = \{e_4\}$ and labeling proceeds until we conclude that no augmentation is possible. Everything in I is labeled and we update the dual solution.

STEP 3: As before, $\delta = \min(t, \delta_w)$ where $\delta_w = \min(\bar{w}_1, \bar{w}_3) = \bar{w}_1 = 2$. We set $v_1 \triangleq v_{sp_1(V_1)}$, where $sp_2(V_1) = \{e_2, e_4\}$ and change t to 1 and $\bar{\mathbf{w}}$ to (0, 0, 2, 0). $\bar{G}$ is now $\{e_1, e_2, e_4\}$.

STEPS 1, 2: We have $\bar{G} \backslash sp_1(I) = \{e_4\}$ and labeling produces the sequence e_4, e_2, e_1 labeled in the stated alternating fashion. The last element e_1 is + labeled and we have that $I \cup \{e_1\} \in \mathscr{I}_2$. Hence, our augmenting sequence is $S = \{e_4, e_2, e_1\}$ with e_2 the departing element. Set updates yield $U_0 = V_0 = \varnothing$, $U_1 = \{e_1\}$, $V_1 = \{e_4\}$ and $U_2 = V_2 = \{e_1, e_4\}$.

At this point, we leave it to the reader to conclude the application whereupon it will be found that a dual revision yields $\delta = t = 1$ and $u_2 \triangleq u_{sp_1(U_2)} = 1$ where $sp_2(U_2) = \{e_1, e_2, e_3, e_4\}$. Thus, the stated $I = \{e_1, e_4\}$ is optimum and has weight 9. ■

In order to address the validity of the algorithm, let us first consider the following pair of lemmas.

Lemma 3.14. Correctness of the Labeling Procedure. *When Step 3 is reached, from 1.1 of the labeling step, the following are satisfied:*

(i) $\overline{G} \subseteq \{e_i\colon D_i^{(1)}$ *is completely unlabeled*$\} \cup \{e_i\colon D_i^{(2)}$ *is completely labeled*$\}$

(ii) $I \cap \{e_i\colon D_i^{(1)}$ *is completely unlabeled*$\} \cap \{e_i\colon D_i^{(2)}$ *is completely labeled*$\} = \varnothing$

(iii) *The update at Step 2.1 changes no* $sp_1(U_i)$

(iv) *The update at Step 2.2 changes no* $sp_2(V_j)$

Proof. If $e_i \in I$, then $D_i^{(1)} = D_i^{(2)} = \{e_i\}$. Clearly, it is all labeled or all unlabeled but not both and (ii) is established.

Now, if $e_i \in \overline{G}\backslash I$ and labeled, then it was scanned by 1.2 and hence, any unlabeled $e_j \in D_i^{(2)}$ were then labeled. Thus, labeled $e_i \in \overline{G}\backslash I$ belongs to $\{e_i\colon D_i^{(2)}$ is completely labeled$\}$. Alternately, if $e_i \in \overline{G}\backslash I$ and unlabeled, it must belong to $sp_1(I)\backslash I$. Step 1.0 labeled all $e_i \in \overline{G}\backslash sp_1(I)$. Had any e_j such that $e_j \in D_i^{(1)}$ been labeled, e_i would have been labeled at 1.3. Thus $D_i^{(1)}$ is completely unlabeled and (i) is established.

In 2.1, $e_{[i]}$ is replaced by $e_{[i+1]}$ where $e_{[i]} \in D_{[i+1]}^{(1)} \triangleq \{sp_1(U_{k[i+1]})\backslash sp_1(U_{k[i+1]-1})\} \cap C_{[i+1]}^{(1)}$. Thus, $e_{[i]} \in U_{k[i+1]}\backslash U_{k[i+1]-1}$ and spans in matroid M_1 are not affected. Similarly, 2.2 replaces $e_{[i]}$ by $e_{[i-1]}$, where $e_{[i]} \in D_{[i-1]}^{(2)} \triangleq \{sp_2(V_{l[i-1]})\backslash sp_2(V_{l[i-1]-1})\} \cap C_{[i-1]}^{(2)}$. But by definition, $e_{[i-1]} \in sp_2(V_{l[i-1]})\backslash sp_2(V_{l[i-1]-1})$ and so, M_2 spans are also unaffected. This completes the proof. ■

Lemma 3.15. Effects of Dual Update. *Let*

$$G^{(1)} \triangleq \left\{ e_i \in sp_1(I)\colon \begin{array}{l} D_i^{(1)} \text{ is completely unlabeled at the} \\ \text{completion of the labeling procedure;} \end{array} \right\}$$

$$G^{(2)} \triangleq \left\{ e_i \in sp_2(I)\colon \begin{array}{l} D_i^{(2)} \text{ is completely labeled at the} \\ \text{completion of the labeling procedure.} \end{array} \right\}$$

Then on each execution of 3.3

(i) $\overline{w}_j$ *decreases by* δ *for* $e_j \notin G^{(1)} \cup G^{(2)}$

(ii) *$\bar{w}_j$ increases by δ for $e_j \in G^{(1)} \cap G^{(2)}$*
(iii) *$\bar{w}_j$ remains unchanged for $e_j \in G^{(1)} \oplus G^{(2)}$*
(iv) $\delta > 0$

Furthermore, after at most $2|G|$ executions, either $t = 0$ or some $\bar{w}_j = 0$ with $e_j \notin \bar{G}$.

Proof. Clearly, the constructions for U and V in 3.1 and 3.2 assure $U_k \backslash U_{k-1}$ is all labeled or all unlabeled for $k = 1, 2, \ldots, p$. Similarly, $V_l \backslash V_{l-1}$ is all labeled or all unlabeled for $l = 1, 2, \ldots, q$.

Let $\bar{k} = \max\{k\colon u_k$ changes in (3-30) and (3-31)$\}$. Then $\bar{w}_j$ for any e_j is impacted only by changes in u_k for $k = k(j), k(j) + 1, \ldots, \bar{k}$. Moreover, since u_p changes only if $U_p \backslash U_{p-1}$ is all unlabeled, we have that $U_{\bar{k}} \backslash U_{\bar{k}-1}$ must be all unlabeled. A consequence is that (3-30) and (3-31) do not impact $\bar{w}_j$ for $e_j \notin sp_1(U_k)$. For e_j with $U_{k(j)} \subset U_{\bar{k}}$, $D_j^{(1)} \subseteq U_{k(j)} \backslash U_{k(j)-1}$ must all be unlabeled or all labeled. If all labeled, (3-30) and (3-31) impact $\bar{w}_j$ an even number of times (half by $+\delta$ and half by $-\delta$). The net effect is zero. If $D_j^{(1)}$ is all unlabeled, (3-30) and (3-31) impact $\bar{w}_j$ an odd number of times with more $+\delta$ then $-\delta$. Overall, (3-30) and (3-31) increase $\bar{w}_j$ by δ when $e_j \in G^{(1)} \triangleq \{e_j \in sp_1(I)\colon D_j^{(1)}$ unlabeled$\}$ and leave all other $\bar{w}_j$ unchanged.

An analogous argument will demonstrate that (3-32) and (3-33) increase $\bar{w}_j$ by δ when $e_j \in G^{(2)} \triangleq \{e_j \in sp_2(I)\colon D_j^{(2)}$ labeled$\}$ and have no impact otherwise. Since the t update reduces all $\bar{w}_j$ by δ we can conclude that (i), (ii), and (iii) hold.

In order to show $\delta > 0$ at each execution of the dual update routine, we must consider the possible ways δ can be determined. One is t. If t were not positive, we would have concluded optimality on the final execution at 3.4. Another possibility is δ_w. The rules ((3-30)–(3-33)) and the update for t, however, decrease $\bar{w}_j$ only when $e_j \notin G^{(1)} \cup G^{(2)}$ and Lemma 3.14 show the final restricted set $\bar{G} \subseteq G^{(1)} \cup G^{(2)}$. Hence, $e_j \notin \bar{G}$ and each $\bar{w}_j$ impacting δ_w is positive.

The final possibility is that δ_u or δ_v determined δ, i.e., that a decreasing u_k or v_l reached zero. Each execution of 3.4 eliminates all sets U_k and V_l with zero u_k or v_l. Further, no constructions create new sets with zero duals. But such $U_k \backslash U_{k-1}$ are unlabeled and $V_l \backslash V_{l-1}$ are labeled. Duals on such sets cannot decrease in (3-30) or (3-32). Thus, any decreasing u_k or v_l began at a positive value and resulting δ_u and δ_v must be positive.

Finally, observe that the nature of the sets created in 3.1 and 3.2 implies that their duals will be monotone nondecreasing until labeling is reinvoked at 3.4. The number of times 3.3 can be executed before 3.4 concludes optimality or reinvokes the labeling portion is bounded by the number of u_k or v_l that might decrease to zero. Since both p and q are no greater than the smaller of the ranks of the two matroids, it follows that the bound is polynomial. ∎

We are now in a position to state and prove the following:

Theorem 3.16. Validity of Algorithm 3C. *Algorithm 3C will produce a maximum weight intersection of two matroids. Moreover, if all w_j are integers, then the number of steps required by the algorithm is bounded by a polynomial in $|G|$, $w^* = \max\{w_j\}$, and the time to check independence in M_1 and M_2.*

Proof. Clearly, the labeling process maintains primal feasibility of the solution

$$x_j = \begin{cases} 1 \text{ if } e_j \in I \\ 0 \text{ otherwise} \end{cases}$$

Dual feasibility depends on $\mathbf{v} \geq \mathbf{0}$, $\mathbf{u} \geq \mathbf{0}$, $t \geq 0$ and $\overline{\mathbf{w}} \geq \mathbf{0}$. If we initialize with

$$t = \max\{w_j\} \tag{3-34}$$

$$\mathbf{u} = \mathbf{0}, \qquad \mathbf{v} = \mathbf{0} \tag{3-35}$$

then $\overline{\mathbf{w}} \geq \mathbf{0}$. Our choice of δ in the dual revision guarantees that no t, $\overline{w}_j$, u_k or v_l is ever allowed to drop below zero.

The more difficult issue has to do with complementary slackness. Obviously complementarity on the constraint (3-16) will be violated until t becomes zero. But all other restrictions are observed. The starting dual solution above, together with $\mathbf{x} = \mathbf{0}$ ($I = \varnothing$) satisfies all complementary slackness limits in x_j, u_k or v_l. Lemmas 3.14 and 3.15 show complementarity is retained. In Lemma 3.14, we observed that labeling simply involved trading elements of the same span sets $sp_1(U_k)$ and $sp_2(V_l)$. Thus, primal constraints that were tight when augmentation begins remain so when it completes. The dual changes (3-31) and (3-33) do make new u_k and v_l positive. Each u_k, however, is the dual multiplier on a constraint

$$\sum_{\{j:\, e_j \in sp_1(U_k)\}} x_j \leq r_1(sp_1(U_k)) \tag{3-36}$$

where $U_k \subseteq I \in \mathscr{I}_1 \cap \mathscr{I}_2$. Since $x_j = 1$ for j with $e_j \in I$, the constraints are obviously tight.

Finally, we come to the condition $x_j > 0 \Rightarrow \overline{w}_j = 0$. The definition of $\overline{G}$ as $\{e_j\colon \overline{w}_j = 0\}$ assures this condition holds for new e_j added to I by augmentation. Moreover, from Lemma 3.14 where $I \subset G^{(1)} \oplus G^{(2)}$, together with Lemma 3.15's conclusion that $\overline{w}_j$ does not change for $e_j \in G^{(1)} \oplus G^{(2)}$, we can see that $I \subseteq \overline{G}$ after duals are updated.

Previously, we saw how labeling requires only polynomially many applications of independence testing procedures for M_1 and M_2. Also, t decreases by δ on every dual change. If all w_j are integers, $\delta \geq 1$ at least when δ_w determines δ. Thus with integer weights, duals can change at most w^* times. ■

We note, in closing the current section, that Theorem 3.16 can be sharpened to avoid the w^* term in the bound and also to escape the integer requirement for $\mathbf{w}$. What is required is to show that only polynomially many e_j can have $\overline{w}_j$ drop to zero in the dual revision update before I is augmented.

3.5.4 *Three Matroid Intersection*

From the developments in the previous section, it is immediate that we could pose the following problem: Given three matroids, $(G, \mathcal{I}_1)$, $(G, \mathcal{I}_2)$, and $(G, \mathcal{I}_3)$ on the same ground set G, find a maximum cardinality subset $I \subseteq G$ that is in $\mathcal{I}_1 \cap \mathcal{I}_2 \cap \mathcal{I}_3$. That is, find a maximum cardinality subset that is simultaneously independent in all three matroids.

It is easy to see that the elegant results we have developed so far do not extend to this *three matroid intersection problem* (or, indeed, to k matroid intersection with $k \geq 3$). For a directed graph $G(V, A)$ let $(A, \mathcal{I}_1)$ be the graphic matroid, $(A, \mathcal{I}_2)$ the partition matroid allowing at most one inbound arc at each vertex, and $(A, \mathcal{I}_3)$ the corresponding partition matroid allowing at most one outbound arc per vertex. A maximum cardinality independent set in $\mathcal{I}_1 \cap \mathcal{I}_2 \cap \mathcal{I}_3$ will be a Hamiltonian path if there is any in G. Thus, a formally efficient algorithm for 3-matroid intersection would provide one for the NP-Complete problem of recognizing directed graphs possessing Hamiltonian paths. It follows that 3-matroid intersection is itself NP-Hard, and in fact it is easy to conclude that the threshold version of the problem is NP-Complete.

3.6 Matroid Parity

Given a single matroid $(G, \mathcal{I})$ and a partition of G into distinct *pairs* $(e_i, \bar{e}_i)$ of elements, the *matroid parity problem* seeks a maximum cardinality or weight subset $I \in \mathcal{I}$ having the property that $e_i \in I$ if and only if $\bar{e}_i \in I$. In various contexts, this problem is also referred to as the *polymatroid matching problem* or the *matchoid problem.* By any name, the problem is of interest in that it yields a common generalization of the problems of matching on finite graphs and matroid intersection.

Example 3.10. Matroid Parity Generalization of 2-Matroid Intersection. To see how matroid parity generalizes 2-matroid intersection, suppose we seek a maximum independent set $I \in \mathscr{I}_1 \cap \mathscr{I}_2$, where $\mathscr{I}_1$ and $\mathscr{I}_2$ are the lists of independent sets for matroids $M_1 \triangleq (G, \mathscr{I}_1)$ and $M_2 \triangleq (G, \mathscr{I}_2)$ respectively. Duplicate G in an identical new set $\hat{G} \triangleq \{\hat{e}\}$ and consider the union matroid $M \triangleq M_1 \vee M_2$ defined in Lemma 3.10. Independent sets in M are those I such that $I \cap G$ is independent in M_1 and $I \cap \hat{G}$ is independent in M_2. By imposing parity constraints that $e_j \in I$, if and only if $\hat{e}_j \in I$, it is easy to see that a maximum parity set I will yield a maximum $I \cap G \in \mathscr{I}_1 \cap \mathscr{I}_2$. ■

Example 3.11. Matroid Parity Generalization of Matching on Graphs. Recall that the classic matching problem on a graph, $G(V, E)$, seeks a maximum cardinality or weight subset $M \subseteq E$ such that no two edges in M are incident to the same vertex in V. While the problem possesses an elegant characterization and an efficient solution procedure, we are interested here in showing that it is generalized by matroid parity.

For a given instance of the matching problem defined on some graph $G(V, E)$ we construct an alternative graph, $\hat{G}(\hat{V}, \hat{E})$, which is a homeomorph of G arising from a single vertex insertion on every edge in E. The construction is demonstrated in Fig. 3.5. Let $(\hat{E}, \mathscr{I})$ be a partition matroid limiting the number of elements of $\hat{E}$ meeting any $v \in V$ (not $\hat{V}$) to at most 1. We then pair elements (edges) in $\hat{E}$ that were derived from the same $e \in E$. Any nonempty, *parity set* $P \subseteq \hat{E}$ will contain an edge $\bar{e}$, if and only if it also has $\bar{\bar{e}}$, where $\bar{e}$ and $\bar{\bar{e}}$ are the adjacent edges created in the homeomorph from G. It is easy to see that if the pair $(\bar{e}, \bar{\bar{e}})$ is independent in $\mathscr{I}$, then no pair, say $(\bar{a}, \bar{\bar{a}})$ could also be independent where $(\bar{a}, \bar{\bar{a}})$ corresponds with an edge in E incident to the same vertex as the edge (also in E) corresponding to $(\bar{e}, \bar{\bar{e}})$. Thus, parity sets in $(\hat{E}, \mathscr{I})$ are in 1 – 1 correspondence with admissible matchings in $G(V, E)$. ■

3.6.1 *Status of Matroid Parity*

We have already developed an algorithm for 2-matroid intersection that is polynomial-time, if independence can be polynomially checked in both matroids. Also, there is a well-known polynomial-time algorithm for weighted matching on arbitrary graphs (see Edmonds (1965c)). It is interesting to ask "What other cases of weighted or cardinality matroid parity have any hope of being polynomially-solved?"

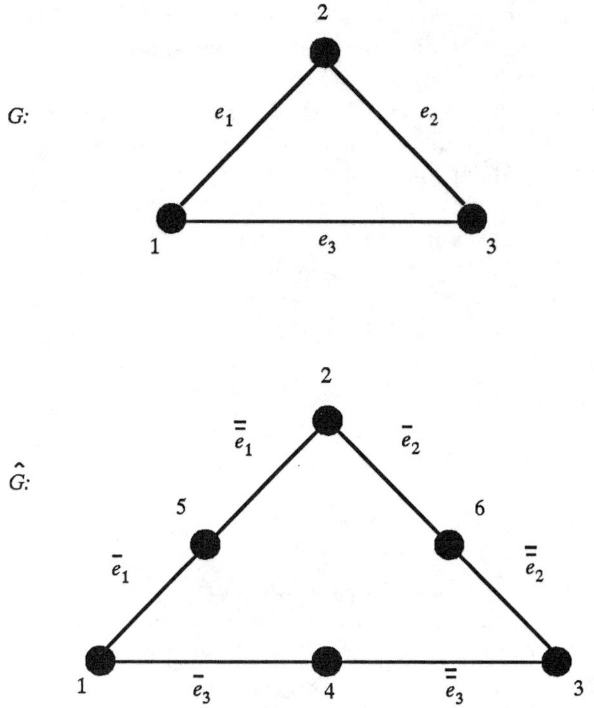

Partition: $\{(\bar{\bar{e}}_1, \bar{e}_2), (\bar{\bar{e}}_2, \bar{\bar{e}}_3), (\bar{e}_1, \bar{e}_3)\}$

Pairs: $\{(\bar{e}_1, \bar{\bar{e}}_1), (\bar{e}_2, \bar{\bar{e}}_2), (\bar{e}_3, \bar{\bar{e}}_3)\}$

Fig. 3.5. Matroid parity generalization of matching on graphs.

Lovász (1978) has shown that even the cardinality version of matroid parity is *NP*-Hard when the underlying matroid is not representable. For representable matroids, it remains an open question at this writing whether a polynomial-time algorithm can exist for weighted matroid parity (even over graphic matroids), but at least two are known for representable cases with a cardinality objective function. The remainder of this section develops the key ideas behind one of them.

3.6.2 Weak and Strong Duality for Matroid Parity

A main element of the cardinality algorithm for representable matroids is a pair of duality results, both due to Lovász, from which we obtain a classic min-max theorem for the matroid parity problem. Our development follows that in Orlin and Vande Vate (1986).

Let $\mathbf{A}$ be a representing matrix specified as part of an instance for the matroid parity problem. We can assume $\mathbf{A}$ to be of full row rank and also that columns are arranged so that $(\mathbf{a}^{2i-1}, \mathbf{a}^{2i})$ are paired. We may call such pairs *lines* and denote them by s_i. Let S be the set consisting of all n lines.

Each column vector $\mathbf{a}$ in $\mathbf{A}$ will be referred to as a *point* and also corresponds to a one-dimensional subspace of $\mathbb{R}^m$. The point, $\mathbf{a}$, induces such a subspace. By a similar interpretation, each line in S induces a two-dimensional subspace in $\mathbb{R}^m$. Such subspaces may be called lines as well.

In terms of these conventions the notion of a matching in a graph can be generalized to these pair lines. Given any $U \subseteq S$, U is said to be a *k-matching*, if $|U| = k$ and $r(\cup_{i \in U}\{\mathbf{a}^{2i-1}, \mathbf{a}^{2i}\}) = 2k$, or informally, $r(U) = (2k)$. That is, a matching of size k is an independent set containing k lines from S. Given any $U \subset S$, we denote by $v(U)$ the maximum cardinality matching contained in U.

If for $U \subseteq S$ we have $r(U) = 2|U| - 1$, U is called a *near-matching*, and if for every proper subset $W \subset U$, W is a matching, U is said to be a *2-circuit*. It is easy to see that if U is a near-matching, it possesses a unique 2-circuit.

For some $U \subseteq \mathbb{R}^m$ and a subspace K of $\mathbb{R}^m$, let $r(U/K) \triangleq r(U \cup K) - r(K)$. Finally, let $v(U/K)$ be the maximum cardinality of a matching $\hat{U}$ in U having $r(\hat{U}/K) = r(\hat{U})$. We are now ready to present the duality results.

For a set U of lines in $\mathbb{R}^m$, let a *cover* of U be a pair (K, Π), where K is a subspace of $\mathbb{R}^m$ and Π is a partition of U. The *capacity* of (K, Π), which we denote by $c(K, \Pi)$, is given by

$$c(K, \Pi) = r(K) + \sum_{i=0}^{t} \lfloor r(U_i/K)/2 \rfloor \tag{3-37}$$

where U_i denotes subset i in the partition Π. We then have

Lemma 3.17. Weak Duality. *Let U be a set of lines in $\mathbb{R}^m$. Then for any matching $\hat{U} \subseteq U$ and any cover (K, Π) of U, $|\hat{U}| \leq c(K, \Pi)$.*

Proof. Denote $\hat{U}$ by $\{(\mathbf{a}^1, \mathbf{a}^2), \ldots, (\mathbf{a}^{2k-1}, \mathbf{a}^{2k})\}$, where $\hat{U}$ is a matching in U and let (K, Π) be a cover of U. Then, for any point $\mathbf{q}$ in K, if $\mathbf{q}$ is in the linear span of $\hat{U}$, say $sp(\hat{U})$, there must be a vector $\mathbf{x}$ different from zero such that $\mathbf{q} = \Sigma_{i=1}^{2k} x_i \mathbf{a}^i$. Removing a pair $(\mathbf{a}^{2i-1}, \mathbf{a}^{2i})$ from $\hat{U}$ with either x_{2i-1} or x_{2i} non-zero, produces another matching U' for which $r(U'/\mathbf{q}) = r(U')$. Observe that if $\mathbf{q} \notin sp(\hat{U})$, any line from $\hat{U}$ can be removed to produce U'. Continuing in this iterative manner, we can determine a matching $U^* \subseteq U$ for which $r(U^*/K) = r(U^*)$ and

$$|\hat{U}| = |U^*| + r(K) \leq v(U/K) + r(K). \tag{3-38}$$

We also have that

$$\nu(U/K) \leq \sum_{i=0}^{t} \nu(U_i/K) \leq \sum_{i=0}^{t} \lfloor r(U_i/K)/2 \rfloor \tag{3-39}$$

and upon combining this with (3-38), we obtain the desired result. ■

The next theorem shows that the relationship between maximum matchings and minimum capacity covers is one of equivalence in the classic min-max sense.

Theorem 3.18. Strong Duality. *Let U be a set of lines in $\mathbb{R}^m$. Then the maximum cardinality of a matching in U is equal to the minimum capacity of a cover of U.*

3.6.3 A Matroid Parity Algorithm—Basic Concepts

A proof of the strong duality result of Theorem 3.18 can be accomplished in a constructive manner by producing an algorithm for the cardinality matroid parity problem on representable matroids. Here, we shall only sketch such a procedure.

The algorithm employs an augmentation scheme where a sequence of matchings of increasing cardinalities are produced in a given subset U^i of S along with a minimum capacity cover in U^i. Let us call the 4-tuple $(U, \hat{U}, K, \Pi)$ a *strong duality set,* if $\hat{U}$ is a maximum cardinality matching in U and (K, Π) is a minimum capacity cover of U. In addition, for a cover (K, Π) of U, let us call the members $U_0, U_1, \ldots, U_t$ of Π, the components of U. Then U_i is an odd or even component depending on the parity of $r(U_i/K)$. In the Orlin-Vande Vate procedure, each even component of the strong duality sets consists of a single line.

If $\hat{U}$ is a matching in an odd component, U_i, we say that $\hat{U}$ is a K-hypomatching of U_i, if $r(\hat{U}/K) = r(\hat{U}) = r(U_i/K) - 1$. Similarly, we may call $\hat{U}$ a K-hypermatching of U_i if $r(\hat{U}/K) = r(\hat{U}) - 1 = r(U_i/K)$. In an even component, if $r(\hat{U}/K) = r(\hat{U}) = r(U_i/K)$, $\hat{U}$ is called a K-perfect matching of U_i.

A procedure can be developed whereby augmentation is sought for some initial, strong duality set by solving a succession of matroid intersection problems with elements corresponding to components. Letting these matroids be $(G, \mathcal{I}_1)$ and $(G, \mathcal{I}_2)$, we construct the first to select linearly independent points and the second, to ensure these points correspond to K-hypomatchings, K-hypermatchings, or K-perfect matchings.

Beginning with some $(U, \hat{U}, K, \Pi)$ and a line $\mathbf{u} \notin U$, we solve the intersection problem defined on $(G, \mathcal{I}_1)$ and $(G, \mathcal{I}_2)$. Let r_1 and r_2 be the rank functions of $\mathcal{I}_1$ and $\mathcal{I}_2$ respectively. There are two possible outcomes:

(i) An $r_1(G)$-intersection results, leading to a $(v(U)+1)$-matching in $U+\mathbf{u}$.

(ii) A strong duality result for matroid intersection showing that the maximum cardinality of an intersection is at most $r_1(G)-1$.

In (i), we find an augmentation that produces a larger matching $\hat{U}'$. In this case, we replace the earlier strong duality set by $(U', \hat{U}', K', \Pi')$ having $U' = \hat{U}'$, $K' = \varnothing$ and Π' consisting of an even component for each line of $\hat{U}'$. In (ii) we restart with $(U+\mathbf{u}, \hat{U}, K', \Pi')$ where Π' is the partition of $U+\mathbf{u}$ derived from Π by combining $\mathbf{u}$ and the components U_i such that either U_i is odd and $U_i \cap G_2 \neq \phi$ or U_i is even and $U_i \subseteq G_2$, into a single component. Here, G_2 is a member of a bipartition of G satisfying $r_1(G_1) + r_2(G_2) \leq r_1(G) - 1$.

3.7 Submodular Functions and Polymatroids

Results for matroids that we developed earlier followed from the numerous elegant properties of rank functions, polyhedra over closed sets, etc. It is natural to ask whether some conclusions remain valid when some of the matroid properties are relaxed. We conclude this chapter with a brief introduction to some results of this type.

3.7.1 Submodular Functions

Consider a rational-valued function, f, which is defined on the class of all subsets of a set G. That is, f is *set-function on G*. We say that f is *submodular* if

$$f(G_1)+f(G_2) \geq f(G_1 \cap G_2)+f(G_1 \cup G_2) \tag{3-40}$$

for all G_1, $G_2 \subseteq G$. If the inequality in (3-40) is reversed, we say that f is *supermodular*. Finally, a set-function f on G is said to be *monotone* if $f(G_1) \leq f(G_2)$ whenever $G_1 \subseteq G_2 \subseteq G$.

The relevance of the submodular function as a generalization of matroids is apparent in the following theorems.

Theorem 3.19. Rank Function Submodularity. *The rank function of any matroid $M \triangleq (G, \mathcal{I})$ is submodular and monotone on subsets of G.*

In fact, characterizations of matroids can be obtained by reference to submodular functions.

Theorem 3.20. Matroid Characterization in Terms of Rank Function (Dillworth (1944)). *Let p be a nonnegative-integer-valued submodular and monotone function defined on* 2^G*, the set of subsets of G, with the property that* $p(A) \leq |A|$ *for all* $A \in 2^G$*. Then* $\mathcal{I}(p) = \{S: S \subseteq G, p(S) = |S|\}$ *is a collection of independent sets of a matroid on G having rank function p.*

Theorem 3.21. Nonnegative Submodular Functions. *Let* $\mathcal{K}$ *be a family of subsets of a ground set G, which is closed under the operations of union and intersection. Also, let p be integer valued, submodular and nonnegative on* $\mathcal{K}\backslash\{\emptyset\}$*. Then,*

$$\mathcal{I}(\mathcal{K}: p) = \{S: p(A) \geq |S \cap A|, A \in \mathcal{K}\backslash\{\emptyset\}\}$$

is the set of independent sets of a matroid on G. Its rank function, $\bar{p}$*, is given by*

$$\bar{p}(X) = \min\left\{|X|, \sum_{i=1}^{t} p(S_i) + |X \cap (G\backslash \cup S_i)|\right\},$$

with the minimum being taken over families $\{S_1, S_2, \ldots, S_t\}$ *of disjoint sets in* $\mathcal{K}\backslash\{\emptyset\}$.

Various corollaries of these theorems can be found in Welsh (1976). Interesting in this regard is the generation of the cycle matroid, M_G, of a graph $G(V, E)$ by the submodular function p defined as $p(Q) = |V(Q)| - 1$, $Q \subseteq E$. Here, $V(Q)$ is the set of vertices in the subgraph of G generated by edges in Q. The reader should try to show this p to satisfy the conditions of Theorem 3.21.

3.7.2 Polymatroids

A polymatroid can be defined in various (equivalent) ways. Here, we shall employ two definitions, each exhibiting obvious relationships to matroids and submodular functions given earlier.

For a finite set G, let $\mathbb{R}_+^G$ denote the space of nonnegative, real valued vectors having coordinates that are indexed by G. A polymatroid is a pair $(G, \mathcal{P})$ where G is nonempty and finite and $\mathcal{P}$ is the set of *independent vectors* forming a nonempty, compact subset of $\mathbb{R}_+^G$ such that

(i) if $\mathbf{x} \in \mathcal{P}$, $\mathbf{y} \in \mathbb{R}_+^G$ and $\mathbf{y} \leq \mathbf{x}$, then $\mathbf{y} \in \mathcal{P}$,
(ii) for every vector $\mathbf{x} \in \mathbb{R}_+^G$, every maximal (component sum) $\bar{\mathbf{x}} \in \mathcal{P}$ with $\bar{\mathbf{x}} \leq \mathbf{x}$ satisfies

$$\sum_{i \in G} \bar{x}_i = \sum_{i \in G} x_i$$

Note that the compactness of $\mathscr{P}$ assures that the notion of maximality is well-defined.

Example 3.12. Polymatroid. It is easy to interpret the basic notions regarding this polymatroid definition considering a simple 2-dimensional case (i.e., $G = \{1, 2\}$). Consider the subset of $\mathbb{R}_+^2$ shown in Figure 3.6a. Denoting the shaded area by $\mathscr{P}_1$, it is not hard to check that $(G, \mathscr{P}_1)$ is a polymatroid. However, $(G, \mathscr{P}_2)$ shown in part (b) is not a polymatroid since condition (ii) is not satisfied. The vector $\mathbf{x}$ shown and the two maximal independent subvectors $\mathbf{x}^1$ and $\mathbf{x}^2$ have, $\Sigma_{i \in G} x_i^1 \neq \Sigma_{i \in G} x_i^2$. From these examples, the reader should be able to characterize the class of polymatroids existing in $\mathbb{R}_+^2$. ■

An alternative definition characterizes a polymatroid as a pair (G, p), p: $2^G \to \mathbb{R}_+$ satisfying

(i) $p(\varnothing) = 0$
(ii) $p(X) \leq p(Y)$ for all $X \subseteq Y \subseteq G$
(iii) $p(X \cup Y) + p(X \cap Y) \leq p(X) + p(Y)$ for all $X \subseteq G$, $Y \subseteq G$.

We shall not take space here to establish the equivalence of these two definitions. This can be found in Welsh (1976).

The new definition has a clear submodularity flavor. Property (ii) is a monotonicity property on the rank function and (iii) asserts its submodularity. Theorem 3.20 can be applied to show that if the rank function is integer-valued and we have $p(\{i\}) = 0$ or 1 for every $i \in G$, then the polymatroid is a matroid.

Suppose we have a polymatroid, (G, p). The vectors $\mathbf{x} \in \mathbb{R}_+^G$ such that

$$\sum_{i \in S} x_i \leq p(S) \text{ for all } S \subseteq G \tag{3-41}$$

are called *independent vectors* in (G, p). The independent vectors of a polymatroid form a convex polyhedron in $\mathbb{R}_+^G$. For a given polymatroid, (G, p), that polyhedron is called the *independence polytope* of (G, p).

Recall that p is assumed monotone and submodular in the definition of a polymatroid. It can also be shown, however, that polymatroids can be constructed from arbitrary nonnegative submodular functions.

Suppose we have the independence polytope, $\mathscr{P}$ of some polymatroid, (G, p), characterized by its (3-41) representation. While, in general, finding the vertices of convex polyhedra is a tedious undertaking at best, the task is not so terrible if our polytope is defined by the independent vectors of a polymatroid.

For $G = \{1, 2, \ldots, n\}$, let (G, p) be a polymatroid where p is nonnega-

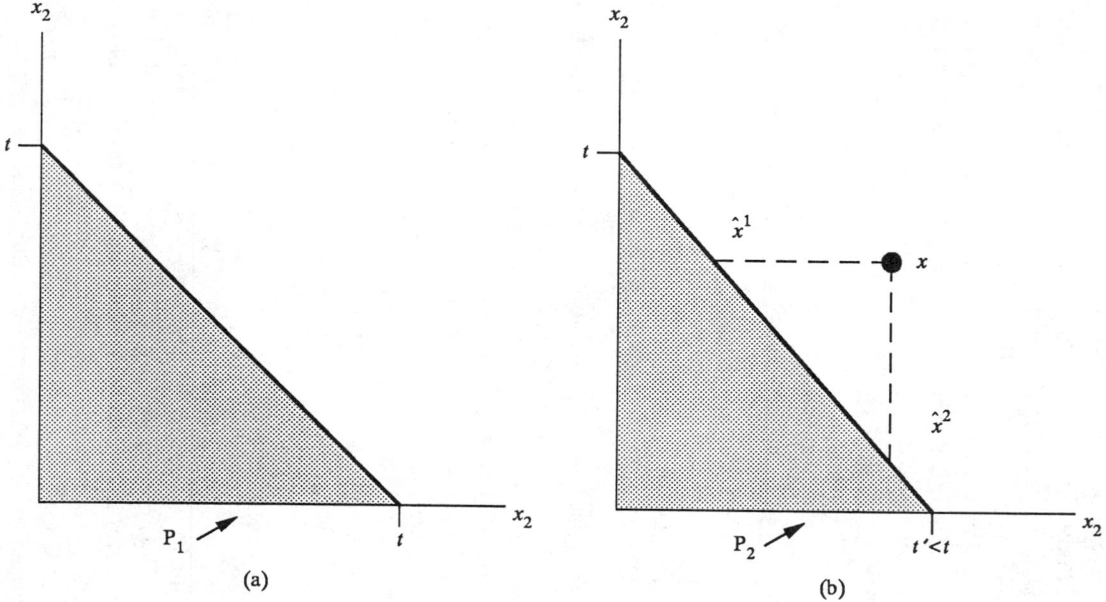

Fig. 3.6. Polymatroid (a) and nonpolymatroid (b) structures in $\mathbb{R}^2_+$.

tive, monotone and submodular. Further, let π be a permutation of the elements in G, where we denote π as $(v_1, v_2, \ldots, v_n)$. Now, define recursively, sets u_π^l such that

$$u_\pi^l = \{v_1, v_2, \ldots, v_l\}; \qquad 1 \leq l \leq n. \tag{3-42}$$

We can now state the following clever result due to Edmonds (1970):

Theorem 3.22. Vertices of the Independence Polytope of a Polymatroid. *Let (G, p) be a polymatroid and let $\mathcal{P}$ be its independence polytope. Then the vertices of $\mathcal{P}$ are points $\mathbf{x}^i \in \mathbb{R}_+^G$ where $\mathbf{x}^i = (x_1^i, x_2^i, \ldots, x_n^i)$ and*

$$x_{v_1}^i = p(U_\pi^1)$$
$$x_{v_2}^i = p(U_\pi^2) - p(U_\pi^1)$$
$$\vdots$$
$$x_{v_k}^i = p(U_\pi^k) - p(U_\pi^{k-1})$$
$$x_{v_{k+1}}^i = x_{v_{k+2}}^i = \cdots = x_{v_n}^i = 0$$

for $0 \leq k \leq n$ and every permutation π of G.

Moreover, it follows that if p is integer-valued, all the vertices of the polymatroid are integers in $\mathbb{R}_+^G$. (Readers may wish to compare to the corresponding Theorem 3.8 for matroids.)

In order to demonstrate the nature of the theorem, let us consider the following example.

Example 3.13. Recursive Determination of the Vertices of the Independence Polytope. We shall expand on a problem given in Welsh. Consider the independence polytope in $\mathbb{R}_+^3$ stated below:

$$x_1 \leq 2, x_2 \leq 3, x_3 \leq 5, x_1 + x_2 \leq 4,$$
$$x_2 + x_3 \leq 6, x_1 + x_3 \leq 6, x_1 + x_2 + x_3 \leq 7$$

and $x_i \geq 0$, $1 \leq i \leq 3$. It is easy to see that the right-hand-sides of these constraints form a submodular function on $G = \{1, 2, 3\}$.

The polytope appears in Figure 3.7. Applying the recursion of Theorem, 3.22 yields the vertices listed in Table 3.1. To illustrate, consider the case of $k = 3$ and $\pi = (2, 1, 3)$. With $v_1 = 2$, we have

$$x_2 = p(\{2\}) = 3$$

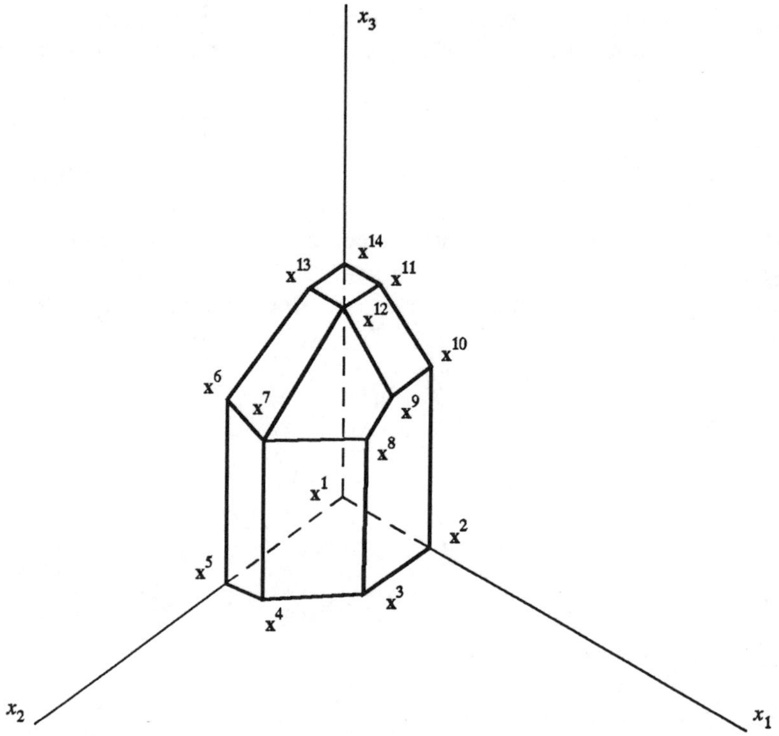

Fig. 3.7. Independence polytope of Example 3.13.

(Observe that this already shows that p is not the rank function of a matroid because $p(\{2\}) > 1$).

$$x_1 = p(\{2, 1\}) - p(\{2\}) = 4 - 3 = 1$$

$$x_3 = p(\{2, 1, 3\}) - p(\{2, 1\}) = 7 - 4 = 3$$

and thus, $\mathbf{x} = (1, 3, 3)$. This is vertex $\mathbf{x}^7$ on the polytope of Figure 3.7. ■

3.7.3 Polymatroid Intersection

With matroids, it proved interesting to consider optimizations over the intersection of two matroids defined on the same ground set. Much of that theory can be seen to generalize directly to polymatroidal intersections $P \triangleq P_1 \cap P_2$ where P_1 is the polymatroid (G, p_1) and P_2 is the polymatroid (G, p_2). As in the case of matroids, the intersection P is not a polymatroid, but does have integrality properties that do not extend to 3-polymatroid intersection. Specifically, Schrijver (1982) shows the following theorem:

TABLE 3.1. Vertices of the Independence Polytope of Example 3.13

π	$\mathbf{k}^a$	$\mathbf{x}^b$	Vertex number on polytope
(1, 2, 3)	3	(2, 2, 3)	8
	2	(2, 2, 0)	3
	1	(2, 0, 0)	2
	0	(0, 0, 0)	1
(1, 3, 2)	3	(2, 1, 4)	9
	2	(2, 0, 4)	10
(2, 1, 3)	3	(1, 3, 3)	7
	2	(1, 3, 0)	4
	1	(0, 3, 0)	5
(2, 3, 1)	2	(0, 3, 3)	6
(3, 1, 2)	3	(1, 1, 5)	12
	2	(1, 0, 5)	11
	1	(0, 0, 5)	14
(3, 2, 1)	2	(0, 1, 5)	13

[a] Note: $\mathbf{x}$ is given as (x_1, x_2, x_3) rather than in π-order.
[b] Values are omitted that generate duplicate vertices.

Theorem 3.23. Vertices of Polymatroid Intersection. *Let (G, p_1) and (G, p_2) be polymatroids on a finite set G. Then, if p_1 and p_2 are integer-valued, all extreme-points of the polytope formed by $\{\boldsymbol{x} \in \mathbb{R}_+^G\}$ satisfying*

$$\sum_{i \in S} x_i \leq p_1(S) \qquad \text{for all } S \subset G$$

$$\sum_{i \in S} x_i \leq p_2(S) \qquad \text{for all } S \subset G$$

are integer-valued $\mathbb{R}_+^G$ vectors.

EXERCISES

3-1. Define some independence systems that are not matroids. Give examples of systems $(G, \mathcal{I})$, where $\mathcal{I}$ is not closed under inclusion.

3-2. Let G be the edge set of a graph $G(V, E)$ and define $\mathcal{I}$ to be the family of edge subsets that form subgraphs of G possessing no subgraphs homeomorphic to K_4. Is $(G, \mathcal{I})$ an independence system? Is it a matroid?

3-3. Consider the graph below and let $(G, \mathcal{I})$ be the corresponding graphic matroid, M_G. Find a circuit of M_G. Let $H = \{e_1, e_3, e_6, e_9, e_{12}, e_{14}\}$. Find $sp(H)$.

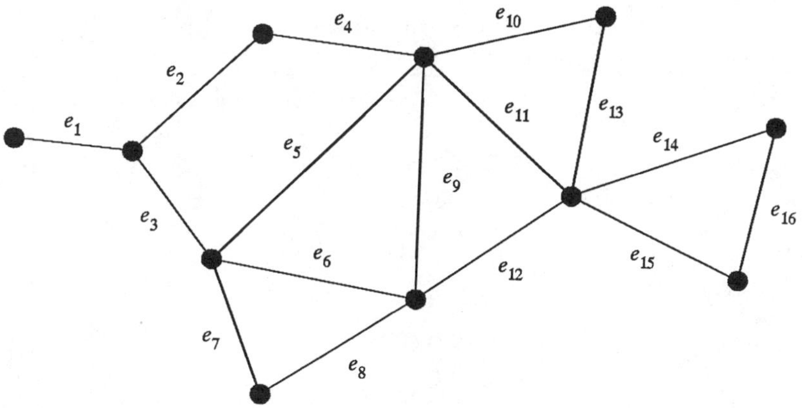

3-4. For any matroid $M \triangleq (G, \mathscr{I})$ show that the *truncated* system $M_k \triangleq (G, \mathscr{I}_k)$ is also a matroid where $\mathscr{I}_k \triangleq \{I \in \mathscr{I}: |I| \leq k\}$.

3-5. Prove Lemma 3.10, i.e., that the union of matroids is a matroid.

3-6. Let M_G be the graphic matroid defined on the graph below. Find the circuits of M_G, the bases of M_G, and the circuits and bases of M_G^*.

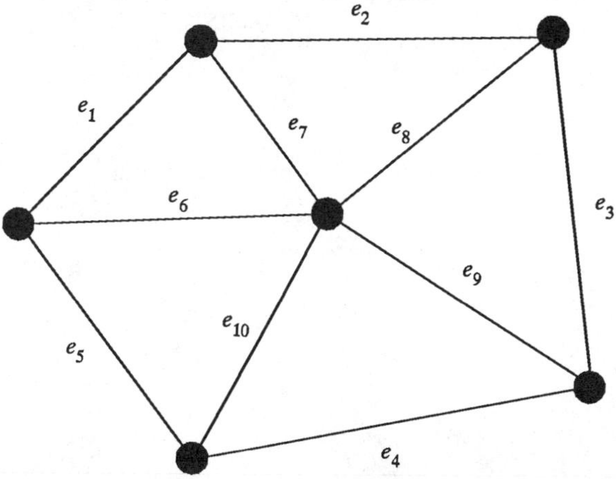

★3-7. (a) Let $E = \{e_i: i \in [1, 2, \ldots, 4n]\}$, $F = \{f_i: i \in [1, 2, \ldots, 4n]\}$ and, for each $i \in [1, 2, \ldots, 2n]$, $S_i = \{e_{2i-1}, e_{2i}, f_{2i-1}, f_{2i}\}$. Let T be the collection of subsets formed by taking either $\{e_{2i-1}, e_{2i}\}$ or $\{f_{2i-1}, f_{2i}\}$ from each subset S_i. For a subset T' of T, define $X \subseteq E \cup F$ to be independent in $M(T')$ if

(i) $|X| \leq 4n$
(ii) $|X \cap S_i| \leq 3$
(iii) $X \notin T'$.

Show that for each $T' \subseteq T$, $M(T')$ is a matroid.

(b) Construct a matroid in which there can be no polynomial procedure for computing rank. Hint: Consider a matroid on $[1, 2, \ldots, 2n]$ defined in terms of $M(T')$ and suppose we are allowed queries of the form "Is X independent in $M(T')$?" (Vande Vate (1984)).

3-8. Apply the Greedy algorithm to the graph in **3-6** and produce a maximum weight forest. Let the edge weights be given by (6, 8, 3, −1, 4, −6, 5, 0, 2, 4) for edges $e_1, e_2, \ldots, e_{10}$ respectively.

3-9. Using the graph in **3-6** (with weights in **3-8**), apply Greedy to the independence system defined in **3-2.**

3-10. For the stated graph in **3-8,** verify the Greedy solution by solving the corresponding system (3-6)–(3-8).

3-11. Consider the directed graph $G(V, A)$ below. Repeat Example 3.5 by defining the graphic and partition matroids on G and finding a maximum cardinality intersection.

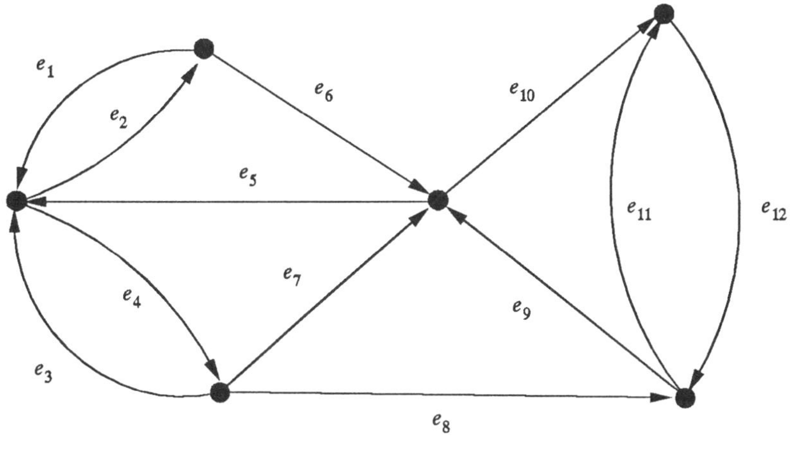

3-12. For $G(V, E)$ below, let $M_1 = M_2 = M_G$. Using Algorithm 3B, find a partition of E each component of which has size 3 and which are independent in their respective matroids.

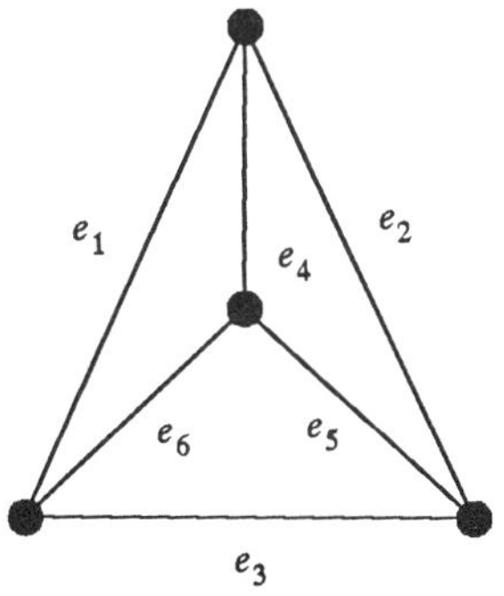

3-13. Find a maximum cardinality matching in the bipartite graph below using the result of Theorem 3.12.

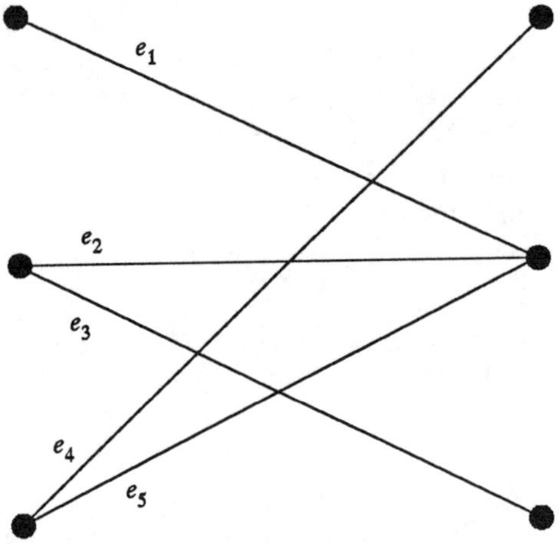

3-14. Formulate and solve **3-13** as a linear program defined by (3-15)–(3-19). Let the edge weights be given by (2, 7, 1, 3, 2) for edges e_1, $e_2, \ldots, e_5$ respectively.

3-15. Solve the weighted intersection problem of **3-14** using Algorithm 3C.

3-16. Let $M=(G, \mathscr{I})$ be a matroid. Prove that for any $H \subseteq G$, $sp(H)$ is unique.

★**3-17.** Let $M=(G, \mathscr{I})$ be a matroid. Then there exists a partition of M into k independent sets if and only if

$$|A| \leq k\, r(A) \text{ for all } A \subseteq G$$

Apply this result to the graphic matroid and prove that the edge set of a graph, $G(V, E)$ can be partitioned into k forests if and only if for every subset $U \subseteq V$, $k(|V|-1) \geq e_U$, where e_U denotes the number of edges with both vertices in U. (Nash-Williams (1964)).

3-18. (Edmonds (1970)). Let $M_1=(G, \mathscr{I}_1)$ and $M_2=(G, \mathscr{I}_2)$ be a pair of matroids on G with rank functions r_1 and r_2 respectively. Then

$$\max_{I \in \mathscr{I}_1 \cap \mathscr{I}_2} |I| = \min_{A \subseteq G} \{r_1(A) + r_2(G \setminus A)\}$$

Now, König's theorem states that for bipartite graphs, the size of a maximum matching is the same as the cardinality of a minimum (vertex) cover. Show that this follows from Edmonds' theorem above.

3-19. For each of the following, show that the indicated combinatorial structure is an independence system and then either show it is a matroid or give an example demonstrating it violates matroid axioms.

(a) G is the arc set of a directed graph. $\mathscr{I}$ is the collection arc subsets with at most one arc inbound at any vertex.
(b) G is the vertex set of an undirected graph. $\mathscr{I}$ is the collection of vertex subsets with no two members incident to the same edge.
(c) G is the edge set of an undirected graph. $\mathscr{I}$ is the collection of matchings, i.e., edge subsets with no two members incident to the same vertex.
(d) G is the vertex set of an undirected graph. $\mathscr{I}$ is the collection of vertex subsets that can be matched (i.e., are spanned by a vertex disjoint collection of edges).
*(e) G is the vertex set of an undirected planar graph with edges viewed as rigid bars and vertices as flexible joints. $\mathscr{I}$ is the collection of vertex/joint subsets that are nonrigid, i.e., when all members are pinned to a location in the plane, some remaining vertices can still move.
(f) G is the set of indices of positive integers $a_1, a_2, \ldots, a_n$. $\mathscr{I}$ is the collection of subsets with sum less than or equal to positive integer b.
(g) G is a finite set of n-vectors (members of $\mathbb{R}^n$). $\mathscr{I}$ is the collection of subsets of G that are affinely independent, i.e., cannot be combined by weights summing to 1 to produce the zero vector (see Appendix A).

3-20. Establish each of the containments in Figure 3.1, i.e., that graphic matroids are regular, regular matroids are binary, and binary matroids are representable.

3-21. Show that both uniform matroids and partition matroids are transversal matroids.

3-22. Establish that the binary matroid, R_{10}, with matrix

$$\begin{bmatrix} 1 & 0 & 0 & 0 & 0 & 1 & 1 & 0 & 0 & 1 \\ 0 & 1 & 0 & 0 & 0 & 1 & 1 & 1 & 0 & 0 \\ 0 & 0 & 1 & 0 & 0 & 0 & 1 & 1 & 1 & 0 \\ 0 & 0 & 0 & 1 & 0 & 0 & 0 & 1 & 1 & 1 \\ 0 & 0 & 0 & 0 & 1 & 1 & 0 & 0 & 1 & 1 \end{bmatrix}$$

is regular but not graphic.

3-23. Establish that the uniform matroid $U_{2,4}$ is representable but not binary.

3-24. Consider the (Fano matroid) system on $G = \{e_1, e_2, e_3, e_4, e_5, e_6, e_7\}$ where all subsets of 3 or fewer elements are independent except $\{e_1, e_2,$

$e_6\}$, $\{e_1, e_4, e_7\}$, $\{e_1, e_3, e_5\}$, $\{e_2, e_3, e_4\}$, $\{e_2, e_5, e_7\}$, $\{e_4, e_5, e_6\}$ and $\{e_3, e_6, e_7\}$.

(a) Establish that this system is a matroid.
(b) Describe the independent sets of the dual of this matroid.
*(c) Show that the matroid and its dual are binary.

3-25. Consider the (Vamos matroid) system on $G = \{a_1, a_2, b_1, b_2, c_1, c_2, d_1, d_2\}$, where all subsets of 4 or fewer elements are independent except $\{a_1, a_2, b_1, b_2\}$, $\{a_1, a_2, c_1, c_2\}$, $\{b_1, b_2, c_1, c_2\}$, $\{b_1, b_2, d_1, d_2\}$ and $\{c_1, c_2, d_1, d_2\}$.

(a) Show that this collection of independent sets forms a matroid on G.
(b) Describe the independent sets of the dual of this matroid.
*(c) Show that this matroid is not representable over any field.

***3-26.** A collection $\{S_j\colon j \in J\}$ of subsets of a finite set S is said to be a d-partition if (i) $|J| \geq 2$; (ii) $|S_j| \geq d$ for all $j \in J$; and (iii) every d-element subset of S is fully contained in a unique S_j. Show that for such a d-partition $\{B \subset S\colon |B| = d+1,\ B \not\subseteq S_j \text{ for any } j \in J\}$ is the set of bases of a matroid (termed a *paving* matroid).

3-27. Given a directed graph $G(V, A)$, two distinguished vertices $s, t \in V$ and nonnegative weights c_{ij} for arcs $(i, j) \in A$,

(a) Formulate as a maximum weight 2-matroid intersection the problem of choosing a subset of arcs that form a shortest (least total weight) path (directed) from s to t. (Hint: introduce complements of arcs.)
(b) Illustrate (a) by applying 2-Matroid Intersection Algorithm 3C on the matroids derived in finding a shortest path from $s = 1$ to $t = 4$ in the following graph.

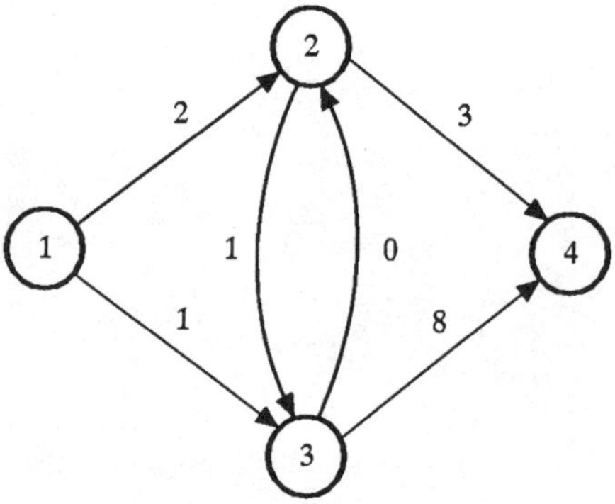

3-28. An even cycle with edges $e_1, \bar{e}_1, e_2, \bar{e}_2, \ldots, e_k, \bar{e}_k$ is obviously a bipartite graph. Why? Suppose weights $w_1, \overline{w}_1, w_2, \overline{w}_2, \ldots, w_k, \overline{w}_k$ are assigned to edges of an even cycle. Formulate a maximum weight 2-matroid intersection problem over the cycle edges and modified weights such that an optimal solution will be either the set $\{e_1, e_2, \ldots, e_k\}$ or the set $\{\overline{e}_1, \overline{e}_2, \ldots, \overline{e}_k\}$, whichever has more (original) total weight.

3-29. The *k-parity* problem is a generalization of the matroid parity problem where elements belong to unique sets of k elements, all of which must be in a solution if any is. As with the usual parity problem, a solution must also be independent in a given matroid (setting $k = 2$ yields the usual parity problem.) Show (by introducing complements) that every maximum total weight k-parity problem can be expressed as one of finding a maximum weight 3-matroid intersection where two of the matroids are based on **3-28.**

3-30. The k-matroid intersection problem is one of finding a maximum total weight subset of a given ground set that is independent in k different matroids on that set. Show (by making k copies of each element of the ground set) that every k-matroid intersection problem can be reduced to the k-parity problem of **3-29.** Then apply **3-29** to conclude every maximum weight k-matroid intersection problem can be written as a maximum weight 3-matroid intersection problem.

3-31. Suppose f_1 and f_2 are rational-valued submodular functions defined on subsets of a finite set G. Show that each of the following is submodular.

(a) $g(S) \triangleq f_1(S) + f_2(S)$
(b) $g(S) \triangleq f_1(S) + k$ (fixed rational k)
(c) $g(S) \triangleq \alpha f_1(S)$ (fixed rational $\alpha \geq 0$)
(d) $g(S) = \min\{k, f_1(S)\}$

3-32. Prove Theorem 3.19, i.e., show that the rank function of a matroid is submodular.

3-33. Following Example 3.12, characterize the polymatroids of $\mathbb{R}^2_+ =$ {nonnegative 2-vectors}.

3-34. Let $T_1, T_2, \ldots, T_n$ be subsets of a finite set T and $N \triangleq \{1, 2, \ldots, n\}$. Show that the function on $S \subseteq N$ defined by $f(S) \triangleq |\bigcup_{i \in S} T_i|$ is submodular. When is it not the rank function of a matroid?

3-35. Consider a bipartite graph with edges in E connecting members of the vertex bipartition S, T, and let weights $w_i \geq 0$ be defined for vertices $i \in S$.

(a) Show that the function on subsets of $\overline{T} \subseteq T$ that sums weights of $\overline{S}$ adjacent to at least one $j \in \overline{T}$ is submodular.
(b) Write the corresponding polymatroid constraints for the graph below.

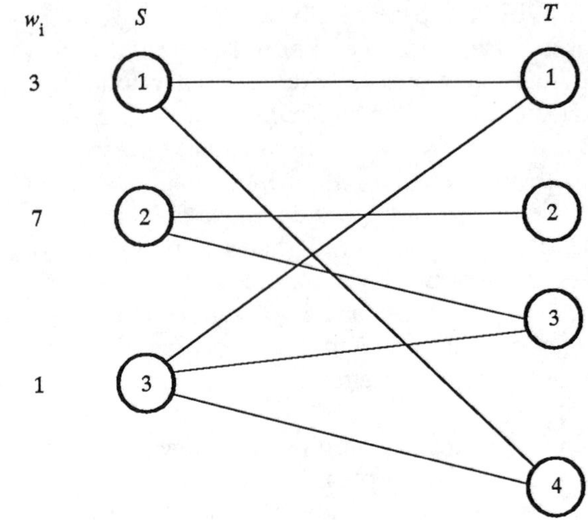

3-36. Let $G(V, A)$ be a directed graph with capacities u_{ij} on arcs $(i, j) \in A$.

(a) Show that the sum of capacities of arcs leaving subsets $\overline{V} \subseteq V$ is a submodular function.

(b) Give the constraints of the corresponding polymatroid for the graph of **3-27**.

4

Polynomial Algorithms—Linear Programming

Modern optimization began in the late 1940s with George Dantzig's (1947) development of the Simplex algorithm for linear programming. In the four decades that have followed, linear programming has remained absolutely central to both applied optimization and theoretical developments. Commercial implementations of the Simplex routinely solve truly massive linear programming models; algorithms like those developed in Chapters 5 and 6 for hard discrete problems use linear programming subroutines; polynomial algorithms for combinatorial problems (for example the matroid and matroid intersection algorithms of Chapter 3) are often proved via a concise linear programming description of the set of solutions.

Although Appendix C provides a brief survey, it is not our purpose in this book to review fundamental linear programming theory. Instead, in this chapter we treat the famous results that establish the polynomial-time solvability of linear programming, together with the related tools that combine with linear programming algorithms to yield polynomial-time procedures for many discrete optimization problems.

4.1 Polynomial-Time Solution of Linear Programs

Readers who have by now become accustomed to the idea that the set of truly well-solved problems is synonymous with the collection of polynomial solvable ones may be surprised to learn that the first proof of polynomial-time solvability of linear programs (the Khachian (1979) algorithm treated in Section 4.1.2) was greeted with great confusion and skepticism. An interesting account of the uproar, which even spilled over into the public media, can be found in Wolfe (1980).

4.1.1 Simplex Anomalies

One source of the confusion was the peculiar complexity theoretic standing of the Simplex algorithm. The enormous success of linear programming is based on the practical fact that the Simplex typically requires a low-order polynomial number of steps to compute an optimal solution. In the worst case, i.e., the one establishing the complexity classification of the algorithm, however, the Simplex is exponential. Thus, with the Simplex we have the very unusual circumstance of an extremely effective practical algorithm that fails the standard formal definition of efficiency. Prior to Khachian's result, the linear programming problem class was in the similarly distressing complexity state of being practically well-solved and replete with the theoretical properties (e.g., strong duality) characteristic of formally tractable cases, yet not known to admit polynomial-time solutions.

A villainous family of instances demonstrating the Simplex's pathological worst case behavior was first provided by Klee and Minty (1972). Their examples are of the form

$$
\begin{array}{lll}
 & \max & x_n \\
(KM_n) & \text{s.t.} & \epsilon \leq x_1 \leq 1 \\
 & & \epsilon x_{j-1} \leq x_j \leq 1 - \epsilon x_{j-1} \qquad \text{for } j = 2, \ldots, n,
\end{array}
$$

where $0 < \epsilon < \frac{1}{2}$. The case of $\epsilon = \frac{1}{4}$, $n = 2$ is illustrated in Figure 4.1.

The polytope of (KM_n) is a twisted version of the 0–1 n-cube. Indeed, when $\epsilon \rightarrow 0$, the limiting polytope is exactly the n-space hypercube and for all $\epsilon \in [0, \frac{1}{2})$ the polytope has 2^n extreme points.

The Simplex algorithm is well known to proceed from extreme-point solution to adjacent extreme-point solution in optimizing a linear program. Certainly, if all 2^n of the extrema in the (KM_n) cubes had to be visited, the algorithm would require time growing exponentially in the n-variable, $2n$-inequality problem input.

The subtle effect of the shifts introduced in (KM_n) when $\frac{1}{2} > \epsilon > 0$ is to order the extreme points of the cube so that there is a strictly improving (with respect to the maximize x_n objective) sequence of adjacent ones including all 2^n extrema. In Figure 4.1, for example, the sequence is the one indicated by extreme point labels $\mathbf{x}^k$. Formalization of the idea that a suitable implementation of the Simplex will follow that sequence yields Klee and Minty's result.

Theorem 4.1. Nonpolynomiality of the Simplex Algorithm. *For every $n \geq 2$, there exist linear programs having n variables, 2n inequality constraints, and integer coefficients bounded by ± 3 such that one implementation of the Simplex algorithm requires $2^n - 1$ pivots.*

Proof. Use $\epsilon = \frac{1}{3}$ in the above system (KM_n) and multiply by 3 to produce integer constraints. See Klee and Minty (1972) for details. ■

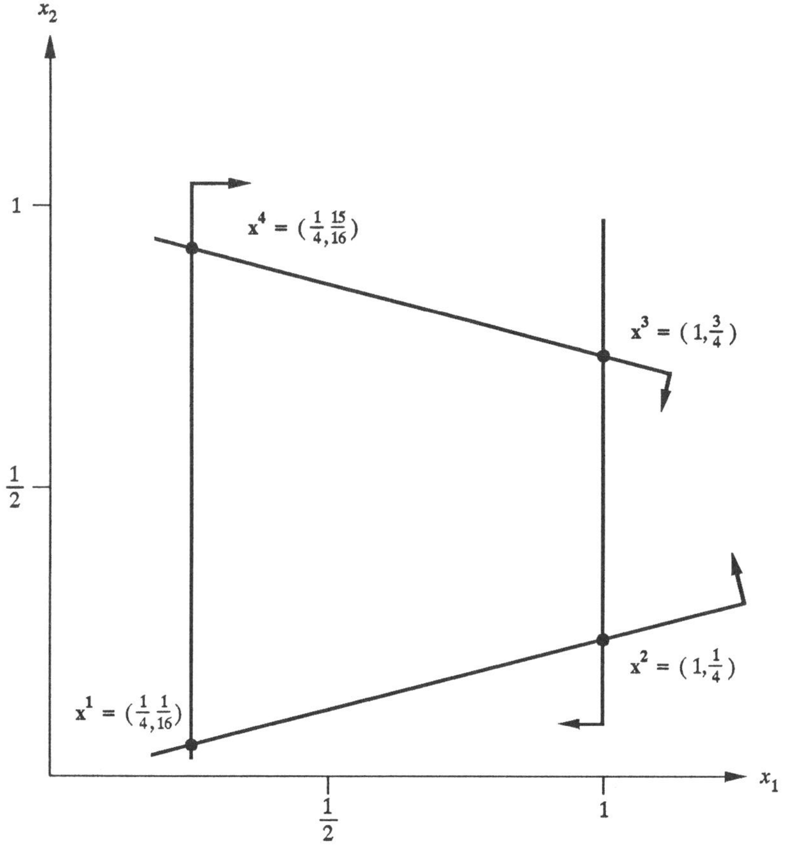

Fig. 4.1. Klee-Minty polytope for $n = 2$, $\epsilon = \frac{1}{4}$.

Although it has long been known that examples like the Klee-Minty problems were extremely pathological, such a proposition is difficult to prove. Still, a variety of the work of the past decade has come very close to doing so by computing the expected number of Simplex steps over various assumed probability distributions on linear programming data. A (partial) list of such contributions is Borgwardt (1982a,b), Smale (1983a,b), Adler (1983), Haimovich (1983) (see also the Tovey results we discuss in Section 7.3). These researchers all derive low order polynomial bounds on the expected number of Simplex pivots, thus confirming practical wisdom with solid theoretical findings. Orders range from $O((1/n + 1/(m-n+1))^{-1})$ to $O((1/n + 1/(m-n+1) - 1/m)^{-1})$ for m inequality constraints and n variables.

4.1.2 Khachian's Ellipsoid Algorithm

We have already noted the considerable uproar produced by the emergence of a provably polynomial time algorithm for linear programming in Khachian's (1979) paper. One reason for the surprise was the very noncombinatorial, nonlinear, nonsimplex underlying structure of this *Ellipsoid* algorithm.

In this section, we present the now classic ellipsoid results. We shall skip over many details of numerical precision, however, in the interest of readability. Readers interested in such matters are referred to Gács and Lovász (1981), and Grötschel, Lovász and Schrijver (1981). (Much of our development is based on the latter.)

Basic Ellipsoid Algorithm

The essential Ellipsoid algorithm is rooted in a nonlinear optimization tradition that can be treated completely independently of linear programming. We begin by taking that approach.

Consider simply a closed convex set $X \subset \mathbb{R}^n$ such that if X is nonempty, it will contain an n-sphere of radius r and be contained in one of radius R, i.e.,

$$X \neq \emptyset \text{ implies } \mathbb{B}(\bar{\mathbf{x}}, r) \subseteq X \subseteq \mathbb{B}(\mathbf{x}^0, R) \tag{4-1}$$

with $\mathbf{x}^0$ known and $\bar{\mathbf{x}} \in X$. In less mathematical terms (4-1) requires, on the one hand, that X be bounded, and on the other, that it be of full dimension.

Our interest is in either finding an $\mathbf{x} \in X$ or proving that none exists. The Ellipsoid algorithm approaches this decision problem by generating a sequence of n-space ellipsoids $\mathbb{B}(\mathbf{x}^0, R) \triangleq E_0 \supseteq E_1 \supseteq E_2 \cdots$. At each iteration, the center of the ellipsoid, $\mathbf{x}^k$, is checked for membership in X. If

$\mathbf{x}^k \in X$, the question is resolved. If not, we generate a *separating* inequality

$$\mathbf{a}^k\mathbf{x} \le a_0^k$$

such that $\mathbf{a}^k\mathbf{x}^k > a_0^k + \epsilon$ for specified error tolerance $\epsilon > 0$ and $\|\mathbf{a}^k\| \ge 1$. That is, we show $\mathbf{x}^k$ lies on one side of $\mathbf{a}^k\mathbf{x} = a_0^k$ and X on the other. The next ellipsoid E_{k+1} is constructed to contain $\{\mathbf{x} \in E_k: \mathbf{a}^k\mathbf{x} \le \mathbf{a}^k\mathbf{x}^k\}$, and the process repeats. Should it continue for so long that some E_l no longer contains enough volume for the inner-ball $\mathbb{B}(\bar{\mathbf{x}}, r)$, we stop with the conclusion that X is empty.

Algebraically, each ellipsoid is represented by

$$E_k \triangleq \{\mathbf{x} \in \mathbb{R}^n: (\mathbf{x} - \mathbf{x}^k)\mathbf{M}_k^{-1}(\mathbf{x} - \mathbf{x}^k) \le 1\} \tag{4-2}$$

for a positive definite matrix $\mathbf{M}_k$. The algorithm is then easily stated (the notation dist($\mathbf{x}$, X) refers to the Euclidean distance from $\mathbf{x}$ to X).

Algorithm 4A. Khachian's Ellipsoid Method. Let X be a compact convex set satisfying (4-1).

STEP 0: *Initialization.* Let $\mathbf{x}^0$ and R be as in (4-1), set $k \leftarrow 0$, and choose ellipsoid matrix $\mathbf{M}_0 \leftarrow R^n(\mathbf{I})$.

STEP 1: *Separation.* Either demonstrate that dist($\mathbf{x}^k$, X) $\le \epsilon$, and stop with an $\mathbf{x}^k$ sufficiently close to X, or exhibit a rational n-vector $\mathbf{a}^k$ with $\|\mathbf{a}^k\| \ge 1$ isolating $\mathbf{x}^k$ in the sense that $\mathbf{a}^k\mathbf{x} \le \mathbf{a}^k\mathbf{x}^k - \epsilon$ for all $\mathbf{x} \in X$. In the latter case, if $k \ge N \triangleq 5n^2\lceil \ln (R/r) \rceil$, stop with the conclusion that X is empty, and if not, proceed to Step 2.

STEP 2: *New Ellipsoid.* Using the $\mathbf{a}^k$ generated at Step 1, update

$$\mathbf{x}^{k+1} \longleftarrow \mathbf{x}^k - \frac{1}{n+1}\left(\mathbf{M}_k \frac{\mathbf{a}^k}{\sqrt{\mathbf{a}^k\mathbf{M}_k\mathbf{a}^k}}\right) \tag{4-3}$$

$$\mathbf{M}_{k+1} \longleftarrow \frac{2n^2+3}{2n^2}\left(\mathbf{M}_k - \frac{2}{n+1}\,\frac{(\mathbf{M}_k\mathbf{a}^k)(\mathbf{M}_k\mathbf{a}^k)^T}{\mathbf{a}^k\mathbf{M}_k\mathbf{a}^k}\right) \tag{4-4}$$

Then set $k \leftarrow k + 1$ and return to Step 1. ■

Example 4.1. Fundamental Ellipsoid Algorithm. Consider the closed convex set

$$X = \{x_1, x_2 \ge 0: x_1 + 2x_2 \le 2, -2x_1 - x_2 \le -1\}$$

It is obvious in Figure 4.2a that this set is contained in the Euclidean ball $\mathbb{B}(\mathbf{0}, 2)$ (i.e., $R = 2$) and contains a ball of say radius $r = \frac{1}{16}$.

Our first, Step 0 ellipsoid has

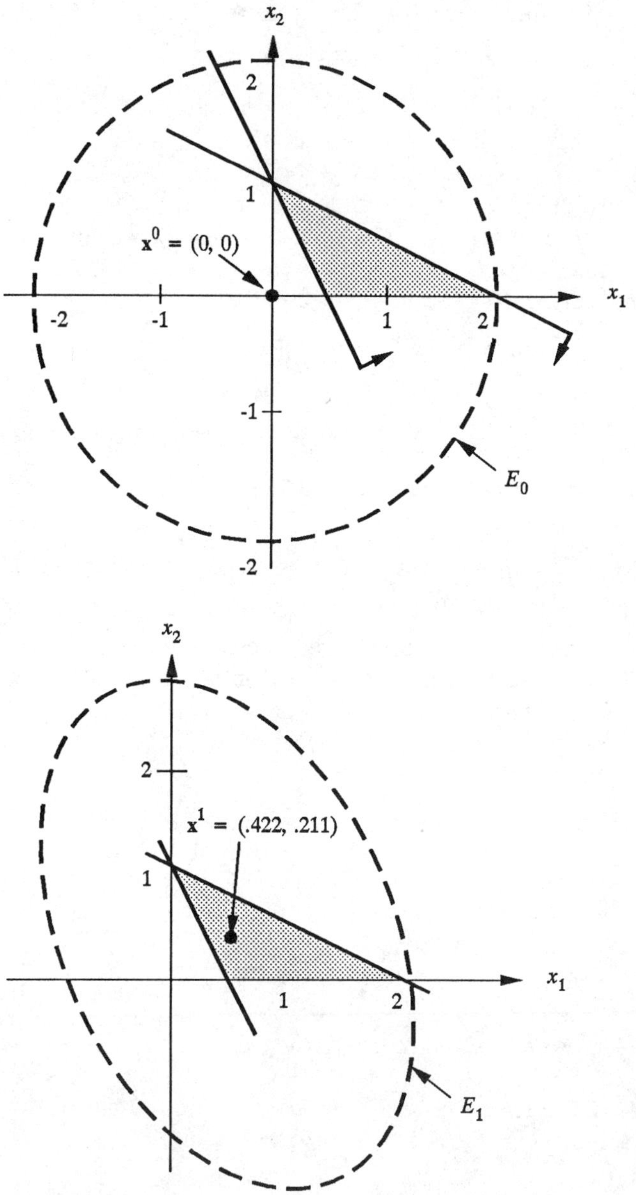

Fig. 4.2. Ellipsoids of Example 4.1; (a) initial ellipsoid E_0, (b) second ellipsoid E_1.

$$\mathbf{M}_0 = R(\mathbf{I}) = \begin{bmatrix} 2 & 0 \\ 0 & 2 \end{bmatrix}.$$

That is, it is exactly the outer ball $\mathbb{B}(\mathbf{0}, 2)$. Using tolerance say $\epsilon = 0$, we can see from Figure 4.2 that $\mathbf{x}^0 = (0, 0) \notin X$. Thus, Step 1 must produce a separating constraint.

The obvious scheme for producing such an inequality with constraints of X given explicitly is to enumerate for one violated by $\mathbf{x}^0$. Clearly, one in this case is $-2x_1 - x_2 \leq -1$ or $\mathbf{a}^1 = (-2, -1)$. Step 2 then updates

$$\mathbf{M}_0\mathbf{a}^0 = \begin{bmatrix} 2 & 0 \\ 0 & 2 \end{bmatrix} \begin{bmatrix} -2 \\ -1 \end{bmatrix} = \begin{bmatrix} -4 \\ -2 \end{bmatrix}$$

$$\mathbf{a}^0\mathbf{M}_0\mathbf{a}^0 = [-2, -1] \begin{bmatrix} -4 \\ -2 \end{bmatrix} = 10$$

$$\mathbf{x}^1 \longleftarrow \begin{bmatrix} 0 \\ 0 \end{bmatrix} - \frac{1}{3\sqrt{10}} \begin{bmatrix} -4 \\ -2 \end{bmatrix} \approx \begin{bmatrix} .422 \\ .211 \end{bmatrix}$$

$$\mathbf{M}_1 \longleftarrow \frac{11}{8}\left(\mathbf{M}_0 - \frac{2}{30}\begin{bmatrix} -4 \\ -2 \end{bmatrix}[-4, -2]\right) \approx \begin{bmatrix} 1.283 & -.733 \\ -.733 & 2.383 \end{bmatrix}$$

(Notice that these calculations involve arithmetic in irrational numbers, raising the sort of numerical questions we are generally ignoring.)

Figure 4.2b shows the new ellipsoid E_1. Notice that it contains all points in E_0 to the feasible side of $\mathbf{a}^0\mathbf{x} \leq \mathbf{a}^0\mathbf{x}^0$. New center $\mathbf{x}^1 \in X$. ■

To completely verify Algorithm 4A, numerous details must be checked. The two most important are summarized in the following lemma. (See the references given before for proofs).

Lemma 4.2. Properties of Algorithm 4A Calculations. *For each iteration k of Algorithm 4A*

(i) Matrix $\mathbf{M}_k$ is positive definite (even after rounding of irrationals to suitable accuracy); and
(ii) $vol\,(E_{k+1})/vol(E_k) < e^{-1/5n}$

where vol (E) denotes the Euclidean volume of E.

Theorem 4.3. Correctness of Algorithm 4A. *Given a compact, convex set X satisfying (4-1), Algorithm 4A either finds an $\mathbf{x}^k$ with dist $(\mathbf{x}^k, X) \leq \epsilon$ or correctly demonstrates that none exists in $N = 5n^2\lceil \ln(R/r)\rceil$ or fewer iterations k.*

Proof. Part (i) of Lemma 4.2 assures that Algorithm 4A will continue computing until one of the stopping rules is attained. For $\mathbf{M}_k$ positive definite, $\mathbf{a}^k\mathbf{M}_k\mathbf{a}^k > 0$, and so $\sqrt{\mathbf{a}^k\mathbf{M}_k\mathbf{a}^k}$ exists.

If we stop with an $\mathbf{x}^k$ having dist $(\mathbf{x}^k, X) \leq \epsilon$, there is nothing more to prove. Suppose instead, that iterations continue until the limit $N = 5n^2\lceil \ln (R/r)\rceil$ is reached, and we stop with the conclusion that X is empty. Then using (ii) of Lemma 4.2, $\text{vol}\,(E_N) < e^{-1/5n}\,\text{vol}\,(E_{N-1}) < e^{-2/5n}\,\text{vol}\,(E_{N-1}) \ldots < e^{-N/5n}\,\text{vol}\,(E_0)$.

Under our choice of $E_0 \triangleq \mathbb{B}(\mathbf{x}^0, R)$, the last volume is $\text{vol}\,(E_0) = (R)^n\,\text{vol}\,(\mathbb{B}(\mathbf{0}, 1))$. Thus,

$$\begin{aligned}\text{vol}\,(E_N) &< \left(e^{\frac{-5n^2\ln(R/r)}{5n}}\right)(R)^n\,\text{vol}\,(\mathbb{B}(\mathbf{0}, 1))\\ &= \left(\frac{r}{R}\right)^n R^n\,\text{vol}\,(\mathbb{B}(\mathbf{0}, 1))\\ &= \text{vol}\,(\mathbb{B}(\mathbf{0}, r))\end{aligned}$$

That is, there is no longer enough volume in E_N to contain the inner ball $\mathbb{B}(\bar{\mathbf{x}}, r)$ and so X must be empty. ■

Khachian and Linear Programming

The general linear program can be stated

$$(LP)\qquad \begin{aligned}\min\quad & \mathbf{cx}\\ \text{s.t.}\quad & \mathbf{Ax} \geq \mathbf{b} \qquad (4\text{-}5)\\ & \mathbf{x} \geq \mathbf{0} \qquad (4\text{-}6)\end{aligned}$$

where $\mathbf{A}$ is m by n and all data are integer. The threshold version of this problem treated in complexity theory (see Chapter 2) is

> Given integer $\mathbf{A}$, $\mathbf{b}$, $\mathbf{c}$, and an integer threshold value v does there exist $\mathbf{x}$ such that $\mathbf{Ax} \geq \mathbf{b}$, $\mathbf{x} \geq \mathbf{0}$, $\mathbf{cx} \leq v$?

Taking $X \triangleq \{\mathbf{x} \geq \mathbf{0}\colon \mathbf{Ax} \geq \mathbf{b},\ \mathbf{cx} \leq v\}$ yields one form to which Algorithm 4A might be addressed.

By applying some standard linear programming theory (see Appendix C), however, we can deal with the full optimization form of (LP). Every linear program (LP) has a dual linear program over the same data

$$(DLP)\qquad \begin{aligned}\max\quad & \mathbf{ub}\\ \text{s.t.}\quad & \mathbf{uA} \leq \mathbf{c} \qquad (4\text{-}7)\\ & \mathbf{u} \geq \mathbf{0} \qquad (4\text{-}8)\end{aligned}$$

The two problems bound each other in that $v(LP) \geq \bar{\mathbf{u}}\mathbf{b}$ for any $\bar{\mathbf{u}}$ satisfying (4-7) and (4-8). Similarly, $v(DLP) \leq \mathbf{c}\bar{\mathbf{x}}$ for any $\bar{\mathbf{x}}$ feasible in (4-5) and (4-6). Optimal solutions to the primal and dual give $v(LP) = v(DLP)$.

From these well-known facts, we may construct an existence question suitable for Algorithm 4A by combining primal and dual feasibility constraints (4-5) through (4-8) with the strong optimality condition $\mathbf{cx} \leq \mathbf{ub}$. Then,

$$X^{LP} = \{(\mathbf{x}, \mathbf{u}) \geq \mathbf{0}: \mathbf{Ax} \geq \mathbf{b}, \mathbf{uA} \leq \mathbf{c}, \mathbf{cx} - \mathbf{ub} \leq 0\} \tag{4-9}$$

The input for our problem is the integer data **A, b, c.** Let us denote the bit length of this data by

$$\begin{aligned} L \triangleq & \sum_{\substack{i=1 \\ b_i \neq 0}}^{m} \lceil 1 + \log_2 |b_i| \rceil + \sum_{\substack{j=1 \\ c_j \neq 0}}^{n} \lceil 1 + \log_2 |c_j| \rceil \\ & + \sum_{i=1}^{m} \sum_{\substack{j=1 \\ a_{ij} \neq 0}}^{n} \lceil 1 + \log_2 |a_{ij}| \rceil + m + n + nm \end{aligned} \tag{4-10}$$

To polynomially apply Algorithm 4A to X^{LP} we must resolve four issues. First X^{LP} must be made full dimension by solving for some variable since $X^{LP} \subset \{(\mathbf{x}, \mathbf{u}): \mathbf{cx} = \mathbf{ub}\}$. We then require radii r and R for the balls of (4-1) and a method for executing separation Step 1 of the algorithm.

The last is straightforward. A finite list of explicitly given constraints defines X^{LP}. We could certainly check feasibility of a given solution $(\mathbf{x}^k, \mathbf{u}^k)$ in time polynomial in L.

Of course, linear programming solutions may be assumed to occur at extreme points of the feasible space. Another lemma will use this fact to fix r and R.

Lemma 4.4. (*LP*) Result Bounds. *Every extreme point solution to X^{LP} is rational with numerators and denominators bounded by 2^L, where L is the length of the (LP) input.*

Proof. The result follows from Cramer's rule on the binding constraints and the fact that 2^L bounds the product of the absolute values of the data, which in turn bounds the determinant of all matrices used in Cramer calculation of numerators and denominators. See Gács and Lovász (1981) for details. ■

Theorem 4.5. Polynomial Linear Programming via the Ellipsoid Algorithm. *Let (LP) be as above and X^{LP} be adjusted to full dimension. Then*

Ellipsoid Algorithm 4A either produces an optimal solution or demonstrates that none exists in time polynomial in m, n and L when $\mathbf{x}^0 \leftarrow \mathbf{0}$, $r \leftarrow 2^{-2L}$, $R \leftarrow 2^L$, $\epsilon \leftarrow 0$, *and separation Step 1 is accomplished by simply testing all constraints.*

Proof. Theorem 4.3 already established that the algorithm would converge in at most $N = 5n^2\lceil \ln(R/r)\rceil$ iterations. We need only check polynomiality.

First, consider the effort step by step. Since Step 1 always produces one of the inequalities of X^{LP} as $\mathbf{a}^k$; it is thus $O(mn)$. Step 2 computations on the result involve only matrix multiplications of size n by n so the effort is $O(n^2)$.

The maximum number of these steps is

$$N = 5n^2\lceil \ln(R/r)\rceil \leq 1 + 5n^2 \ln(2)\log_2(R/r). \tag{4-11}$$

Lemma 4.4 implies that if X^{LP} is nonempty and full dimension, it contains a ball of radius $r = 2^{-2L}$ and has a solution within $\mathbb{B}(\mathbf{0}, R)$, where $R = 2^L$. Thus, by (4-11)

$$\begin{aligned} N &\leq 1 + 5n^2 \ln(2)\log_2(2^L/2^{-2L}) \\ &= O(n^2 L) \end{aligned}$$

Multiplying this value by the effort per step gives an $O((nm + n^2)n^2L)$ bound on all computation. ∎

Sliding Objective Version

Although theoretically sound, the existence approach of simultaneously seeking a primal and dual solution to (*LP*) is not very appealing computationally. Many variants have been proposed (see the survey by Bland, Goldfarb and Todd (1981)).

A particularly appealing one is a *sliding objective* scheme in which the objective function is viewed as a moving constraint in $X^{\bar{v}} = \{\mathbf{x} \geq \mathbf{0}: \mathbf{A}\mathbf{x} \geq \mathbf{b}, \mathbf{c}\mathbf{x} \leq \bar{v}\}$. Here $\bar{v}$ is an *incumbent* solution value, i.e., $\bar{v} = \mathbf{c}\bar{\mathbf{x}}$ for the least cost feasible $\bar{\mathbf{x}}$ so far encountered.

So long as the Ellipsoid's center $\mathbf{x}^k$ violates one of the constraints $\mathbf{x} \geq \mathbf{0}$, $\mathbf{A}\mathbf{x} \geq \mathbf{b}$ or $\mathbf{c}\mathbf{x} \leq \bar{v}$ that constraint provides the $\mathbf{a}^k$ for Step 2. If all constraints of $X^{\bar{v}}$ are satisfied by $\mathbf{x}^k$, $\bar{v} \leftarrow \mathbf{c}\mathbf{x}^k$ gives a new incumbent value. We then take $\mathbf{a}^k \leftarrow \mathbf{c}$ on the next iteration. That is, we assume $\mathbf{x}^k$ violates a constraint of the form $\mathbf{c}\mathbf{x} \leq \mathbf{c}\mathbf{x}^k - \epsilon$. The effect is to either find a still better incumbent or prove none exists.

With some modifications in the stopping limit, this version of the Ellipsoid algorithm provides a direct linear programming optimization method that runs in polynomial time. Grötshel, Lovász, and Schrijver (1981) give a thorough treatment.

Computation

By resolving the open question of whether linear programs could be solved in polynomial time, Khachian's Ellipsoid algorithm certainly represented a theoretical breakthrough. Some quite important complexity results have also been shown to arise as side benefits of Ellipsoid results (see Sections 4.3 and 4.4).

Unfortunately, however, practical experience with Khachian's approach at solving applied linear programs has been poor. Since the algorithm works by slow reduction of feasible volume, it gains little from special structure of the underlying linear program. Thus, unlike the Simplex algorithm, the Ellipsoid method seems to require something like its worst case computational bound. Length L becomes so large on even tiny applied examples that such limits are far more than could be computationally feasible. In addition, matrices $\mathbf{M}_k$ become very dense, creating large numerical burdens.

4.1.3 Karmarkar's Projective Scaling Algorithm

Although of great theoretical importance, Khachian's Ellipsoid method has (to date) proved a failure as a practical linear programming algorithm. Still, its feasible-volume-reducing, nonlinear, numerical concept opened many new research paths.

In 1984, N. Karmarkar announced a different *projective scaling* algorithm descended from Khachian's in its broad perspective. At this writing it is unsettled whether Karmarkar's method will prove a bonafide competitor to the Simplex method on practical linear programming. It is clear, however, that the algorithm is both polynomially-bounded in the worst case and computationally viable on typical data. Thus, it certainly does represent another fundamental achievement in the long and important quest for a completely satisfactory linear programming algorithm.

In this section we develop the important properties of the projective transformation algorithm. Our discussion is motivated, in part, by excellent sets of notes due to Vijaya Chandru (1984) and Richard Stone (1985).

Strategy of the Algorithm

As we saw in the last section, Khachian's Ellipsoid method may be viewed as multidimensional binary search. At each iteration of the algorithm the search narrows to the feasible half of the current ellipsoid. The objective function of the underlying (*LP*) is introduced only via a threshold constraint or duality.

Suppose that, like the Khachian case, our current solution $\mathbf{x}^k$ can be viewed as the center of an ellipsoid. But this time the entire volume of the ellipsoid is feasible and we are optimizing. Our course is obvious: step to the boundary of the ellipsoid along the radius vector of most rapid objective function improvement.

Such steps along radii of feasible ellipsoids form the core of Karmarkar's method. At each iteration a large inscribed ellipsoid is fit into the interior of the projectively transformed (*LP*) feasible space. A radius vector of rapid objective function improvement is then selected, and the solution progresses along that vector toward the ellipsoid boundary. In the original space the projective scaling algorithm is optimizing over inscribed paraboloids.

Problem Form and Projection

It should be clear that success with the above strategy requires a canonical linear programming format into which it is computationally easy to inscribe maximal ellipsoids centered at a current solution. Karmarkar selected just such a form

$$\min \quad \mathbf{hy} \tag{4-12}$$

$$\text{s.t.} \quad \mathbf{Ky} = \mathbf{0} \tag{4-13}$$

$$(TLP) \qquad \mathbf{1y} = 1 \tag{4-14}$$

$$\mathbf{y} \geq \mathbf{0} \tag{4-15}$$

with solution

$$\mathbf{y}^k \triangleq (1/n)\mathbf{1} \tag{4-16}$$

(We leave until later to show that this form is general.)

Ignoring for a moment the main requirements $\mathbf{Ky} = \mathbf{0}$, observe that constraints (4-14) and (4-15) describe a very simple *LP* polytope. It is the regular set (ironically called a *simplex*) $\mathbb{S} \triangleq \{\mathbf{x} \geq \mathbf{0}\colon \mathbf{1x} = 1\}$ having the n-space unit vectors as vertices.

Figure 4.3a illustrates for $n = 3$. Notice that in this simple space the current solution $\mathbf{y}^k = (\frac{1}{3}, \frac{1}{3}, \frac{1}{3})$ lies exactly at the center. Regularity of the simplex's shape renders the maximum volume inscribed ellipsoid, no more than a $(n-1)$-sphere (here a circle) of radius

$$r = \frac{1}{\sqrt{n(n-1)}}$$

or about 0.41 for $n = 3$.

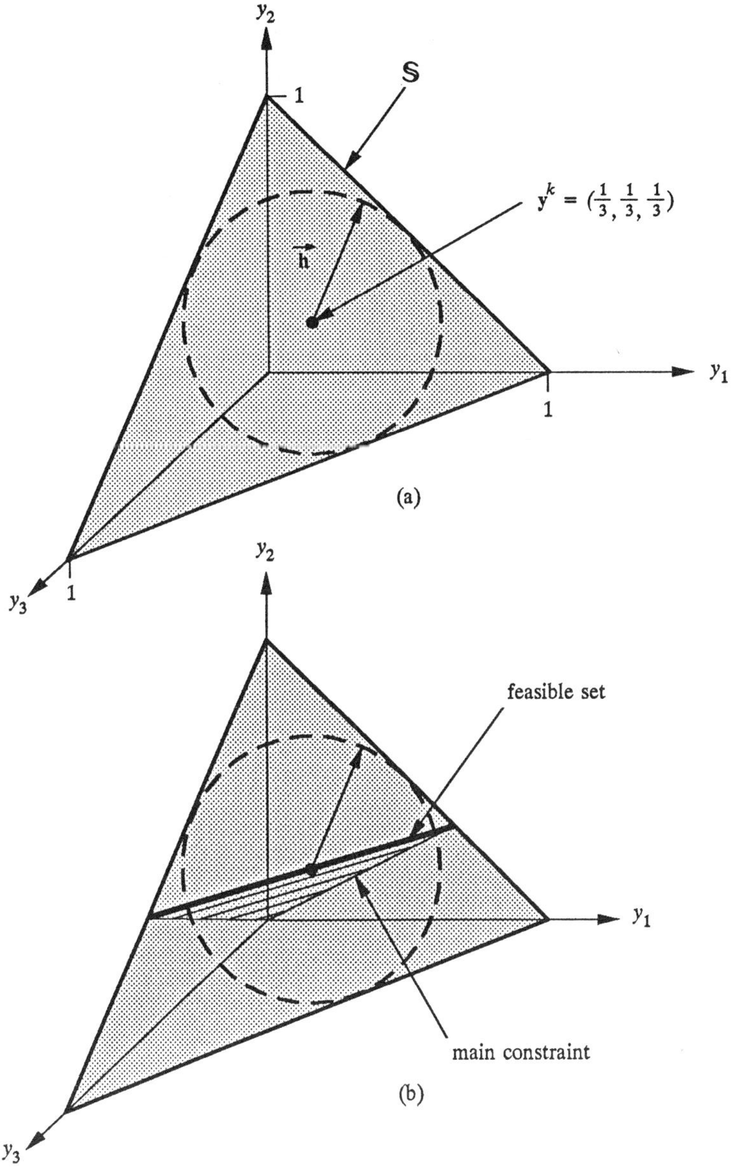

Fig. 4.3. Solution spaces in Karmarkar problem form; (a) simplex with inscribed sphere, (b) simplex with main constraint.

To minimize **hy** over the simplex, we would merely need to take the step indicated in Figure 4.3a—a step in the direction $-\vec{\mathbf{h}} \triangleq -\mathbf{h}$. Normalizing for the length of $\vec{\mathbf{h}}$ and using a step parameter $\alpha > 0$, the update becomes

$$\mathbf{y}^{k+1} \leftarrow \mathbf{y}^{k} - \frac{\alpha}{\sqrt{n(n-1)}} \frac{\vec{\mathbf{h}}}{\|\vec{\mathbf{h}}\|}. \qquad (4\text{-}17)$$

If the entire simplex (and thus, the entire inscribed sphere) is feasible, $\mathbf{y}^{k+1}$ will remain so for $0 < \alpha < 1$.

Now consider adding a $\mathbf{Ky} = \mathbf{0}$ system satisfied by the current $\mathbf{y}^k$. In Figure 4.3b, we illustrate with $\mathbf{K} = [1, -3, 2]$ or

$$y_1 - 3y_2 + 2y_3 = 0.$$

The feasible space of the full linear program is the intersection of $\{\mathbf{y}\colon \mathbf{Ky} = \mathbf{0}\}$ with the simplex. Furthermore, the intersection of $\{\mathbf{y}\colon \mathbf{Ky} = \mathbf{0}\}$ with the $(n-1)$-sphere inscribed within the simplex contains the center of the sphere and is therefore a sphere of the same radius in lower dimension. For example, in the case of Figure 4.3b this intersection is the indicated $(n-2)$-sphere or line segment.

How are we to improve now with respect to the objective function? In the misleadingly simple case of 3-space problems with one $\mathbf{Ky} = \mathbf{0}$ constraint, only two directions are feasible and we merely pick the one of the two that improves upon $\mathbf{hy}^k$. More effort is required, however, in higher dimensional space. We must compute a direction

$$\vec{\mathbf{h}} \in \{\mathbf{y}\colon \mathbf{Ky} = \mathbf{0}, \mathbf{1y} = 1\} \qquad (4\text{-}18)$$

such that step (4-16) will yield objective function improvement. Some straightforward calculus will show that the $\vec{\mathbf{h}}$ with most rapid objective function increase per unit step is obtained by *projection,* of the true cost vector **h** on $\{\mathbf{y}\colon \mathbf{Ky} = \mathbf{0}, \mathbf{1y} = 1\}$. The needed arithmetic is

$$\vec{\mathbf{h}} \longleftarrow (\mathbf{I} - \mathbf{P}^T(\mathbf{P}\,\mathbf{P}^T)^{-1}\,\mathbf{P})\mathbf{h} \qquad (4\text{-}19)$$

where

$$\mathbf{P} \triangleq \begin{bmatrix} \mathbf{K} \\ \hline \mathbf{1} \end{bmatrix}. \qquad (4\text{-}20)$$

$\mathbf{P}^T$ denotes the transpose of **P**.

Centering Transformation

A major gap remains if the above movement along cost projections is to form the main algorithm step. After a single such step, the solution is no longer at the center of $\{\mathbf{y} \geq \mathbf{0}\colon \mathbf{1y} = 1\}$ and so the spheres no longer have their elegant form.

To produce an algorithm that always steps from the center of the simplex, Karmarkar introduced a clever data transformation. At each iteration, calculations are performed in a transformed space, where the current solution is the simplex center, and then mapped back to true space.

To distinguish the spaces, let us denote the original problem

$$(KLP) \qquad \begin{aligned} \min \quad & \mathbf{cx} && (4\text{-}21)\\ \text{s.t.} \quad & \mathbf{Ax} = \mathbf{0} && (4\text{-}22)\\ & \mathbf{1x} = 1 && (4\text{-}23)\\ & \mathbf{x} \geq \mathbf{0} && (4\text{-}24) \end{aligned}$$

and suppose the current feasible solution $\mathbf{x}^k > \mathbf{0}$ (i.e., positive in every component). The (*TLP*) form above, with **y**'s and **K**'s, will refer to transformed entities.

Karmarkar's centering transformation is the vector-valued function $\theta: \mathbb{S} \to \mathbb{S}$ where $\mathbb{S} \triangleq \{\mathbf{x} \geq \mathbf{0}: \mathbf{1x} = 1\}$ denotes the simplex and

$$y_j \triangleq \theta_j(\mathbf{x}) \triangleq \frac{(x_j/x_j^k)}{\sum_{l=1}^{n} (x_l/x_l^k)} \qquad j = 1, 2, \ldots, n \qquad (4\text{-}25)$$

Several properties of the θ transformation are easily established. (Here let $\text{Diag}(\mathbf{x}^k)$ denote a diagonal matrix with x_j^k, the jth diagonal element.)

Lemma 4.6. Transform Properties. *Let (KLP) and transformation θ be as above with $\mathbf{x}^k$ a strictly positive feasible solution. Then*

(i) $\theta(\mathbf{x}^k) = 1/n(\mathbf{1})$

(ii) *θ is a 1:1 mapping from $\mathbb{S}$ onto $\mathbb{S}$*

(iii)
$$\theta_j^{-1}(\mathbf{y}) = \frac{x_j^k y_j}{\sum_{l=1}^{n} x_j^k y_j} \qquad \text{for } j = 1, 2, \ldots, n \qquad (4\text{-}26)$$

(iv) $\{\mathbf{x} \in \mathbb{S}: A\mathbf{x} = \mathbf{0}\} \triangleq \{\theta^{-1}(\mathbf{y}) \in \mathbb{S}: A[\text{Diag}(\mathbf{x}^k)]\mathbf{y} = \mathbf{0}\}$

Proof.

(i)
$$\theta_j(\mathbf{x}^k) \triangleq \frac{(x_j^k/x_j^k)}{\sum_{l=1}^{n} (x_l^k/x_l^k)} = 1/n$$

(ii) Choose $\overline{\mathbf{x}} \in \mathbb{S}$ and define $\overline{\mathbf{y}} = \theta(\overline{\mathbf{x}})$. Clearly, $\overline{\mathbf{y}} \geq \mathbf{0}$. To finish establishing $\overline{\mathbf{y}} \in \mathbb{S}$, we verify the sum,

$$\sum_{t=1}^{n} \overline{y}_t \triangleq \sum_{t=1}^{n} \frac{(\overline{x}_t/x_t^k)}{\sum_{l=1}^{n} (\overline{x}_l/x_l^k)} = 1.$$

(iii) To establish the 1 : 1 part of (ii) and verify the expression for the inverse, observe

$$\theta_j^{-1}(\theta(\bar{\mathbf{x}})) = \frac{x_j^k\,\theta_j(\bar{\mathbf{x}})}{\sum_{l=1}^{n} x_l^k\,\theta_l(\bar{\mathbf{x}})}$$

$$= \frac{(x_j^k(\bar{x}_j/x_j^k))\Big/\left(\sum_{l=1}^{n}(\bar{x}_l/x_l^k)\right)}{\left(\sum_{l=1}^{n} x_l^k(\bar{x}_l/x_l^k)\right)\Big/\left(\sum_{l=1}^{n}(\bar{x}_l/x_l^k)\right)}$$

$$= \frac{\bar{x}_j}{\sum_{l=1}^{n}\bar{x}_l} = \frac{\bar{x}_j}{1}$$

the last by $\bar{\mathbf{x}} \in \mathbb{S}$.

(iv) Pick $\bar{\mathbf{x}} \in \{\mathbf{x} \in \mathbb{S}\colon \mathbf{A}\mathbf{x} = \mathbf{0}\}$ and define $\bar{\mathbf{y}} \triangleq \theta(\bar{\mathbf{x}})$. Then, $[\mathrm{Diag}(\mathbf{x}^k)]\bar{\mathbf{y}}$ has j^{th} entry

$$x_j^k\left(\frac{\bar{x}_j/x_j^k}{\sum_{l=1}^{n}\bar{x}_l/x_l^k}\right) = \frac{\bar{x}_j}{\sum_{l=1}^{n}(\bar{x}_l/x_l^k)}.$$

Thus,

$$\mathbf{A}[\mathrm{Diag}(\mathbf{x}^k)]\bar{\mathbf{y}} = \left[\frac{\mathbf{A}\bar{\mathbf{x}}}{\sum_{l=1}^{n}(\bar{x}_l/x_l^k)}\right] = \mathbf{0}$$

because the denominator is positive for $\mathbf{x}^k > 0$, $\bar{\mathbf{x}} \in \mathbb{S}$. ■

Example 4.2. Karmarkar Transformation. Consider the specific problem

$$\begin{array}{lrl}
\min & 3x_1 - x_2 + x_3 & \\
\text{s.t.} & -x_1 \qquad + 2x_3 & = 0 \\
 & x_1 + x_2 + x_3 & = 1 \\
 & x_1, x_2, x_3 & \geq 0.
\end{array}$$

Figure 4.4 shows the original space, and the transformed one associated with positive feasible solution $\mathbf{x}^k \triangleq (\frac{2}{5}, \frac{2}{5}, \frac{1}{5})$.

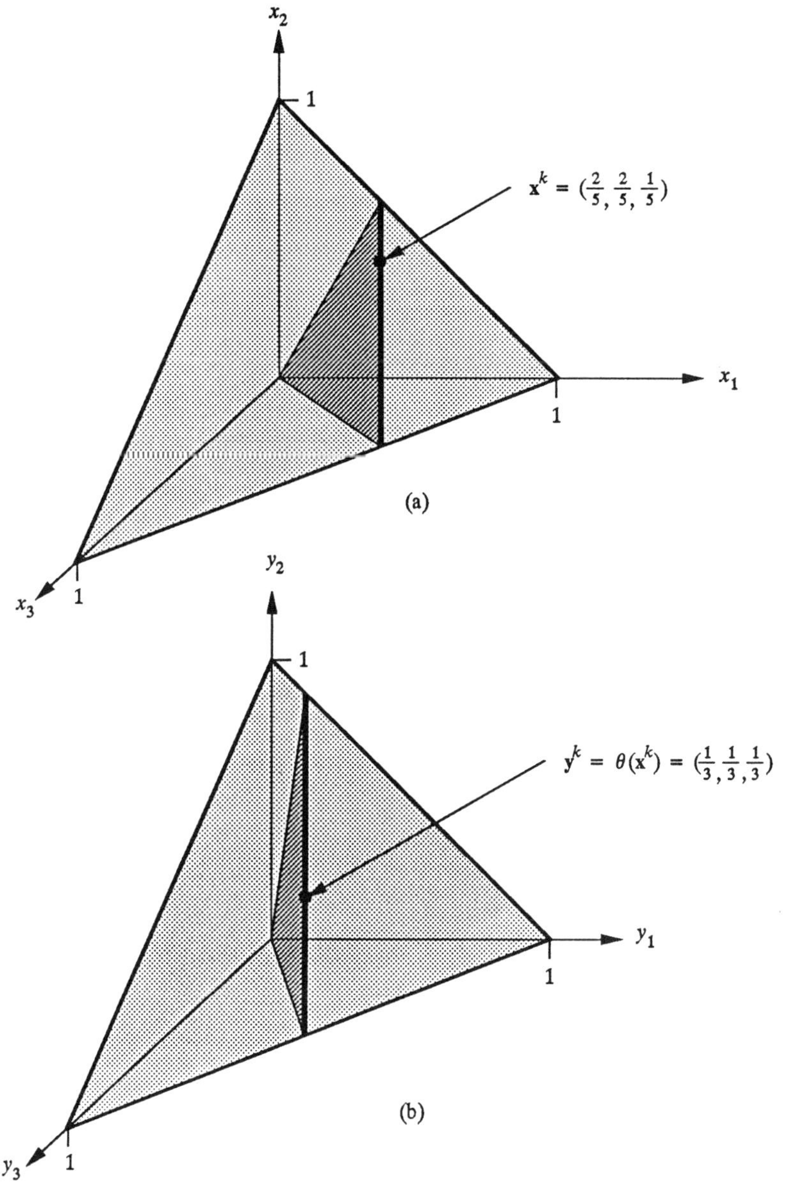

Fig. 4.4. Original and transform space for Example 4.2.

Observe first that $\mathbf{x}^k$ transforms to $\mathbf{y}^k \triangleq (\frac{2}{5}/\frac{2}{5}, \frac{2}{5}/\frac{2}{5}, \frac{1}{5}/\frac{1}{5})/3 = (\frac{1}{3})\mathbf{1}$, the center of the transformed set. Each other $\mathbf{x} \in \mathbb{S}$ maps to a $\mathbf{y}$ in the simplex of transformed space. For example, with $\bar{\mathbf{x}} = (\frac{3}{10}, \frac{5}{10}, \frac{2}{10})$, $\theta(\bar{\mathbf{x}}) \triangleq \bar{\mathbf{y}} = (\frac{3}{10}/\frac{2}{5}, \frac{5}{10}/\frac{2}{5}, \frac{2}{10}/\frac{1}{5})$ divided by its component sum or $(\frac{1}{4}, \frac{5}{12}, \frac{1}{3})$. It is easy to also verify that $\theta^{-1}(\bar{\mathbf{y}}) = (\frac{3}{10}, \frac{5}{10}, \frac{2}{10}) \triangleq \bar{\mathbf{x}}$.

Finally, consider (iv) of Lemma 4.6 and $\tilde{\mathbf{x}} = (\frac{1}{3}, \frac{1}{2}, \frac{1}{6})$. Certainly for $\mathbf{A} = [-1, 0, 2]$, $\mathbf{A}\tilde{\mathbf{x}} = 0$. Transformed value $\tilde{\mathbf{y}} \triangleq \theta(\tilde{\mathbf{x}}) = (\frac{2}{7}, \frac{3}{7}, \frac{2}{7})$. Notice that $\mathbf{A}\tilde{\mathbf{y}} \neq 0$. Rescaling with Diag($\mathbf{x}^k$), however, gives

$$\begin{aligned}\mathbf{A}[\text{Diag}(\mathbf{x}^k)]\tilde{\mathbf{y}} &= [-1, 0, 2]\begin{bmatrix} \frac{2}{5} & 0 & 0 \\ 0 & \frac{2}{5} & 0 \\ 0 & 0 & \frac{1}{5} \end{bmatrix}\begin{bmatrix} \frac{2}{7} \\ \frac{3}{7} \\ \frac{2}{7} \end{bmatrix} \\ &= [-1, 0, 2]\begin{bmatrix} \frac{4}{35} \\ \frac{6}{35} \\ \frac{2}{35} \end{bmatrix} = 0. \end{aligned}$$

■

Complete Algorithm

Having introduced the key notions, we are now able to fully state the Projective Transformation Algorithm.

Algorithm 4B. Karmarkar's Projective Transformation. Let (*KLP*) be as above such that $(1/n)\mathbf{1}$ is feasible and $\mathbf{cx}$ is minimized at $\mathbf{0}$ over feasible $\mathbf{x}$.

STEP 0: *Initialization.* Pick $\mathbf{x}^0 \leftarrow (1/n)\mathbf{1}$, fix tolerance $q > 0$, and set $k \leftarrow 0$.

STEP 1: *Stopping.* If $\mathbf{cx}^k/\mathbf{cx}^0 < 2^{-q}$ (q is an error tolerance constant), stop; $\mathbf{x}^k$ may be rounded to an optimal solution. Otherwise, proceed to Step 2.

STEP 2: *Projection.* Compute the projection onto the feasible transformed set of the objective function as

$$\vec{\mathbf{c}} \longleftarrow (\mathbf{I} - \mathbf{P}^T(\mathbf{P}\mathbf{P}^T)^{-1}\mathbf{P})[\text{Diag}(\mathbf{x}^k)]\mathbf{c} \tag{4-27}$$

where

$$\mathbf{P} \triangleq \begin{bmatrix} \mathbf{A}[\text{Diag}(\mathbf{x}^k)] \\ \hdashline \mathbf{1} \end{bmatrix}. \tag{4-28}$$

STEP 3: *New Point.* Update

$$\mathbf{y}^{k+1} \longleftarrow (1/n)\mathbf{1} - \frac{\alpha}{\sqrt{n(n-1)}} \frac{\vec{\mathbf{c}}}{\|\vec{\mathbf{c}}\|} \tag{4-29}$$

in transformed space (Here α is a constant in the interval (0, 1), $\alpha = \frac{1}{4}$ will serve for convergence). Then compute

$$\mathbf{x}^{k+1} \longleftarrow \theta^{-1}(\mathbf{y}^{k+1}),$$

advance $k \leftarrow k + 1$, and return to Step 1. ■

Example 4.3. Karmarkar's Algorithm. Consider the simple example problem

$$\begin{aligned} \min \quad & 6x_1 - 6x_2 + x_3 \\ \text{s.t.} \quad & x_1 - x_2 = 0 \\ & x_1 + x_2 + x_3 = 1 \\ & x_1, x_2, x_3 \geq 0. \end{aligned}$$

It is obvious that $\mathbf{x}^0 \triangleq (\frac{1}{3}, \frac{1}{3}, \frac{1}{3})$ is feasible, and Figure 4.5 demonstrates that the unique optimal solution is $\mathbf{x} = (\frac{1}{2}, \frac{1}{2}, 0)$ with value 0. Thus, the assumed conditions for Karmarkar's algorithm are fulfilled.

For $k = 0$, $q = 4$, stopping does not occur at Step 1 and we proceed to project. Here,

$$\mathbf{P} = \begin{bmatrix} 1 & -1 & 0 \\ 1 & 1 & 1 \end{bmatrix}$$

so

$$(\mathbf{I} - \mathbf{P}^T(\mathbf{P}\mathbf{P}^T)^{-1}\mathbf{P}) = \begin{bmatrix} 1 & 0 & 0 \\ 0 & 1 & 0 \\ 0 & 0 & 1 \end{bmatrix}$$

$$- \begin{bmatrix} 1 & 1 \\ -1 & 1 \\ 0 & 1 \end{bmatrix} \left(\begin{bmatrix} 1 & -1 & 0 \\ 1 & 1 & 1 \end{bmatrix} \begin{bmatrix} 1 & 1 \\ -1 & 1 \\ 0 & 1 \end{bmatrix} \right)^{-1} \begin{bmatrix} 1 & -1 & 0 \\ 1 & 1 & 1 \end{bmatrix}$$

$$= \begin{bmatrix} \frac{1}{6} & \frac{1}{6} & -\frac{1}{3} \\ \frac{1}{6} & -\frac{1}{6} & -\frac{1}{3} \\ -\frac{1}{3} & -\frac{1}{3} & \frac{2}{3} \end{bmatrix}.$$

Thus, the projected (scaled) $\mathbf{c}$ vector is

$$\vec{\mathbf{c}} \longleftarrow \begin{bmatrix} \frac{1}{6} & \frac{1}{6} & -\frac{1}{3} \\ \frac{1}{6} & \frac{1}{6} & -\frac{1}{3} \\ -\frac{1}{3} & -\frac{1}{3} & \frac{2}{3} \end{bmatrix} \begin{bmatrix} \frac{1}{3} & 0 & 0 \\ 0 & \frac{1}{3} & 0 \\ 0 & 0 & \frac{1}{3} \end{bmatrix} \begin{bmatrix} 6 \\ -6 \\ 1 \end{bmatrix} = \begin{bmatrix} -\frac{1}{9} \\ -\frac{1}{9} \\ +\frac{2}{9} \end{bmatrix}.$$

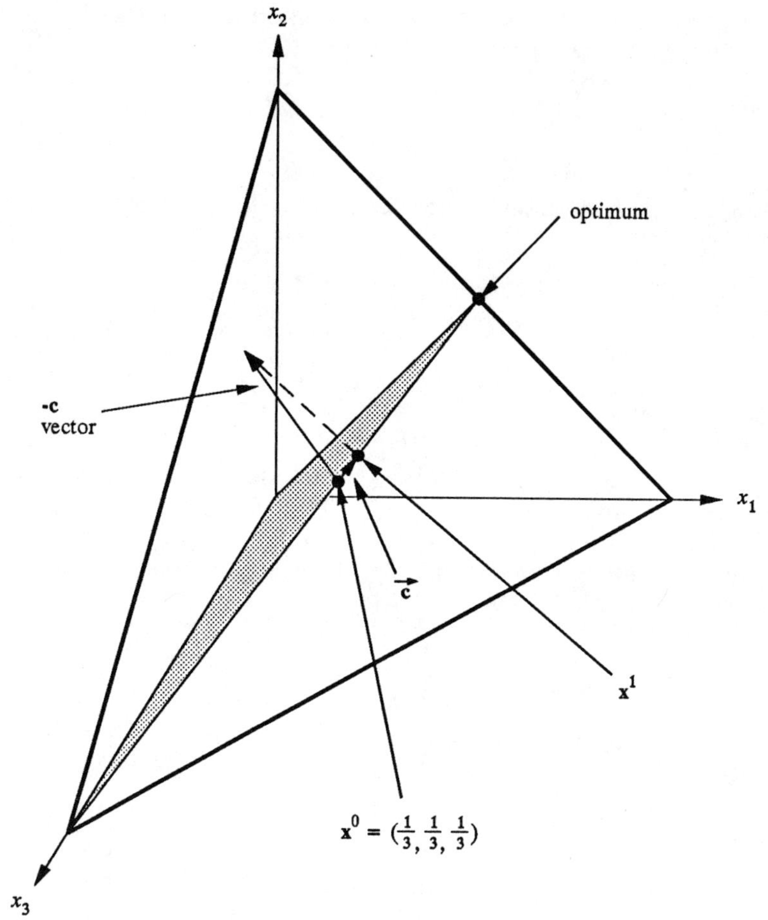

Fig. 4.5. Karmarkar Example 4.3.

Note that for reasons to which we must return, it is $(\text{Diag}(\mathbf{x}^0))\mathbf{c}$ that is being projected, not $\mathbf{c}$ itself.

Using $\alpha = \frac{1}{4}$, the new point in transformed space is

$$\mathbf{y}^1 = \begin{bmatrix} \frac{1}{3} \\ \frac{1}{3} \\ \frac{1}{3} \end{bmatrix} - \frac{1}{4\sqrt{3(2)}} \begin{bmatrix} -\frac{1}{9} \\ -\frac{1}{9} \\ \frac{2}{9} \end{bmatrix} \approx \begin{bmatrix} .3444 \\ .3444 \\ .3111 \end{bmatrix}.$$

The corresponding point in original space is

$$\mathbf{x}^1 = \theta^{-1}(\mathbf{y}^1) = [.3444, .3444, .3111]$$

(the same as $\mathbf{y}^1$ in this first iteration only).

Cost $\mathbf{cx}^1 = .3111 < .3333 = \mathbf{cx}^0$, so that the objective has improved. Iteration now continues with $k = 1$. ■

Starting Form

It is time now to confront the issue of placing linear programs into the somewhat unintuitive starting format for Algorithm 4B.

Suppose we are supplied a problem in standard form for the Simplex algorithm, i.e.,

$$\begin{aligned} \min \quad & \mathbf{cx} \\ \text{s.t.} \quad & \mathbf{Ax} = \mathbf{b} \\ & \mathbf{x} \geq \mathbf{0}. \end{aligned}$$

By the same arguments as those used to form the initial ellipsoid for Khachian's algorithm, we know there is an upper limit, say R, on each component of any solution. Thus, $\mathbf{1x} \leq nR$.

Adding this constraint to the formulation with slack x_0 gives

$$\begin{aligned} \min \quad & \mathbf{cx} \\ \text{s.t.} \quad & \mathbf{Ax} = \mathbf{b} \\ & x_0 + \mathbf{1x} = nR \\ & x_0 \geq 0,\ \mathbf{x} \geq \mathbf{0} \end{aligned}$$

Then, multiplying the objective and all rows by $(nR)^{-1}$ and substituting $\mathbf{x}' \triangleq (nR)^{-1}\mathbf{x}$, $x_0' \triangleq (nR)^{-1}x_0$, $\mathbf{b}' \triangleq (nR)^{-1}\mathbf{b}$

$$\begin{aligned} \min \quad & \mathbf{cx}' \\ \text{s.t.} \quad & \mathbf{Ax}' = \mathbf{b}' \\ & x_0' + \mathbf{1x}' = 1 \\ & x_0' \geq 0,\ \mathbf{x}' \geq \mathbf{0}. \end{aligned}$$

Subtracting a multiple b_i' of the last row from each other completes the derivation of the standard form for the projective scaling method

$$\begin{aligned} \min \quad & \mathbf{cx}' \\ \text{s.t.} \quad & \mathbf{A}' \begin{bmatrix} x_0' \\ \mathbf{x}' \end{bmatrix} = \mathbf{0} \\ & \mathbf{1} \begin{bmatrix} x_0' \\ \mathbf{x}' \end{bmatrix} = 1 \\ & x_0' \geq 0,\ \mathbf{x}' \geq \mathbf{0}, \end{aligned}$$

where $\mathbf{A}' = [-\mathbf{b}' \mid \mathbf{A} - \mathbf{b}'(\mathbf{1}^T)]$.

As stated, however, Algorithm 4B requires more than standard form. It must also be true that $\mathbf{x}^0 = (1/n)\mathbf{1}$ is a feasible solution and the objective is minimized at $\mathbf{cx} = 0$.

Theoretically speaking, these limits are easily confronted by dealing with an existence rather than an optimization form of linear programming. We saw in Khachian Section 4.1.2 that every *LP* optimization can be posed as an existence question by listing primal feasibility, dual feasibility and strong duality constraints. Thus, the optimization form of a linear program can be reduced to the issue of existence of a solution $\mathbf{x}''$ to a new constraint system

$$(LPX)\qquad \begin{aligned} \mathbf{A}''\mathbf{x}'' &= \mathbf{0} \\ \mathbf{1x}'' &= 1 \\ \mathbf{x}'' &\geq \mathbf{0} \end{aligned}$$

To obtain starting conditions for Algorithm 4B, we need only add an artificial variable w with main constraint column equal to the negative sum of the columns of $\mathbf{A}''$. Then the linear program,

$$\begin{aligned} \min \quad & w \\ \text{s.t.} \quad & \mathbf{A}''\mathbf{x}'' - (\mathbf{A}''\mathbf{1})w = \mathbf{0} \\ & \mathbf{1x}'' + w = 1 \\ & \mathbf{x}'' \geq \mathbf{0},\ w \geq 0, \end{aligned}$$

has initial feasible solution $\mathbf{x}'' = (n+1)^{-1}\mathbf{1}$, $w = (n+1)^{-1}$, and its optimal value is 0 exactly when system (*LPX*) has a solution.

In computational practice, one would not, of course, wish to expand the problem statement so much. As with Khachian, there is a sliding objective alternative. We refer the reader to Karmarkar's (1984) paper for details.

Transformed Objective

A more subtle detail of Algorithm 4B we have not addressed relates to projection of the scaled cost $(\text{Diag}(\mathbf{x}^k))\mathbf{c}$, rather than the $\mathbf{c}$ itself, in choosing the update direction at Step 2. To see why some change is needed, note that the true objective function value $\mathbf{c}(\theta^{-1}(\bar{\mathbf{y}}))$ of a point in transform space is

$$\mathbf{c}[\theta^{-1}(\bar{\mathbf{y}})] = \frac{\sum_{j=1}^{n} c_j(x_j^k \bar{y}_j)}{\sum_{j=1}^{n} (x_j^k \bar{y}_j)}$$

Thus, the linear function $((\text{Diag}(\mathbf{x}^k))\mathbf{c})\bar{\mathbf{y}} = \Sigma_{j=1}^{n} (x_j^k c_j) y_i$ gives the correct numerator, and so is a reasonable local approximation to use in seeking cost improvement within transform space.

The greater difficulty is that the transformation of **c** is inherently nonlinear. The consequence is that no straight line movement in transform space can be guaranteed to improve the true objective value for a very great distance.

Karmarkar's analysis of the delicate issue of how large a step is possible is posed in terms of *potential functions*—approximations to true and transformed cost.

In particular, he proposes

$$f(\mathbf{x}) \triangleq \sum_{j=1}^{n} \ln\left(\frac{\mathbf{cx}}{x_j}\right) \tag{4-30}$$

to approximate **cx** in true space and

$$g(\mathbf{y}) \triangleq \sum_{j=1}^{n} \ln\left(\frac{(\text{Diag}(\mathbf{x}^k))\mathbf{cy}}{y_j}\right) - \sum_{j=1}^{n} \ln(x_j^k) \tag{4-31}$$

to approximate cost in transform space. Direct substitution will establish that

$$f(\mathbf{x}) = g(\theta(\mathbf{x})), \tag{4-32}$$

i.e., that one potential function measures the same progress in real space that the other measures in transform space.

It is also important to see that

$$f(\mathbf{x}) = n \ln(\mathbf{cx}) - \ln\left(\prod_{j=1}^{n} x_j\right)$$

Since $\Sigma_{j=1}^{n} x_j = 1$ for any feasible **x**, the last term will be relatively stable for small changes in **x**. A similar argument motivates $g(\mathbf{y})$ as a good local cost approximation in transform space. A more precise statement is given in another Karmarkar lemma.

Lemma 4.7. Potential Function Improvement. *Let $\mathbf{x}^k$, $\mathbf{c}$, α and $\mathbf{y}^{k+1}$ be as in Algorithm 4B above. Then either*

(i) $(\text{Diag}(\mathbf{x}^k))\mathbf{c}\mathbf{y}^{k+1} = 0$, or
(ii) $g(n^{-1}\mathbf{1}) - g(\mathbf{y}^{k+1}) \geq \delta$,

where $\delta > 0$ is a constant depending on α. ($\alpha = \frac{1}{4}$ implies $\delta = \frac{1}{8}$)

Polynomiality Argument

Using the above lemmas, we are now ready to establish polynomiality of Karmarkar's algorithm.

Theorem 4.8. Karmarkar's Algorithm. *Any linear program can be polynomially reduced to one in form (KLP) specified for Algorithm 4B. Moreover, then in time polynomial in the size of the (KLP) input, the algorithm either computes a solution or proves none exists.*

Proof. We have already shown how problems reduce to Algorithm 4B format. For polynomiality, consider the implications of the two cases in Lemma 4.7. When outcome (i) holds

$$\mathbf{c}\mathbf{x}^{k+1} = \frac{\sum_{j=1}^{n} c_j x_j^k y_j^{k+1}}{\sum_{j=1}^{n} x_j^k y_j} = 0. \tag{4-33}$$

On the other hand, if $g(n^{-1}\mathbf{1}) - g(\mathbf{y}^{k+1}) \geq \delta$, then relation (4-32) assures

$$g(\theta(\mathbf{x}^k)) - g(\theta(\mathbf{x}^{k+1})) = f(\mathbf{x}^k) - f(\mathbf{x}^{k+1}) \geq \delta \tag{4-34}$$

If at any iteration condition (4-33) arises, the algorithm will clearly stop at Step 1. If it runs k iterations without stopping in that way, then (4-34) yields

$$f(\mathbf{x}^k) \leq f(\mathbf{x}^0) - k\delta$$

Applying the definition of f, this implies

$$\sum_{j=1}^{n} \ln\left(\frac{\mathbf{c}\mathbf{x}^k}{x_j^k}\right) \leq \sum_{j=1}^{n} \ln\left(\frac{\mathbf{c}\mathbf{x}^0}{x_j^0}\right) - k\delta$$

or, after some rearranging,

$$n \ln\left(\frac{\mathbf{c}\mathbf{x}^k}{\mathbf{c}\mathbf{x}^0}\right) \leq \sum_{j=1}^{n} \ln(x_j^k) - \sum_{j=1}^{n} \ln(x_j^0) - k\delta.$$

Now, $x_j^0 = 1/n$ for all j, so that we may further group

$$n \ln\left(\frac{\mathbf{c}\mathbf{x}^k}{\mathbf{c}\mathbf{x}^0}\right) \leq \sum_{j=1}^{n} \ln(n x_j^k) - k\delta.$$

Then, since $x_j^k \leq 1$ for all j

$$n \ln\left(\frac{\mathbf{c}\mathbf{x}^k}{\mathbf{c}\mathbf{x}^0}\right) \leq n \ln(n) - k\delta. \tag{4-35}$$

Now choose $\bar{k} = \delta^{-1}(nq + n\ln(n))$. Then, substituting in (4-35)

$$n \ln\left(\frac{\mathbf{c}\mathbf{x}^{\bar{k}}}{\mathbf{c}\mathbf{x}^0}\right) \leq -nq$$

or dividing by n and exponentiating yields

$$\frac{\mathbf{c}\mathbf{x}^{\bar{k}}}{\mathbf{c}\mathbf{x}^0} \leq 2^{-q}. \tag{4-36}$$

That is, after $\bar{k} = O(nq + n\ln(n))$ iterations, the algorithm will stop at Step 1. Since each iteration clearly requires polynomial effort, the theorem is established. ■

Using a rank-one update rule, each iteration can actually be shown to require $O(n^{2.5})$ computations on average to do necessary projections (see Karmarkar (1984)). Thus, the entire effort is polynomial $O(n^{3.5}q + n^{3.5}\ln n)$. Using results of Section 4.1.2, that needed accuracy is limited by 2^{-L}, where L is the length of the (KLP) input, we obtain the bound $O(n^{3.5}(L + \ln n))$. Recent improvements have brought the complexity down to $O((m+n)n + (m+n)^{1.5}nL)$ where m denotes the number of constraints and n the number of variables (Vaidya (1987)).

4.2 Integer Solvability of Linear Programs

Although the algorithms for linear programming are significant in their own right, our main interest lies with how they can be used to solve discrete optimization problems. Certainly a minimum requirement for linear programs to yield discrete solutions is that they admit a linear programming formulation. That is, we require a linear program with integer extrema.

When do linear programs have integer optimal solutions? In this section we present the classic results.

4.2.1 Unimodularity

Much of standard linear programming theory (see Appendix C) relates to *basis* submatrices, i.e., maximal sets of linearly independent columns of a linear program in equality format

$$\begin{aligned} \min \quad & \mathbf{c}\mathbf{x} \\ \text{s.t.} \quad & \mathbf{A}\mathbf{x} = \mathbf{b} \\ & \mathbf{x} \geq \mathbf{0}. \end{aligned} \tag{4-37}$$

Extreme points of the corresponding feasible space arise as solutions with only variables of a basis nonzero.

The constraint matrix **A** is said to be *unimodular* if every basis matrix **B** of **A** has determinant, det(**B**) $= \pm 1$. The following classic result of Veinott and Dantzig (1968) shows the implications for integer solvability.

Theorem 4.9. Unimodularity and Equality Linear Programs. *Let A be an integer matrix with linearly independent rows. Then the following are equivalent:*

(i) *A is unimodular.*
(ii) *Extreme-points of* $S^{=} \triangleq \{\boldsymbol{x}: \boldsymbol{A}\boldsymbol{x} = \boldsymbol{b},\ \boldsymbol{x} \geq \boldsymbol{0}\}$ *are integer for any integer right-hand-side* ***b***.
(iii) *Every basis submatrix* ***B*** *of* ***A*** *has integer inverse* $\boldsymbol{B}^{-1}$.

Proof. We proceed by showing (i) $\Rightarrow$ (ii) $\Rightarrow$ (iii) $\Rightarrow$ (i).

For (i) $\Rightarrow$ (ii), we apply the definition of unimodularity. Every extreme point of $S^{=}$ must correspond to a nonbasic subvector $\mathbf{x}^N$ fixed at zero and a basic subvector $\mathbf{x}^B$ determined by $\mathbf{B}\mathbf{x}^B = \mathbf{b}$ for basis submatrix **B**. If **b** is integer, then by Cramer's rule, components of the $\mathbf{x}^B$ solution will be of the form

$$x_j^B = \frac{\det(\text{integer matrix})}{\det(\mathbf{B})}.$$

Unimodularity of **A** thus assures such extreme point solutions are integer.

For (ii) $\Rightarrow$ (iii), let **B** be a basis submatrix of **A**. **B** is nonsingular so it must have an inverse $\mathbf{B}^{-1}$. We shall show all columns $\mathbf{B}^{-1}\mathbf{e}^j$ are integer when (ii) holds ($\mathbf{e}^j$ is the jth unit vector).

Choose an integer vector **u**, so that $\mathbf{v} \triangleq \mathbf{u} + \mathbf{B}^{-1}\mathbf{e}^j \geq \mathbf{0}$. Premultiplying by **B** gives $\mathbf{Bv} = \mathbf{Bu} + \mathbf{e}^j$, and we see that **Bv** is the sum of integer vectors and thus, integer itself. Choosing $\mathbf{b} \leftarrow \mathbf{Bv}$ and applying (ii), we know solution **v** of the system $\mathbf{Bv} = \mathbf{b}$, $\mathbf{v} \geq \mathbf{0}$ must be integer. But then $\mathbf{B}^{-1}\mathbf{e}^j \triangleq \mathbf{v} - \mathbf{u}$ is integer.

To prove (iii) $\Rightarrow$ (i), we consider an (integer) basis submatrix **B** of **A** and its inverse $\mathbf{B}^{-1}$. By (iii), $\mathbf{B}^{-1}$ is integer, and by definition of inverses,

$$1 = \det(\mathbf{I}) = \det(\mathbf{B}\mathbf{B}^{-1}) = \det(\mathbf{B})\det(\mathbf{B}^{-1}).$$

The only nonzero integer choices for the final two determinants are ± 1, and so det(**B**) $= \pm 1$ and **A** is unimodular. ■

The unimodularity property relates to maximal submatrices of **A**. The stronger total unimodularity property considers all submatrices. In particular, a matrix **A** is said to be *totally unimodular* if every square submatrix of **A** has determinant ± 1 or 0. Hoffman and Kruskal's (1956) classic result on total unimodularity can be viewed as a corollary to Theorem 4.9.

Corollary 4.10. Total Unimodularity and Inequality Linear Programs. *Let* $\boldsymbol{A}$ *be an integer matrix. Then the following are equivalent.*

(i) *Every submatrix of* $\boldsymbol{A}$ *has determinant* ± 1 *or* 0.
(ii) *Extreme points of* $S^{\geq} = \{\boldsymbol{x}\colon \boldsymbol{A}\boldsymbol{x} \geq \boldsymbol{b},\ \boldsymbol{x} \geq \boldsymbol{0}\}$ *are integer for any integer right-hand-side* $\boldsymbol{b}$.
(iii) *Every nonsingular submatrix of* $\boldsymbol{A}$ *has an integer inverse.*

Proof. Given any integer matrix **A**, we may create an instance to which Theorem 4.9 may be applied by appending a negative identity matrix. Recognizing that any nonsingular submatrix **P** of **A** corresponds to a basis

$$\begin{pmatrix} \mathbf{P} & \mathbf{0} \\ \mathbf{Q} & -\mathbf{I} \end{pmatrix}$$

of the equality system, $[\mathbf{A}, -\mathbf{I}]$ yields the corollary upon direct application of the theorem. ∎

The most prominent class of totally unimodular matrices are those that arise as the vertex-arc incidence matrix **E** of a directed graph. Specifically, **E** is a *vertex-arc incidence matrix,* if each column of **E** contains exactly two nonzero entries, $+1$ and -1. Rows correspond to vertices of the graph, and columns have the $+1$, where arc (i, j) leaves and the -1 where it enters. Figure 4.6 provides an example, and the next theorem establishes total unimodularity.

Theorem 4.11. Total Unimodularity of Vertex-Arc Incidence Matrices. *Every vertex-arc incidence matrix,* $\mathbf{E}$*, of a directed graph is totally unimodular.*

Proof. We proceed by induction on the size k of submatrices **M**. For $k = 1$, the theorem follows from the fact that all elements of **E** are 0, $+1$ or -1.

Now, assume the result holds for $k < \sigma$ and consider a σ by σ submatrix **M**. If any column of **M** has no nonzero entries, $\det(\mathbf{M}) = 0$ and the theorem follows. Similarly, if every column of **M** has two nonzero entries, they must be one $+1$ and one -1. Summing rows yields the zero vector, so $\det(\mathbf{M}) = 0$.

We conclude that if $\det(\mathbf{M}) \neq 0$, there is a column of **M** with exactly one

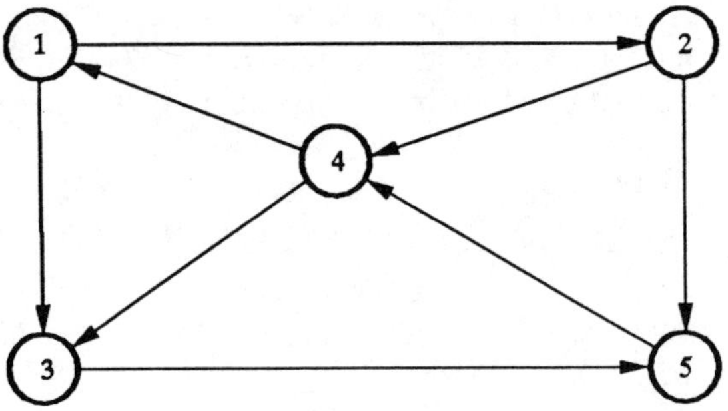

	(1, 2)	(4, 1)	(1, 3)	(2, 4)	(4, 3)	(2, 5)	(5, 4)	(3, 5)
vertex 1	1	−1	1	0	0	0	0	0
vertex 2	−1	0	0	1	0	1	0	0
vertex 3	0	0	−1	0	−1	0	0	1
vertex 4	0	1	0	−1	1	0	−1	0
vertex 5	0	0	0	0	0	−1	1	−1

Fig. 4.6. A directed graph and its vertex-arc incidence matrix.

nonzero entry $+1$ or -1. Expanding by minors on that element yields $\det(\mathbf{M}) = \pm 1 \det(\mathbf{M}')$ where $\mathbf{M}'$ is a $(\sigma - 1)$ by $(\sigma - 1)$ matrix. It follows from the inductive hypothesis that $\det(\mathbf{M}) = \pm 1$ or 0. ■

4.2.2 Testing Total Unimodularity

It is obvious that totally unimodular matrices are highly desirable ones in discrete optimization, because they assure integer solvability (for integer right-hand-sides). When the matrices are known to arise as a vertex-arc incidence matrix, we have already seen (Theorem 4.11) that total unimodularity is guaranteed.

What if the matrix's origin is not known to be a network one? It may still be possible to transform to a vertex-arc incidence matrix by suitable row operations. It may also be that the matrix is totally unimodular but not of network origin. Figure 4.7 illustrates the latter case.

In matroid terms (see Chapter 3), totally unimodular matrices are those that produce *regular* matroids under real arithmetic. Matroids arising from vertex-arc incidence matrices are called *graphic* matroids.

Recent and elegant results by Seymour (1980) have both shown constructively the relation between these two classes of matroids and provided a formally efficient mechanism for checking if a matroid is regular. That is, Seymour has developed an elegant scheme for checking if the corresponding matrix is totally unimodular.

$$\begin{bmatrix} 1 & 0 & 0 & 0 & 0 & -1 & 1 & 0 & 0 & 1 \\ 0 & 1 & 0 & 0 & 0 & 1 & -1 & 1 & 0 & 0 \\ 0 & 0 & 1 & 0 & 0 & 0 & 1 & -1 & 1 & 0 \\ 0 & 0 & 0 & 1 & 0 & 0 & 0 & 1 & -1 & 1 \\ 0 & 0 & 0 & 0 & 1 & 1 & 0 & 0 & 1 & -1 \end{bmatrix}$$

Fig. 4.7. A totally unimodular matrix not arising from any graph.

We shall briefly sketch these results. The key notion is *sums* of matroids. Consider, specifically, two binary matroids $M_1 \triangleq (G_1, \mathscr{I}_1)$, $M_2 \triangleq (G_2, \mathscr{I}_2)$. Then we may form a new ground set $G \triangleq G_1 \oplus G_2 \triangleq (G_1 \cup G_2) \backslash (G_1 \cap G_2)$ and define sum matroid $M \triangleq M_1 \oplus M_2$ on G as the one with circuits (minimal dependent sets) C of the form $C_1 \oplus C_2$, where C_1 and C_2 are circuits of M_1 and M_2, respectively.

If G_1 and G_2 are nonempty, but $G_1 \cap G_2 = \varnothing$, $M_1 \oplus M_2$ is said to be a *1-sum* of matroids M_1 and M_2. If $G_1 \cap G_2 = \{e\}$ where e is a dependent singleton in neither M_1, M_2 nor their duals M_1^*, M_2^*, we call $M_1 \oplus M_2$ a *2-sum*. Finally, if $G_1 \cap G_2 = H$ with $|H| = 3$ and H is a circuit of both M_1 and M_2, we term $M_1 \oplus M_2$ a *3-sum*.

Figure 4.8 illustrates these ideas for graphic matroids on a connected graph G. In part (a), G is the 1-sum of G_1 and G_2 obtained by identifying a vertex. Part (b) shows the 2-sum operation corresponds to identifying a common edge. A 3-sum arises as in part (c) by doing the same operation on a common 3-circuit.

Seymour's theorem constructively characterizes regular matroids in terms of such matroid sums.

Theorem 4.12. Constructive Characterization of Regular Matroids. *Any regular matroid M can be constructed as 1-, 2-, and 3-sums of graphic matroids, cographic matroids, and matroids isomorphic to the binary matroid R_{10} of the matrix in Figure 4.7.*

We know graphic matroids arise from vertex-arc incidence matrices and cographic ones are duals of graphics. In a rather precise sense, therefore, Theorem 4.12 shows R_{10} is the only non-network building block of totally unimodular matrices.

To use Seymour's results for recognizing totally unimodular matrices, we reverse the sequence. A matroid M is said to *1-, 2- or 3-separate* into M_1 and M_2, respectively, if M is the 1-, 2- or 3-sum of matroids M_1 and M_2 of suitable rank. To check whether a representable matroid is regular, we can thus proceed by finding 1-, 2- and 3-separations of the matroid and checking whether the resulting component matroids are graphic, cographic, or

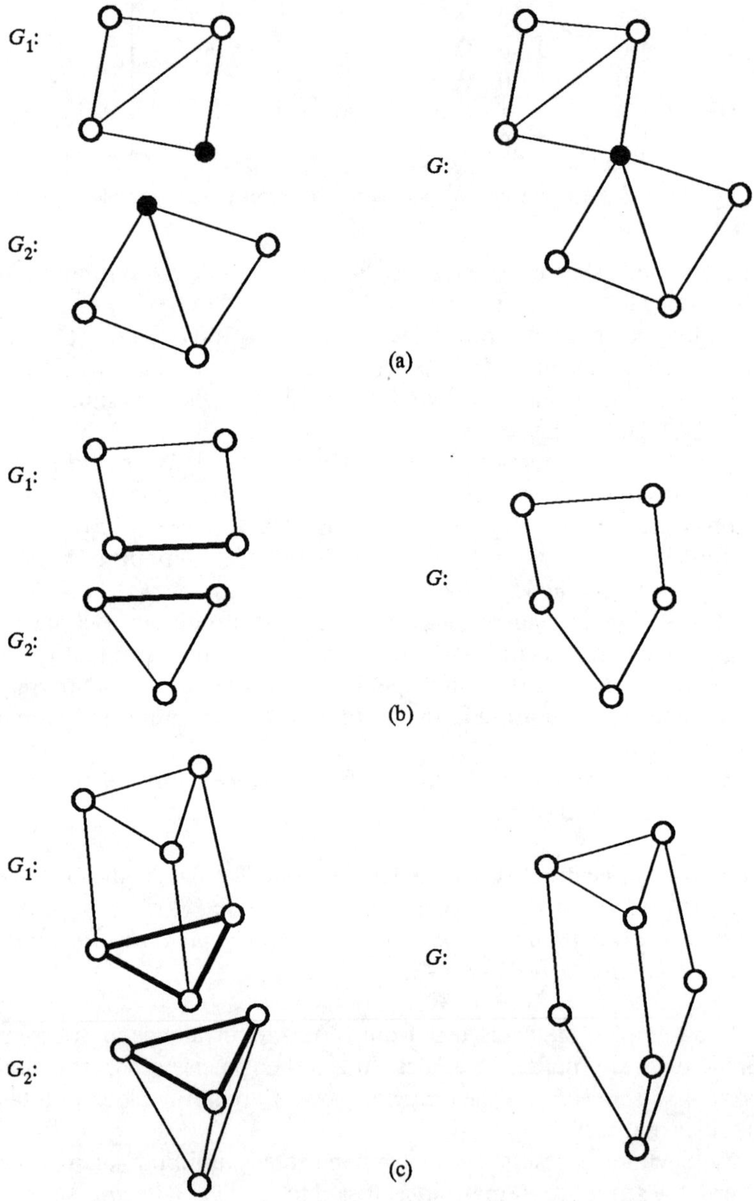

Fig. 4.8. Sums of graphic matroids; (a) 1-sum, (b) 2-sum, (c) 3-sum.

isomorphic to R_{10}. All such separations can be produced in polynomial time, and whether a matroid is graphic, cographic or isomorphic to R_{10} can also be checked in polynomial time. Thus, regularity of a matroid can be determined in polynomial time by concatenation of these procedures. We refer the interested reader to Bixby (1981) for further elaboration.

4.2.3 Total Dual Integrality

Corollary 4.10 showed total unimodularity is sufficient for any inequality-constrained system $S^{\geq} \triangleq \{\mathbf{x} \geq \mathbf{0}: \mathbf{A}\mathbf{x} \geq \mathbf{b}\}$ to have integer extreme points when $\mathbf{b}$ is integer. Total unimodularity is also necessary if the result must hold for all right-hand-sides $\mathbf{b}$. If the problem form restricts $\mathbf{b}$, however, it is easy to find cases where weaker properties will do.

Example 4.4. Total Unimodularity and Matching. One famous case known to have integer extreme points arises in the computation of maximum weight *perfect matchings* (edge sets meeting each vertex exactly once) in an undirected graph $G(V, E)$ with $|V|$ even. If we denote for $\bar{V} \subset V$

$$F(\bar{V}) \triangleq \{e \in E\text{: one end is } \bar{V}\text{, one is not}\}$$

$$E(\bar{V}) \triangleq \{e \in E\text{: both ends of } e \text{ belong } \bar{V}\},$$

and associate variable x_e with edge $e \in E$, the maximum weight perfect matching problem is

$$\begin{aligned} \max \quad & \sum_{e \in E} w_e x_e & & \\ \text{s.t.} \quad & \sum_{e \in F(v)} x_e = 1 & \text{for all } v \in V \quad & \text{(4-38)} \\ & x_e \geq 0 & \text{for all } e \in E \quad & \text{(4-39)} \\ & x_e \text{ integer} & \text{for all } e \in E \quad & \text{(4-40)} \end{aligned}$$

Edmonds' (1965c) classic investigation of matching polyhedra and algorithms showed that this integer optimization problem can be solved by linear programming, if (4-40) is replaced by

$$\sum_{e \in E(\bar{V})} x_e \leq \frac{|\bar{V}| - 1}{2} \qquad \text{for all } \bar{V} \subset V \text{ with } |\bar{V}| \text{ odd.} \qquad \text{(4-41)}$$

That is, all extreme point solutions of constraints (4-38), (4-39), (4-41) are integer.

Still, it is easy to construct cases where these constraints are not totally unimodular. Consider, for example, constraints (4-38) on a G that is an odd cardinality cycle. The corresponding submatrix will have the form

$$\begin{bmatrix} 1 & 0 & \cdots & 1 \\ 1 & 1 & & 0 \\ 0 & 1 & & \cdot \\ \cdot & 0 & & \cdot \\ \cdot & \cdot & & 0 \\ 0 & 0 & \cdots & 1 \end{bmatrix}$$

with |determinant| = 2. ■

The Edmonds algorithm shows that the matching constraints above have integer solutions for all weight sets $\{w_e: e \in E\}$. A primal variant of Edmonds' algorithm due to Cunningham and Marsh (1978) produces an LP-optimal dual solution that is integer whenever weights $\mathbf{w}$ are. The latter property led Edmonds and others to investigate the notion of *total dual integrality.*

Total dual integrality can be defined for a linear program

$$(P) \qquad \begin{aligned} \min \quad & \mathbf{cx} \\ \text{s.t.} \quad & \mathbf{Ax} \geq \mathbf{b} \\ & \mathbf{x} \geq \mathbf{0} \end{aligned}$$

posed over rational **A, b, c.** The dual of that linear program is

$$(D) \qquad \begin{aligned} \max \quad & \mathbf{ub} \\ \text{s.t.} \quad & \mathbf{uA} \leq \mathbf{c} \\ & \mathbf{u} \geq \mathbf{0} \end{aligned}$$

Constraints of (P) are said to be *totally dual integer* (TDI) if (D) has an integer optimal solution for all integer **c** for which it has any optimal solution.

We have just specified why matching problems are TDI. The primal blossom algorithm constructs the needed integer dual optimum for any integer weight vector.

The main importance of the TDI property is its sufficiency for integer solvability.

Theorem 4.13. Integer Solvability of TDI Systems. *If a rational constraint system* $S^{\geq} \triangleq \{\boldsymbol{Ax} \geq \boldsymbol{b},\ \boldsymbol{x} \geq \boldsymbol{0}\}$ *is totally dual integer, and if* $\boldsymbol{b}$ *is integer, then all extreme point solutions of the system are integer.*

Proof. We proceed by showing that if $S^{\geq}$ has a noninteger extreme point solution $\bar{\mathbf{x}}$ when $\mathbf{b}$ is integer, then there exists an integer cost vector $\bar{\mathbf{c}}$ for which the corresponding dual linear program (D) has a fractional optimal solution (contradicting total dual integrality). For this purpose, we may take $\mathbf{A}$ integer, as fractions could easily be cleared without changing the solution space.

Choose such a fractional extremum $\bar{\mathbf{x}}$. Standard linear programming theory (see the Appendix C) shows $\bar{\mathbf{x}}$ must be a nonnegative basic solution to

$$\{\mathbf{Ax} - \mathbf{Is} = \mathbf{b},\ \mathbf{x} \geq \mathbf{0},\ \mathbf{s} \geq \mathbf{0}\}.$$

That is, $\bar{\mathbf{x}}$ and the corresponding $\bar{\mathbf{s}}$ must be expressible as $(\bar{\mathbf{x}}^B, \bar{\mathbf{x}}^N)$, $(\bar{\mathbf{s}}^B, \bar{\mathbf{s}}^N)$, and $\mathbf{A}$, $\mathbf{b}$ partitionable accordingly, so that

$$\begin{pmatrix} \mathbf{A}_1^B & \mathbf{0} \\ \mathbf{A}_2^B & -\mathbf{I} \end{pmatrix} \begin{pmatrix} \bar{\mathbf{x}}^B \\ \bar{\mathbf{s}}^B \end{pmatrix} = \begin{pmatrix} \mathbf{b}^1 \\ \mathbf{b}^2 \end{pmatrix}$$

with $\mathbf{A}_1^B$ square. ($\mathbf{A}_1^B$ is nonvacuous because otherwise $\bar{\mathbf{x}}$ is integer.)

Expressed this way, $\bar{\mathbf{x}}^B$ is the unique real solution to the integer equation system $\mathbf{A}_1^B\mathbf{x} = \mathbf{b}^1$. Standard results from number theory show such an integer equation system has an integer solution if and only if the greatest common divisor (g.c.d.) of the coefficients in each row of $\mathbf{A}_1^B$ divides the corresponding right-hand-side in $\mathbf{b}^1$.

Here there are no integer solutions because the unique real solution is the fractional $\bar{\mathbf{x}}_1^B$. Thus, there is a row of $\mathbf{Ax} \geq \mathbf{b}$, say $\mathbf{a}^B\mathbf{x}^B + \mathbf{a}^N\mathbf{x}^N \geq \beta$, such that the g.c.d. of elements of $\mathbf{a}^B$, say γ, does not divide β, yet $\mathbf{a}^B\bar{\mathbf{x}}^B = \beta$.

We pick $\bar{\mathbf{c}} \triangleq (\bar{\mathbf{c}}^B, \bar{\mathbf{c}}^N) \leftarrow ((1/\gamma)\mathbf{a}^B, \lceil(1/\gamma)\mathbf{a}^N\rceil)$. The definition of γ assures this is indeed an integer cost row. Furthermore, the (P) solution value for solution $\bar{\mathbf{x}}$ is $((1/\gamma)\mathbf{a}^B)\bar{\mathbf{x}}^B = (\beta/\gamma)$.

Now consider the dual with this $\bar{\mathbf{c}}$. One obviously feasible dual solution is $\bar{\mathbf{u}}$, with $\bar{u}_i = (1/\gamma)$ on row $\mathbf{a}^B\mathbf{x}^B + \mathbf{a}^N\mathbf{x}^N \geq \beta$ and $\bar{u}_i = 0$ otherwise. This solution has dual objective function value β/γ. Since this equals the primal, we know both $\bar{\mathbf{u}}$ and $\bar{\mathbf{x}}$ are optimal for their respective problems. But β/γ is fractional because γ does not divide β. Thus, every optimal solution $\mathbf{u}^*$ to (D) for this $\bar{\mathbf{c}}$ will have fractional solution value $\mathbf{u}^*\mathbf{b} = \beta/\gamma$. With $\mathbf{b}$ integer, we may conclude all such $\mathbf{u}^*$ are fractional and $S^{\geq}$ is not TDI. This completes the proof. ■

Many famous (polynomially solvable) models in discrete optimization can be shown to have totally dual integer formulations like the matching case of Example 4.4. Interested readers are referred to Schrijver (1981).

Giles and Pulleyblank (1979) also prove a form of converse to Theorem 4.13. It shows every polytope with integer extreme points has a TDI representation. We conclude our treatment by stating this result.

Theorem 4.14. Necessity of a TDI Form. *Let $\bar{S}$ be a polyhedron with all extreme points $\boldsymbol{x}$ integer. Then there exists a unique, minimal, totally dual integer system $S = \{\boldsymbol{x} \geq \boldsymbol{0}\colon \boldsymbol{A}\boldsymbol{x} \geq \boldsymbol{b}\}$, with $\boldsymbol{A}$ rational, $\boldsymbol{b}$ integer such that $S = \bar{S}$.*

4.3 Equivalences between Linear and Polynomial Solvability

In Section 4.1, we presented the polynomial procedures available for linear programming. We explored, in Section 4.2, special properties of linear programming problems that cause an *LP* solution to solve an associated discrete optimization problem.

One of the oldest conjectures in modern discrete optimization is that there is an equivalence between linear and polynomial solvability. That is, any discrete problem that can be succinctly expressed as a linear program can probably be polynomial-time solved and vice versa. Now we will introduce the growing body of literature that both proves a form of this conjecture and gives it practically implementable value.

4.3.1 Separation and Optimization

To begin, we must return to Khachian's Ellipsoid method of Section 4.1.2. Recall that the procedure addresses a closed convex set $X \subset \mathbb{R}^n$ such that (perhaps after translating the axes)

$$\mathbb{B}(\hat{\mathbf{x}}, r) \subseteq X \subseteq \mathbb{B}(\mathbf{0}, R) \tag{4-42}$$

for given inner and outer radii r and R.

The sliding objective version of the algorithm introduced earlier addresses the *optimization* problem

$$\begin{array}{lll} & \min & \mathbf{cx} \\ (Opt) & \text{s.t.} & \mathbf{x} \in X, \end{array}$$

where $\mathbf{c}$ is a rational vector. Adjusting for error tolerances $\epsilon > 0$, the Ellipsoid algorithm actually solves the *weak optimization, problem*

(WOpt) Find a rational n-vector $\bar{\mathbf{x}}$ such that $\text{dist}(\bar{\mathbf{x}}, X) \leq \epsilon$ and $\min\{\mathbf{cx}\colon \mathbf{x} \in X\} \leq \mathbf{c}\bar{\mathbf{x}} - \epsilon$.

At most, N iterations are required, where N depends only on n and the length (logarithms) of data items R, r, $\|\mathbf{c}\|$ and ϵ.

Each iteration must solve a *separation* problem of the form

(Sep) Given a rational n-vector $\mathbf{x}^k$ either show $\mathbf{x}^k \in X$ or find a rational vector $\mathbf{a}^k$ such that $\max\{\mathbf{a}^k\mathbf{x}: \mathbf{x} \in X\} < \mathbf{a}^k\mathbf{x}^k$.

(Note that the direction of the inequalities has been reversed from Section 4.1.2.) The corresponding *weak separation* form with tolerances added is

(WSep) Given a rational n-vector $\mathbf{x}^k$, either show $\text{dist}(\mathbf{x}^k, X) \leq \epsilon$ or find a rational vector $\mathbf{a}^k$ with $\|\mathbf{a}^k\| \geq 1$ such that $\max\{\mathbf{a}^k\mathbf{x}: \mathbf{x} \in X\} \leq \mathbf{a}^k\mathbf{x}^k + \epsilon$.

The interesting fact is that almost all the difficulty in producing a polynomial algorithm for (*Opt*) (or (*WOpt*)) reduces to (*Sep*) (respectively (*WSep*)). We may make this explicit in the following lemma.

Lemma 4.15. Weak Separation Implies Weak Optimization. *Let (WSep) and (WOpt) be as above over sets X satisfying (4-42). Then if (WSep) can be solved in time polynomial in the length of its input, so can (WOpt).*

Proof. Considering n, r, R and ϵ as given parameters, we already know Algorithm 4A gives a (*WOpt*) solution after polynomially many invocations of (*WSep*). If each (*WSep*) call requires polynomial time at most, the total effort will therefore require only polynomially many algorithmic steps.

The only remaining issue is to show that vectors $\mathbf{a}^k$ generated by (*WSep*) will be sufficiently well-behaved that ellipsoid update calculations will require only polynomial time. That this is true is inherent in the fact that (*WSep*) is polynomial. It must follow that $\mathbf{a}^k$ is polynomial in length or (*WSep*) could not even write it out. ■

Lemma 4.15 is a fairly straightforward insight about the Ellipsoid algorithm. The interesting and less expected result, demonstrated in the seminal paper of Grötschel, Lovász and Schrijver (1981) (and also in Padberg and Rao (1980)), is that the Lemma 4.15 has a converse. That is, *(WOpt)* and *(WSep)* are polynomially equivalent.

To develop their proof of the converse, we require the notion of a *polar set*. Denoted X^*, the polar of a convex set X is the convex set

$$X^* \triangleq \{\mathbf{v}: \mathbf{v}\mathbf{x} \leq 1 \text{ for all } \mathbf{x} \in X\}.$$

Polars have many elegant properties vis-à-vis their defining sets. Important ones for our discussion are

- $(X^*)^* = X$ (4-43)
- $\mathbb{B}(\mathbf{x}, r) \subseteq X \subseteq \mathbb{B}(\mathbf{0}, R)$ implies $\mathbb{B}(\mathbf{0}, 1/R) \subseteq X^* \subseteq \mathbb{B}(\mathbf{0}, 1/r)$. (4-44)

In light of (4-44), we can imagine applying the Ellipsoid algorithm to the polar problem of any appropriate instance set X in the earlier discussion. Natural polar forms of weak optimization and separation problems are then

(WOpt)* Given a rational vector $\mathbf{d}$, find a rational n-vector $\bar{\mathbf{v}}$ such that $\text{dist}(\mathbf{v}, X^*) \le \epsilon$ and $\min\{\mathbf{dv}: \mathbf{v} \in X^*\} \le \mathbf{d}\bar{\mathbf{v}} - \epsilon$,

and

(WSep)* Given a rational n-vector $\mathbf{v}^k$, either show $\text{dist}(\mathbf{v}^k, X^*) \le \epsilon$ or find a rational vector $\mathbf{a}^k$ with $\|\mathbf{a}^k\| \ge 1$ such that $\max\{\mathbf{a}^k\mathbf{v}: \mathbf{v} \in X^*\} \le \mathbf{a}^k\mathbf{v}^k + \epsilon$.

Lemma 4.16. Weak Optimization Implies Weak Polar Separation. *Let (WOpt) be the weak optimization form defined above over sets X satisfying (4-42) and (WSep*) the corresponding weak separation problem over polars of the X's. Then if (WOpt) can be solved in time polynomial in the length of its input, so can (WSep*).*

Proof. Consider a polar set X^*, a vector $\mathbf{v}^k$, and a tolerance $\epsilon^* > 0$. We wish to show *(WSep*)* can be polynomially solved with this input.

Imagine first solving the corresponding *(WOpt)* problem with cost $(-\mathbf{v}^k)$ and error tolerance $\epsilon \leftarrow \epsilon^* r$. By hypothesis, we will obtain in polynomial time a vector $\mathbf{z}$ such that $\text{dist}(\mathbf{z}, X) \le \epsilon^* r$ and $-\mathbf{v}^k\mathbf{z} \le \min\{-\mathbf{v}^k\mathbf{x}: \mathbf{x} \in X\} + \epsilon^* r$, i.e., $\mathbf{v}^k\mathbf{z} \ge \max\{\mathbf{v}^k\mathbf{x}: \mathbf{x} \in X\} - \epsilon^* r$.

If it happens that $\mathbf{v}^k\mathbf{z} \le 1$, then $\mathbf{v}^k\mathbf{x} \le \mathbf{v}^k\mathbf{z} \le 1 + \epsilon^* r$ for all $\mathbf{x} \in X$, and $\hat{\mathbf{v}} \triangleq (1 + \epsilon^* r)^{-1}\mathbf{v}^k \in X^*$. It follows that $\|\hat{\mathbf{v}}\| \le 1/R$, and so,

$$\begin{aligned}\text{dist}(\mathbf{v}^k, X^*) &\le \|\mathbf{v}^k - \hat{\mathbf{v}}\| \triangleq \|(1 + \epsilon^* r)\hat{\mathbf{v}} - \hat{\mathbf{v}}\| \\ &= \epsilon^* r \|\hat{\mathbf{v}}\| \le (\epsilon^* r)(1/r) = \epsilon^*\end{aligned}$$

That is, the first outcome possibility of *(WSep*)* is fulfilled.

On the other hand, suppose $\mathbf{v}^k\mathbf{z} > 1$. Then we may assert $\mathbf{a}^k \leftarrow \mathbf{z}$ satisfies the second *(WSep*)* case. To see that this is true, pick $\hat{\mathbf{x}} \in X$ close enough to $\mathbf{z}$ that

$$\|\mathbf{z} - \hat{\mathbf{x}}\| \le \epsilon^* r. \tag{4-45}$$

Then for every $\mathbf{v} \in X^*$, $\hat{\mathbf{x}} \in X$, $\mathbf{v}^k\mathbf{z} > 1$ gives

$$\mathbf{vz} = \mathbf{v}(\mathbf{z} - \hat{\mathbf{x}}) + \mathbf{v}\hat{\mathbf{x}} \le \mathbf{v}(\mathbf{z} - \hat{\mathbf{x}}) + 1 < \mathbf{v}(\mathbf{z} - \hat{\mathbf{x}}) + \mathbf{v}^k\mathbf{z}.$$

Then, by the Cauchy-Schwartz inequality, we have that

$$\mathbf{v}\mathbf{z} < \|\mathbf{v}\|\,\|\mathbf{z} - \hat{\mathbf{x}}\| + \mathbf{v}^k\mathbf{z}. \qquad (4\text{-}46)$$

But $\mathbf{v} \in X^*$ implies $\|\mathbf{v}\| \leq 1/r$. Combining this with (4-46) and (4-45) gives $\mathbf{z}\mathbf{v} < (\epsilon^* r)(1/r) + \mathbf{z}\mathbf{v}^k$ as required. ■

We are now ready for the main theorem.

Theorem 4.17. Equivalence of Weak Separation and Weak Optimization. *Let (WSep) and (WOpt) be the weak separation and optimization problems defined above over sets X satisfying (4-42). Then (WSep) is polynomially solvable in the length of its input if and only if (WOpt) is.*

Proof. (⇒) Follows directly from Lemma 4.15.

(⇐) Under Lemma 4.16, if *(WOpt)* is polynomially solvable, then so is (*WSep**) in the polar. But then by Lemma 4.15, applied on the polars, (*WOpt**) is polynomially solvable. Now again applying Lemma 4.16 the weak separation problem ((*WSep**)*) on the polar of the polar must be polynomially solvable. But since $(X^*)^* = X$, the latter is exactly *(WSep)*. ■

The above theorem holds for quite general convex sets X (once the length of X is decided). In discrete optimization, our interest is almost always in *rational polytopes,* i.e., in bounded sets expressable as the intersection of finitely many half-space sets of the form $\{\mathbf{x}: \mathbf{a}\mathbf{x} \geq a_0\}$ with $\mathbf{a}$ a rational n-vector, and a_0 a rational scalar.

Although the proof is too technical to include here, Grötschel *et al.* (1981) extended Theorem 4.17 over rational polytopes to eliminate the weak approximation elements needed for more general X's. The following theorem summarizes their extension.

Theorem 4.18. Equivalence of Separation and Optimization on Rational Polytopes. *Suppose problems (Opt) and (Sep) above are posed over rational polytopes X satisfying (4-42). Then (Opt) is solvable in time polynomial in the length of its input if and only if (Sep) is.*

4.3.2 Combinatorial Significance of the Separation Implies Optimization

Both directions of Theorem 4.18 have great significance in discrete optimization, and we treat them separately.

First consider the polynomial separation implies polynomial optimization direction (for rational polytopes). This result can be viewed as a general skeleton for proving polynomial solvability of any discrete optimization problem. We need only two elements:

(i) a concise linear programming description of the set of discrete solutions to the problem; and
(ii) a polynomial separation scheme for either showing that every inequality of the linear system is satisfied, or exhibiting a violated one.

Item (i) is, of course, a very difficult task. In fact, results we shall survey in Section 6.3.2 suggest it is quite unlikely that a suitable linear programming formulation will ever be found for *NP*-Complete problems.

Still, many of the most elegantly solved problems in discrete optimization have been shown to admit simple linear programming forms. For example, return to the perfect matching problem of Example 4.4. Recall that equations (4-38) through (4-40) provide a correct integer programming formulation of the problem on a given graph $G(V, E)$; replacing (4-40) by (4-41) gives a linear programming characterization. This linear programming form settles issue (i).

For (ii), we require a polynomial time scheme for checking a specified $\mathbf{x}^k \in \mathbb{R}^{|E|}$. We wish to either show $\mathbf{x}^k$ satisfies all constraints or exhibit a violated one.

At first glance, this may appear a trivial assignment. Why not merely check each constraint for violation as we did in the example of Ellipsoid in Section 4.1.2?

A little thought, however, will show such a procedure would not combine with the Ellipsoid algorithm to produce a polynomial time scheme for perfect matching. The subtlety lies with input length. Constraint system (4-41) is exponential in the size of the original graph because it has rows for each odd vertex subset. A naïve separation scheme that explicitly checked each such row would thus yield polynomial Ellipsoid solution of the specified linear program, but would be exponential in the original matching input.

We see that more ingenuity is required. In particular, we require an algorithm running in time polynomial with respect to the original graph input that implicitly checks all rows of the linear programming form.

Given an $\mathbf{x}^k \in \mathbb{R}^{|E|}$, we could check whether constraints (4-38) or (4-39) were violated in the naïve way of explicitly trying each one. There are only polynomially many to try. If a violated one is found, it provides the needed separation vector $\mathbf{a}^k$.

If all the easy constraints are satisfied by $\mathbf{x}^k$, we move on to (4-41). Subtraction of one half the sum of degree constraints (4-38) for $v \in \overline{V}$ from each constraint of (4-41) gives the more suggestive form

$$\sum_{e \in F(\bar{V})} x_e \geq 1 \qquad \text{for all } \bar{V} \subset V, |\bar{V}| \text{ odd.} \tag{4-47}$$

We see that the constraints merely require that the **x**-total across any *odd cut set* (i.e., edge set leaving an odd vertex set) should be at least 1.

To readers familiar with network flows, this form will immediately suggest a separation scheme. Ford and Fulkerson (1956) presented a classic, polynomial-time method for finding the minimum capacity cutset separating a specified vertex $s \in \bar{V}$ from another $t \notin \bar{V}$ where capacity is defined as the sum over $e \in F(\bar{V})$ of given edge values u_e.

Given an $\mathbf{x}^k$ to separate, we could proceed by employing the Ford–Fulkerson procedure with values $u_e \leftarrow x_e^k$. If we try all the $O(|V|^2)$ vertex pairs $s, t \in V$, and record the minimum cut capacity, we will certainly find a cut with $\mathbf{x}^k$ total less than 1 if such exists. With one reservation, this is exactly what is needed for polynomial separation of constraints (4-47). The only problem is that our scheme checks all cut sets, and (4-47) constrains only odd cut sets. Still, the notion we have outlined can be adapted to check only odd cut sets. See Padberg and Rao (1982) for a full development.

Including Ellipsoid calculation with this separation method will certainly produce a scheme that is less efficient than the best known algorithm for perfect matching. The point is, however, that it is one derived via a broadly applicable tool. Moreover, the Theorem 4.18 approach has produced polynomial algorithms for some problems not known to be efficiently solvable in any other way. A difficult example is treated in Grötschel *et al.* (1981).

4.3.3 Combinatorial Significance of Optimization Implies Separation

Although it does not provide an obvious construction, the converse of Theorem 4.18 may be the direction of broadest theoretical importance. To see its significance, observe that virtually every discrete optimization problem

$$\begin{array}{lll} & \min & \mathbf{cx} \\ (DOpt) & \text{s.t.} & \mathbf{x} \in D \end{array}$$

over some discrete set D can be expressed in the *(Opt)* linear programming format with a suitable choice of rational polytope. We reserve the details for Chapter 6, but the idea is that the smallest convex set containing all $\mathbf{x} \in D$ is usually a rational polytope. That *convex hull of solutions,* denoted $[D]$, thus provides, in principle, a linear program formulation of *(DOpt).*

We introduced this section by mentioning the conjecture that polynomially solvable *(DOpt)*'s always admit succinct linear programming characterizations. Simply knowing that the convex hull exists is not enough to fulfill this conjecture. Instead, we would like to be able to write down all needed constraints in a fashion similar to say (4-47) for matching.

The key difference is in the notion of a *succinct* linear programming description. Again reserving the details until Section 6.3, the essential requirement is to describe $[D]$ with a system of linear inequalities that have short (i.e., polynomial length) derivations.

The converse of Theorem 4.18 demonstrates that, at least for cases where $[D]$ is bounded and full dimension, the separation problem over $[D]$ must be polynomial-time solvable for any test vector $\mathbf{x}^k \in \mathbb{R}^n$. By carefully choosing an appropriate $\mathbf{x}^k$, we can make any nonredundant constraint of $[D]$ the only one with suitable rational coefficients that can separate $\mathbf{x}^k$. Thus, it follows that all such constraints must be derivable in polynomial time because the polynomial time separation algorithm must be able to produce each one. With a great deal more precision, this line of argument can thus give a form of proof to the classic conjecture above as a corollary to Theorem 4.18.

The weakness of the converse of Theorem 4.18 is that it is not very constructive. Some recent research, however, by Martin (1987), Martin, Rardin and Campbell (1987), Jeroslow (1986), and Balas and Pulleyblank (1983) seems to provide constructions for many cases.

4.4 Integer Programs with a Fixed Number of Variables

Consider the pure nonnegative integer program

$$(IP) \qquad \begin{array}{ll} \min & \mathbf{cx} \\ \text{s.t.} & \mathbf{Ax} \geq \mathbf{a} \\ & \quad \mathbf{x} \geq \mathbf{0} \\ & \quad \mathbf{x} \text{ integer} \end{array}$$

with integer $\mathbf{A}$, $\mathbf{a}$ and $\mathbf{c}$, and $\mathbf{x}$ an n-vector. It is very easy to see that when $\mathbf{x}$ is limited to $0-1$ values (i.e., $\mathbf{x} \leq \mathbf{1}$ is also a constraint), (IP)'s with $n \leq k$ fixed can, in principle, be solved in polynomial time. Simply trying all $0-1$ combinations will require a number of computations bounded by a polynomial in the size of the data. Effort is at most 2^k times the computation to evaluate and check a particular $\mathbf{x}$ solution. Of course, this bound is exponential in k, but we have assumed k fixed.

In exercises we ask the reader to show this polynomial bound on integer programs with fixed numbers of variables easily extends via binary expansion to any (*IP*) with explicit integer upper bounds $\mathbf{x} \leq \mathbf{u}$ as part of the data. But the general case, with no explicit bounds given, is far more complex. In this section we briefly sketch Lenstra's (1983) resolution.

Theorem 4.19. Polynomial Solution of Integer Programs with Fixed Numbers of Variables (Lenstra). *Let (IP) be the general integer program above. Then for any fixed n, the associated decision problem*

Does there exist an integer n-vector $\boldsymbol{x} \geq \boldsymbol{0}$ such that $\boldsymbol{Ax} \geq \boldsymbol{a}$, $\mathbf{cx} \leq v$?

*can be answered in time bounded by a polynomial in the length of **A, a, c** and a threshold v.*

4.4.1 Lattices and Basis Reduction

A *lattice* of Euclidean n-space is a set L generated by a basic set of vectors $\{\mathbf{b}^i\colon 1 = 1, 2, \ldots, n\}$. Specifically,

$$L \triangleq \left\{\sum_{i=1}^{n} z_i\mathbf{b}^i\colon z_i \text{ integer } i = 1, 2, \ldots, n\right\}.$$

One obvious example is the set of integer n-vectors. All elements of that lattice are generated by the basis of unit vectors $\{\mathbf{e}^i\colon 1 = 1, 2, \ldots, n\}$.

For any given lattice L there will be many alternatives for the generating basis set. Some straightforward algebra, however, shows they have constant determinant.

Lemma 4.20. Invariant Determinants. *For any lattice L, $d_L = \det[\boldsymbol{b}^1, \boldsymbol{b}^2, \ldots, \boldsymbol{b}^n]$ for generating basic vectors $\boldsymbol{b}^i$ is independent of the generator set chosen. Furthermore, for each generator set*

$$d_L \leq \prod_{i=1}^{n} \|\boldsymbol{b}^i\|. \tag{4-48}$$

Solutions to (IP) above are elements of the intersection of an integer lattice and $\{\mathbf{x} \geq \mathbf{0}\colon \mathbf{Ax} \geq \mathbf{a}\}$. One of the critical arguments in Lenstra's results is the observation that there exists a *reduced basis* of every lattice with the product of (4-48) upper bounded.

Lemma 4.21. Reduced Basis. *For every lattice L of $\mathbb{R}^n$ there exists a reduced basis $\{\bar{\boldsymbol{b}}^i\colon i = 1, \ldots, n\}$ such that*

$$\prod_{i=1}^{n} \|\bar{\boldsymbol{b}}^i\| \leq c_1[n]d_L$$

where $c_1[n]$ is a constant depending only on n and d_L is the determinant of a generating set for L.

It is not a simple matter to efficiently exhibit the reduced lattice basis promised by Lemma 4.21. Still, polynomial algorithms have appeared. The reader is referred to Lenstra, Lenstra, and Lovász (1982), and Kannan (1983a,b).

4.4.2 Polynomial Solution

Like the theoretical developments of Section 4.3, Lenstra's proof of Theorem 4.19 begins with a result based on Ellipsoid Algorithm 4A.

Lemma 4.22. Bounding Transformation. *Let (IP) be as above, a general integer program, and v an objective function threshold. Then there is a polynomial transformation τ of $F \triangleq \{\boldsymbol{x} \geq 0\colon\ \boldsymbol{Ax} \geq \boldsymbol{a},\ \boldsymbol{cx} \leq v\}$, a point $\boldsymbol{p} \in \tau(F)$ and a radius r such that $\mathbb{B}(\boldsymbol{p}, r) \subset \tau(F) \subset \mathbb{B}(\boldsymbol{p}, c_2[n]r)$, where $\mathbb{B}(\boldsymbol{p}, r)$ denotes the Euclidean ball of radius r about $\boldsymbol{p}$, and $c_2[n]$ is a constant depending only on n.*

Our interest is in whether $F \cap \{\text{integer } \mathbf{x}\}$ is empty. Using the last lemma, we may consider the equivalent question of whether $\tau(F) \cap L$ is empty, where L is the lattice obtained when τ is applied to the set of integer n-vectors.

We shall proceed recursively using the reduced basis $\{\bar{\mathbf{b}}^1, \bar{\mathbf{b}}^2, \ldots, \bar{\mathbf{b}}^n\}$ of L guaranteed by Lemma 4.21. At each stage we either obtain a $\boldsymbol{y} \in \tau(F) \cap L$ or demonstrate that the number of integer multipliers (lattice levels) of $\bar{\mathbf{b}}^n$ that must be tried in finding such a $\mathbf{y}$ is bounded by constants depending only on n. Trying each of these multipliers in turn yields a bounded number of problems in dimension $n-1$. Continuing through $n-1, n-2, \ldots, 1$ produces an answer to the main question in a number of steps polynomial in **A, b, c** and v (but of course exponential in our fixed n).

To be more explicit we require one more of Lenstra's propositions.

Lemma 4.23. Distance to a Lattice Point. *For a lattice L of $\mathbb{R}^n$, any $\boldsymbol{p} \in \mathbb{R}^n$ and any basis $\{\boldsymbol{b}^i\colon i = 1, 2, \ldots, n\}$ generating L, there exists a $\boldsymbol{y} \in L$ such that*

$$\|\boldsymbol{p} - \boldsymbol{y}\| \leq \left(\frac{\sqrt{n}}{2}\right) \max_{i=1,\ldots,n} \{\|\boldsymbol{b}^i\|\} \tag{4-49}$$

We may now sketch the whole proof of Theorem 4.19. Suppose, for simplicity, that the basis vectors of our reduced basis of L are numbered so that $\|\bar{\mathbf{b}}^1\| \leq \|\bar{\mathbf{b}}^2\| \leq \cdots \leq \|\bar{\mathbf{b}}^n\|$, and apply Lemma 4.23 with $\mathbf{p}$ as in the transformation Lemma 4.22 to obtain a $\mathbf{y} \in L$ with $\|\mathbf{p} - \mathbf{y}\| \leq (\sqrt{n}/2)\|\bar{\mathbf{b}}^n\|$. If $\|\mathbf{p} - \mathbf{y}\| \leq r$ (the inner radius in Lemma 4.22), then $\mathbf{y} \in \mathbb{B}(\mathbf{p}, r) \subseteq \tau(F)$, and $\mathbf{y}$ is the needed element of $L \cap \tau(F)$.

If $\|\mathbf{p} - \mathbf{y}\| > r$, then since $\tau(F) \subseteq \mathbb{B}(p, c_2[n]r)$,

$$\|\mathbf{p} - \mathbf{x}\| \leq c_2[n]r < c_2[n]\|\mathbf{p} - \mathbf{y}\| \leq c_2[n]\frac{\sqrt{n}}{2}\|\bar{\mathbf{b}}^n\| \tag{4-50}$$

for all $\mathbf{x} \in \tau(F)$.

Consider now sub-lattices of the form

$$L_k \triangleq \left\{k\bar{\mathbf{b}}^n + \sum_{i=1}^{n-1} z_i\bar{\mathbf{b}}^i\colon z_i \text{ integer } i = 1, 2, \ldots, n-1\right\}$$

with k a fixed integer. If d_{L_0} denotes the basis determinant of the lattice when $k = 0$, we can apply Lemmas 4.21 and 4.20 in turn to establish

$$\prod_{i=1}^{n}\|\bar{\mathbf{b}}^i\| \leq c_1[n]d_L = c_1[n]\left(\frac{d_L}{d_{L_0}}\right)d_{L_0} \leq c_1[n]\left(\frac{d_L}{d_{L_0}}\right)\prod_{i=1}^{n-1}\|\bar{\mathbf{b}}^i\|$$

Dividing by $\Pi_{i=1}^{n-1}\|\bar{\mathbf{b}}^i\|$ gives

$$\|\bar{\mathbf{b}}^i\| \leq c_1[n]\left(\frac{d_L}{d_{L_0}}\right) \tag{4-51}$$

Combining (4-51) and (4-50), we see that every element of $\mathbb{B}(\mathbf{p}, c_2[n]r) \supseteq \tau(F)$ lies within $c_2[n](\sqrt{n}/2)c_1[n](d_L/d_{L_0})$ of $\mathbf{p}$. It can be shown that (d_L/d_{L_0}) is the distance between parallel hyperplanes containing the sub-lattices L_k. Thus, $\mathbf{x} \in \tau(F)$ that also belongs to L will lie on one of the $c_2[n](\sqrt{n})(c_1[n])$ sub-lattices L_k of such hyperplanes that intersect $\mathbb{B}(\mathbf{p}, c_2[n]r)$. This completes the sketch of the proof.

EXERCISES

4-1. Prove Theorem 4.1, i.e., establish that the Klee-Minty examples do make the Simplex an exponential algorithm.

4-2. Consider applying the primal simplex to the linear program associated with a minimum cost flow problem on a directed graph $G(V, A)$ having unit costs c_{ij} on arcs $(i, j) \in A$ and net supply b_i at vertex i (see Appendix B). In particular consider the (Zadeh (1973)) family of problems (N_n) with $V_1 \triangleq \{s, t, 1, \bar{1}\}$,

$$V_k \triangleq V_{k-1} \cup \{k, \bar{k}\}, k \geq 2. A_1 \triangleq \{(1, s), (1, \bar{1}), (t, \bar{1}), (s, t)\}$$

$$A_k \triangleq A_{k-1} \cup \{(k, s), (t, \bar{k})\} \cup \{(k, \bar{i}), (i, \bar{k}): 1 \leq i \leq k-1\} \; k \geq 2$$

$$b_1 \triangleq 1, b_2 \triangleq 3, b_k \triangleq 2^{k-1} + 2^{k-3} \text{ for } k \geq 3$$

$$b_{\bar{1}} \triangleq -2, b_{\bar{2}} \triangleq -2, b_{\bar{k}} \triangleq -2^{k-1} - 2^{k-3} \text{ for } k \geq 3$$

$$b_s \triangleq b_k \triangleq 0$$

$$c_{st} \triangleq \infty, c_{is} \triangleq c_{t\bar{i}} \triangleq 0 \text{ for all } i, c_{i\bar{i}} \triangleq 0$$

$$c_{i\bar{j}} \triangleq c_{j\bar{i}} \triangleq 2^{i-1} - 1, i \geq 2, j < i$$

and the starting basic feasible solution $\bar{x}_{is} \triangleq b_i$ for all i, $\bar{x}_{t\bar{i}} \triangleq -b_{\bar{i}}$ for all $\bar{i}$, $\bar{x}_{st} \triangleq \Sigma_{i=1}^{n} b_i$, $\bar{x}_{1\bar{1}} = 0$, $\bar{x}_{i\bar{j}} \triangleq \bar{x}_{j\bar{i}} \triangleq 0$ for all $i \geq 2$ and $j < i$. Show that application of the primal simplex (with the most negative c_{ij} entering) to (N_n) requires at least $2^n + 2^{n-2} - 2$ pivots when started from $\bar{\mathbf{x}}$, i.e., that the simplex can be exponential even if restricted to minimum cost network flow problems.

4-3. Consider the linear program

$$\begin{array}{ll} \min & 6x_1 + 3x_2 + 5x_3 \\ \text{s.t.} & 2x_1 + 1x_2 + 4x_3 \geq 9 \end{array}$$

$$0 \leq x_1 \leq 2, \qquad 0 \leq x_2 \leq 4, \qquad 0 \leq x_3 \leq 1$$

(a) Place the problem in format (4-9) suitable for solution by Khachian's Ellipsoid Algorithm.
(b) Execute two iterations of Khachian Algorithm 4A on the input of (a). Start with an ellipsoid centered at zero.
(c) Place the above problem in the form (4-12)–(4-15) assumed in Karmarkar's algorithm

4-4. Continue the Karmarkar Algorithm 4B calculations of Example 4.3 for two additional iterations. For each iteration

(a) Draw figures showing progress in both original (i.e., **x**) space and transformed (i.e., **y**) space.
(b) Write explicitly the transformation θ and its inverse employed at each iteration, and verify Lemma 4.6 holds for each step.
(c) Compute potential functions (4-30)–(4-31), and verify that Lemma 4.7 is satisfied.

4-5. Prove Lemma 4.4, i.e., establish that numerators and denominators of (LP) optima are bounded by 2^L.

4-6. Derive Karmarkar projection calculation (4-19) and (4-20) by showing that the closest point in $\{\mathbf{y}: \mathbf{Ky} = \mathbf{0}, \mathbf{1y} = \mathbf{1}\}$ to a given vector $\mathbf{h}$ is the $\vec{\mathbf{h}}$ of (4-19).

4-7. Show that the intersection of $\{\mathbf{x} \in \mathbb{R}^n: \mathbf{A}\mathbf{x} = \mathbf{0}\}$ with the n-sphere is a sphere of lower dimension. Assume the center of the sphere belongs to $\{\mathbf{x} \in \mathbb{R}^n: \mathbf{A}\mathbf{x} = \mathbf{0}\}$.

4-8. The Assignment Problem of Exercise **1-2(c)** can be formulated

$$\begin{array}{lll} \min & \sum_i \sum_j w_{ij} x_{ij} & \\ \text{s.t.} & \sum_i x_{ij} = 1 & \text{for every } j \\ & \sum_i x_{ij} = 1 & \text{for every } i \\ & x_{ij} \geq 0 & \text{for every } i \text{ and } j \\ & x_{ij} \text{ integer} & \text{for every } i \text{ and } j \end{array}$$

(a) Show that the linear programming constraints of this problem (those other than "x_{ij} integer") are totally unimodular.

(b) Apply (a) to conclude that the "x_{ij} integer" constraint in the above formulation is redundant (unneeded).

4-9. The constraint matrix $\mathbf{A}$ of an equality-constrained linear program (4-37) is said to have the *consecutive-ones* property if $\mathbf{A}$ is binary and the rows of $\mathbf{A}$ can be sequenced so that all 1's in each column lie in consecutive rows.

(a) Show that such consecutive-ones matrices are totally unimodular. (Hint: transform to a vertex-arc incidence matrix and apply Theorem 4.11).

(b) Give an example of a nonunimodular matrix with the corresponding *circular ones* property that 1's in a column are consecutive under the circular notion the first row follows the last.

4-10. Establish that the R_{10} matrix of Figure 4.7 is totally unimodular.

4-11. The two-commodity flow problem on a directed graph $G(V, A)$ has the form

$$\begin{array}{ll} \min & \mathbf{c}^1\mathbf{x}^1 + \mathbf{c}^2\mathbf{x}^2 \\ \text{s.t.} & \mathbf{E}\mathbf{x}^1 = \mathbf{b}^1 \\ & \mathbf{E}\mathbf{x}^2 = \mathbf{b}^2 \\ & \mathbf{x}^1 + \mathbf{x}^2 \leq \mathbf{u} \\ & \mathbf{x}^1, \mathbf{x}^2 \geq \mathbf{0}, \end{array}$$

where $\mathbf{E}$ is the vertex-arc incidence matrix of G, $\mathbf{u}$ is a vector of positive joint capacities, $\mathbf{b}^1$, $\mathbf{b}^2$ are requirements vectors with component sum zero, and $\mathbf{c}^1$, $\mathbf{c}^2$ are cost vectors.

(a) Give an example showing that the constraints of this problem are not always unimodular.

*(b) Show that the constraints are unimodular when the undirected graph underlying G contains no subgraph homeomorphic to K_4, the complete graph on 4 vertices (Truemper and Soun (1979)).

4-12. Refer to the *1-separations* and *2-separations* on graphs introduced in Section 4.2.2.

(a) Show that the following graph can be 1- and 2-separated into components that are either cycles or parallel edges connecting a pair of vertices.

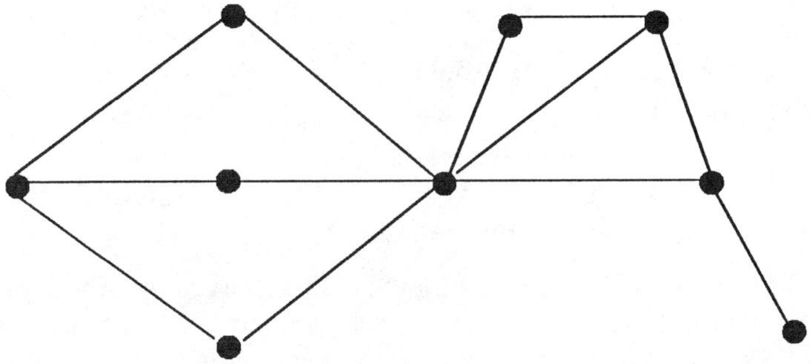

*(b) Show that such a decomposition is always possible if the given graph contains no subgraph homeomorphic to K_4, the complete graph on 4 vertices.

4-13. One form of the minimum cost flow problem on a directed graph $G(V, A)$ is

$$\begin{aligned} \min \quad & \mathbf{cx} \\ \text{s.t.} \quad & \mathbf{Ex} = \mathbf{0} \\ & \mathbf{u} \geq \mathbf{x} \geq \mathbf{l}, \end{aligned}$$

where $\mathbf{E}$ is the vertex-arc incidence matrix of G, and $\mathbf{l}$ and $\mathbf{u}$ are given arc lower and upper bounds. Establish that the constraint systems for such linear programs are totally unimodular.

4-14. An equivalent form of the problem of **4-13** is

$$\begin{aligned} \min \quad & (\mathbf{cP})\mathbf{w} \\ \text{s.t.} \quad & \mathbf{l} \leq \mathbf{Pw} \leq \mathbf{u}, \\ & \mathbf{w} \geq \mathbf{0} \end{aligned}$$

where $\mathbf{P}$ is the incidence matrix of (directed) circuits of G (i.e., columns corresponding to circuits, rows to arcs).

(a) Give an example of a graph for which the **P** matrix is not totally unimodular.
(b) State the linear programming dual of the above path form of the min cost flow problem.
(c) Use **4-13** to show that the dual of (b) is totally dual integral.

★4-15. Edmonds (1965c) famous result is that the *matching problem* on a graph $G(V, E)$ (see Section 1.1) can be solved by solving the linear program

$$\begin{aligned} \max \quad & \sum_{i}\sum_{j>i} w_{ij}x_{ij} \\ \text{s.t.} \quad & \sum_{i<k} x_{ik} + \sum_{j>k} x_{kj} \le 1 && \text{for all } k \in V \\ & \sum_{i\in S}\sum_{\substack{j\in S\\ j>i}} x_{ij} \le \left\lceil \frac{|S|-1}{2} \right\rceil && \text{for all } S \subset V \\ & x_{ij} \ge 0 && \text{for all } (i, j) \in E. \end{aligned}$$

Show that this constraint system is totally dual integral.

4-16. Section 3.4.2 introduced the polytope of a matroid $M \triangleq (G, \mathscr{I})$ as (3-7) $\{\mathbf{x} \ge \mathbf{0}\text{: } \mathbf{A}\mathbf{x} \le \mathbf{r}\}$, where **A** is the incidence matrix of the closed sets of M and **r** is the vector of closed set ranks.

(a) Establish that such (**A**, **r**) pairs are totally dual integral.
(b) Devise a polynomial time separation algorithm for these constraints, i.e., an algorithm that either exhibits an inequality violated by a given vector $\bar{\mathbf{x}}$ or proves none exists. (Hint: use Greedy Algorithm 3A).

4-17. Section 3.5.3 presents the two matroid intersection polytope (3-17)–(3-19) generalizing the case of **4-16.**

(a) Establish that $\left(\begin{array}{c|c} \mathbf{A} & \mathbf{r} \\ \mathbf{B} & \mathbf{s} \end{array}\right)$ systems are totally dual integral.
(b) Devise a polynomial time separation algorithm for these constraints, i.e., an algorithm that either exhibits an inequality violated by a given vector $\bar{\mathbf{x}}$ or proves none exists. (Hint: use Greedy Algorithm 3A).

4-18. The *NP*-Hard *Steiner tree problem* on a given graph $G(V, E)$ is to find a minimum total weight tree connecting vertices of a designated vertex set $S \subset V$. If decision variables $x_e = 1$ if edge $e \in E$ is part of the tree and $= 0$ otherwise, an obviously valid family of inequalities is

$\sum_{e\in D} x_e \ge 1$ for every cut D indexing a set of edges dividing the graph into at least two components, no one of which contains all vertices of S.

Devise a polynomial-time separation algorithm for this family of constraints, i.e., an algorithm that either exhibits an inequality violated by a given $|E|-$ vector $\bar{\mathbf{x}} \geq 0$ or proves none exists. (Hint: use the fact that a minimum capacity cut set separating any two specified vertices can be computed in polynomial time.)

4-19. Consider the famous *NP*-Hard *Traveling Salesman Problem* (see Section 1.1) on a complete graph over vertices V. Let variable $x_{ij} = 1$ if edge (i, j) is selected for the salesman's tour and $= 0$ otherwise, and consider a given vector $\bar{\mathbf{x}}$ of length $\frac{1}{2}|V|(|V|-1)$. For each of the following families of valid inequalities or equalities for the problem, devise a polynomial time separation algorithm to exhibit a member violated by $\bar{\mathbf{x}}$ or prove none exists.

(a) $x_{ij} \geq 0$ for all $i, j \in V, j > i$

(b) $\sum_{j>i} x_{ij} + \sum_{j<i} x_{ji} = 2$ for all $i \in V$ (assume $\bar{\mathbf{x}}$ failed to violate any constraint of (a)).

(c) $\sum_{i \in S} \sum_{\substack{j \in S \\ j>i}} x_{ij} \leq |S| - 1$ for all $S \subset V, |S| \geq 3$. (Assume $\bar{\mathbf{x}}$ failed to violate any constraint of (a) or (b)). (Hint: use the fact that the minimum capacity cut separating any pair of vertices in a graph can be determined in polynomial time.)

4-20. The *NP*-Hard *minimum 3-cut* problem (see Dahlhaus, Johnson, Papadimitriou, Seymour and Yannakakis (1984), and Cunningham (1987)) is to find a minimum total weight subset of edges $\bar{E}$ in a given graph $G(V, E)$ such that given vertices in $T \subset V, |T| = 3$, cannot reach one another when edges of $\bar{E}$ are removed from the graph. Let decision variable $x_e = 1$ if edge e belongs to $\bar{E}$ and $= 0$ otherwise, and consider an arbitrary $|E|$-vector $\bar{\mathbf{x}} \geq \mathbf{0}$. For each of the following families of valid inequalities, devise a polynomial time separation algorithm, i.e., produce an algorithm that exhibits an inequality violated by $\bar{\mathbf{x}} \geq \mathbf{0}$ or proves none exists. (Hint: use the fact that a shortest path between any two vertices can be found in polynomial time).

(a) $\sum_{e \in P} x_e \geq 1$ for every $P \subset E$ indexing edges of a path between two vertices in T.

(b) $\sum_{t \in T} \sum_{e \in P_t} x_e \geq 2$ for every $v \notin T$ and $P_t \subset E$ indexing edges of a path from $t \in T$ to v.

4-21. Consider the set of incidence vectors of edges in a given graph $G(V, E)$ such that removal of those edges divides the graph into two components. If decision variable $x_e = 1$ when edge $e \in E$ belongs to such a *cut set* and $= 0$ otherwise, one family of valid inequalities is given by

$$\sum_{e \in \langle D \rangle} x_e - \sum_{e \in \langle C \setminus D \rangle} x_e \leq |D| - 1$$

for every $C \subseteq E$ indexing a cycle of G and every $D \subseteq C$, $|D|$ odd. Here $\langle S \rangle$ denotes the subgraph induced by the vertex set S. Consider an arbitrary $|E|$-vector $\bar{\mathbf{x}} \geq \mathbf{0}$. Devise a polynomial time separation algorithm that will either exhibit a member of this set of constraints violated by $\bar{\mathbf{x}}$ or prove none exists. (Hint: create an auxiliary graph with two copies of each vertex and use the fact that the shortest path between all pairs of vertices in a graph can be found in polynomial time.)

4-22. Given a graph $G(V, E)$ it can be shown that the polytope of incidence vectors of perfect (exact) matchings must satisfy

(i) $x_e \geq 0$ for every $e \in E$,

(ii) $\sum_{e \text{ meeting } v} x_e = 1$ for every $v \in V$

★(iii) $\sum_{e \in D} x_e \geq 1$ for every $D \subset E$, that indexes a cut set dividing the graph into components with oddly many vertices.

Devise a polynomial time separation algorithm for these families of constraints, i.e., an algorithm that either exhibits a constraint violated by a specified $|E|$-vector $\bar{\mathbf{x}}$ or proves none exists. (Padberg and Grötshel, Chapter 9 in Lawler, Lenstra, Rinnooy Kan and Shmoys (1985)).

4-23. An integer matrix $\mathbf{L}$ is said to be *Leontief* if its rows and columns can be arranged so that (i) the first nonzero entry in each column is $+1$, (ii) there exists at least one column with its $+1$ in the last row, (iii) any nonzero entries in columns other than the $+1$ are negative and in rows below the $+1$. For the system $\{\mathbf{x} \geq \mathbf{0},\ \mathbf{L}\mathbf{x} = \mathbf{b}\}$ where $\mathbf{L}$ is restricted Leontief

(a) Give an example where $\mathbf{L}$ is not unimodular.
(b) Develop a polynomial algorithm for constructing optimal primal and dual linear programs posed over such systems with $\mathbf{b}$ a feasible right-hand-side.
(c) Use the results of (b) to show that such systems are totally dual integral for all $\mathbf{b}$ for which they are feasible.

4-24. Prove Lemma 4.20, i.e., show that the determinant of an integer lattice L is independent of the set of generators chosen to represent the lattice.

4-25. Consider the bounded linear integer program

$$\begin{aligned} \min \quad & \mathbf{cx} \\ \text{s.t.} \quad & \mathbf{Ax} \geq \mathbf{b} \\ & \mathbf{u} \geq \mathbf{x} \geq \mathbf{0} \\ & \mathbf{x} \text{ integer,} \end{aligned}$$

where **c**, **A**, **b** and **u** are integer and **x** is an n-vector. Show that for any fixed n this problem can be solved in time polynomial in the length of its data. (Hint: represent each x_j as a sum $\Sigma_{l=0}^{\lfloor \log_2 u_j \rfloor} 2^l y_j^l$ and enumerate.)

5

Nonpolynomial Algorithms—Partial Enumeration

A discrete optimization problem expressed in the form

$$P(S) \qquad \begin{aligned} \min \quad & \mathbf{cx} \\ \text{s.t.} \quad & \mathbf{Ax} \geq \mathbf{b} \\ & \mathbf{x} \in S = \{\mathbf{x} \geq \mathbf{0}\colon x_j = 0 \text{ or } 1 \text{ for } j \in I\}, \end{aligned}$$

quite clearly has only a finite number of solutions that need to be considered in identifying an optimum. Any choice of 0 or 1 values for the x_j with $j \in I$ leaves a linear program that can be solved to find the best completion of the solution vector (if any exists); there are only $2^{|I|}$ 0 and 1 combinations to check.

One could think of solving $P(S)$ by *total enumeration,* i.e., by trying all those finitely many possibilities. It is easy, however, to see that the computational effort involved in total enumeration would soon become prohibitive. A problem with even a modest two hundred 0–1 variables would have 2^{200} or about 10^{60} cases to consider; one with 201 variables would have twice as many.

We have seen in Chapters 3 and 4 how some discrete optimization problems are amenable to solution by polynomially-bounded algorithms that proceed to optimal solutions through a direct, constructive sequence of improving solutions. We have also seen that many, in fact most, discrete optimization problems (actually their threshold versions) fall in the class *NP*-Complete—a class for which such algorithms are not likely to be found.

What alternatives then are available (short of total enumeration) when direct, polynomially-bounded algorithms are impossible? If the problem sizes we deal with are not too large, there are many. In this and the following chapter, we shall develop the theory of the two main classes of strategies: *partial enumeration* and *polyhedral description.* Both yield algorithms that are in the worst case nonpolynomial. Thus, in a pure complexity theoretic sense, neither is an improvement on total enumeration. Still, both may provide effective practical procedures for commonly observed instances of important discrete optimization problems, and recent work by Crowder, Johnson, and Padberg (1983), Crowder and Padberg (1980), and others has demonstrated the value of combining the techniques.

5.1 Fundamentals of Partial Enumeration

Although total enumeration is a hopeless strategy for most discrete optimization problems, it does not follow that the notion of enumeration should be completely rejected; fixing variables does leave us with a smaller, and thus easier problem.

Suppose we pick some $p \in I$ of $P(S)$ and *partially enumerate* the set S by dividing it into two subsets

$$S_1 = \{\mathbf{x} \in S\colon x_p = 0\}$$

and

$$S_2 = \{\mathbf{x} \in S\colon x_p = 1\}.$$

If we could establish that one of these classes of solutions contain no specific solution that needs to be considered in our search for an optimum, then the whole class could be discarded without explicit enumeration of its membership.

Such consideration of solutions in classes is the core of partial enumeration algorithms. At any stage of such an algorithm, the set S is divided into subset, S_0, of already considered solutions and a list of other subsets S_1, $S_2, \ldots, S_q$ that collectively exhaust S, i.e.,

$$S = S_0 \cup S_1 \cup S_2 \cup \ldots \cup S_q.$$

Each subset S_k characterizes a *candidate problem* (denoted $P(S_k)$) derived from $P(S)$ by restricting the feasible space to $S_k \cap \{\mathbf{x}\colon \mathbf{A}\mathbf{x} \geq \mathbf{b}\}$. The algorithm selects some candidate problem to examine and evaluates whether any solution to that candidate need be considered further in identifying an optimal solution to $P(S)$. If not, $P(S_k)$ is *fathomed,* i.e., it is removed from the *candidate list* and $S_0 \leftarrow S_0 \cup S_k$. Unhappily, there remains the possibility that this evaluation fails: we cannot prove that the current candidate can be fathomed and in such a case, S_k must be further enumerated. $P(S_k)$ is replaced in the candidate list by one or more new candidate problems, each of which is derived by adding further restrictions to S_k.

5.1.1 Branch and Bound

How does one go about proving that no solution to a candidate problem can be optimal in the overall problem? One possibility is to check *feasibility.* If it can be shown that a candidate problem $P(S_k)$ has no feasible solution, it certainly has no solution that is optimal for $P(S)$. A second idea is to evaluate the solutions to $P(S_k)$ relative to any known feasible solutions for $P(S)$. Some feasible solutions may be known before we begin our computations; others will be generated during processing, e.g., when we happen on to a $P(S_k)$ that is easy enough to solve explicitly. Define the *incumbent solution* to be the best (in terms of the objective function) feasible solution to $P(S)$ uncovered so far. Also, let v^* denote its objective function value. Then, a candidate problem $P(S_k)$ can be fathomed if it can be shown that no solution to $P(S_k)$ can have a lower value of the objective function than v^*. Since $P(S_k)$ is a discrete optimization problem, it will usually be quite difficult to know exactly how good a solution to $P(S_k)$ actually exists. Instead, we attempt to prove $P(S_k)$ has no solution superior to the incumbent by showing that some relatively easily computed lower bound $\beta(S_k)$ on the optimal solution value of $P(S_k)$ (lower bound because $P(S_k)$ is a minimization problem) is at least as great as the incumbent solution value, v^*.

Branch-and-bound algorithms for discrete optimization problems can be defined as those partial enumeration procedures employing tests of feasibility and comparison to an incumbent solution to fathom candidate problems. Recall that the problem value function $v(\cdot)$ is defined

$$v(\cdot) \triangleq \begin{cases} \text{the value of an optimal solution to the} \\ \text{problem } (\cdot)\text{, if one exists,} \\[1ex] +\infty \text{ if } (\cdot) \text{ is a minimizing and infeasible} \\ \text{problem or a maximizing and unbounded one,} \\[1ex] -\infty \text{ if } (\cdot) \text{ is a minimizing and unbounded} \\ \text{problem or a maximizing and infeasible one.} \end{cases}$$

If $v^* \leftarrow \infty$ until a feasible solution is obtained, then we can simply say we fathom candidate problems whenever we have that $v(P(S_k)) \geq v^*$.

The partial enumeration approach is conveniently represented with a *tree* (from which branch and bound derives its name). Nodes of the tree represent candidate problems and branches show new restrictions added when a candidate cannot be fathomed. The notions developed thus far, as well as this tree representation, are illustrated by the following example:

Example 5.1. Elementary Branch and Bound on a Knapsack Problem. Consider the following 0–1 knapsack problem:

$$
\begin{array}{lll}
 & \min & 4x_1 + 5x_2 + 16x_3 \\
KNP(S) & \text{s.t.} & 20x_1 + x_2 + 2x_3 \geq 2 \\
 & & (x_1, x_2, x_3) \in S = \{\text{binary } (x_1, x_2, x_3)\}
\end{array}
$$

We require a scheme for defining candidate problems $KNP(S_k)$ and a procedure for computing a bound $\beta(S_k)$ on $v(KNP(S_k))$. Problems $KNP(S_k)$ may be formed from $KNP(S)$ by fixing some x_j at 1, other x_j at 0, and leaving the rest free. Let κ_k be the built in sum of cost coefficients for variables fixed at 1 in the candidate problem $KNP(S_k)$. Similarly, let γ_k be the sum of constraint coefficients for variables fixed at 1 in $KNP(S_k)$. Then, a bound on $v(KNP(S_k))$ is given by

$$
\beta(S_k) \triangleq \begin{cases} +\infty \text{ if } 2 > \gamma_k + (\text{the sum of constraint coefficients on unfixed variables}) \\ \kappa_k \quad \text{otherwise.} \end{cases}
$$

That is, $KNP(S_k)$ is proved infeasible if $2 >$ maximum possible constraint sum, and otherwise, we know only that $v(KNP(S_k)) \geq \kappa_k$. Furthermore, whenever $\gamma_k \geq 2$, a feasible solution can be derived by setting all unfixed variables at 0.

Processing of the example begins by placing the entire problem in the candidate list. Since that problem is now the only member of the candidate list, it becomes the first $KNP(S_1)$ chosen. $KNP(S_1)$ is represented by node 1 in the tree of Figure 5.1. For that candidate, $\gamma_1 = 0 < 2$. This makes the problem too difficult to solve directly, so processing moves to the bound calculation. Since $\gamma_1 + 20 + 1 + 2 \geq 2$, $\beta(KNP(S_1)) = \kappa_1 = 0$. Comparison to the incumbent solution value $v^* = +\infty$ shows we must further partition. This can be accomplished by fixing the value of some previously free variable. All three variables are free in S_1 and in the absence of any better rule, we shall simply choose x_1. The candidate problems of nodes 2 and 3 in Figure 5.1 result. Again, there is an option, which of the two problems in

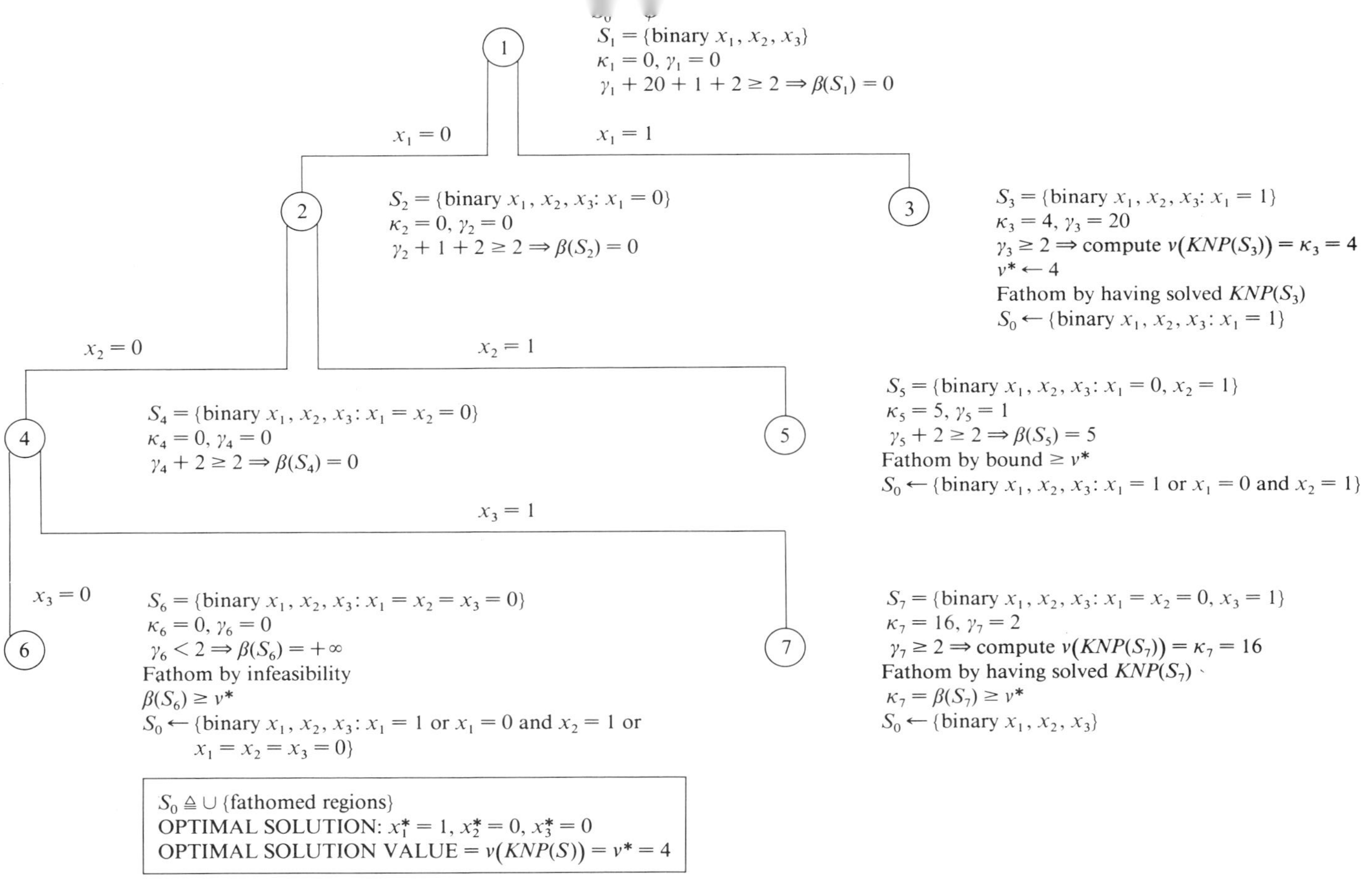

Fig. 5.1. Branch and bound tree for knapsack Example 5.1.

the candidate list should be explored next? Suppose we pick $KNP(S_2)$. In this case, efforts to solve the problem or to bound it out of contention also fail. The candidate problems of nodes 4 and 5 in Figure 5.1 are the consequence. Here, x_2 was chosen and we partition S_2 into subsets S_4 and S_5. Proceeding next to the (arbitrarily selected) candidate problem $P(S_3)$, we are more successful. Since $\gamma_3 = 20 \geq 2$, $v(KNP(S_3)) = \kappa_3 = 4$. This feasible solution to $KNP(S)$ both provides an incumbent solution $\mathbf{x}^* = (1, 0, 0)$ with $v^* = 4$ and allows us to fathom the entire set S_3.

The remainder of the solution process is similar. $P(S_4)$ can neither be solved nor excluded on the basis of bounds, and so it is partitioned into $P(S_6)$ and $P(S_7)$. Although the partial solution of candidate $P(S_5)$ is not yet a feasible solution, its bound of 5 already exceeds the incumbent solution value, 4. Thus, $P(S_5)$ is fathomed. Consideration of $P(S_6)$ yields a bound of $+\infty$ because the problem is infeasible; $P(S_7)$ can be solved completely, but the resulting feasible solution is not superior to our present incumbent. As a consequence, both $P(S_6)$ and $P(S_7)$ are fathomed in turn. The algorithm stops with the now optimal incumbent solution $x_1^* = 1$, $x_2^* = 0$, $x_3^* = 0$. ■

With the general notions in mind, we are now ready to formalize a branch and bound algorithm (Figure 5.2 provides a flow chart):

Algorithm 5A. Fundamental Branch and Bound for P(S)

STEP 0: *Initialization.* Establish the candidate list by making the full problem $P(S)$ its sole entry and set $v^* \leftarrow +\infty$.

STEP 1: *Candidate Selection.* Choose some member $P(S_k)$ from the current candidate list.

STEP 2: *Bounding.* Compute one or more lower bounds $\beta(S_k)$ on $v(P(S_k))$. If any $\beta(S_k) \geq v^*$, go to Step 6 and fathom. Otherwise, go to Step 3.

STEP 3: *Feasible Solutions.* If convenient, generate one or more feasible solutions $\tilde{\mathbf{x}}$ to $P(S_k)$. If all such $\tilde{\mathbf{x}}$ have $\mathbf{c}\tilde{\mathbf{x}} \geq v^*$ or if none are convenient to generate, go to Step 5. Otherwise, go to Step 4.

STEP 4: *Incumbent Saving.* Save the solution $\tilde{\mathbf{x}}$ from Step 3 with minimum $\mathbf{c}\tilde{\mathbf{x}}$ by setting $v^* \leftarrow \mathbf{c}\tilde{\mathbf{x}}$, $\mathbf{x}^* \leftarrow \tilde{\mathbf{x}}$. If v^* is now $-\infty$, stop; $P(S)$ is unbounded. If v^* is finite, and $\tilde{\mathbf{x}}$ can be proved to solve $P(S_k)$ (i.e., min $\mathbf{c}\tilde{\mathbf{x}} = \max \beta(S_k)$), go to Step 6 and fathom. Otherwise, go to Step 5.

STEP 5: *Branching.* Replace $P(S_k)$ in the candidate list by one or more further restricted candidate problems $P(S_{k1}), P(S_{k2}), \ldots, P(S_{kp})$. Then, return to Step 1.

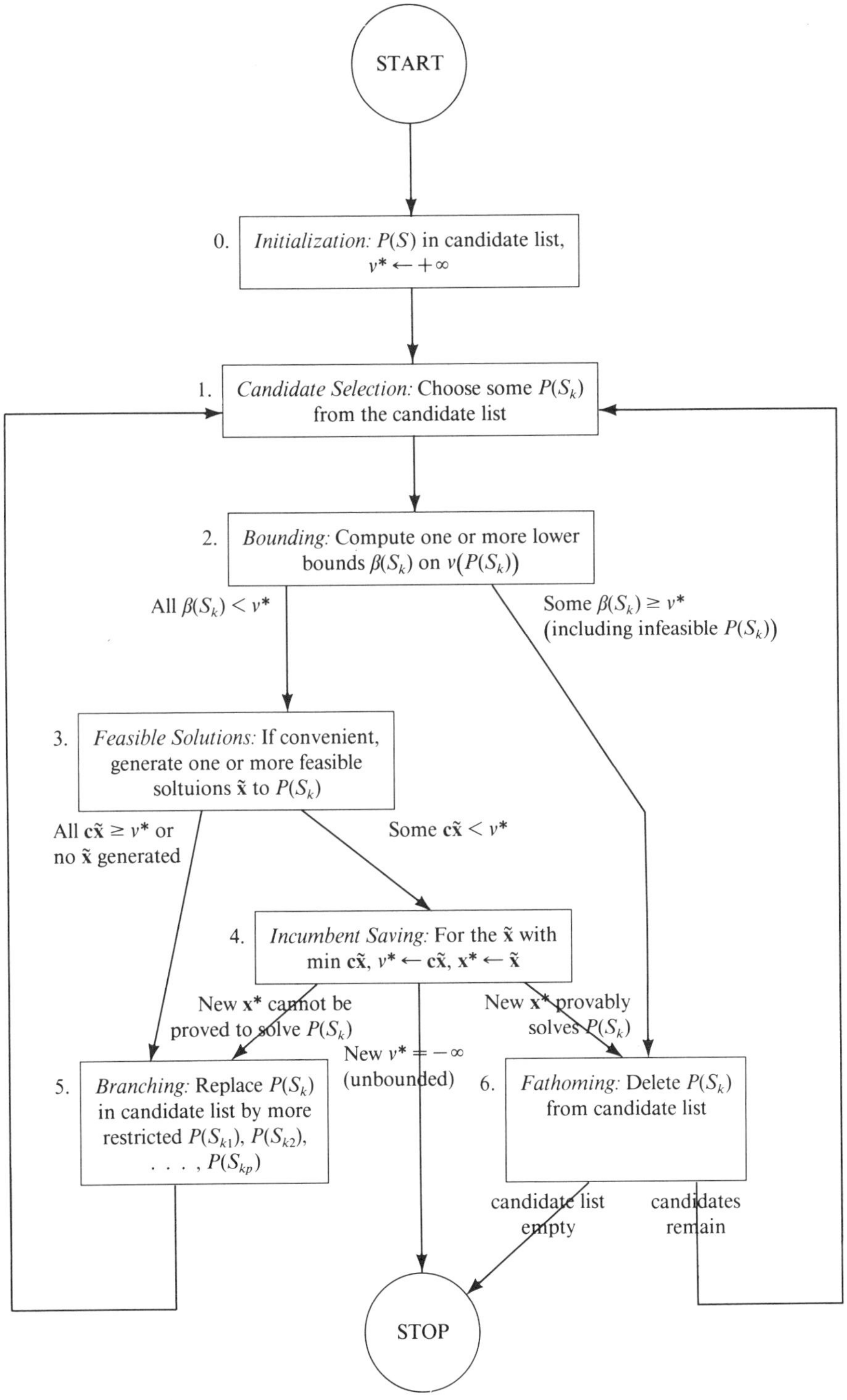

Fig. 5.2. Fundamental branch and bound algorithm 5A.

STEP 6: *Fathoming.* Fathom $P(S_k)$, i.e., delete $P(S_k)$ from the candidate list. If the candidate list is now empty, stop; the incumbent solution, $\mathbf{x}^*$, is optimal unless none exists ($v^* = +\infty$) in which case $P(S)$ is infeasible. If candidates remain, return to Step 1. ■

Theorem 5.1. Convergence of Branch and Bound. *Suppose Algorithm 5A is operated so that*

(i) there are only finitely many candidate problems $P(S_k)$ that can be created by any sequence of executions of Step 5,
(ii) Step 5 never creates a candidate problem that has previously belonged to the candidate list,
(iii) at the completion of any execution of Step 5,

$$S = \left(\bigcup_{\left\{ \substack{k \text{ in candidate} \\ \text{list}} \right\}} S_k \right) \cup \left(\bigcup_{\left\{ \substack{l \text{ fathomed} \\ \text{at Step 6}} \right\}} S_l \right)$$

Then the algorithm stops after finitely many executions of its steps with $v^* = v(P(S))$.

Proof. Unless Algorithm 5A stops at Step 4, it considers one candidate problem after another until it stops at Step 6. Consideration of each problem begins at Step 1, moves through Step 2, Step 3, Step 4, and ends with deletion of the candidate at Step 5 or Step 6. By (ii), candidate problems are not recreated and by (i), there are only finitely many to consider. Thus, Algorithm 5A clearly stops after executing finitely many of its steps.

It remains to show that it stops with the correct answer, i.e., with $v^* = v(P(S))$. If termination occurs at Step 4, $v^* = -\infty$ because we have encountered a candidate problem $P(S_u)$ with $v(P(S_u)) = -\infty$. Since the definition of a candidate problem implies

$$\{\mathbf{x} \text{ feasible for } P(S)\} \supseteq \{\mathbf{x} \text{ feasible for } P(S_u)\},$$

$$-\infty \le v(P(S)) \le v(P(S_u)) = -\infty.$$

Now, suppose we stop at Step 6. When such a stop occurs, the candidate list is empty. But the list begins nonempty at Step 0 and (iii) assures it remains so until $S = S_0$, the union of all fathomed regions. If we denote by v_l^* the incumbent solution value at the time candidate $P(S_l)$ was fathomed, then validity of bounds used at Step 2 and incumbent selection rules of Step 4 together give

$$v(P(S_l)) \ge \beta(S_l) \ge v_l^* \ge v^*$$

Thus $v(P(S_0)) \geq v^*$, and since v^* is the value of a feasible solution to $P(S)$, we will stop only when

$$v^* \geq v(P(S)) = v(P(S_0)) \geq v^*. \qquad \blacksquare$$

5.1.2 *Candidate Problems from Partial Solutions*

Algorithm 5A is a very general procedure, stated so that it encompasses both elementary and sophisticated branch and bound schemes. Still, Theorem 5.1 shows that very little is required to make even the general procedure finite. We must only show that Step 5 (i) can produce only finitely many candidate problems, (ii) never duplicates candidates, and (iii) always maintains an exhaustive classification of S.

By far, the most common scheme for satisfying these requirements is to subdivide the set S of $P(S)$ by fixing values of x_j for j in some index set as a *partial solution.* Such a scheme subdivides a candidate problem at Step 5 by fixing the value of an integer variable, x_p, not assigned a specific value in the current S_k. If we reach a point where no such variable exists, i.e., S_k is fully enumerated, we merely fathom after solving the implied linear program at Step 3 and processing at Step 4.

Example 5.1 illustrated exactly such a branching scheme. Corollary 5.2 shows that such simple branching notions are enough to fulfill the requirements of Theorem 5.1.

Corollary 5.2. Convergence of Partial Solution Branch and Bound. *Let $P(S)$ be above with I the set of variable subscripts subject to 0–1 constraints. Also, suppose candidate problems $P(S_k)$ of Algorithm 5A are specified by partial solution, i.e., by fixing the values of some x_j with $j \in J_k \subseteq I$. More specifically, assume that Step 3 of Algorithm 5A solves any linear $P(S_k)$ (i.e., candidates with $J_k = I$) and that Step 5 creates candidate problems by fixing the value of some new x_p with $p \in I \backslash J_k$. Then, such an Algorithm 5A stops after finitely many executions of its steps with $v^* = v(P(S))$.*

Proof. We shall apply Theorem 5.1. Note first that the number of partial solutions of length l is at most the number of ways of choosing l variables to fix, multiplied by the number of possible values of the l variables or $[(|I|)!/(l!(|I| - l)!)][2^l]$. Our assumption that Step 3 solves (and fathoms at Step 4) any fully enumerated solutions shows that the length of solutions created by Step 5 is bounded by $|I|$. Thus, at most

$$\sum_{l=0}^{|I|} \frac{(|I|)!}{l!(|I| - l)!} 2^l$$

candidate problems can ever be created; (i) of Theorem 5.1 holds. Now the branching procedure of Step 5 in a partial solution approach divides the current candidate problem set S_k into the two sets, $S_k \cap \{\mathbf{x}\colon x_p = 0\}$ and $S_k \cap \{\mathbf{x}\colon x_p = 1\}$, where x_p is a discrete variable not fixed in S_k. Clearly, this fulfills the exhaustiveness required of (iii) in Theorem 5.1; 1 and 0 are the only feasible values for x_p. For nonrepeat property (ii), observe that all candidate problems have the first, $P(S)$ as a common ancestor. Thus, tracing the evolution in the branch and bound tree for any two candidates will identify a first branching where their paths diverge. For one the partition variable will have $x_p = 1$ and for the other $x_p = 0$. Since these possibilities are mutually exclusive the two candidates must be distinct. This completes the proof. ■

5.1.3 Bounds and Feasible Solutions from Relaxations

To have an Algorithm 5A that can actually be coded for a computer, we must detail more than how branching occurs to create candidate problems at Step 5. It is equally important to examine how bounds are calculated at Step 2 and how feasible solutions are generated at Step 3.

The design of suitable bounding and solution generating procedures is at the heart of most branch and bound research, but there is a common thread. Define a *relaxation* of the minimization problem (P) as any problem $(\tilde{P})$ such that

(i) every feasible solution to (P) is feasible for $(\tilde{P})$,
(ii) the objective function of problem (P), evaluated at any feasible point for (P) is greater than or equal to the objective function for $(\tilde{P})$ evaluated at the same point.

Relaxations provide an obvious source of bounds.

Lemma 5.3. Relaxation Bounds. *Let (P) be any optimization problem (minimize form) and $(\tilde{P})$ be a relaxation of (P). Then $v(\tilde{P}) \le v(P)$.*

Proof. Any $\mathbf{x}$ feasible for (P) is also feasible for $(\tilde{P})$. Thus, if (P) is infeasible ($v(P) = +\infty$) the stated inequality holds and if it is unbounded ($v(P) = -\infty$), so is $v(\tilde{P})$. Also, if $v(P)$ has a finite optimum its objective function value in (P), which is greater than or equal to its objective function value in $(\tilde{P})$, provides an upper bound on $v(\tilde{P})$ because that (P) optimum is feasible in $(\tilde{P})$. ■

We see that we can create bounds for a branch and bound procedure by structuring relaxations of our discrete problem $P(S)$. The value of an optimal solution to any such relaxation provides a bound function for Step

2 of Algorithm 5A. Certainly, any optimal solution $\tilde{\mathbf{x}}$ to a relaxation of $P(S)$ may also provide a feasible solution for Step 3, if it satisfies all constraints of the original problem. If the objective functions of the original and relaxed problems agree, we can make a stronger statement.

Lemma 5.4. Optimal Solutions from Relaxations. *Let $(\tilde{P})$ be a relaxation of the minimization problem (P), and $\tilde{\mathbf{c}}\mathbf{x}$ and $\mathbf{cx}$ the objective functions of the two problems respectively. Then if $\tilde{\mathbf{x}}$ is optimal in $(\tilde{P})$ and both*

(i) $\tilde{\mathbf{x}}$ is feasible in (P)
(ii) $\mathbf{c}\tilde{\mathbf{x}} = \tilde{\mathbf{c}}\tilde{\mathbf{x}}$

$\tilde{\mathbf{x}}$ is optimal in (P).

Proof. An $\tilde{\mathbf{x}}$ satisfying the hypotheses of the lemma is necessarily feasible in (P). This implies $\mathbf{c}\tilde{\mathbf{x}} \geq v(P)$. But then by (ii) and Lemma 5.3,

$$\tilde{\mathbf{c}}\tilde{\mathbf{x}} \triangleq v(\tilde{P}) \leq v(P) \leq \mathbf{c}\tilde{\mathbf{x}} = \tilde{\mathbf{c}}\tilde{\mathbf{x}}$$

It follows that $\tilde{\mathbf{x}}$ is a feasible solution to (P) that achieves $v(P)$, and so $\tilde{\mathbf{x}}$ is optimal in (P). ■

Example 5.2. Relaxations of the Knapsack Problem of Example 5.1. Return to the knapsack problem

$$KNP(S) \qquad \begin{array}{ll} \min & 4x_1 + 5x_2 + 16x_3 \\ \text{s.t.} & 20x_1 + x_2 + 2x_3 \geq 2 \\ & (x_1, x_2, x_3) \in S \\ & = \left\{ x_1, x_2, x_3 : \begin{array}{r} 0 \leq x_j \leq 1 \text{ for } j = 1, 2, 3 \\ x_j \text{ integer for } j = 1, 2, 3 \end{array} \right\} \end{array}$$

The enumeration scheme used in Example 5.1 produced candidate problems $KNP(S_k)$ that were identical to $KNP(S)$ except that one or more x_j were fixed in value. One way to form a relaxation of such a $KNP(S_k)$ is to zero the objective function coefficients of variables still free (unfixed) in S_k. For example $P(S_5)$ in Example 5.1 was

$$KNP(S_5) \qquad \begin{array}{ll} \min & 4x_1 + 5x_2 + 16x_3 \\ \text{s.t.} & 20x_1 + x_2 + 2x_3 \geq 2 \\ & (x_1, x_2, x_3) \in S \\ & x_1 = 0 \\ & x_2 = 1 \end{array}$$

Our *free variable zeroing* relaxation would be

$$
\widetilde{KNP}(S_5) \qquad \begin{array}{ll} \min & 4x_1 + 5x_2 + 0x_3 \\ \text{s.t.} & 20x_1 + x_2 + 2x_3 \geq 2 \\ & (x_1, x_2, x_3) \in S \\ & x_1 = 0 \\ & x_2 = 1 \end{array}
$$

Constraints are the same as in $KNP(S_5)$, but the objective function is now an underestimate. Solution of the relaxation gives $v(\widetilde{KNP}(S_5)) = 5$, equal to the bound $\beta(S_5)$ computed in Example 5.1. This is no coincidence. The function $\beta(S_5)$ used in Example 5.1 is exactly $v(\widetilde{KNP}(S_5))$ where $\widetilde{KNP}(S_k)$ is derived from $KNP(S_k)$ by zeroing cost coefficients on free variables.

There are a host of other relaxations of $KNP(S_k)$'s that an algorithm designer might choose to employ. One possibility is to relax upper bounds, i.e., replace S by

$$
\tilde{S} = \left\{ x_1 x_2 x_3 : \begin{array}{l} x_j \geq 0 \text{ for } j = 1, 2, 3 \\ x_j \text{ integer for } j = 1, 2, 3 \end{array} \right\}.
$$

Another is to relax integrality, i.e., replace S by $\bar{S} = \{x_1, x_2, x_3 : 0 \leq x_j \leq 1$ for $j = 1, 2, 3\}$.

The latter approach gives a linear program relaxation. To illustrate on the candidate problem $KNP(S_5)$ in Example 5.1, the corresponding relaxation $KNP(\bar{S}_5)$ has optimal solution $x_1 = 0$, $x_2 = 1$, $x_3 = \frac{1}{2}$ with $v(KNP(\bar{S}_5)) = 13 \leq v(KNP(S_5)) = 21$. The relaxation has provided the lower bound guaranteed by Lemma 5.3.

To illustrate Lemma 5.4, consider this linear programming relaxation of Example 5.1 candidate problem $KNP(S_4)$ with $S_4 = \{(x_1, x_2, x_3) \in S\colon x_1 = x_2 = 0\}$. The corresponding linear programming relaxation, $KNP(\bar{S}_4)$, has optimal solution that happens also to be feasible for $KNP(S_4)$. Thus, since cost functions agree, we can conclude that $(x_1, x_2, x_3) = (0, 0, 1)$ solves $KNP(S_4)$. Candidate problems $KNP(S_6)$ and $KNP(S_7)$ could have been avoided in Example 5.1. ■

5.1.4 Branch and Bound and Dynamic Programming

The reader may recognize the knapsack problem of Examples 5.1 and 5.2 as a member of the 0–1 knapsack problem class treated by dynamic programming in elementary texts. Since informally, dynamic programming involves trying a number of variable combinations, it is reasonable to ask whether it can properly be considered a partial enumeration strategy.

The answer is definitely *yes*. To see the connection, however, we must introduce a new process for discarding candidate problems. Suppose (as in Example 5.1) κ_k is the constant term in the objective function of the candidate problem $P(S_k)$ that results from the extra constraints distinguishing $P(S_k)$ from $P(S)$. Then, one candidate problem, $P(S_j)$, is said to *dominate* another, $P(S_k)$, if $\kappa_j \leq \kappa_k$ and every feasible solution for $P(S_k)$ is also feasible for $P(S_j)$. In other words, $P(S_j)$ dominates $P(S_k)$ if any committed portion κ_j, of its objective function value is at least as small as that of $P(S_k)$, and its feasible space is at least as large.

It is obvious that dominated candidate problems need not be considered in searching for an optimal solution to a discrete optimization problem. The following corollary states this more precisely.

Corollary 5.5. Convergence of a Partial Enumeration Algorithm Using Dominance. *Suppose Algorithm 5A is operated as specified in Theorem 5.1 except that candidate problems that are known to be dominated by other candidates are never chosen at Step 1. Then, the algorithm stops after finitely many executions of its steps with* $v^* = v(P(S))$.

Having defined the notion of dominance, we can now fit dynamic programming algorithms into the structure of partial enumeration procedures. *Dynamic programming* schemes for discrete optimization problems, $P(S)$, may be viewed as partial enumeration algorithms that rely primarily on dominance to avoid evaluating all possible candidate problems.

Although discrete dynamic programming algorithms are actually performing partial enumeration, the branching rules and candidate problems are usually somewhat hidden in the state/stage structure. Before further discussing their connection with branch and bound, we shall re-solve Example 5.1 by elementary dynamic programming.

Example 5.3. Dynamic Programming Enumeration of Example 5.1. Consider again, the knapsack problem $KNP(S)$ of Example 5.1, and let *states* of the dynamic program be levels of satisfaction, ρ, of the constraint row $20x_1 + x_2 + 2x_3 \geq 2$. Clearly, the only possible states are $\rho = 0-$, 1, $2+$ (the first denoting 0 or less and the last denoting 2 or more). We define value function $\omega^*(\cdot)$ as

$$\omega_t^*(\rho) = v \begin{pmatrix} \min & \sum_{j=1}^{t} c_j x_j & \\ \text{s.t.} & \sum_{j=1}^{t} a_j x_j = \rho & \\ & x_j = 0 \text{ or } 1 & \text{for } j = 1, \ldots, t \end{pmatrix}$$

where c_j and a_j are the objective function and constraint coefficients of x_j, respectively. In terms of this notation, we have the simple dynamic programming recursion

$$\omega_0^*(\rho) = \begin{cases} 0 & \text{if } \rho = 0- \\ +\infty & \text{otherwise} \end{cases}$$

$$\omega_t^*(\rho) = \min_{x_t = 0 \text{ or } 1} \{c_t x_t + \omega_{t-1}^*(\rho - a_t x_t)\}.$$

The following tables apply this recursion to solve *KNP(S)*.

STAGE $t = 1$

State ρ	x_t	Partial solution	Node	Objective value	$\omega_t^*(\rho)$
$0-$	0	$x_1 = 0$	2	0	0
	1	$x_1 = 1$	3	4	
1	0	$x_1 = 0$	2	$+\infty$	4
	1	$x_1 = 1$	3	4	
$2+$	0	$x_1 = 0$	2	$+\infty$	4
	1	$x_1 = 1$	3	4	

STAGE $t = 2$

State ρ	x_t	Partial solution	Node	Objective value	$\omega_t^*(\rho)$
$0-$	0	$x_2 = 0, x_1 = 0$	4	0	0
	1	$x_2 = 1, x_1 = 0$	5	5	
1	0	$x_2 = 0, x_1 = 1$	8	4	4
	1	$x_2 = 1, x_1 = 0$	5	5	
$2+$	0	$x_2 = 0, x_1 = 1$	8	4	4
	1	$x_2 = 1, x_1 = 1$	9	9	

STAGE $t = 3$

State ρ	x_t	Partial solution	Node	Objective value	$\omega_t^*(\rho)$
$2+$	0	$x_3 = 0, x_2 = 0, x_1 = 1$	10	4	4
	1	$x_3 = 1, x_2 = 0, x_1 = 0$	7	16	

From the last table, an optimal solution is $(x_1^*, x_2^*, x_3^*) = (1, 0, 0)$ with value 4. ■

For comparative purposes, we have added columns to the tables showing the full partial solution being evaluated when we fix x_t and then use the corresponding optimal setting of $x_1, x_2, \ldots, x_{t-1}$. We have also drawn an enumeration tree of these partial solutions in Figure 5.3 and noted the tree node numbers in the dynamic programming tables. For example, when $t = 2$ and $\rho = 1$, we examined the settings $x_2 = 0$ and $x_2 = 1$ in carrying out the dynamic programming calculations. In the $x_2 = 0$ case, previous results tell us the corresponding optimal choice for x_1 is 1; $\omega^*_{t-1}(\rho - a_t x_t) = \omega^*_1(1)$ was achieved with $x_1 = 1$. In a similar fashion, we can deduce that trying $x_2 = 1$ at this point in the dynamic program is evaluating the partial solution $x_1 = 0$, $x_2 = 1$. The enumeration tree node numbers for these two cases are 8 and 5 respectively.

To complete the comparison to Example 5.1, we have numbered the nodes in Figure 5.3 to match those of Figure 5.1. In addition, the committed objective function and constraint values κ_i and γ_i are provided in Figure 5.3 just as they were in Figure 5.1.

It is clear that the enumeration depicted in Figure 5.3 is not total. For example, there is room for two additional nodes below node 5 with $x_1 = 0$, $x_2 = 1$ and $x_3 = 0$ or 1. How could we omit these nodes, labeled as say 11 and 12, without incurring the risk of overlooking the optimal solution? The answer is that both are dominated. Had we investigated nodes 11 and 12, we would have computed $\kappa_{11} = +\infty$, $\gamma_{11} = 1$, $\kappa_{12} = 21$, $\gamma_{12} = 3$. Since node 11 is infeasible, it is dominated by any feasible node. Alternately, node 12 does correspond to a feasible solution, but it uses no less resource than node 7, and it is more costly. By Corollary 5.5, we have lost nothing by not explicitly considering nodes 11 and 12.

In examining Figures 5.1 and 5.3, we can see that comparing the branch and bound algorithm to the dynamic program yields no clear winner in terms of efficiency. By adhering to dominance, the dynamic program avoided consideration of node 6 ($x_1 = x_2 = x_3 = 0$), which in time was checked in Figure 5.1. On the other hand, by failing to look ahead at as-yet-unconsidered stages, as the branch and bound algorithm did, the dynamic programming procedure was forced to enumerate the extra nodes 8, 9 and 10 that were not checked in Example 5.1. It is also clear that the dynamic programming structure required a certain amount of redundant computation that was avoided with branch and bound. Although such duplication could have been reduced had we used a more sophisticated dynamic programming structure, the tables of Example 5.3 show nodes 2 and 3 being evaluated three times and node 8 twice. In Example 5.1, these nodes were checked only once.

The knapsack problem of Examples 5.1 and 5.3 is only a tiny, contrived illustration. Still it is generally true that any discrete dynamic program-

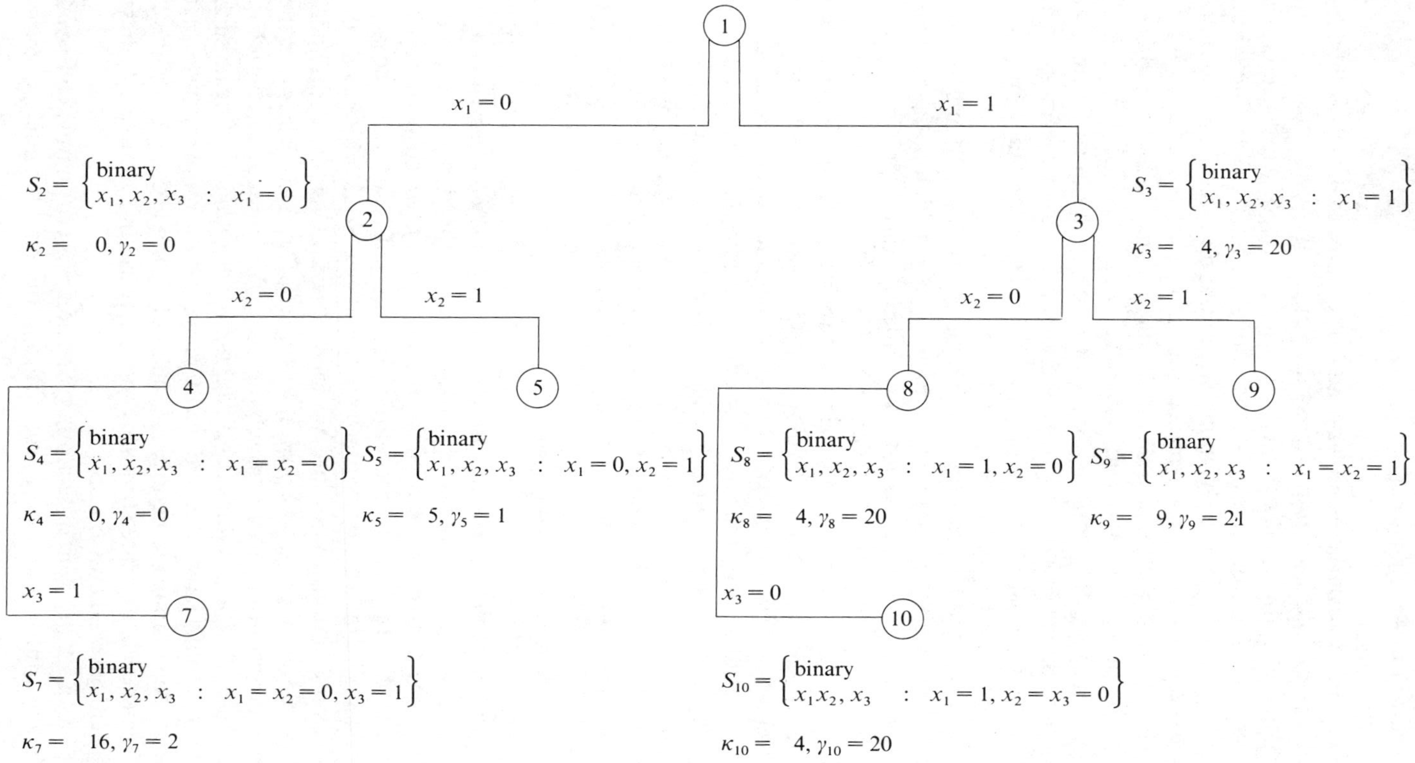

Fig. 5.3. Dynamic programming enumeration tree for knapsack Example 5.3.

ming procedure can be viewed as partial enumeration via dominance. Where such dominance tests can be efficiently introduced, they will reduce the number of solutions explicitly enumerated (see Ibarki (1976) or Marsten and Morin (1977) for more details). The forward looking feasibility and optimality checks of branch and bound, however, may do at least as well when the dynamic program does not yield a polynomially-bounded algorithm. Moreover, the branch and bound structure can be employed in any $P(S)$, not just those for which dominance can be efficiently checked. Thus, the most effective strategy for approaching difficult discrete optimization problems seems to be one where dominance (when efficiently checked), feasibility and bounds are all taken into account.

5.1.5 Exponential Order of Partial Enumeration

It is clear that a partial enumeration scheme might, in particularly unfortunate circumstances, result in total enumeration, and thereby require computation time growing more than polynomially with problem size. It is not particularly easy, however, to contrive a system of examples that exhibit such worst case behavior.

Chvátal (1980) has identified just such a class of problems. The problems are special knapsacks of the following very simple form:

$$\begin{aligned} &\min && \sum_{j=1}^{n} a_j x_j \\ CKP(S)\quad &\text{s.t.} && \sum_{j=1}^{n} a_j x_j \geq b = \left\lceil \frac{1}{2} \sum_{j=1}^{n} a_j \right\rceil \\ & && (x_1, x_2, \ldots, x_n) \in S = \left\{ \mathbf{x}: \begin{array}{c} 0 \leq x_j \leq 1 \text{ and } x_j \text{ integer for} \\ j = 1, 2, \ldots, n \end{array} \right\} \end{aligned}$$

where coefficients a_j satisfy

(i) $a_j > 0$ for $j = 1, 2, \ldots, n$

(ii) $\sum_{j \in J} a_j \geq b$ for every $J \subseteq \{1, 2, \ldots, n\}$ with $|J| \geq \frac{9}{10} n$

(iii) $\sum_{j \in J_1} a_j \neq \sum_{j \in J_2} a_j$ for any distinct $J_1, J_2 \subseteq \{1, 2, \ldots, n\}$

(iv) there exists no $J \subseteq \{1, 2, \ldots, n\}$ such that $\sum_{j \in J} a_j = b$.

If such problems are solved via partial enumeration with candidate problems being specified by partial solution, then the characterizing set of a candidate problem $CKP(S_k)$ would have the form

$$S_k = S \cap \{\mathbf{x}\colon x_j = 0 \text{ for } j \in J_k^- \text{ and } x_j = 1 \text{ for } j \in J_k^+\}.$$

Here J^- and J^+ are nonoverlapping subsets of $\{1, 2, \ldots, n\}$. A simplified version of Chvátal's result is given by the following theorem.

Theorem 5.6. Exponential Order of Partial Enumeration on Chvátal's Knapsack Problems. *Consider the class of problems $CKP(S)$, and suppose such problems are to be solved by Algorithm 5A with candidate problems being specified by partial solution (as in Section 5.1.2) and candidate problems being fathomed whenever dominated (as in Section 5.1.4) or bounded by the linear programming bound $v(P(\bar{S}_k)) \geq v^*$ with $\bar{S}_k = \{\mathbf{x}\colon 0 \leq x_j \leq 1 \text{ for } j = 1, 2, \ldots, n\} \cap \{\mathbf{x}\colon x_j = 0 \text{ for all } j \in J_k^- \text{ and } x_j = 1 \text{ for all } j \in J_k^+\}$. Then the number of candidate problems that must be explored in solving any particular problem of the class is at least $2^{n/10}$, i.e., exponential in n.*

Proof. Any partial solution enumeration of $CKP(S)$ would require investigation of at least $2^{n/10}$ candidate problems if no candidate $CKP(S_k)$ with $|J_k^-| + |J_k^+| \leq n/10$ can be fathomed. We shall show that (i) through (iv) imply exactly that.

First, consider dominance. If a candidate $CKP(S_d)$ is to be dominated, there must be another candidate $CKP(S_p)$ such that $J_d^- \cup J_d^+ \supseteq J_p^- \cup J_p^+$, $b - \Sigma_{j \in J_d^+} a_j \geq b - \Sigma_{j \in J_p^+} a_j$, and $\Sigma_{j \in J_d^+} a_j \geq \Sigma_{j \in J_p^+} a_j$. Clearly, this is only possible when $\Sigma_{j \in J_d^+} a_j = \Sigma_{j \in J_p^+} a_j$, which is precluded by (iii). Thus, no candidate of interest can be passed over because it is dominated.

Turning to fathoming by bounding, observe that the main constraint of $CKP(S_k)$ will be

$$\sum_{j \notin J_k^- \cup J_k^+} a_j x_j \geq b - \sum_{j \in J_k^+} a_j.$$

Now (i) assumes the right side of this expression is at most b. Thus, by (ii), the linear program $CKP(\bar{S}_k)$ always has a feasible solution $x_j = 1$ for all $j \notin J_k^- \cup J_k^+$ when $|J_k^+| + |J_k^-| \leq n/10$. Moreover, at linear optimality its main constraint will be satisfied as an equality, so that $v(CKP(\bar{S}_k)) = (b - \Sigma_{j \in J_k^+} a_j) + (\Sigma_{j \in J_k^+} a_j) = b$. Since (iv) assures $v(CKP(S)) > b$, it cannot be that $b = v(CKP(\bar{S}_k)) \geq v^* \geq v(CKP(S))$. Thus, no $CKP(S_k)$ with $|J_k^+| + |J_k^-| \leq n/10$ can be fathomed by the bound $v(CKP(\bar{S}_k))$ and the theorem holds. ■

Although the class of $CKP(S)$ is obviously contrived, Chvátal (1980) has shown it is not small. In fact, the probability that $\{a_j\}$ randomly chosen in the range $1 \leq a_j \leq m$, lead to a $CKP(S)$ approaches 1 as m and n become large.

5.2 Elementary Bounds

Having completed our discussion of fundamental notions in partial enumeration, we are now ready to begin studying practical procedures for carrying out the very general strategy embodied in Algorithm 5A. It should be clear that the most important single element in a branch and bound algorithm is an adequate bounding procedure. Only through strong bounds can we expect to fathom candidate problems rapidly enough to avoid being overcome by the exponential growth in the number of potential candidates.

5.2.1 Implicit Enumeration

The most easily computed bounds experiencing widespread use in discrete optimization are simple feasibility and cost comparisons obtained by direct inspection of a problem's constraints and objective function. The many schemes of this type are known collectively as *implicit enumeration* procedures.

Since they depend on observations about a problem's structure, many implicit enumeration schemes are adapted specifically to particular problem forms. One general structure is the $P(S)$ where all variables, x_j are restricted to be 0 or 1 and all costs, c_j, are nonnegative. We shall call such a problem $IEP(S)$. Any pure 0–1 problem is easily placed in this form by substituting the expression $(1 - x_j')$ for any x_j having a negative objective function coefficient; x_j' replaces x_j as a decision variable.

Now suppose that we are confronted with a problem $IEP(S)$ and that our current candidate, $IEP(S_k)$, has fixed the values of all variables except those with indices in the free set F_k. Then the current candidate problem k can be written as

$$IEP(S_k) \qquad \begin{aligned} \min \quad & \sum_{j \in F_k} c_j x_j + \kappa_k \\ \text{s.t.} \quad & \sum_{j \in F_k} a_{ij} x_j + \gamma_i^k \geq b_i \qquad \text{for all } i \\ & x_j = 0 \text{ or } 1 \qquad \text{for all } j \in F_k. \end{aligned} \tag{5-1}$$

Here, κ_k is the constant term obtained from objective function coefficients of fixed x_j (i.e., $j \notin F_k$), and the γ_i^k are the corresponding fixed terms of the constraint rows.

Nonnegativity of costs, together with these definitions gives an obvious way of discovering feasible solutions for Step 3 of Algorithm 5A.

Lemma 5.7. Feasible Solutions from Implicit Enumeration. *Suppose for $IEP(S_k)$ as above,*

$$\gamma_i^k \geq b_i \qquad \textit{for all } i$$

Then the solution $x_j = 0$ *for all* $j \in F_k$ *solves IEP*(S_k)*, and IEP*(S_k) *can be fathomed.*

We also need bounds on $v(IEP(S_k))$. One way is to create a series of one-row relaxations and take the greatest of the corresponding solution values. That is, to consider bounds of the form

$$\beta_0(S_k) \triangleq \max_i v\begin{pmatrix} \min & \sum_{j\in F_k} c_j x_j + \kappa_k & \\ \text{s.t.} & \sum_{j\in F_k} a_{ij}x_j + \gamma_i^k \geq b_i & \\ & x_j = 0 \text{ or } 1 & \text{for all } j \in F_k \end{pmatrix}.$$

Notice, however, that we are still obligated to solve a discrete knapsack problem for each i in order to evaluate bound $\beta_0(S_k)$. Knapsack problems are relatively easy discrete problems, but still far too difficult for quick and dirty implicit enumeration computation. More must be relaxed.

The possibility tried in Example 5.1 was to zero the (nonnegative) objective function coefficients c_j. The resulting maximum of one-row relaxation bounds is

$$\beta_1(S_k) \triangleq \max_i v\begin{pmatrix} \min & \kappa_k & \\ \text{s.t.} & \sum_{j\in F_k} a_{ij}x_j + \gamma_i^k \geq b_i & \\ & x_j = 0 \text{ or } 1 & \text{for all } j \in F_k \end{pmatrix}.$$

In each one-row optimization, the objective function is constant, so that any feasible solution is optimal at value κ_k. The only issue is whether some one-row relaxation is infeasible, i.e., whether there is an i such that $\Sigma_{j\in F_k} \max\{0, a_{ij}\} + \gamma_i^k < b_i$.

Other schemes retain the objective function of β_0's one-row problems but relax $x_j = 0$ or 1 to $x_j \geq 0$ or $1 \geq x_j \geq 0$. The former gives

$$\beta_2(S_k) \triangleq \max_i v\begin{pmatrix} \min & \sum_{j\in F_k} c_j x_j + \kappa_k & \\ \text{s.t.} & \sum_{j\in F_k} a_{ij}x_j + \gamma_i^k \geq b_i & \\ & x_i \geq 0 & \text{for all } j \in F_k \end{pmatrix}.$$

It is very easy to see that each one-row relaxation in β_2 can be solved by picking j to minimize ratios c_j/a_{ij}. Using this idea and deleting already feasible rows gives the compact form

$$\beta_2(S_k) = \kappa_k + \max_{\{i:\, \gamma_i^k < b_i\}} \left((b_i - \gamma_i^k) \min_{\substack{j \in F_k \\ a_{ij} > 0}} \{c_j / a_{ij}\} \right).$$

Bound statement 5B summarizes the above bounds along with a β_3 derived as in β_2 except that upper bounds $x_j \leq 1$ are retained.

Bounds 5B. One-Row Implicit Enumeration Bound Functions. Let *IEP*(S) be as above and *IEP*(S_k) an associated candidate problem.

Row feasibility

$$\beta_1(S_k) \triangleq \begin{cases} +\infty & \text{if for any row } i, \\ & \sum_{j \in F_k} \max\{0, a_{ij}\} + \gamma_i^k < b_i \\ \kappa_k & \text{otherwise} \end{cases}$$

Row cost ratio

$$\beta_2(S_k) \triangleq \kappa_k + \max_{\{i:\, \gamma_i^k < b_i\}} \left\{ (b_i - \gamma_i^k) \min_{\{j \in F_k:\, a_{ij} > 0\}} \{c_j / a_{ij}\} \right\}$$

Row linear program

$$\beta_3(S_k) \triangleq \kappa_k + \max_{\{i:\, \gamma_i^k < b_i\}} v \begin{pmatrix} \min & \sum_{j \in F_k} c_j x_j & \\ \text{s.t.} & \sum_{j \in F_k} a_{ij} x_j \geq b_i - \gamma_i^k & \\ & 0 \leq x_j \leq 1 & \text{for all } j \in F_k \end{pmatrix}$$ ■

Example 5.4. Implicit Enumeration. Consider the simple *IEP*

$$\begin{array}{llllll} \min & 10x_1 & + 2x_2 & - \;\; x_3 & + 3x_4 & \\ \text{s.t.} & 4x_1 & - 2x_2 & - \;\; x_3 & + \;\; x_4 & \geq 1 \\ & 3x_1 & + \;\; x_2 & + 3x_3 & & \geq 4 \\ & x_1, & x_2, & x_3, & x_4 = 0 \text{ or } 1. & \end{array}$$

To apply implicit enumeration techniques, let us first rewrite the problem in standard form. Substituting $x_3 = 1 - x_3'$. The result is

$$\begin{array}{lll} & \min & 10x_1 + 2x_2 + \;\; x_3' + 3x_4 - 1 \\ IEX(T) & \text{s.t.} & 4x_1 - 2x_2 + \;\; x_3' + \;\; x_4 \geq 2 \\ & & 3x_1 + \;\; x_2 - 3x_3' \qquad\quad \geq 1 \end{array}$$

$$(x_1, x_2, x_3', x_4) \in T = \{\text{binary } (x_1, x_2, x_3', x_4)\}.$$

Following Algorithm 5A, we begin enumerating by setting $v^* \leftarrow +\infty$, and making our first candidate problem $IEX(T_1)$ with $T_1 = T$. For this candidate, $\kappa_1 = -1$, $\gamma_1^1 = 0$, $\gamma_2^1 = 0$, $F_1 = \{1, 2, 3, 4\}$

$$\sum_{j \in F_1} \max\{0, a_{1j}\} + \gamma_1^1 = 4 + 0 + 1 + 1 + 0 \geq 2$$

$$\sum_{j \in F_1} \max\{0, a_{2j}\} + \gamma_2^1 = 3 + 1 + 0 + 0 + 0 \geq 1.$$

Thus, $\beta_1(T_1) = \kappa_1 = -1$.

$$\begin{aligned} \beta_2(T_1) &= \kappa_1 + \max_{\{i:\, \gamma_i^k < b_i\}} \left\{ (b_i - \gamma_i^k) \min_{\{j \in F_1:\, a_{ij} > 0\}} \{c_j/a_{ij}\} \right\} \\ &= -1 + \max\{(2\text{-}0) \min\{\tfrac{10}{4}, \tfrac{1}{1}, \tfrac{3}{1}\}, (1\text{-}0) \min\{\tfrac{10}{3}, \tfrac{2}{1}\}\} \\ &= 1 \end{aligned}$$

Now,

$$v \begin{pmatrix} \min & 10x_1 + 2x_2 + 1x_3' + 3x_4 \\ \text{s.t.} & 4x_1 - 2x_2 + x_3' + x_4 \geq 2 \\ & 0 \leq x_1, x_2, x_3', x_4 \leq 1 \end{pmatrix} = 3.5$$

$$v \begin{pmatrix} \min & 10x_1 + 2x_2 + 1x_3' + 3x_4 \\ \text{s.t.} & 3x_1 + x_2 - 3x_3' \geq 1 \\ & 0 \leq x_1, x_2, x_3', x_4 \leq 1 \end{pmatrix} = 2$$

and it follows that $\beta_3(T_1) = -1 + \max\{3.5, 2\} = 2.5$.

Since none of these bounds exceeds $v^* = +\infty$, and the conditions of Lemma 5.7 for a feasible solution are not met, we branch. Let us create the two new candidates $IEX(T_2)$ and $IEX(T_3)$ with

$$T_2 = \{(x_1, x_2, x_3', x_4) \in T: x_1 = 1\}$$

$$T_3 = \{(x_1, x_2, x_3', x_4) \in T: x_1 = 0\}$$

Consider $IEX(T_2)$ first. Here, $\kappa_2 = 9$, $\gamma_1^2 = 4$, $\gamma_2^2 = 3$, $F_2 = \{2, 3, 4\}$. Since $\gamma_1^2 \geq b_1$ and $\gamma_2^2 \geq b_2$, $\beta_1(T_2) = \beta_2(T_2) = \beta_3(T_2) = \kappa_2$. We cannot fathom at Step 2 of Algorithm 5A. $IEX(T_2)$, however, does satisfy Lemma 5.7. We can obtain the incumbent solution $x_1^* = 1$, $x_2^* = 0$, $x_3'^* = 0$, $x_4^* = 0$ with $v^* \leftarrow \kappa_2 = 9$. This leads to a fathom at Step 4.

The remaining candidate is $IEX(T_3)$. There, $\kappa_3 = -1$, $\gamma_1^3 = 0$, $\gamma_2^3 = 0$, $F_3 = \{2, 3, 4\}$, and computation of our three bounds yields $\beta_1(T_3) = -1$, $\beta_2(T_3) = 1$, $\beta_3(T_3) = 3$. We cannot fathom or obtain a feasible solution; we must further subdivide T_3. By continuing our use of β_1, β_2 and β_3, we will eventually conclude that no solution in T_3 is feasible, i.e., that our incumbent $x_1^* = 1$, $x_2^* = 0$, $x_3'^* = 0$, $x_4^* = 0$ is optimal. In terms of the original variables, this solution is $x_1^* = 1$, $x_2^* = 0$, $x_3^* = 1$, $x_4^* = 0$ at value 9. ■

All the bounding rules of 5B consider one row at a time. Clearly, we would expect stronger bounds if we considered several rows.

One technique is to consider nonnegative combinations of the rows of (5-1). It is easy to verify that any $\mathbf{x}$ satisfying (5-1) also satisfies such a nonnegative sum or *surrogate constraint,*

$$\sum_{j \in F_k} (\mathbf{u}\mathbf{a}^j)x_j + \mathbf{u}\gamma^k \geq \mathbf{u}\mathbf{b} \tag{5-2}$$

Here $\mathbf{u} \geq \mathbf{0}$ is the weight vector and $\mathbf{a}^j$ is the *jth* column of (5-1), γ^k is the vector of γ_i^k and $\mathbf{b}$ is the vector of b_i. Once we have the valid row (5-2) we can then apply the same sorts of analyses implicit in the one-row Bounds 5B.

A different approach for combining rows is to compare *completion cardinality,* i.e., minimums and maximums on the number of $j \in F_k$ that can be 1 in any feasible completion of the current partial solution. If the minimum for one row exceeds the maximum for another, $IEP(S_k)$ is infeasible.

Both the surrogate and this completion cardinality notion are detailed in Bounds 5C.

Bounds 5C. Implicit Enumeration Multi-row Bound Functions. Let $IEP(S)$ be as above, $IEP(S_k)$ an associated candidate problem, $\mathbf{u} \geq \mathbf{0}$ a vector of row multipliers, and $\mathbf{a}^j$ the *jth* column of $\mathbf{A}$.

Surrogate without cost

$$\beta_4(S_k) \triangleq \begin{cases} +\infty & \text{if } \sum_{j \in F_k} \max\{0, \mathbf{u}\mathbf{a}^j\} + \mathbf{u}\gamma^k < \mathbf{u}\mathbf{b} \\ \kappa_k & \text{otherwise} \end{cases}$$

Surrogate with cost

$$\beta_5(S_k) \triangleq \kappa_k + \mathbf{u}(\mathbf{b} - \gamma^k) + \sum_{j \in F_k} \min\{0, c_j - \mathbf{u}\mathbf{a}^j\}$$

Completion cardinality

$$\beta_6(S_k) \triangleq \begin{cases} +\infty & \text{if } \max_i \{q_i^{\min}\} > \min_i \{q_i^{\max}\} \\ \kappa_k & \text{otherwise} \end{cases}$$

where $q_i^{\min} \triangleq$ minimum cardinality of a subset Q of F_k such that

$$\sum_{j \in Q} a_{ij} + \gamma_i^k \geq b_i$$

$q_i^{\max} \triangleq$ maximum cardinality of a subset Q of F_k such that

$$\sum_{j \in Q} a_{ij} + \gamma_i^k \geq b_i. \qquad \blacksquare$$

The stated $\beta_4(S_k)$ is clearly just a weighted row extension of $\beta_1(S_k)$. Any $\beta_6(S_k)$ formalizes the notion of a $+\infty$ bound (i.e., proof of infeasibility) when the least number of $x_j = 1$ for $j \in F_k$ tolerable in one row is less than the greatest number feasible in another.

It is not nearly so obvious that $\beta_5(S_k)$ is obtained by weighting constraint rows without zeroing the cost row as in β_4. The next theorem proves β_5's validity.

Theorem 5.8. Validity of $\boldsymbol{\beta_5}$. *Let $IEP(S_k)$ be as above and $\beta_5(S_k)$ as defined in Bounds 5C. Then*

$$\beta_5(S_k) = v\begin{pmatrix} \text{min} & \kappa_k + \sum_{j \in F_k} (c_j - \mathbf{u}\mathbf{a}^j)x_j + \mathbf{u}(\mathbf{b} - \gamma^k) \\ \text{s.t.} & x_j = 0 \text{ or } 1 \qquad \text{for all } j \in F_k \end{pmatrix} \tag{5-3}$$

and the latter is a valid relaxation of $IEP(S_k)$.

Proof. First note that $\beta_5(S_k)$ does indeed compute the value of the optimization problem (5-3). The computation shown in Bound 5C merely adds the constants $\kappa_k + \mathbf{u}(\mathbf{b} - \gamma^k)$ to the sum of optimal cost contributions for $x_j = 0$ or 1.

To show the problem of (5-3) is a relaxation of $IEP(S_k)$, observe first that its feasible space certainly contains that of $IEP(S_k)$. It remains only to show the objective in (5-3) underestimates the $IEP(S_k)$ objective function at $x_j = 0$ or 1 feasible in $IEP(S_k)$.

When $\{x_j\colon j \in F_k\}$ is feasible it certainly satisfies any nonnegative weighted surrogate of main rows as in (5-2). That is,

$$0 \geq \mathbf{u}\mathbf{b} - \sum_{j \in F_k} (\mathbf{u}\mathbf{a}^j)x_j - \mathbf{u}\gamma^k.$$

Adding this (nonpositive) term to $IEP(S_k)$ objective function, and regrouping

$$\sum_{j \in F_k} c_j x_j + \kappa_k + \mathbf{u}\mathbf{b} - \sum_{j \in F_k} (\mathbf{u}\mathbf{a}^j)x_j - \mathbf{u}\gamma^k$$
$$= \sum_{j \in F_k} (c_j - \mathbf{u}\mathbf{a}^j)x_j + \kappa_k + \mathbf{u}(\mathbf{b} - \gamma^k).$$

This completes the proof that (5-3) is a relaxation. $\blacksquare$

Return to the third candidate problem of Example 5.4. After eliminating the x_1 fixed at 0 we have

$$
\begin{array}{llr}
 & \min & 2x_2 + x_3' + 3x_4 - 1 \\
IEX(T_3) & \text{s.t.} & -2x_2 + x_3' + x_4 \geq 2 \\
 & & x_2 - 3x_3' \qquad \geq 1 \\
 & & x_2, x_3', x_4 = 0 \text{ or } 1
\end{array}
$$

with $\kappa_3 = -1$, $\gamma_1^3 = 0$, $\gamma_2^3 = 0$, $F_3 = \{2, 3, 4\}$.

To illustrate surrogate constraint bounds, we must pick row multipliers u_1 and u_2. Arbitrarily, choose $\mathbf{u} = (u_1, u_2) = (1, 2)$. The resulting surrogate constraint of β_4 is

$$(1(-2) + 2(1))x_2 + (1(1) + 2(-3))x_3' + (1(1) + 2(0))x_4 \geq 1(2) + 2(1)$$

or

$$-5x_3 + x_4 \geq 4.$$

Since $\Sigma_{j \in F_3} \max\{\mathbf{ua}^j, 0\} = 0 + 1 \ngeq 4$, $\beta_4(T_3) = +\infty$. In contrast to the weaker Bounds 5B, we would fathom.

The surrogate combination of β_5 includes cost. In $IEX(T_3)$ with arbitrary $\mathbf{u} = (1, 2)$, β_5 computes as

$$
\begin{aligned}
\beta_5(T_3) &= -1 + (1, 2)(2\text{-}0, 1\text{-}0) + \min\{0, 2\text{-}0\} \\
&\qquad + \min\{0, 1 + 5\} + \min\{0, 3\text{-}1) \\
&= 3
\end{aligned}
$$

Although β_5 is greater than or equal to β_1, β_2 and β_3 for this case, we would not be able to fathom on the $+3$ value.

Turning finally to the completion cardinality notion of β_6, we can see that at least and at most two of x_2, x_3' and x_4 can be positive if row 1 of $IEX(T_3)$ is to be satisfied. Thus, $q_1^{\min} = q_1^{\max} = 2$. Similarly $q_2^{\min} = 1$ and $q_2^{\max} = 2$ (do not forget x_4). Since $\max_i\{q_i^{\min}\} \leq \min_i\{q_i^{\max}\}$, we would not fathom; $\beta_6(T_3) = \kappa_3 = -1$.

There are numerous variants of the six implicit enumeration schemes we have presented that can add strength to bounds. In the interest of space, we shall omit these extensions (see Garfinkel and Nemhauser (1972, Chapter 4) for a discussion). In comparing the tests we have presented, or their extensions, however, it is important that the reader keep in mind a fundamental trade-off. Stronger and stronger bounds can be obtained, if the algorithm designer is willing to invest more and more effort in their computation. For example, we saw that bounds β_4 and β_5 dominated β_1, β_2

and β_3 in our example. The former required multipliers $\mathbf{u} \geq \mathbf{0}$, however. We have not addressed how one chooses such multipliers (see Geoffrion (1969) for an approach), but it should be clear that finding satisfactory $\mathbf{u}$ will add considerable computation. The stronger, surrogate bounds could be justified only if experience shows the enumeration they allow us to skip is more costly than the computational effort they add.

5.2.2 The Linear Programming Relaxation

The implicit enumeration bounds of the previous section were derived from rather simple conditions defined on single constraint rows or combinations of rows. We know from Section 5.1.3, however, that any relaxation of a problem $P(S)$ (i.e., any easier problem having a feasible space containing that of $P(S)$ and an objective function underestimating that of $P(S)$) can serve as a source of bounds.

Linear programming is, without doubt, the most successful branch of optimization. Although most discrete optimization problems are not (as far as is known) linear programs, it is clear that each $P(S)$ is closely associated with a linear program; each $P(S)$ has a linear objective function and mostly linear constraints.

An exact formulation of the linear program associated with $P(S)$ is

$$P(\bar{S}) \qquad \begin{array}{ll} \min & \mathbf{cx} \\ \text{s.t.} & \mathbf{Ax} \geq \mathbf{b} \\ & \mathbf{x} \in \bar{S} = \{\mathbf{x} \geq \mathbf{0}\text{: } 1 \geq x_j \text{ for } j \in I\} \end{array}$$

We shall refer to $P(\bar{S})$ as the *linear programming relaxation* of $P(S)$. Notice that it is formed from the *linear programming relaxation* of the set S, i.e., the set $\bar{S}$ obtained by replacing constraints $x_j = 0$ or 1 by $1 \geq x_j \geq 0$.

The core of the power of $P(\bar{S})$ follows from the fact that linear programs are relatively easy optimization problems to solve and from the obvious properties summarized in the following lemma.

Lemma 5.9. Linear Programming Relaxations. *For $P(S)$ and $P(\bar{S})$ defined above*

(i) $v(P(\bar{S})) \leq v(P(S))$

and (ii) If $\bar{\mathbf{x}}$ solves $P(\bar{S})$ and its components $\bar{x}_j$ satisfy $\bar{x}_j = 0$ or 1 for each $j \in I$, then $\bar{\mathbf{x}}$ solves $P(S)$.

We see that solving $P(\bar{S})$ yields both a source of solved problems at Step 3 of Algorithm 5A and a source of bounds for Step 2. Since solution procedures involving linear programming relaxations are so common, we shall formally state and illustrate a version of Algorithm 5A that employs them with the partial solutions concept of Section 5.1.2.

Algorithm 5D. Linear Programming Based Branch and Bound

STEP 0: *Initialization.* Establish the candidate list by making the full problem $P(S)$ its sole entry. Set $v^* \leftarrow +\infty$.

STEP 1: *Candidate Selection.* Choose some member $P(S_k)$ of the current candidate list.

STEP 2: *Bounding.* Attempt to solve the linear programming relaxation $P(\bar{S}_k)$. If $v(P(\bar{S}_k)) = -\infty$, stop; $P(S)$ is unbounded. If $v(P(\bar{S}_k)) \geq v^*$, go to Step 6 and fathom. Otherwise, let $\bar{\mathbf{x}}$ be the $P(\bar{S}_k)$ optimum obtained and go to Step 3.

STEP 3: *Feasible Solutions.* If $\bar{x}_j = 0$ or 1 for all $j \in I$, i.e., $\bar{\mathbf{x}}$ satisfies all discrete constraints, set $\tilde{\mathbf{x}} \leftarrow \bar{\mathbf{x}}$. Otherwise, attempt to construct an $\tilde{\mathbf{x}}$ feasible in $P(S_k)$, beginning from $\bar{\mathbf{x}}$. If $\mathbf{c}\tilde{\mathbf{x}} \geq v^*$ or no $\tilde{\mathbf{x}}$ was formed, proceed to Step 5. Otherwise, go to Step 4.

STEP 4: *Incumbent Saving.* Save $\tilde{\mathbf{x}}$ as a new incumbent solution by $\mathbf{x}^* \leftarrow \tilde{\mathbf{x}}$, $v^* \leftarrow \mathbf{c}\tilde{\mathbf{x}}$. If $\mathbf{c}\tilde{\mathbf{x}} = \mathbf{c}\bar{\mathbf{x}}$, go to Step 6 and fathom; $\tilde{\mathbf{x}}$ solves $P(S_k)$. Otherwise, go to Step 5.

STEP 5: *Branching.* Choose some fractional component x_p with $p \in I$ satisfying $0 < \bar{x}_p < 1$. Replace $P(S_k)$ in the candidate list with two new problems, one characterized by $S_k \cap \{\mathbf{x}\colon x_p = 0\}$ and the other by $S_k \cap \{\mathbf{x}\colon x_p = 1\}$. Then return to Step 1.

STEP 6: *Fathoming.* Fathom $P(S_k)$, i.e., delete it from the candidate list. If the candidate list is now empty, stop; $\mathbf{x}^*$ is optimal unless v^* still equals $+\infty$ in which case $P(S)$ is infeasible. If candidates remain, return to Step 1. ■

Example 5.5. Linear Programming Based Branch and Bound. Consider the discrete optimization problem

$$\begin{array}{lll} & \min & 10x_1 - 2x_2 + 6x_3 + x_4 + 18x_5 \\ LPX(S) & \text{s.t.} & 7x_1 + x_2 + x_3 + 4x_5 \geq 4 \\ & & x_2 + 3x_3 + 5x_4 - 4x_5 \geq 3 \\ & & (x_1, x_2, x_3, x_4, x_5) \in S = \{\mathbf{x}\colon x_1, x_2, x_3, x_4 = 0 \text{ or } 1, x_5 \geq 0\} \end{array}$$

Here the linear relaxation of S is $\bar{S} = \{\mathbf{x} \geq \mathbf{0}\colon x_1, x_2, x_3, x_4 \leq 1\}$. The linear programming relaxation $LPX(\bar{S})$ has solution $x_1 = \frac{3}{7}$, $x_2 = 1$, $x_4 = \frac{2}{5}$, $x_3 = x_5 = 0$ with value $\frac{94}{35}$ or about 2.7. An optimal solution to the full problem is $x_1 = x_2 = x_4 = 1$, $x_3 = x_5 = 0$ with value 9. Details of how this solution is computed by Algorithm 5D are presented in Table 5.1. ■

TABLE 5.1. Solution of Example 5.5

Algorithm 5D step	Action/result
0	$v^* \leftarrow +\infty$, $S_1 \leftarrow S$
1	Select candidate $LPX(S_1)$
2	Solve $LPX(\bar{S}_1)$ for $v(LPX(\bar{S}_1)) = \frac{94}{35} \approx 2.7$ $\bar{\mathbf{x}} \leftarrow (\frac{3}{7}, 1, 0, \frac{2}{5}, 0)$.
3	Conclude $\bar{\mathbf{x}}$ is not feasible for $LPX(S_1)$
5	Replace $LPX(S_1)$ in the candidate list by $LPX(S_2)$ and $LPX(S_3)$ with $S_2 \leftarrow S_1 \cap \{\mathbf{x}: x_1 = 0\}$, $S_3 \leftarrow S_1 \cap \{\mathbf{x}: x_1 = 1\}$.
1	Select candidate $LPX(S_2)$
2	Solve $LPX(\bar{S}_2)$ for $v(LPX(\bar{S}_2)) = \frac{25}{2} = 12.5$ $\bar{\mathbf{x}} \leftarrow (0, 1, 0, 1, \frac{3}{4})$
3	Since $\bar{\mathbf{x}}$ is feasible for $LPX(S_2)$ set $\tilde{\mathbf{x}} \leftarrow \bar{\mathbf{x}}$
4	Save a new incumbent $v^* \leftarrow \mathbf{c}\tilde{\mathbf{x}} = 12.5$ $\mathbf{x}^* \leftarrow \tilde{\mathbf{x}} = (0, 1, 0, 1, \frac{3}{4})$
6	Fathom $LPX(S_2)$ because $\bar{\mathbf{x}}$ solved it.
1	Select candidate $LPX(S_3)$
2	Solve $LPX(\bar{S}_3)$ for $v(LPX(\bar{S}_3)) = \frac{42}{5} = 8.4$ $\bar{\mathbf{x}} \leftarrow (1, 1, 0, \frac{2}{5}, 0)$
3	Conclude $\bar{\mathbf{x}}$ is not feasible for $LPX(S_3)$
5	Replace $LPX(S_3)$ in the candidate list by $LPX(S_4)$ and $LPX(S_5)$ with $S_4 \leftarrow S_3 \cap \{\mathbf{x}: x_4 = 0\}$ $S_5 \leftarrow S_3 \cap \{\mathbf{x}: x_4 = 1\}$
1	Select candidate $LPX(S_4)$
2	Solve $LPX(\bar{S}_4)$ for $v(LPX(\bar{S}_4)) = 12$ $\bar{\mathbf{x}} \leftarrow (1, 1, \frac{2}{3}, 0, 0)$
3	Conclude $\bar{\mathbf{x}}$ is not feasible for $LPX(S_4)$
5	Replace $LPX(S_4)$ in the candidate list by $LPX(S_6)$ and $LPX(S_7)$ with $S_6 \leftarrow S_4 \cap \{\mathbf{x}: x_3 = 0\}$ $S_7 \leftarrow S_4 \cap \{\mathbf{x}: x_3 = 1\}$
1	Select candidate $LPX(S_6)$
2	Attempt to solve $LPX(\bar{S}_6)$ and conclude it is infeasible, i.e., $v(LPX(\bar{S}_6)) = +\infty$.
6	Fathom $LPX(S_6)$ because $v(LPX(\bar{S}_6)) \geq v^* = 12.5$
1	Select candidate $LPX(S_7)$
2	Solve $LPX(\bar{S}_7)$ for $v(LPX(\bar{S}_7)) = 14$
6	Fathom $LPX(S_7)$ because $v(LPX(\bar{S}_7)) \geq v^* = 12.5$
1	Select candidate $LPX(S_5)$
2	Solve $LPX(\bar{S}_5)$ for $v(LPX(\bar{S}_5)) = 9$ $\bar{\mathbf{x}} \leftarrow (1, 1, 0, 1, 0)$
3	Since $\bar{\mathbf{x}}$ is feasible for $LPX(S_5)$ set $\tilde{\mathbf{x}} \leftarrow \bar{\mathbf{x}}$
4	Save a new incumbent $v^* \leftarrow \mathbf{c}\tilde{\mathbf{x}} = 9$ $\mathbf{x}^* \leftarrow \tilde{\mathbf{x}} = (1, 1, 0, 1, 0)$
6	Fathom $LPX(S_5)$ because $\bar{\mathbf{x}}$ solved it. Since the candidate list is now empty, conclude v^* and $\mathbf{x}^*$ are optimal

5.2.3 *Linear Programming versus Implicit Enumeration*

Although it may not be clear at first glance, most of the implicit enumeration bounds 5B and 5C have a close relationship with linear programming. This connection can be formalized to yield a comparison of the relative fathoming power of most bounds from implicit enumeration versus those from linear programming relaxations.

Theorem 5.10. Strength of Implicit Enumeration versus Linear Programming Bounds. *Let $IEP(S_k)$ be a candidate problem derived as in Section 5.2.1 from a pure 0–1 integer program with nonnegative cost coefficients. Then, the linear programming relaxation $IEP(\bar{S}_k)$ satisfies*

$$v(IEP(\bar{S}_k)) \geq \max_{1 \leq l \leq 5} \{\beta_l(S_k)\}$$

where $\beta_1, \ldots, \beta_5$ are the implicit enumeration schemes in Bounds 5B and 5C.

Proof. The linear programming relaxation of $IEP(S_k)$ is

$$IEP(\bar{S}_k) \qquad \begin{array}{lll} \min & \sum_{j \in F_k} c_j x_j + \kappa_k & \\ \text{s.t.} & \sum_{j \in F_k} a_{ij} x_j + \gamma_i^k \geq b_i & \text{for all } i \\ & 1 \geq x_j \geq 0 \quad \text{for all } j \in F_k & \end{array}$$

Each $\beta_3(S_k)$ is a relaxation obtained by considering only one main constraint row, i. Thus, $v(IEP(\bar{S}_k)) \geq \beta_3(S_k)$. It also follows that $v(IEP(\bar{S}_k)) \geq \beta_2(S_k)$ because β_2 can be written (as in Section 5.2.1)

$$\beta_2(S_k) = \max_{\{i:\, \gamma_i^k < b_i\}} v \begin{pmatrix} \min & \sum_{j \in F_k} c_j x_j + \kappa_k & \\ \text{s.t.} & \sum_{j \in F_k} a_{ij} x_j \geq b_i - \gamma_i^k & \\ & x_j \geq 0 & \text{for all } j \in F_k \end{pmatrix}$$

Now each optimization in β_1 is the special β_4 where $\mathbf{u}$ is a unit vector. If the linear programming relaxation dominates β_4 it certainly dominates β_1. When $\beta_4(S_k) = \kappa_k$, dominance is clear. If $\beta_4(S_k) = +\infty$, there exists some $\mathbf{u} \geq \mathbf{0}$ such that

$$\sum_{j \in F_k} \max\{0, \mathbf{u}\mathbf{a}^j\} + \mathbf{u}\gamma^k < \mathbf{u}\mathbf{b}. \tag{5-4}$$

Thus (5-4) implies

$$\sum_{j\in F_k} \mathbf{u}\mathbf{a}^j x_j + \mathbf{u}\gamma^k \geq \mathbf{u}\mathbf{b}$$

$$0 \leq x_j \leq 1 \qquad \text{for all } j \in F_k$$

has no solutions. That is, $v(IEP(\bar{S}_k)) = +\infty$.

To prove β_5 is dominated by the linear programming relaxation, we shall introduce a new bound. Geoffrion (1969) has defined the *strongest* surrogate constraint function β_G as

$$\beta_G(S_k) \triangleq \max_{\mathbf{u}\geq\mathbf{0}} \left\{\kappa_k + \mathbf{u}(\mathbf{b} - \gamma^k) + \sum_{j\in F_k} \min\{0, c_j - \mathbf{u}\mathbf{a}^j\}\right\}.$$

Clearly, $\beta_5(S_k) \leq \beta_G(S_k)$. But introducing new variables, $w_j = -\min\{0, c_j - \mathbf{u}\mathbf{a}^j\}$, yields

$$\beta_G(S_k) = v\begin{pmatrix} \max & \kappa_k + \mathbf{u}(\mathbf{b} - \gamma^k) - \sum_{j\in F_k} w_j & \\ \text{s.t.} & c_j - \mathbf{u}\mathbf{a}^j \geq -w_j & \text{for all } j \in F_k \\ & w_j \geq 0 & \text{for all } j \in F_k \\ & \mathbf{u} \geq \mathbf{0} & \end{pmatrix}.$$

The linear programming dual of the problem on the right is exactly $IEP(\bar{S}_k)$ above. Thus,

$$\beta_5(S_k) \leq \beta_G(S_k) = v(IEP(\bar{S}_k)).$$ ■

Two notes need to be added to Theorem 5.10. First, such a theorem does not hold for all implicit enumeration bounds. The simple $IEP(S)$

$$\begin{array}{ll} \min & 6x_1 \\ \text{s.t.} & 3x_1 \geq 2 \\ & -3x_1 \geq -2 \\ & x_1 = 0 \text{ or } 1 \end{array}$$

illustrates that while $v(IEP(\bar{S})) = 4$, the completion cardinality bound, $\beta_6(IEP(S)) = +\infty$. The second point is one we have mentioned before. Theorem 5.10 demonstrates that the linear programming relaxation bound dominates simple implicit enumeration bounds. It does not necessarily suggest that the additional computational effort incurred in determining the linear programming bound is justified.

5.2.4 *Alternative Formulations*

It is very often the case that the same discrete optimization problem admits a number of different formulations, i.e., there are a number of different combinations of variables and constraints that lead to the same optimal solution set. It is important to realize that such alternative formulations may lead to quite different bound values—even when exactly the same bounding strategy is employed. We shall have more to say about this phenomenon in later sections of this book. Indeed all the theory of cutting planes treated in Chapter 6 can be viewed as identifying formulations with linear programming relaxations that more and more closely approximate the integer feasible space.

At this point, we shall only illustrate the above issue with a simple example.

Example 5.6. Alternative Linear Relaxations. Consider

$$P1(S) \qquad \begin{array}{ll} \min & 2x_1 + 26x_2 + 45x_3 + 60x_4 - 100x_5 \\ \text{s.t.} & x_1 + x_2 + x_3 + x_4 \geq 4x_5 \\ & \mathbf{x} \in S = \{(x_1, x_2, x_3, x_4, x_5) \text{: } \mathbf{x} \text{ binary}\} \end{array}$$

Quite clearly, the unique optimal solution is to make all $x_j = 0$. That solution is cheaper than others with any $x_j = 0$, and the only other feasible solution is all $x_j = 1$. Thus, $v(P1(S)) = 0$. The linear programming relaxation $P1(\bar{S})$ yields the fractional optimal solution $(1, 0, 0, 0, \frac{1}{4})$ with solution value -23. Still, $P1(S)$ is not the only way to formulate our problem. It is just as correct to write the main constraints of the problem in any of the four ways shown in Table 5.2. Notice, however, that each of the four yields a different bound value although all are derived via linear programming relaxation. Of course bound improvements are obtained at the cost of solving more complex linear programs. The fact remains that integer-equivalent formulations, dealt with by the same bounding strategy, can yield substantially different bound values. ■

5.3 Conditional Bounds and Penalties

Regardless of the bounding scheme employed in a branch and bound algorithm, the bound is likely to improve if the feasible space of the problem can be more tightly described. The previous section illustrated how changes in formulation can produce tighter bounds.

TABLE 5.2. Illustrative Variation of Bounds with Formulation in Example 5.6

Formulation name	Constraints (other than $\mathbf{x} \in S$)	Optimal Solution to linear programming relaxation	Linear relaxation solution value
$P1(S)$	$x_1 + x_2 + x_3 + x_4 \geq 4x_5$	$(1, 0, 0, 0, \frac{1}{4})$	-23
$P2(S)$	$x_1 \geq x_5$ $x_2 + x_3 + x_4 \geq 3x_5$	$(\frac{1}{3}, 1, 0, 0, \frac{1}{3})$	$-6\frac{2}{3}$
$P3(S)$	$x_1 + x_2 \geq 2x_5$ $x_3 + x_4 \geq 2x_5$	$(1, 0, 1, 0, \frac{1}{2})$	-3
$P4(S)$	$x_1 \geq x_5$ $x_2 \geq x_5$ $x_3 \geq x_5$ $x_4 \geq x_5$	$(0, 0, 0, 0, 0)$	0

A similar improvement can be achieved if we can more closely describe the defining set S_k associated with a candidate problem $P(S_k)$ of a particular branch and bound algorithm. One approach is to compute *range restrictions,*[1] i.e., tighter upper and lower limits on discrete variable values. For example, we might be able to conclude at a certain stage of branch and bound enumeration that, while the current candidate problem $P(S_k)$ cannot be fathomed, any optimal solution to $P(S_k)$ will have $x_q = 0$. Setting $S_k \leftarrow S_k \cap \{\mathbf{x}: x_q = 0\}$ will tend to improve bounds computed from relaxations of $P(S_k)$ and facilitate direct solution of $P(S_k)$.

How does one show that the range of a variable can be tightened? The approach is exactly like the one employed to demonstrate that a candidate problem cannot lead to an improvement over an incumbent solution: we compute bounds over portions of the feasible set S_k.

More specifically, range restriction procedures are implemented via *conditional bounds,* i.e., bounds that are valid for only a portion of the feasible space of a candidate problem. For any 0–1 variable x_j, denote $P(S_k \cap \{\mathbf{x}: x_j = 0\})$ and $P(S_k \cap \{\mathbf{x}: x_j = 1\})$ by $P(S_k: x_j = 0)$ and $P(S_k: x_j = 1)$, respectively. Then, from our previous development of branch and bound, we know that whenever $v(P(S_k: x_j = 1)) \geq v^*$, we can set $S_k \leftarrow S_k \cap \{\mathbf{x}: x_j = 0\}$. Similarly, if $v(P(S_k: x_j = 0)) \geq v^*$, we can set $S_k \leftarrow S_k \cap \{\mathbf{x}: x_j = 1\}$. Of course, if $\min\{v(P(S_k: x_j = 0)), v(P(S_k: x_j = 1))\} \geq v^*$, we can fathom $P(S_k)$.

As with main fathoming, range restriction is usually accomplished through computing bounds on conditional candidate problems rather than by actually solving the problems. In principle, any bound relationship can

[1] Some authors use the term *pegging* for what we shall call *range restriction.*

be used. Still, we would not wish to invest great amounts of computation in range restriction; the effort would probably be better devoted to computing main bounds and fathoming.

5.3.1 *Conditional Bounds from Implicit Enumeration*

One obvious source of such quick and dirty relationships is the collection of implicit enumeration bounds developed in Section 5.2.1. All the Bounds 5B and 5C could be used, but the most commonly employed is the one presented as β_1. Specifically, for pure 0–1 problems, $IEP(S_k)$, with nonnegative objective function coefficients,

$$\beta_1(S_k) \triangleq \begin{cases} +\infty & \text{if for any row } i \\ & \sum_{j \in F_k} \max\{0, a_{ij}\} + \gamma_i^k < b_i \\ \kappa_k & \text{otherwise} \end{cases}$$

where F_k is the index set of free variables, γ_i^k is the committed component in row i arising from variables fixed in S_k, and κ_k is the corresponding committed component of the objective function.

Assuming $\beta_1(S_k) < v^*$, application of β_1 in the conditional bound context leads immediately to the following three rules:

(i) For each $p \in F_k$ such that $c_p + \kappa_k \geq v^*$, $S_k \leftarrow S_k \cap \{\mathbf{x}: x_p = 0\}$.

(ii) For each $p \in F_k$ such that there is a row i with

$$\sum_{\substack{j \in F_k \\ j \neq p}} \max\{0, a_{ij}\} + a_{ip} + \gamma_i^k < b_i,$$

$$S_k \longleftarrow S_k \cap \{\mathbf{x}: x_p = 0\}.$$

(iii) For each $p \in F_k$ such that there is a row i with

$$\sum_{\substack{j \in F_k \\ j \neq p}} \max\{0, a_{ij}\} + \gamma_i^k < b_i$$

$$S_k \longleftarrow S_k \cap \{\mathbf{x}: x_p = 1\}.$$

Rule (i) is derived from the observation that the new committed part of the objective function has $v(IEP(S_k\colon x_p = 1)) \geq v^*$; rules (ii) and (iii) follow from infeasibility of $IEP(S_k\colon x_p = 1)$ and $IEP(S_k\colon x_p = 0)$ respectively.

Example 5.7. Implicit Enumeration Conditional Bounds. The rules are illustrated by the candidate problem

$$\begin{array}{lll} & \min & x_{13} + 10x_{29} + x_{30} + \ \ x_{37} + 82 \\ IEP(S_k) & \text{s.t.} & -10x_{13} + \ \ x_{29} + x_{30} + 5x_{37} + 15 \geq 18 \\ & & (x_{13}, x_{29}, x_{30}, x_{37}) \in S_k = \{\text{binary 4-vectors}\} \end{array}$$

with $v^* = 85$. Rule (i) assures $x_{29} = 0$ in any helpful solution to $IEP(S_k)$ because $10 + 82 \geq 85$. Rules (ii) and (iii) show $IEP(S_k\text{: } x_{13} = 1)$ and $IEP(S_k\text{: } x_{37} = 0)$ are infeasible. It follows that we can replace S_k by $\{(x_{13}, x_{29}, x_{30}, x_{37})\text{: } x_{13} = x_{29} = 0, x_{37} = 1, x_{30} = 0 \text{ or } 1\}$. ■

5.3.2 *Post-Linear Programming Penalties*

When linear relaxations $P(\overline{S}_k)$ are solved in the main bound procedure of a branch and bound algorithm, linear programming theory provides us with another class of conditional bounds. Let us assume surplus variables are added to the original decision vector, $\mathbf{x}$, in the usual way, so that $P(S)$ is expressed in the equivalent, equality form

$$\begin{array}{lll} & \min & (\mathbf{c}, \mathbf{0})\mathbf{x} \\ P(S) & \text{s.t.} & (\mathbf{A}, -\mathbf{I})\mathbf{x} = \mathbf{b} \\ & & \mathbf{x} \in S = \{\mathbf{x} \geq \mathbf{0}\text{: } x_j = 0 \text{ or } 1 \text{ for all } j \in I\} \end{array}$$

For simplicity, let us also deal only with the root candidate problem $P(S)$, although any $P(S_k)$ could be treated in the same way.

If $P(\overline{S})$ is solved by any simplex procedure, we will end with both an LP-optimal solution, $\bar{\mathbf{x}}$, and a revised representation of $P(S)$ in terms of the nonbasic variables at LP-optimality. Following Appendix C, that representation can be stated

$$\begin{array}{llll} & \min & \sum_{j \in N} \bar{c}_j x_j + v(P(\overline{S})) & \\ P_N(S) & \text{s.t.} & x_k = \bar{x}_k - \sum_{j \in N} \bar{a}_{kj} x_j & \text{for all } k \notin N \\ & & \mathbf{x} \in S & \end{array}$$

where N is the index set of nonbasic variables, $\bar{a}_{kj}$ is the coefficient of nonbasic variable x_j in the representation of basic variable x_k, and $\bar{c}_j$ is the adjusted cost of x_j after basic variables have been eliminated from the original cost row $(\mathbf{c}, \mathbf{0})$.

The representation $P_N(S)$ immediately allows us to compute *penalties* for forcing binary variables down to 0 from their $\bar{x}_j$ values, or up to 1. The next theorem formalizes these penalty calculations of Beale and Small (1965) and Driebeek (1966).

Theorem 5.11. Linear Relaxation Penalties. *Let $P(S)$ be represented in the equivalent form $P_N(S)$ obtained at simplex optimality of $P(S)$. Then for any $k \in I$*

$$v(P(S\colon x_k = 0)) \geq v(P(\bar{S}\colon x_k = 0)) \geq v(P(\bar{S})) + \begin{cases} 0 & \text{if } k \in N \\ \min\limits_{\substack{j \in N \\ \bar{a}_{kj} > 0}} \left\{ \dfrac{\bar{x}_k \bar{c}_j}{\bar{a}_{kj}} \right\} & \text{if } k \notin N \end{cases}$$

$$v(P(S\colon x_k = 1)) \geq v(P(\bar{S}\colon x_k = 1)) \geq v(P(\bar{S})) + \begin{cases} \bar{c}_k & \text{if } k \in N \\ \min\limits_{\substack{j \in N \\ \bar{a}_{kj} < 0}} \left\{ \dfrac{(1 - \bar{x}_k)\bar{c}_j}{-\bar{a}_{kj}} \right\} & \text{if } k \notin N \end{cases}$$

Proof. Consider first some nonbasic, binary x_k with $k \in I \cap N$. Since such $\bar{x}_k = 0$, it is clear that $v(P(\bar{S})) = v(P(\bar{S}\colon x_k = 0))$. Moreover, from linear programming we know $\bar{c}_k$ to be an underestimate of the unit cost of increasing x_k. Thus, $v(P(\bar{S}\colon x_k = 1)) \geq v(P(\bar{S})) + \bar{c}_k$.

To deal with basic, binary x_k, i.e., $k \in I \backslash N$, consider the following relaxation of $P_N(\bar{S})$.

$$\begin{aligned} \min \quad & \sum_{j \in N} \bar{c}_j x_j + v(P(\bar{S})) \\ \text{s.t.} \quad & x_k = \bar{x}_k - \sum_{j \in N} \bar{a}_{kj} x_j \qquad\qquad (5\text{-}5) \\ & x_j \geq 0 \qquad \text{for all } j \in N \end{aligned}$$

If the constraint $x_k = 0$ is added, $\bar{x}_k = \Sigma_{j \in N} \bar{a}_{kj} x_j$. Noting that all $\bar{c}_j \geq 0$, the optimal solution value of such a relaxation can then be obtained by inspection as $v(P(\bar{S})) + \min\{\bar{x}_k \bar{c}_j / \bar{a}_{kj}\colon \bar{a}_{kj} > 0\}$. Similarly, if $x_k = 1$ the (5-5) relaxation has value $v(P(\bar{S})) + \min\{-(1 - \bar{x}_k)\bar{c}_j / \bar{a}_{kj}\colon \bar{a}_{kj} < 0\}$. The remainder of the theorem follows from the fact that (5-5) is a relaxation of $P_N(\bar{S})$, which is in turn equivalent to $P(\bar{S})$ and a relaxation of $P(S)$. ■

Example 5.8. Up and Down Penalties. Consider the simple knapsack example

$$\begin{aligned} & \min \quad && 5x_1 + x_2 + 3x_3 \\ KNX(S) \quad & \text{s.t.} \quad && 20x_1 + x_2 + 2x_3 \geq 2 \\ & && (x_1, x_2, x_3) \in S = \{\text{integer } \mathbf{x}\colon \mathbf{0} \leq \mathbf{x} \leq \mathbf{1}\}. \end{aligned}$$

After including a surplus variable, x_4, and solving $KNX(\bar{S})$ we obtain the equivalent form

$$\begin{array}{lll} & \min & \frac{3}{4}x_2 + \frac{5}{2}x_3 + \frac{1}{4}x_4 + \frac{1}{2} \\ KNX_N(S) & \text{s.t.} & x_1 = \frac{1}{10} - \left[\frac{1}{20}x_2 + \frac{1}{10}x_3 - \frac{1}{20}x_4\right] \\ & & (x_1, x_2, x_3) \in S,\ x_4 \geq 0 \end{array}$$

Here the nonbasic set $N = \{2, 3, 4\}$.

For the nonbasic variable, x_3, $v(KNX(S\colon x_3 = 0)) \geq \frac{1}{2} + 0$ and $v(KNX(S\colon x_3 = 1)) \geq \frac{1}{2} + \frac{5}{2}$. For the basic variable x_1, a bound on $v(KNX(S\colon x_1 = 0))$, i.e., the cost with x_1 moved down to 0 is $\frac{1}{2} + \min\{(\frac{1}{10})(\frac{3}{4})/(\frac{1}{20}), (\frac{1}{10})(\frac{5}{2})/(\frac{1}{10})\} = 2$. The bound for moving x_1 up to 1 that underestimates $v(KNX(S\colon x_1 = 1))$ is $\frac{1}{2} + \frac{9}{10}(\frac{1}{4})/\frac{1}{20} = 5$. ■

Theorem 5.11 addressed the computation of simple penalties or conditional bounds on basic binary variables via one-row relaxations, (5-5), of $P_N(\bar{S})$. Many extensions are possible if additional constraints are added to the one-row relaxations. One simple notion is to add to the relaxations, upper bounds on x_j for $j \in N$. For example, (5-5) might be solved with the additional constraints.

$$x_j \leq 1 \qquad \text{for all } j \in N \cap I$$

The solution value of the resulting one-row linear program is not as easily written down as was the case for Theorem 5.11, but it is elementary to compute. In the case of x_1 in Example 5.8, this extension would show

$$v(KNX(S\colon x_1 = 0)) \geq \frac{1}{2} + \frac{3}{4} + \frac{1}{20}\left(\frac{5}{2}\right) \Big/ \frac{1}{10} = 2\frac{1}{2}.$$

Tomlin (1971) proposed a more subtle extension that incorporates the constraints x_j integer for all $j \in N \cap I$ into the relaxations (5-5).

Theorem 5.12. Tomlin Penalties. *Let $P(S)$ be represented in the equivalent form $P_N(S)$ obtained at simplex optimality of $P(\bar{S})$. Then for any $k \in I \backslash N$ with $0 < \bar{x}_k < 1$*

$$v(P(S)) \geq v(P(\bar{S})) + \min_{\substack{j \in N \\ g_{kj} > 0}} \left\{\frac{\bar{c}_j}{g_{kj}}\right\},$$

where

$$g_{kj} = \begin{cases} \dfrac{-\bar{a}_{kj}}{1-\bar{x}_k} & j \notin I,\ \bar{a}_{kj} \le 0 \\[2ex] \dfrac{\bar{a}_{kj}}{\bar{x}_k} & j \notin I,\ \bar{a}_{kj} < 0 \\[2ex] \min\left\{\dfrac{1-(\bar{a}_{kj}-\lfloor \bar{a}_{kj} \rfloor)}{1-\bar{x}_k}, \dfrac{\bar{a}_{kj}-\lfloor \bar{a}_{kj} \rfloor}{\bar{x}_k}\right\} & j \in I. \end{cases}$$

Proof. Clearly the theorem will be true if

$$\begin{aligned} \min \quad & \sum_{j \in N} \bar{c}_j x_j + v(P(\bar{S})) \\ \text{s.t.} \quad & \sum_{j \in N} g_{kj} x_j \ge 1 \qquad\qquad (5\text{-}6) \\ & x_j \ge 0 \qquad \text{for all } j \in N \end{aligned}$$

is a relaxation of $P(S)$. That this is the case follows since the constraint (5-6) is the classic mixed-integer cutting plane of Gomory (1960a), and does hold for all $\mathbf{x}$ feasible in $P(S)$. We shall reserve the proof of this fact, however, for the development of cutting planes in Section 6.2. ∎

Although it was convenient to develop them at this point, the reader should observe that the penalties of Theorem 5.12 are not conditional bounds valid only in parts of the feasible region. Instead, they globally lower bound the cost of achieving integrality on a particular basic variable. Thus, while they may improve our bound on $v(P(S))$, they do not provide range restrictions.

To illustrate, return to x_1 of Example 5.8. The constraint (5-5) obtained from $KNX_N(S)$ is

$$x_2 \min\left\{\frac{1-\frac{1}{20}}{1-\frac{1}{10}}, \frac{\frac{1}{20}}{\frac{1}{10}}\right\} + x_3 \min\left\{\frac{1-\frac{1}{10}}{1-\frac{1}{10}}, \frac{\frac{1}{10}}{\frac{1}{10}}\right\} + x_4\left(\frac{\frac{1}{20}}{1-\frac{1}{10}}\right) \ge 1$$

Thus, $v(KNX(S)) \ge \frac{1}{2} + \min\{\frac{3}{4}/\frac{1}{2}, \frac{5}{2}/1, \frac{1}{4}/\frac{1}{18}\} = 2$. Since 2 is a valid lower bound for the whole problem $KNX(S)$, it is greater than or equal to the minimum of the up and down bounds computed in Example 5.8. Note however that the up bound of 5 exceeds the one we have obtained from

Tomlin penalties. The higher bound is possible because it is valid only in the restricted region $S \cap \{\mathbf{x}: x_1 = 1\}$.

As we have already noted several times, the relative usefulness of any bounding—including the above penalties—depends on whether enough enumeration is avoided by a bound to recover its computation cost. Although exceptions are known, post-linear programming penalties have often not justified themselves in this sense. Forrest *et al* (1974) show that serious difficulty arises from the high degree of degeneracy often found in linear relaxations of discrete optimization problems. In such cases, the penalties of Theorems 5.11 and 5.12 are very poor estimates of the conditional solution values they lower bound. A second difficulty follows from the effort to compute the $\bar{a}_{kj}$. Although these values are consequences of the optimal basis, most simplex codes do not routinely compute them. The revised simplex methods used in large procedures require $\bar{a}_{kj}$ for only one j per iteration. Thus, their design usually makes computation of all $\bar{a}_{kj}$ rather bulky and time consuming.

5.4 Heuristic Aspects of Branch and Bound

Even if candidate problems are being characterized by partial solution, and the bounding procedures to be used at Step 2 of Algorithm 5A have been selected, various details of the procedure remain to be specified. Most such details are essentially heuristic in nature; only empirical experience with a particular class of problems can indicate which alternative is really best. The range of alternatives and considerations in selecting among them, however, can be set out fairly generally.

5.4.1 Candidate Problem Selection

At any point in the evolution of a branch and bound Algorithm 5A, there may be many active candidate problems among which to choose at Step 1. One important heuristic aspect of such an algorithm is the rule used to make that selection.

What information is available to aid candidate selection? One source is bounding computations. As we process a selected candidate problem $P(S_k)$ through Algorithm 5A, we compute lower bounds $\beta(S_k)$. When the candidate cannot be fathomed, S_k will be subdivided via branching. But the original $P(S_k)$ is a relaxation of all the new candidates. Thus, $\beta(S_k)$ is a valid lower bound on the solution values of the new candidates. If conditional bounds were computed in processing the original $P(S_k)$, they may provide more precise information. Each conditional bound corresponds to a subset of the feasible region for $P(S_k)$. Should that subset subsequently be chosen as the feasible region for one of the new candidate problems, the computed conditional bound is a readily available lower bound on the new candidate.

We shall denote by $\sigma(S_l)$ the best such *a priori stored bound* on a stored candidate problem $P(S_l)$. Before we actually choose a candidate to process, $\sigma(S_l)$ is the greatest known lower bound on $v(P(S_l))$.

Let us now define two specific candidate selection alternatives. A *depth first* (also called *last in first out*) rule selects the candidate that was most recently added to the candidate list, or equivalently, the one of greatest branching depth. A *best bound* procedure picks the candidate $P(S_k)$ with least stored bound, $\sigma(S_k)$.

Figure 5.4 illustrates these alternatives for the first ten candidate problems of an enumeration. Numbers within nodes of the branch and bound tree indicate the sequence in which the nodes (or the corresponding candidate problems) are considered. Hypothetical values for bounds $\sigma(S_k)$ available when each problem was selected, and bounds $\beta(S_k)$ computed at Step 2 are shown beside the node. Since $\sigma(S_l)$ of candidates derived from a parent candidate $P(S_k)$ exceed $\beta(S_k)$, we may presume conditional bounds of Section 5.3 were employed in setting $\sigma(S_l)$'s. We assume that v^* remains at value 90 throughout both cases and that all aspects of the algorithms except Step 1 are identical.

From Figure 5.4, we see the depth first rule proceeds ever deeper in the tree, partitioning the same problem until it can be fathomed. On the other hand, best bound skips from location to location, generally having more breadth and less depth at any given point.

How can the alternatives be compared? We can identify four measures:

- *candidate storage* is the maximum number of candidate problems that can ever be simultaneously active (stored).
- *calculation restart* is the degree to which bound and branching calculations for one candidate problem can use results from work on previous candidates.
- *number of candidate problems* is the number of candidate problems for which bounds must be explicitly calculated in obtaining a provable optimal solution to $P(S)$.
- *quality of incumbent solutions* is the rate at which values of the (feasible) incumbent solutions approach $v(P(S))$.

Considerable technical detail must be added to state precise theorems ranking candidate selection rules according to these measures. It is not hard to see, however, that depth first will usually dominate all other procedures on the first two.

Consider candidate storage. Since depth first will further and further

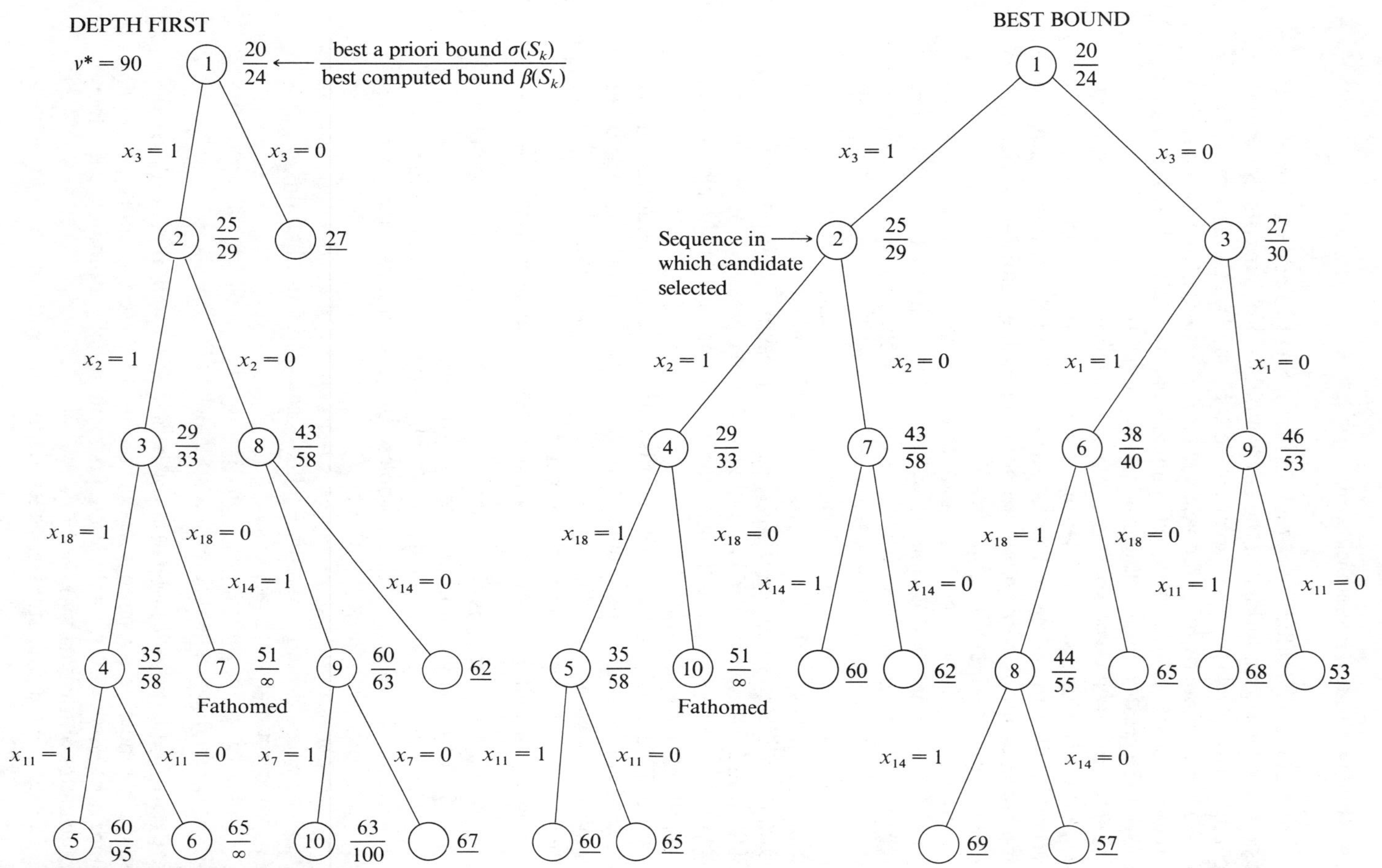

DEPTH FIRST
v* = 90
best a priori bound σ(S_k)
best computed bound β(S_k)
Fathomed
Fathomed
BEST BOUND
Sequence in which candidate selected

enumerate the same candidate until it can be fathomed, all simultaneously stored candidates (except possibly the last two) will have different numbers of fixed discrete variables. Thus, there will never be more than $1 + |I|$ candidates to store at one time (1 plus one for each possible length). Any reasonable alternative to the depth first rule will still have to provide for storing those same $|I| + 1$ candidates. But if any but the longest candidate has been partitioned, there may be many others to store simultaneously.

Calculation restart is even more intuitive. Bound and other computations per candidate problem will generally be saved if the candidate most like the previous one is chosen at each execution of Step 1. For example, if bounds are computed by simplex linear programming, one may only have to make a few pivots to move from the optimal solution to say $P(\bar{S}_k)$ to that of $P(S_k\colon x_j = 1)$. There are contrived counter-examples, but depth first generally maximizes the frequency of such easy cases. Any method that skips around in the branch and bound tree will require more bound computation effort.

If depth first were dominant on all four effectiveness measures, it would be universally employed. Unfortunately, it ranks poorly on our third criterion—the number of candidate problems. To see why this is so, study the convergence plots of Figure 5.5. These graphs show the value of the incumbent solution, v^*, and the stored lower bound, $\sigma(S_k)$, of candidates selected in fully solving the example of Figure 5.4.

In both graphs, the upper bound, v^*, is a monotone, nonincreasing step function, although there is no reason the two functions should be identical. Changes in v^* occur when new incumbents are uncovered.

The difference between best bound and depth first pertains to stored lower bounds. Since best bound always picks the active candidate with least lower bound, selected $\sigma(S_k)$ form a monotone nondecreasing function of the selection number. Depth first will also keep monotone $\sigma(S_k)$ so long as we partition the same candidate. But monotonicity may be lost when we backtrack, i.e., fathom the current candidate and move backwards in the tree.

How can we compare, on such plots, the total number of candidates required under various rules? Note that the plot of $\sigma(S_k)$ under best bound never exceeds $v(P(S))$. This occurs because our chosen $\sigma(S_k)$, the *least* stored lower bound, is always an overall lower bound on $v(P(S))$. It follows that (if bound calculations are independent of v^*) no candidate considered under best bound could have been avoided by other rules. Such candidates may be fathomed after we begin computations on them, but the bound we had when we selected candidates satisfied $\sigma(S_k) \leq v(P(S)) \leq v^*$.

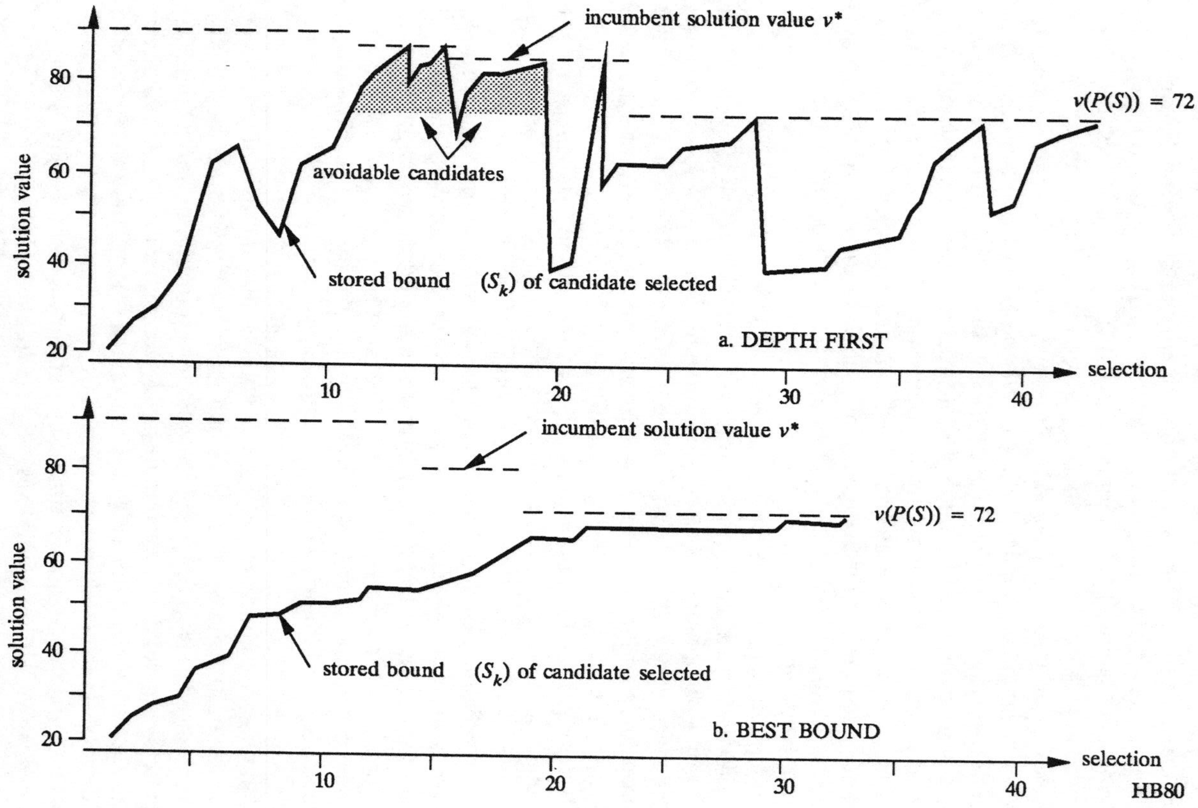

incumbent solution value v^*
80
60
40
20
solution value
avoidable candidates
stored bound (S_k) of candidate selected
$v(P(S)) = 72$
a. DEPTH FIRST
10
20
30
40
selection
incumbent solution value v^*
80
60
40
20
solution value
stored bound (S_k) of candidate selected
$v(P(S)) = 72$
b. BEST BOUND
10
20
30
40
selection
HB80

By contrast, notice in Figure 5.5 that $\sigma(S_k)$ often exceeds $v(P(S))$ under depth first. Therein is the extra enumeration cost of the depth first rule. The indicated candidates (those with $\sigma(S_k) > v(P(S))$) could have been avoided had we used the best bound alternative. For algorithms where bounding and branching are independent of v^*, best bound minimizes the number of candidate problems processed.

So far, we have depth first dominant with respect to storage and computation per candidate, best bound dominates on the number of candidates. The picture becomes even less clear when we turn to the incumbent solution quality criterion. That measure is often the most important because it is typical (as indicated in Figure 5.5) that branch and bound algorithms obtain good incumbent solutions fairly early in processing. The time-consuming part of the solution process comes in proving the quality of such solutions by slowly improving the lower bounds. Thus, we would prefer rules that rapidly decrease v^*; in case, our computational resources are exhausted before optimality can be formally proved.

Unfortunately, no simple rule like depth first or best bound is known to dominate alternatives in producing good incumbent solutions. Thus, we are left with conflict on some of our four measures and uncertainty on the single most important. Any actual code will be forced to adopt an empirical compromise. Forecasting schemes like those of Section 5.4.3 appear to be the most viable basis for such compromises.

5.4.2 *Feasible Solutions*

In developing both Algorithm 5A and its implicit enumeration and linear programming variants, we have moved rather quickly over Step 3, generation of feasible solutions. That is because very little is required of Step 3 to prove convergence of branch and bound (as in Theorems 5.1 and 5.2). The only real requirement is that a candidate problem $P(S_k)$ should be completely solved if it is fully enumerated, i.e., has an assigned value for each x_j with $j \in I$.

In the practical world of actual branch and bound codes, feasible solutions play a much more central role. A great deal of computation is often invested in nonexact, rounding algorithms to generate solutions at Step 3. To see why, we need only to consider again the measures of candidate selection effectiveness treated in the previous section. Obviously, if we expend more effort in generating feasible solutions, the quality of our incumbent solution should improve more rapidly. But there are other gains as well. We have already noted that the number of candidate problems to be stored may grow rapidly as the algorithm proceeds. Good incumbent solutions can help to control that growth. If a candidate can be fathomed, it need never be stored. Even if a candidate could not be fathomed when it was set up, its stored bound $\sigma(S_k)$ may be high enough to delete $P(S_k)$ without explicitly computing its Step 2 bounds. The incumbent solution saving Step 4 of Algorithms 5A and 5D can be restated:

STEP 4: *Incumbent Saving.* Save the new feasible solution $\tilde{\mathbf{x}}$ as the incumbent by $\mathbf{x}^* \leftarrow \tilde{\mathbf{x}}$, $v^* \leftarrow \mathbf{c}\tilde{\mathbf{x}}$. If v^* is now $-\infty$, stop; $P(S)$ is unbounded. If not, delete from the candidate list any $P(S_i)$ with saved bound $\sigma(S_i) \geq$ the new v^*. Then go to Step 6 if $\tilde{\mathbf{x}}$ solves the current $P(S_k)$ and to Step 5 otherwise.

Rapidly decreasing incumbent solution values v^* will eliminate many stored problems in this step.

Another look at the plot for best bound candidate selection in Figure 5.5 will show how good feasible solutions can also reduce the number of candidate problems processed through the bounding effort of Step 2. Lower incumbent solutions values help to minimize the effort spent on avoidable candidates with stored bound $\sigma(S_k) > v(P(S))$.

5.4.3 *Branching Variable Selection and Pseudo-Costs*

We have noted above that a forecast of the true candidate solution values $v(P(S_k))$ could assist in selecting among candidate problems at Step 1 of Algorithm 5A. Forecasts can also assist with another heuristic part of the procedure. At Step 5 of the algorithm, we must (assuming we are using a partial solution approach) select a previously free variable to fix. Two new candidates will be created, one with that variable fixed at 0 and the other with it fixed at 1.

What variable do we pick? Many strategies have been proposed. We might try to branch so that uncertainty is resolved most rapidly, i.e., branch on variables where 1 and 0 seem equally acceptable values. On the other hand, we might try to make the most obvious decisions—branch on the variable where it is most clear that 1 is preferable to 0 (or vice versa).

Either way, we require forecasts of $v(P(S_k))$ or more precisely $v(P(S_k\colon x_p = 0))$ and $v(P(S_k\colon x_p = 1))$. In the literature of discrete optimization, such forecasts are called *pseudo-costs.* Although they estimate true costs (e.g. $v(P(S_k))$, $v(P(S_k\colon x_p = 0))$, $v(P(S_k\colon x_p = 1))$) pseudo-costs need not be any sort of bound. Their role is in guiding heuristic decisions in the algorithm, not in generating incumbent solutions or in proving fathomability.

Since our goal in pseudo-costing is to estimate solution values, there are numerous techniques that might be attempted. Most practical work, however, has concentrated on a single notion—*pricing infeasibility.* Beginning from the (infeasible if we do not fathom) solution of the relaxation solved for bounding, peudo-cost procedures typically estimate optimal solution cost by adding an adjustment for the amount of infeasibility.

To provide a more precise context, we shall assume for the remainder of this section that bounds are obtained by solving linear programming relaxations $P(\bar{S}_k)$. (See, for example, Land and Powell (1977) for a more complete discussion.) In such cases, the optimality linear relaxation solution $\bar{\mathbf{x}}^k$ will be feasible for all constraints except the one requiring x_j integer for $j \in I$.

We can define the *integer infeasibility* of the $\bar{\mathbf{x}}^k$ solution of $P(\bar{S}_k)$ by

$$\xi(S_k) = \sum_{j \in I} \min\{\bar{x}_j^k, 1 - \bar{x}_j^k\}.$$

The quantity $\xi(S_k)$ is the distance to the nearest integer choice of $x_j, j \in I$. Depending on how many pricing parameters we wish to employ, the infeasibility pricing notion leads immediately to several estimates of true solution values.

Suppose x_p is selected as the variable on which to partition $P(S_k)$. Possible estimators for the value of the two new candidate problems are

$$v(P(S_k\colon x_p = 0)) \approx v(P(\bar{S}_k)) + \mu \bar{x}_p^k + \mu \sum_{\substack{j \in I \\ j \neq p}} \min\{\bar{x}_j^k, 1 - \bar{x}_j^k\} \tag{5-7}$$

$$v(P(S_k\colon x_p = 1)) \approx v(P(\bar{S}_k)) + \mu(1 - \bar{x}_p^k) + \mu \sum_{\substack{j \in I \\ j \neq p}} \min\{\bar{x}_j^k, 1 - \bar{x}_j^k\} \tag{5-8}$$

$$v(P(S_k\colon x_p = 0)) \approx v(P(\bar{S}_k)) + \mu_p \bar{x}_p^k + \sum_{\substack{j \in I \\ j \neq p}} \mu_j \min\{\bar{x}_j^k, 1 - \bar{x}_j^k\} \tag{5-9}$$

$$v(P(S_k\colon x_p = 1)) \approx v(P(\bar{S}_k)) + \mu_p(1 - \bar{x}_p^k) + \sum_{\substack{j \in I \\ j \neq p}} \mu_j \min\{\bar{x}_j^k, \bar{x}_j^k\} \tag{5-10}$$

$$v(P(S_k\colon x_p = 0)) \approx v(P(\bar{S}_k)) + \mu_p^- \bar{x}_p^k + \sum_{\substack{j \in I \\ j \neq p}} \min\{\mu_j^- \bar{x}_j^k, \mu_j^+(1 - \bar{x}_j^k)\} \tag{5-11}$$

$$v(P(S_k\colon x_p = 1)) = v(P(\bar{S}_k)) + \mu_p^+(1 - \bar{x}^k) + \sum_{\substack{j \in I \\ j \neq p}} \min\{\mu_j^- \bar{x}_j^k, \mu_j^+(1 - \bar{x}_j^k)\} \tag{5-12}$$

The first estimate pair weights all infeasibility by the single price μ; the second uses a separate price μ_j for each $j \in I$; the last assesses different prices μ_j^- and μ_j^+ for moving x_j down to 0 or up to 1.

There remains the issue of how we obtain the prices μ, μ_j, or μ_j^- and μ_j^+. One source would be the penalties of Section 5.3.2. The Tomlin penalties of Theorem 5.12 lower bound the cost of integrality on particular variables, exactly what we wish to estimate by $(\mu_j \min\{\bar{x}_j^k, 1 - \bar{x}_j^k\})$. The earlier penalties of Theorem 5.11 distinguish between up and down directions as we must to approximate μ_j^+ and μ_j^-.

We have noted earlier, however, that degeneracy often makes penalties rather weak estimators. Better information can be obtained directly from branch and bound enumeration. Each time we solve relaxations of candidates where x_p is forced down to 0 or up to 1, we learn about the cost of making x_p integer. As the enumeration proceeds, more and more of this information is available for pseudo-costing. The next example shows a complete sequence.

Example 5.9. Pseudo-Costs. Figure 5.6 shows the branch and bound tree resulting from the early phases of solving some hypothetical example $P(S)$. As usual, numbers on the nodes (candidate problems) show the order in which the nodes were selected for evaluation. The linear programming relaxation bounds obtained when candidates were processed by Algorithm 5D's Step 2 are noted next to each of the first seven nodes.

Processing has now been completed at node 7, and we wish to estimate the optimal solution values for $P(S_8) = P(S_7{:}\ x_2 = 0)$ and $P(S_9) = P(S_7{:}\ x_2 = 1)$. First consider the one-parameter, (5-7) and (5-8) approach. Each time we solve a candidate problem with a newly fixed variable, the increase in bound over the predecessor node for which that variable was free is a measure of the cost of making it integer. For example, comparison of bounds at nodes 1 and 3 shows an increment of $(5.0 - 3.6) = 1.4$ was required to move x_3 from the fractional $\frac{8}{9}$ to the integer 1. The implied estimate of μ is $(1.4)/(\frac{1}{9}) = 12.6$.

At the moment when we are constructing nodes 8 and 9, six such values are available. Averaging gives the estimate

$$\mu \approx \frac{1}{6}\left(\frac{4.2 - 3.6}{\frac{8}{9}} + \frac{5.0 - 3.6}{\frac{1}{9}} + \frac{5.2 - 5.0}{\frac{1}{9}} + \frac{5.8 - 5.0}{\frac{8}{9}} + \frac{5.3 - 4.2}{\frac{8}{9}} + \frac{5.1 - 4.2}{\frac{1}{9}}\right) \approx 4.22.$$

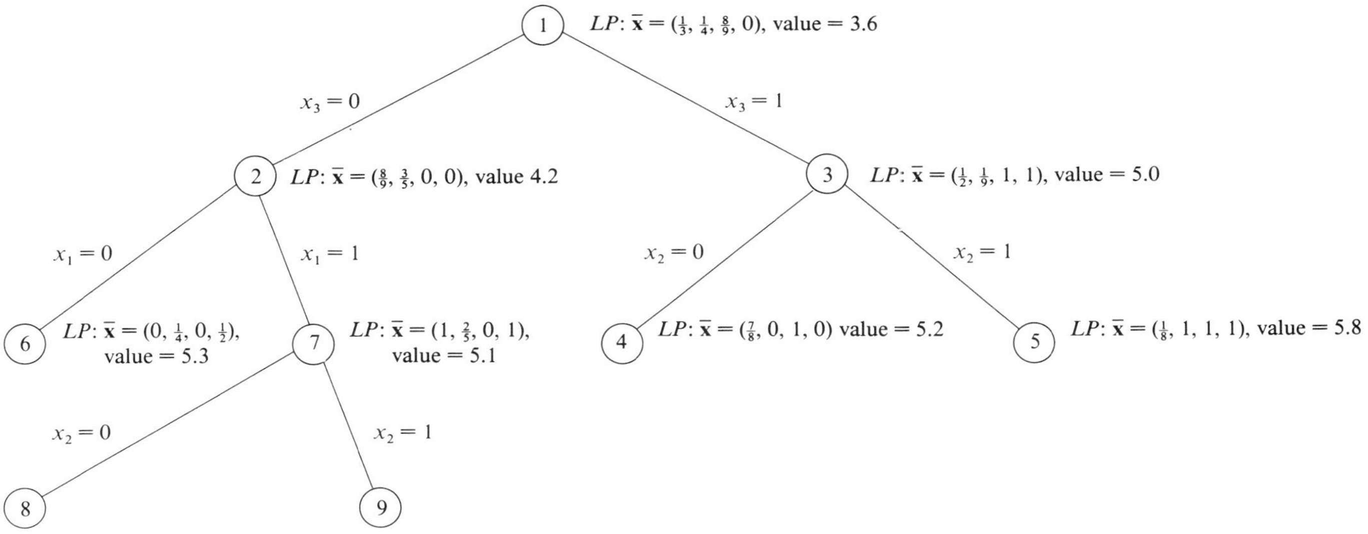

Fig. 5.6. Pseudo-cost Example 5.9.

Using this value, we would forecast

$$v(P(S_8)) \approx 5.1 + 4.22\left(\frac{2}{5}\right) + 4.22(0 + 0 + 0) \approx 6.79$$

$$v(P(S_9)) \approx 5.1 + 4.22\left(\frac{3}{5}\right) + 4.22(0 + 0 + 0) \approx 7.63.$$

Pseudo-cost schemes (5-9) through (5-12) use separate weights for each variable. Our partial solution branching scheme offers a natural way to approximate those weights. The fixed variable of one candidate is made integer in the next linear programming relaxation. Incremental cost of the relaxation can be attributed to integerizing the variable. Thus, the μ_2 needed at nodes 8 and 9 could be estimated after nodes 4 and 5 as

$$\mu_2 \approx \frac{1}{2}\left(\frac{5.2 - 5.0}{\frac{1}{9}} + \frac{5.8 - 5.0}{\frac{8}{9}}\right) \approx 1.35.$$

The corresponding pseudo-cost forecasts are

$$v(P(S_8)) \approx 5.1 + 1.35\left(\frac{2}{5}\right) \approx 5.64$$

$$v(P(S_9)) \approx 5.1 + 1.35\left(\frac{3}{5}\right) \approx 5.91.$$

By an analogous approach we could estimate separate up and down weights μ_j^+ and μ_j^-. Required values are

$$\mu_2^+ \approx \frac{5.8 - 5.0}{\frac{8}{9}} \approx .90$$

$$\mu_2^- \approx \frac{5.2 - 5.0}{\frac{1}{9}} \approx 1.80$$

and resulting pseudo-costs become

$$v(P(S_8)) \approx 5.1 + 1.80\left(\frac{2}{5}\right) \approx 5.82$$

$$v(P(S_9)) \approx 5.1 + .90\left(\frac{3}{5}\right) \approx 5.64.$$

■

5.5 Constructive Dual Techniques

Duality is one of the most powerful concepts in optimization. One might define a *dual* (D) to a given *primal* problem (P) as a related optimization problem having variables corresponding to the constraints of (P) and satisfying the *weak duality* relationship $v(D) \leq v(P)$. The effectiveness of almost all the conquering algorithms that have rendered particular optimization problems well solved can be traced to the direct or indirect exploitation of such a dual. In some cases, we are able to deal with a relatively complex primal by solving a somewhat easier dual; other times, the existence of a dual provides us with a test for optimality that permits stopping without searching through all possible solutions; in still other settings, the dual variables can be assigned an economic interpretation that assists in understanding the full implications of an optimal primal solution.

We shall see that, with the exception of problems in the class P, the application of duality in discrete optimization problems is quite difficult. In general, it is not possible to solve a primal problem by solving its dual. Still, duals do exist for discrete problems, and they do satisfy the fundamental relationship $v(D) \leq v(P)$. Thus, the various dual values are at least candidates for bound functions in a branch and bound algorithm. If we can find dual problems that are reasonably easy to solve and have solution values that form tight bounds on the values of candidate problems $v(P(S_k))$, we will have the core of a potentially valuable enumerative algorithm.

Every candidate problem, $P(S_k)$, of such a branch and bound procedure is a separate primal with its own dual. Thus, any algorithm may actually solve many duals—one at each node of the branch and bound tree. It will simplify notation, however, if we discuss only the root candidate problem, $P(S)$. Thus, all exposition of this section assumes we wish to bound (or if we are lucky, solve) $P(S)$. Adjustments for any other candidate are straightforward, requiring only substitution for variables assigned fixed values in S_k.

We shall also rewrite our standard discrete optimization problem $P(S)$ as

$$(P) \qquad \begin{aligned} \min \quad & \mathbf{cx} \\ \text{s.t.} \quad & \mathbf{Rx} \geq \mathbf{r} \\ & \mathbf{x} \in T. \end{aligned}$$

Here we imagine T to be a subset of S containing only points that satisfy specified rows of the system $\mathbf{Ax} \geq \mathbf{b}$. Aspects of $\mathbf{Ax} \geq \mathbf{b}$ not imbedded in T have been rewritten as $\mathbf{Rx} \geq \mathbf{r}$. It will turn out below that deciding what constraints to enforce in the set T critically affects resulting bounds. At this point it is only necessary to think of T as the easily enforced constraints and $\mathbf{Rx} \geq \mathbf{r}$ as the complicating ones.

5.5.1 Lagrangean Duals

To further develop the idea of duality in enumerative algorithms, we must concentrate on a specific dual. Define the *Lagrangean relaxation* (P_u), of our primal problem (P) as

$$(P_u) \qquad \begin{array}{ll} \min & \mathbf{cx} + \mathbf{u}(\mathbf{r} - \mathbf{Rx}) \\ \text{s.t.} & \mathbf{x} \in T \end{array}$$

where $\mathbf{u}$ is a nonnegative vector with one component for each row of $\mathbf{R}$. The *Lagrangean dual,* (D_L), is then the optimization problem in $\mathbf{u}$

$$(D_L) \qquad \begin{array}{ll} \max & v(P_u) \\ \text{s.t.} & \mathbf{u} \geq \mathbf{0}. \end{array}$$

Notice that we have obtained (P_u) by partially relaxing the constraints $\mathbf{Rx} \geq \mathbf{r}$ or $\mathbf{r} - \mathbf{Rx} \leq \mathbf{0}$. They have not been dropped entirely but instead are rolled up with weights u_i and added to the objective function. The next theorem demonstrates that this process leaves (P_u) a relaxation in the sense of Section 5.1.3 and thus that (D_L) satisfies weak duality.

Theorem 5.13. Weak Lagrangean Duality. *Let (P), (P_u) and (D_L) be as defined above. Then, for all $\mathbf{u} \geq \mathbf{0}$, $v(P_u) \leq v(P)$. Consequently, $v(D_L) \leq v(P)$.*

Proof.

$$v(P) \geq v\begin{pmatrix} \min & \mathbf{cx} \\ \text{s.t.} & \mathbf{x} \in T \\ & \mathbf{u}(\mathbf{r} - \mathbf{Rx}) \leq 0 \end{pmatrix} \quad \begin{array}{l}\text{because any } \mathbf{x} \text{ satisfying} \\ (\mathbf{r} - \mathbf{Rx}) \leq \mathbf{0} \text{ in } (P) \text{ satisfies} \\ \mathbf{u}(\mathbf{r} - \mathbf{Rx}) \leq 0 \text{ for } \mathbf{u} \geq \mathbf{0}.\end{array}$$

$$\geq v\begin{pmatrix} \min & \mathbf{cx} + \mathbf{u}(\mathbf{r} - \mathbf{Rx}) \\ & \mathbf{x} \in T \\ & \mathbf{u}(\mathbf{r} - \mathbf{Rx}) \leq 0 \end{pmatrix} \quad \begin{array}{l}\text{because we have added a} \\ \text{nonpositive quantity to} \\ \text{the objective function}\end{array}$$

$$\geq v\begin{pmatrix} \min & \mathbf{cx} + \mathbf{u}(\mathbf{r} - \mathbf{Rx}) \\ & \mathbf{x} \in T \end{pmatrix} \quad \begin{array}{l}\text{by Lemma 5.3, i.e., because} \\ \text{the latter is a relaxation}\end{array}$$

The last problem is exactly (P_u), and we have the first part of the theorem. Now since $v(P_u) \leq v(P)$ holds for all $\mathbf{u}$'s feasible in (D_L), it certainly follows that $\sup_{u \geq 0} v(P_u) \triangleq v(D_L) \leq v(P)$ and the theorem is proved. ∎

We have now proved that (P_u) is indeed a relaxation of (P). The next theorem shows when it yields a solution to (P).

Theorem 5.14. Strong Lagrangean Duality. *Let (P), (P_u) and (D_L) be as defined above. If $\hat{\mathbf{x}}$ solves $(P_{\hat{u}})$ for some $\hat{\mathbf{u}} \geq \mathbf{0}$, and in addition*

$$\mathbf{R}\hat{\mathbf{x}} \geq \mathbf{r} \tag{5-13}$$

$$\hat{\mathbf{u}}(\mathbf{r} - \mathbf{R}\hat{\mathbf{x}}) = 0 \tag{5-14}$$

then $\hat{\mathbf{x}}$ solves (P).

Proof. An $\hat{\mathbf{x}}$ satisfying the hypotheses of the theorem is feasible in (P) because of (5-13) and the fact that any $\hat{\mathbf{x}}$ solving $(P_{\hat{u}})$ has $\hat{\mathbf{x}} \in T$. In particular, this implies $\mathbf{c}\hat{\mathbf{x}} \geq v(P)$. But weak duality (Theorem 5.13) shows

$$v(P) \geq v(P_{\hat{u}}) = \mathbf{c}\hat{\mathbf{x}} + \hat{\mathbf{u}}(\mathbf{r} - \mathbf{R}\hat{\mathbf{x}}).$$

Under (5-14) the last expression is $\mathbf{c}\hat{\mathbf{x}}$, and we can conclude $v(P) \geq \mathbf{c}\hat{\mathbf{x}}$. Thus, since $\hat{\mathbf{x}}$ is feasible in (P) and $\mathbf{c}\hat{\mathbf{x}} = v(P)$, $\hat{\mathbf{x}}$ is optimal. ■

Only in comparatively rare circumstances will discrete problems satisfy the requirements of Theorem 5.14. In fact, Nemhauser and Ullman (1968) show that implicit enumeration formulations *IEP*(S) (see Section 5.2.1) can be solved via Theorem 5.14 only when they can be solved directly by linear programming. Still, $v(D_L)$ is a lower bound on $v(P)$. How strong a bound it is depends on which constraints are dualized (i.e., represented as $\mathbf{Rx} \geq \mathbf{r}$) and which constraints are retained in the set T. Some examples will illustrate.

Example 5.10. Lagrangean Duality with All Linear Constraints Dualized. Consider the class of discrete problems

$$(Q) \qquad \begin{array}{ll} \min & \mathbf{cx} \\ \text{s.t.} & \mathbf{Ax} \geq \mathbf{b} \\ & \mathbf{x} \in B = \{\text{binary } \mathbf{x}\}, \end{array}$$

where all linear constraints are to be dualized. The Lagrangean relaxation for $\mathbf{u} \geq \mathbf{0}$ is

$$(Q_u) \qquad \begin{array}{ll} \min & \mathbf{cx} + \mathbf{u}(\mathbf{b} - \mathbf{Ax}) \\ \text{s.t.} & \mathbf{x} \in B. \end{array}$$

If we rewrite the objective function as $\Sigma_j(c_j - \mathbf{ua}^j)x_j + \mathbf{ub}$, where $\mathbf{a}^j$ is the jth column of $\mathbf{A}$, (Q_u) can be solved by inspection.

$$v(Q_u) = \mathbf{ub} + \sum_j \min\{0, c_j - \mathbf{ua}^j\}.$$

One specific case is

$$\begin{aligned} \min \quad & 3x_1 + 2x_2 \\ \text{s.t.} \quad & 2x_1 + 5x_2 \geq 3 \\ & 5x_1 + 2x_2 \geq 3 \\ & (x_1, x_2) \in \{\text{binary 2-vectors}\}. \end{aligned}$$

Clearly, the unique optimal solution is the only feasible solution $\mathbf{x}^* = (1, 1)$; $v(Q) = 3 + 2 = 5$. For $\mathbf{u} = (1, 1)$

$$v(Q_u) = (1, 1)(3, 3) + \min\{0, 3 - 7\} + \min\{0, 2 - 7\} = -3,$$

which is a lower bound on $v(Q)$. The best such bound occurs for the $\mathbf{u}^* = (\frac{4}{21}, \frac{11}{21})$ that solves (D_L). (We defer the question of how to find these multipliers until Section 5.5.2). At $\mathbf{u}^*$

$$v(Q_{u^*}) = v(D_L) = (\tfrac{4}{21}, \tfrac{11}{21})(3, 3) + \min\{0, 3 - \tfrac{63}{21}\} + \min\{0, 2 - \tfrac{42}{21}\} = \tfrac{15}{7}.$$

We note that $\frac{15}{7}$ is also $v(\overline{Q})$, the bound obtainable from the linear programming relaxation solution $\overline{\mathbf{x}} = (\frac{3}{7}, \frac{3}{7})$. ■

Example 5.11. Lagrangean Duals of Uncapacitated Warehouse Location Problems. Warehouse location problems are ones that choose which of a given set of potential warehouses, i, to build in order to supply demand points j at minimum total cost. Costs include both a fixed cost $f_i > 0$ for building warehouse i and transportation costs $\Sigma_i \Sigma_j c_{ij} x_{ij}$, where x_{ij} is the amount shipped from i to j and c_{ij} is the unit transportation cost. By introducing variables x_i that equal 1 when warehouse i is built and 0 otherwise, we obtain the formulation

$$\begin{aligned} (UW) \qquad \min \quad & \sum_i \sum_j c_{ij} x_{ij} + \sum_i f_i x_i \\ \text{s.t.} \quad & \sum_i x_{ij} \geq d_j & \text{for all } j \qquad & (5\text{-}15) \\ & \sum_j x_{ij} \leq \left(\sum_j d_j\right) x_i & \text{for all } i \qquad & (5\text{-}16) \\ & d_j \geq x_{ij} \geq 0 & \text{for all } i \text{ and } j \qquad & (5\text{-}17) \\ & x_i = 0 \text{ or } 1 & \text{for all } i. \qquad & (5\text{-}18) \end{aligned}$$

Here, d_j is the demand at point j.

Several alternative Lagrangean dualizations present themselves. We could dualize (5-15), or (5-16), or both. Suppose we investigate the case of dualizing (5-15). Lagrangean relaxations are then

$$(UW_u)\qquad \begin{array}{ll} \min & \sum_i \sum_j c_{ij}x_{ij} + \sum_i f_i x_i + \sum_j u_j\left(d_j - \sum_i x_{ij}\right) \\ \text{s.t.} & \text{(5-16) (5-17) and (5-18).} \end{array}$$

After some rearrangement, it is easy to see that (UW_u) can be solved by separately solving one trivial problem for each i. Specifically,

$$v(UW_u) = \sum_j d_j u_j + \sum_i v\left(\begin{array}{lll} \min & f_i x_i + \sum_j (c_{ij} - u_j)x_{ij} & \\ \text{s.t.} & \sum_j x_{ij} \leq \left(\sum_j d_j\right) x_i & \text{for all } i \\ & d_j \geq x_{ij} \geq 0 & \text{for all } j \\ & x_i = 0 \text{ or } 1 & \end{array}\right). \qquad (5\text{-}19)$$

To solve the separate i problems, we consider $x_i = 0$, which implies all $x_{ij} = 0$, versus $x_i = 1$, in which case

$$x_{ij} = \begin{cases} d_j & \text{if } (c_{ij} - u_j) < 0 \\ 0 & \text{otherwise} \end{cases}$$

One specific case with $i = 1, 2, 3$ and $j = 1, 2$ is

$$(UWX)\qquad \begin{array}{ll} \min & 8x_{11} + 7x_{12} + 5x_{21} + 4x_{22} + 1x_{31} + 3x_{32} + 36x_1 + 12x_2 + 36x_3 \\ \text{s.t.} & x_{11} + x_{21} + x_{31} \geq 6 \\ & x_{12} + x_{22} + x_{32} \geq 6 \\ & \mathbf{x} \in T = \left\{ \begin{array}{ll} \mathbf{x}: & x_{11} + x_{12} \leq 12x_1 \\ & x_{21} + x_{22} \leq 12x_2 \\ & x_{31} + x_{32} \leq 12x_3 \\ & 6 \geq x_{ij} \geq 0 \quad \text{for } i = 1, 2, 3;\ j = 1, 2 \\ & x_i = 0 \text{ or } 1 \quad \text{for } i = 1, 2, 3 \end{array} \right\}. \end{array}$$

The optimal solution to (UWX) is to open only warehouse 3, i.e., $x_3 = 1$, $x_{31} = x_{32} = 6$, $v(UWX) = 36 + 1(6) + 3(6) = 60$. All other variables are 0.

Now consider (5-19) for $u_1 = 4$, $u_2 = 6$. The subproblem for $i = 2$ is

$$\begin{array}{ll} \min & 12x_2 + (5-4)x_{21} + (4-6)x_{22} \\ \text{s.t.} & x_{21} + x_{22} \le 12x_2 \\ & 6 \ge x_{21}, x_{22} \ge 0 \\ & x_2 = 0 \text{ or } 1. \end{array}$$

If $x_2 = 0$, $x_{21} = x_{22} = 0$ at cost 0. If $x_2 = 1$, $x_{21} = 0$, but $x_{22} = 6$; cost is $12 + (-2)6 = 0$, so x_2 may be either 0 or 1. A check of similar problems for $i = 1$ and 3 shows that $x_1 = x_3 = 0$ is optimal. Thus, $v(UWX_u) = 6(4) + 6(6) + 0 + 0 + 0 = 60$, equal to $v(UWX)$ and thus $v(D_L)$. This compares to the linear programming relaxation solution $\bar{x}_{22} = \bar{x}_{31} = 6$, $\bar{x}_2 = \bar{x}_3 = \frac{1}{2}$ with cost $v(\overline{UWX}) = 54$. ■

In discussing both examples, we have compared the Lagrangean dual value $v(D_L)$ to the linear programming relaxation $v(\bar{P})$. With our first example it happened that $v(D_L) = v(\bar{P})$; in the second $v(D_L) > v(\bar{P})$. Geoffrion (1974) has provided a characterization of $v(D_L)$ that allows us to see why this is the case. That characterization centers on the *convex hull* (denoted $[T]$) of points in T. That convex hull is the intersection of all convex sets containing T. (Refer to Appendix A for further detail if needed).

Theorem 5.15. Characterization of $v(D_L)$. *Let (P), (P_u) and (D_L) be as defined above. Then*

$$v(D_L) = v\begin{pmatrix} \min & \boldsymbol{cx} \\ \text{s.t.} & \boldsymbol{Rx} \ge \boldsymbol{r} \\ & \boldsymbol{x} \in [T] \end{pmatrix}$$

where $[T]$ denotes the convex hull of points in T.

Proof.

$$v(D_L) \triangleq v\left(\max_{\mathbf{u}\ge\mathbf{0}} v\left(\min_{\mathbf{x}\in T} \mathbf{cx} + \mathbf{u}(\mathbf{r} - \mathbf{Rx})\right)\right)$$

$$= v\left(\max_{\mathbf{u}\ge\mathbf{0}} v\left(\min_{\mathbf{x}\in[T]} \mathbf{cx} + \mathbf{u}(\mathbf{r} - \mathbf{Rx})\right)\right),$$

because an optimization problem with a linear objective function achieves an optimum (if any exists) on the boundary of its convex hull. Whenever v functions are nested and constraint sets are convex, the order of nesting can be interchanged. Interchanging gives

$$v(D_L) = v\left(\min_{\mathbf{x}\in[T]} \mathbf{cx} + v\left(\max_{\mathbf{u}\ge\mathbf{0}} \mathbf{u}(\mathbf{r} - \mathbf{Rx})\right)\right).$$

But since the inner optimization can be bounded only if $\mathbf{Rx} \geq \mathbf{r}$, we can conclude

$$v(D_L) = v\left(\begin{array}{ll} \min & \mathbf{cx} + v\left(\begin{array}{ll} \max & \mathbf{u}(\mathbf{r} - \mathbf{Rx}) \\ & \mathbf{u} \geq \mathbf{0} \end{array}\right) \\ \text{s.t.} & \mathbf{Rx} \geq \mathbf{r} \\ & \mathbf{x} \in [T] \end{array}\right).$$

The theorem now follows because $\mathbf{r} - \mathbf{Rx} \leq \mathbf{0}$ implies $\mathbf{u} = \mathbf{0}$ is always an optimal solution to the inner optimization. ∎

Corollary 5.16. Lagrangean Dual versus Linear Programming Relaxation. *Let (P) and (D_L) be as defined above and ($\bar{P}$) be the linear programming relaxation of (P). Then $v(D_L) \geq v(\bar{P})$.*

Proof. The problem ($\bar{P}$) is defined as

$$(\bar{P}) \qquad \begin{array}{ll} \min & \mathbf{cx} \\ \text{s.t.} & \mathbf{Rx} \geq \mathbf{r} \\ & \mathbf{x} \in \bar{T}, \end{array}$$

where $\bar{T}$ is the linearization of the set T. The Corollary follows directly from Theorem 5.15, since the convex hull is the smallest convex set containing T, i.e., $[T] \subseteq \bar{T}$. ∎

To understand the implications of the above results, let us first look at Corollary 5.16. It demonstrates that there was no accident when we had $v(D_L) \geq v(\bar{P})$ in both Examples 5.10 and 5.11. The Lagrangean dual will always give a bound value that is at least as large as the one obtained from the linear programming relaxation.

But our examples showed one case where the Lagrangean dual value was strictly greater than the linear relaxation and one where the two were equal. For an explanation we must turn to Theorem 5.15, which establishes that we can see how strong $v(D_L)$ can be by understanding something about the convex hull $[T]$. Certainly $[T] \subseteq \bar{T} \triangleq$ the linear programming relaxation of T. When $[T] = \bar{T}$, i.e., when subproblems (P_u) can always be solved by solving their linear programming relaxations, we say T has the *integrality property*. The consequence is easy to see.

Corollary 5.17. Integrality Property. *Let (P), (P_u) and (D_L) be as defined above. If T has the integrality property, $v(D_L) = v(\bar{P})$.*

Proof. Immediate from Theorem 5.15 and the definition of the integrality property. ∎

It is quite easy to write the convex hull of the set B in Example 5.10:

$$[B] = \overline{B} = \{\mathbf{x}: 0 \le x_1, x_2 \le 1\}.$$

We can see immediately from Corollary 5.17 that $v(D_L)$ must always equal the value of the linear programming relaxation. Whenever $[T]$ is indistinguishable from $\overline{T}$, $v(D_L)$ cannot yield a better bound than $v(\overline{P})$.

The convex hull in Example 5.11 is more complex. It can be written,

$$[T] = \{\mathbf{x}: x_{ij} \le d_j x_i \text{ for all } i \text{ and } j \text{ and } 0 \le x_i \le 1 \text{ for all } i\}.$$

Here, the single constraints $\Sigma_j x_{ij} \le (\Sigma d_j)x_i$ have been replaced by the more restrictive ones $x_{ij} \le d_j x_i$ for all j. Since $\overline{T} \ne [T]$, we were able to have $v(D_L) > v(\overline{UW})$.

What can we conclude from this discussion? One missing piece of information is how one finds good Lagrange multipliers (we shall turn to that issue in the next section). For the moment, it will suffice to say that the available processes are usually complex and require frequent solution of problems (P_u). Corollary 5.16 has made it clear that we can always achieve a bound from such a process that is at least as strong as the one from the corresponding linear programming relaxation. If our choice, however, of constraints to explicitly enforce (i.e., our choice of the set T) is so simple that $[T] = \overline{T}$, Corollary 5.17 tells us the Lagrangean procedure cannot improve on the linear relaxation.

Whether that is satisfactory depends on the linear relaxation. It is not hard to find examples where the treatment of some troublesome constraints via lagrange multipliers leaves $\overline{T} = [T]$, but allows us to evaluate $v(\overline{P})$ when $(\overline{P})$ is so large that it would be hopeless to approach it by usual linear programming techniques. In such cases, the Lagrangean approach is satisfactory in spite of its having the $\overline{T} = [T]$ property of Corollary 5.17.

If the linear programming bound is not likely to be adequate, we must be more clever in selecting a dualization scheme. Specifically, we must choose constraints to keep in the set T and others to roll into the objective function in such a way that assures the Lagrangean subproblems (P_u) are tractable, yet not linear programs. Example 5.11 exhibited just such a formulation. Such approaches can lead to efficient procedures with bounds far better than those resulting from the linear relaxation.

5.5.2 Lagrangean Dual Search

To this point, we have observed something of the potential power of Lagrangean dual approaches, but we have not confronted the matter of finding good dual multipliers. To determine optimal multipliers, we must solve the nonlinear program

$$(D_L) \qquad \begin{array}{ll} \sup & v(P_u) \\ \text{s.t.} & \mathbf{u} \geq \mathbf{0} \end{array}$$

Its constraints $\mathbf{u} \geq \mathbf{0}$ are quite simple ones. But what sort of objective function is $v(P_u)$? We know it is a rather costly one to evaluate; to determine $v(P_u)$ for a given $\mathbf{u}$ we must solve the Lagrangean relaxation, (P_u).

Still, the function has some useful properties. Before reviewing them we shall make one mild assumption:

> For the remainder of Section 5.5, we shall take the set T to be nonempty and finite, i.e., assume that only finitely many $\mathbf{x} \in T$ may solve relaxations (P_u).

Of course, one case satisfying this assumption arises when all decision variables are 0–1. Most cases with bounded mixed-integer solution spaces, however, also conform to the assumption; a (P_u) solution will occur at one of the finitely many extreme points of the linear programs that results when 0–1 variables are fixed. Unbounded cases can also be treated, but they only obfuscate the discussion.

Theorem 5.18. Piecewise-Linear Concavity of the Lagrangean Dual Function. *Let (P) and (P_u) be the primal and Lagrangean relaxation problems defined above over a finite set T. Then $v(P_u)$ is a piecewise-linear concave function of* $\mathbf{u}$.

Proof. By definition, $(P_u) = \min_{\mathbf{x} \in T}\{\mathbf{cx} + \mathbf{u}(\mathbf{r} - \mathbf{Rx})\}$. Thus, (under our assumption that T is nonempty and finite), the function $v(P_u)$ is merely the minimum of finitely many linear functions, $\mathbf{cx} + \mathbf{u}(\mathbf{r} - \mathbf{Rx})$. There is one for each $\mathbf{x} \in T$. The consequence is that $v(P_u)$ is piecewise-linear. Moreover, since the finite minimum of concave functions is concave, and every linear function is concave, we can also conclude that $v(P_u)$ is concave. ■

Example 5.12. Dual Function. Consider the simple example problem

$$(P) \qquad \begin{array}{ll} \min & 3x_1 + 2x_2 \\ \text{s.t.} & 5x_1 + 2x_2 \geq 3 \\ & 2x_1 + 5x_2 \geq 3 \end{array}$$

$$(x_1, x_2) \in T = \left\{ \begin{array}{l} \text{integer} \\ (x_1, x_2) \end{array} : \begin{array}{l} 0 \leq x_1 \leq 1, 0 \leq x_2 \leq 2 \\ 8x_1 + 8x_2 \geq 1 \end{array} \right\}.$$

Figure 5.7 graphs the solution. The set $T = \{(0, 1), (0, 2), (1, 0), (1, 1), (1, 2)\}$, because $(0, 0)$ violates the linear constraint $8x_1 + 8x_2 \geq 1$. Clearly,

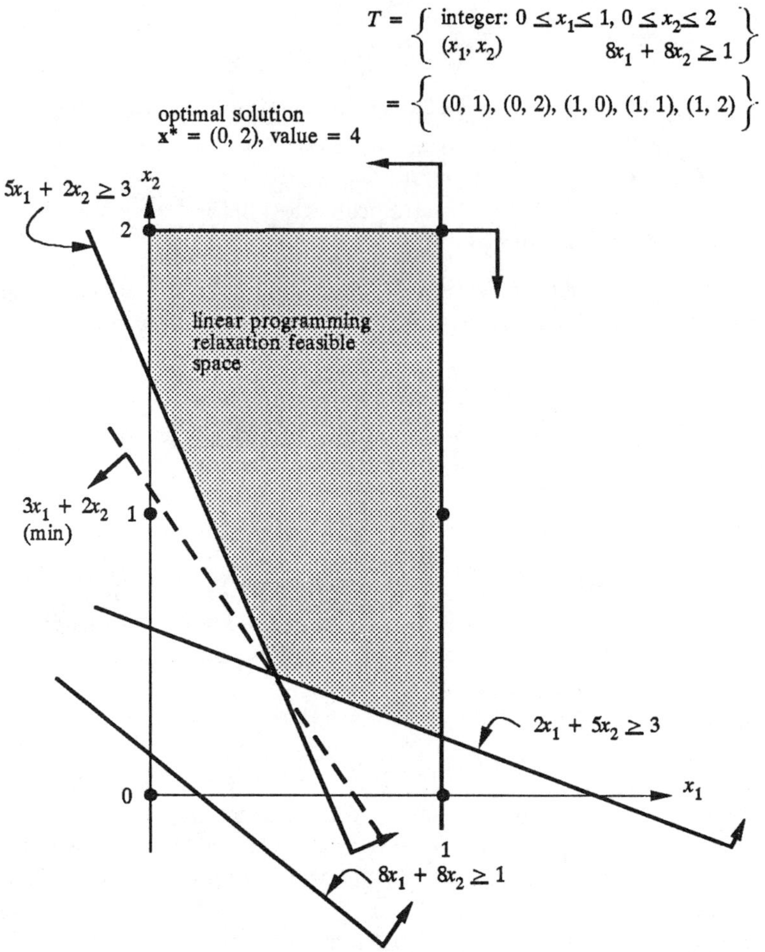

Fig. 5.7. Graphic solution of Example 5.12.

$x_1 = 0$, $x_2 = 2$ is the optimal solution to (P) because points (0, 1) and (1, 0) in T violate main constraints.

With multipliers u_1 and u_2, Lagrangean relaxations are

$$(P_u)\qquad \begin{array}{ll} \min & 3x_1 + 2x_2 + u_1(3 - 5x_1 - 2x_2) + u_2(3 - 2x_1 - 5x_2) \\ \text{s.t.} & (x_1, x_2) \in \{(0, 1), (0, 2), (1, 0), (1, 1), (1, 2)\}. \end{array}$$

Direct substitution of the five points in T, in the objective function yields

$$v(P_u) = \min\{2 + u_1 - 2u_2, 4 - u_1 - 7u_2, 3 - 2u_1 + u_2,\\ 5 - 4u_1 - 4u_2, 7 - 6u_1 - 9u_2\}.$$

Figure 5.8 illustrates the concave nature of this Lagrangean function. It is formed as the minimum of five linear functions in (u_1, u_2), one for each $(x_1, x_2) \in T$. Only four show because $5 - 4u_1 - 4u_2$ is dominated. The optimal Lagrange multipliers are the nonnegative ones maximizing $v(P_u)$. Here, they are $(u_1, u_2) = (\frac{1}{3}, 0)$ with $v(D_L) = 2\frac{1}{3}$. ■

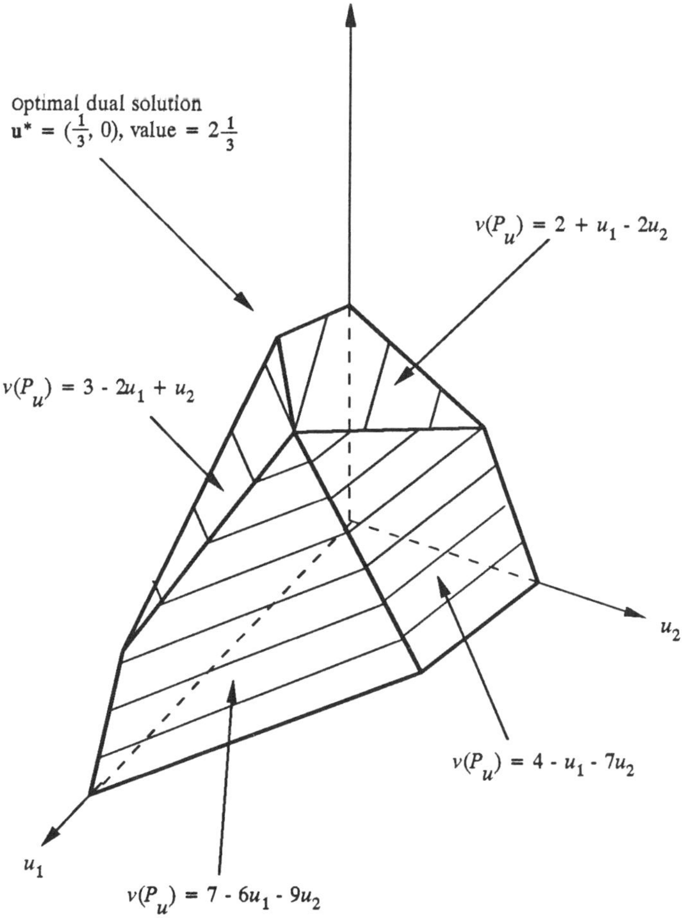

Fig. 5.8. Lagrangean dual surface for Example 5.12.

Subgradient Methods

The most straightforward methods of nonlinear programming (see, for example, Bazaraa and Shetty (1979)) employ gradient[2] search. Beginning at some specified point, they advance along the gradient of the objective function at that point until progress is no longer satisfactory. The gradient at the new point is then computed, and the search continues. If the function being maximized is concave, such a process will allow us to approach an optimum (at least in the absence of constraints).

Theorem 5.18 showed that $v(P_u)$ is indeed concave. However, its piecewise-linear form is troublesome. At points $\hat{\mathbf{u}}$, where $(P_{\hat{u}})$ is solved by a unique $\hat{\mathbf{x}} \in T$, $v(P_u)$ is differentiable; a gradient exists and equals $(\mathbf{r} - \mathbf{R}\hat{\mathbf{x}})$. Where several $\hat{\mathbf{x}} \in T$ solve $(P_{\hat{u}})$, however, $v(P_u)$ is not differentiable.

If we are to deal with this lack of differentiability, we need a generalization of the gradient. To this purpose, we define a vector $\mathbf{s}$ to be a *subgradient* of a (concave) function $\theta(\mathbf{u})$ at $\hat{\mathbf{u}}$ if for all $\mathbf{u}$, $\theta(\hat{\mathbf{u}}) + \mathbf{s}(\mathbf{u} - \hat{\mathbf{u}}) \geq \theta(\mathbf{u})$.

Theorem 5.19. Subgradients of the Lagrangean Dual Function. *Let (P_u) be the Lagrangean relaxation of the above discrete optimization problem (P) where T is nonempty and finite. Then the collection of subgradients of $v(P_u)$ at any given $\boldsymbol{u}$ has the form*

$$\left\{ \sum_{\boldsymbol{x} \in T(\boldsymbol{u})} \lambda_x(\boldsymbol{r} - \boldsymbol{R}\boldsymbol{x}) : \sum_{\boldsymbol{x} \in T(\boldsymbol{u})} \lambda_x = 1,\ \lambda_x \geq 0 \text{ for all } \boldsymbol{x} \in T(\boldsymbol{u}) \right\}, \tag{5-20}$$

where $T(\boldsymbol{u}) \triangleq \{\boldsymbol{x} \in T: \boldsymbol{x} \text{ solves } (P_u)\}$.

Proof. Consider any given $\hat{\mathbf{u}}$ and let

$$\mathbf{s} = \sum_{\mathbf{x} \in T(\hat{\mathbf{u}})} \lambda_x(\mathbf{r} - \mathbf{R}\mathbf{x}) \qquad \text{where all } \lambda_x \geq 0$$

and

$$\sum_{\mathbf{x} \in T(\hat{\mathbf{u}})} \lambda_x = 1.$$

Then, noting that for all $\mathbf{x} \in T(\hat{\mathbf{u}})$, the value of $\mathbf{cx} + \hat{\mathbf{u}}(\mathbf{r} - \mathbf{Rx})$ is identical.

$$\begin{aligned} v(P_{\hat{u}}) + \mathbf{s}(\mathbf{u} - \hat{\mathbf{u}}) &= \sum_{\mathbf{x} \in T(\hat{\mathbf{u}})} \lambda_x[\mathbf{cx} + \hat{\mathbf{u}}(\mathbf{r} - \mathbf{Rx})] + \left[\sum_{\mathbf{x} \in T(\hat{\mathbf{u}})} \lambda_x(\mathbf{r} - \mathbf{Rx}) \right](\mathbf{u} - \hat{\mathbf{u}}) \\ &= \sum_{\mathbf{x} \in T(\hat{\mathbf{u}})} \lambda_x[\mathbf{cx} + \hat{\mathbf{u}}(\mathbf{r} - \mathbf{Rx}) + \mathbf{u}(\mathbf{r} - \mathbf{Rx}) - \hat{\mathbf{u}}(\mathbf{r} - \mathbf{Rx})] \end{aligned}$$

[2] A gradient is a vector of partial derivatives with respect to each component of the decision vector $\mathbf{u}$.

$$\geq \min_{\mathbf{x} \in T(\hat{\mathbf{u}})} [\mathbf{cx} + \mathbf{u}(\mathbf{r} - \mathbf{Rx})]$$

$$\geq \min_{\mathbf{x} \in T} [\mathbf{cx} + \mathbf{u}(\mathbf{r} - \mathbf{Rx})] = v(P_u).$$

This demonstrates that every element of the set of (5-20) is a subgradient of $v(P_u)$.

A proof that every subgradient has the (5-20) form requires notions from convex analysis that are beyond the scope of this book (the reader is referred to Bazaraa and Shetty (1979), page 192). Generally, the argument depends on the facts that the subgradients form a convex set with extreme points, $\mathbf{s}$, that can be written $\mathbf{s} = (\mathbf{r} - \mathbf{Rx})$ for some $\mathbf{x} \in T(\hat{\mathbf{u}})$. ■

One immediate consequence of Theorem 5.19 is that any optimal $\hat{\mathbf{x}}$ to some $(P_{\hat{u}})$ yields a subgradient, $(\mathbf{r} - \mathbf{R}\hat{\mathbf{x}})$, for $v(P_u)$ at $\hat{\mathbf{u}}$. Figure 5.9 shows the Lagrangean surface of Figure 5.8 from a more direct, top view. At points such as $\tilde{\mathbf{u}}$, where a unique $\tilde{\mathbf{x}}$ solves $(P_{\tilde{u}})$, there is only one subgradient; at $\tilde{\mathbf{x}}$, $v(P_u) = \mathbf{c}\tilde{\mathbf{x}} + \mathbf{u}(\mathbf{r} - \mathbf{R}\tilde{\mathbf{x}})$ and differentiation with respect to $\mathbf{u}$ gives $(\mathbf{r} - \mathbf{R}\tilde{\mathbf{x}})$. In other cases, e.g., point $\hat{\mathbf{u}}$ in Figure 5.9, there are several subgradients. When both $\mathbf{x}^1$ and $\mathbf{x}^2$ solve $(P_{\hat{u}})$, $(\mathbf{r} - \mathbf{Rx}^1)$, or $(\mathbf{r} - \mathbf{Rx}^2)$ or any convex combination of the two is a subgradient.

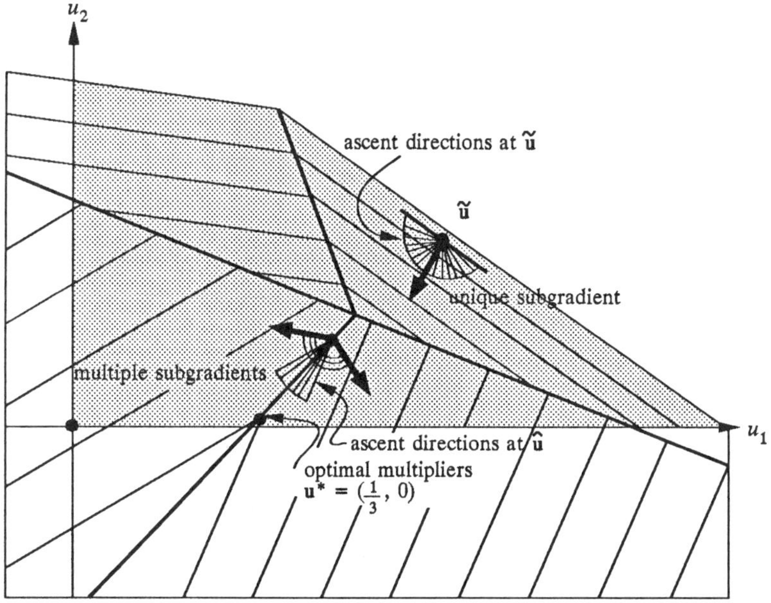

Fig. 5.9. Subgradients of the Lagrangean dual function of Example 5.12.

A direction, $\mathbf{d}$, is said to be an *ascent* direction of the function $\theta(\mathbf{u})$ at $\hat{\mathbf{u}}$, if there exists some $\delta > 0$ such that $\theta(\hat{\mathbf{u}} + \lambda\mathbf{d}) > \theta(\hat{\mathbf{u}})$ for all $\lambda \in (0, \delta)$. The appeal of a gradient (when it exists) is that it is always an ascent direction. Unfortunately, this property does not extend to all subgradients. Consider, for example, the $\hat{\mathbf{u}}$ of Figure 5.9. Movement in either of the two extreme subgradient directions decreases $v(P_u)$ from $v(P_{\hat{u}})$. To find a subgradient that is an ascent direction, we must take a combination of the two extremes.

In spite of the fact that subgradients may not be ascent directions, they do take us closer in a Euclidean distance sense to an optimal solution. For example, small steps in either of the extreme subgradient directions at $\hat{\mathbf{u}}$ in Figure 5.9, although yielding a decrease in $v(P_u)$, do move us closer in distance to the optimal $\mathbf{u}^* = (\frac{1}{3}, 0)$. The next lemma formalizes this idea.

Lemma 5.20. Subgradient Distance Improvement. *Let $\theta(\boldsymbol{u})$ be a continuous, concave function of $\boldsymbol{u}$. Also, define for each α the set $U(\alpha) = \{\boldsymbol{u}: \theta(\boldsymbol{u}) \geq \alpha\}$. Then, for any given α, $\hat{\boldsymbol{u}} \notin U(\alpha)$ and $\boldsymbol{s}$ subgradient to θ at $\hat{\boldsymbol{u}}$ there exist $\hat{\delta}$ such that*

$$\|\boldsymbol{u} - (\hat{\boldsymbol{u}} + \lambda\boldsymbol{s})\| < \|\boldsymbol{u} - \hat{\boldsymbol{u}}\|$$

for all $\boldsymbol{u} \in U(\alpha)$ and $\lambda \in (0, \hat{\delta})$.

Proof. Select α. If $U(\alpha)$ is empty, the lemma is vacuous. If not, choose $\mathbf{u} \in U(\alpha)$. Then,

$$\|\mathbf{u} - (\hat{\mathbf{u}} + \lambda\mathbf{s})\|^2 = \|\mathbf{u} - \hat{\mathbf{u}}\|^2 - 2\lambda\mathbf{s}(\mathbf{u} - \hat{\mathbf{u}}) + \lambda^2 \|\mathbf{s}\|^2. \qquad (5\text{-}21)$$

But since $\hat{\mathbf{u}} \notin U(\alpha)$, $\theta(\mathbf{u}) > \theta(\hat{\mathbf{u}})$. Moreover, $\mathbf{s}$, a subgradient, implies

$$\theta(\mathbf{u}) \leq \theta(\hat{\mathbf{u}}) + \mathbf{s}(\mathbf{u} - \hat{\mathbf{u}}).$$

Thus, it must be that $\mathbf{s}(\mathbf{u} - \hat{\mathbf{u}}) > 0$ and $-2\lambda\mathbf{s}(\mathbf{u} - \hat{\mathbf{u}}) < 0$. For $\lambda \in (0, \hat{\delta})$, where $\hat{\delta} = \mathbf{s}(\mathbf{u} - \hat{\mathbf{u}})/\|\mathbf{s}\|^2$, it follows from (5-21) that $\|\mathbf{u} - (\hat{\mathbf{u}} + \lambda\mathbf{s})\|^2 < \|\mathbf{u} - \hat{\mathbf{u}}\|^2$ or $\|\mathbf{u} - (\hat{\mathbf{u}} + \lambda\mathbf{s})\| < \|\mathbf{u} - \hat{\mathbf{u}}\|$. ■

Before formally stating a subgradient search procedure, we must deal with one more issue. Our dual problem (D_L) is *constrained* by the restriction $\mathbf{u} \geq \mathbf{0}$. How can we incorporate such constraints into our search strategy? One answer is to *project* points $\mathbf{u}$ on the feasible set $U = \{\mathbf{u} \geq \mathbf{0}\}$ as we proceed, i.e., replace them by their closest neighbors in U.

We can state the fundamental result from convex analysis that justifies such projection as follows:

Lemma 5.21. Distance Improvement by Projection on Convex Sets. *Let U be a closed convex set of $\mathbb{R}^m$, $\boldsymbol{u}^* \in U$ and $\boldsymbol{v} \in \mathbb{R}^m \backslash U$. Then*

$$\|u^* - u\| \leq \|u^* - v\|$$

where u *satisfies*

$$\|u - v\| = \inf_{w \in U} \|w - v\|$$

Although projection on a convex feasible set theoretically advances a search, computation of the projection can be quite complex. The closed convex set of interest to us, however, is $U = \{\mathbf{u} \geq 0\}$. Given any $\hat{\mathbf{u}} \ngeq \mathbf{0}$, it is easy to show the projection of $\hat{\mathbf{u}}$ on $\{\mathbf{u} \geq \mathbf{0}\}$ is obtained by zeroing negative components. Specifically

$$\hat{u}_i \leftarrow \begin{cases} \hat{u}_i & \text{if } \hat{u}_i \geq 0 \\ 0 & \text{otherwise.} \end{cases}$$

Algorithm 5E. Subgradient Lagrangean Search

STEP 0: *Initialization.* Pick any $\mathbf{u}^1 \geq \mathbf{0}$, and set $k \leftarrow 1$, $v^0 \leftarrow -\infty$.

STEP 1: *Lagrangean Relaxation.* Solve the Lagrangean relaxation (P_{u^k}) for an optimal $\mathbf{x}^k \in T$. If $(\mathbf{r} - \mathbf{R}\mathbf{x}^k) \leq \mathbf{0}$ and $\mathbf{u}^k(\mathbf{r} - \mathbf{R}\mathbf{x}^k) = 0$, stop; $\mathbf{x}^k$ solves the primal problem (P) and $v(P_{u^k}) = v(D_L) = v(P)$.

STEP 2: *Incumbent Savings.* If $v^{k-1} < v(P_{u^k}) = \mathbf{c}\mathbf{x}^k + \mathbf{u}^k(\mathbf{r} - \mathbf{R}\mathbf{x}^k)$, save a new dual incumbent solution by $v^k \leftarrow v(P_{u^k})$. Otherwise, set $v^k \leftarrow v^{k-1}$.

STEP 3: *Subgradient Step.* Compute a new point

$$\mathbf{u}^{k+1} \longleftarrow \mathbf{u}^k + \lambda_k(\mathbf{r} - \mathbf{R}\mathbf{x}^k)/\|\mathbf{r} - \mathbf{R}\mathbf{x}^k\|, \tag{5-22}$$

λ_k is a step size satisfying

$$\lambda_k > 0 \qquad \text{for all } k \tag{5-23}$$

$$\lim_{k \to \infty} \lambda_k = 0 \tag{5-24}$$

$$\sum_{k=1}^{\infty} \lambda_k = +\infty. \tag{5-25}$$

STEP 4: *Projection.* Project the new $\mathbf{u}^{k+1}$ on $\{\mathbf{u} \geq \mathbf{0}\}$ by setting

$$u_j^{k+1} \longleftarrow \max\{0, u_j^{k+1}\} \qquad \text{for all } j. \tag{5-26}$$

Then replace $k \leftarrow k + 1$ and return to Step 1. ■

Two details of Algorithm 5E require comment. First, we see that an incumbent solution value, v^k, is maintained at Step 2. This is necessary because subgradient steps do not guarantee improvements in $v(P_u)$. The other new element is the stepsize rule of (5-23) to (5-25). After giving an example, we shall reproduce Polyak's (1967) proof that these very general ground rules are enough to guarantee convergence.

Example 5.13. Subgradient Lagrangean Search on Example 5.12. Return to Example 5.12 and pick $\mathbf{u}^1 = (.400, .200)$. To apply Algorithm 5E, we first solve (P_{u^1}) to obtain the optimal $\mathbf{x}^1 = (0, 1)$. There $(\mathbf{r} - \mathbf{R}\mathbf{x}^1) = (3\text{-}0\text{-}2, 3\text{-}0\text{-}5) = (1, -2)$, and $v(P_{u^1}) = 2 + (.400, .200)(1, -2) = 2.000$. Since $(\mathbf{r} - \mathbf{R}\mathbf{x}^1) \not\leq \mathbf{0}$, we do not stop at Step 1. Still, $v(P_{u^1})$ does provide a new incumbent solution $v^1 \leftarrow 2.000$ at Step 2.

To compute the next **u** via (5-22), we need a stepsize rule satisfying the quite general conditions (5-23) through (5-25). We shall arbitrarily choose the scheme

$$\lambda_k = 1/(2k). \tag{5-27}$$

Applying this rule,

$$\mathbf{u}^2 \longleftarrow (.400, .200) + (\tfrac{1}{2})(1, -2)/\|(1, -2)\| \approx (.624, -.247).$$

Projection revises $\mathbf{u}^2$ to $(\max\{0, .624\}, \max\{0, -.247\}) = (.624, 0.)$. There, $\mathbf{x}^2 = (1, 0)$ solves the relaxation with $v(P_{u^2}) = 1.752$. Since this new value does not improve the dual incumbent, $v^2 \leftarrow v^1 = 2.000$.

The procedure continues by applying (5-22) and (5-27) again such that

$$\mathbf{u}^3 \longleftarrow (.624, 0.) + (\tfrac{1}{4})(-2, 1)/\|(-2, 1)\| \approx (.400, .112).$$

Here, no projection is necessary and the new relaxation, (P_{u^3}), improves our incumbent:

$$v^3 \longleftarrow v(P_{u^3}) = 2.176.$$

Points $\mathbf{u}^4$, $\mathbf{u}^5$, . . . are computed similarly with the sequence converging (in the limit) to our optimal $\mathbf{u}^* = (.333, 0.)$. Since convergence is not finite, an actual code for the algorithm would terminate via some *ad hoc* stopping rule.

Figure 5.10 shows this sequence of points. Note that the points approach $\mathbf{u}^*$ in space although the function value does not always increase. ■

Theorem 5.22. Convergence of Subgradient Lagrangean Search. *Let (P_u) be the Lagrangean relaxation of the discrete optimization problem (P), where the constraint set, T, is nonempty and finite. Then, if Algorithm 5E is applied to solve the Lagrangean dual (D_L), the algorithm either stops at Step 1 with $v(P_{u^k}) = v(D_L) = v(P)$, or generates a sequence $\{v^k\}$ satisfying, $\lim_{k\to\infty} v^k = v(D_L)$.*

Proof. If Algorithm 5E stops at Step 1, the strong duality conditions of Theorem 5.14 have been fulfilled. It follows that $v(P_{u^k}) = v(D_L) = v(P)$.

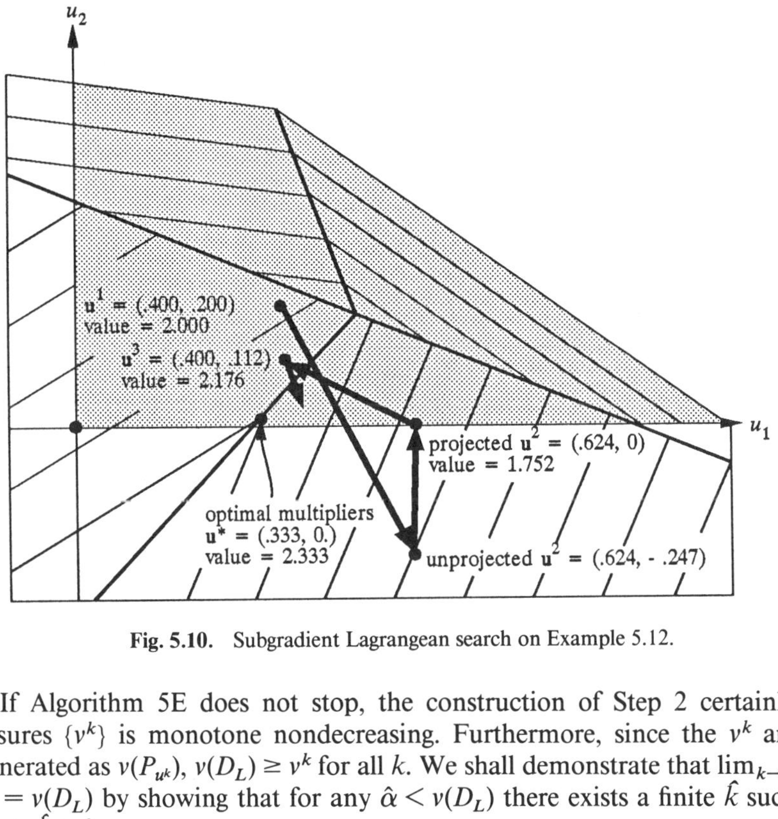

Fig. 5.10. Subgradient Lagrangean search on Example 5.12.

If Algorithm 5E does not stop, the construction of Step 2 certainly assures $\{v^k\}$ is monotone nondecreasing. Furthermore, since the v^k are generated as $v(P_{u^k})$, $v(D_L) \geq v^k$ for all k. We shall demonstrate that $\lim_{k\to\infty} v^k = v(D_L)$ by showing that for any $\hat{\alpha} < v(D_L)$ there exists a finite $\hat{k}$ such that $v^{\hat{k}} \geq \hat{\alpha}$.

Pick $\hat{\alpha} < v(D_L)$ and $\hat{\mathbf{u}} \geq \mathbf{0}$ with $v(P_u) > \hat{\alpha}$. We have shown $v(P_u)$ is a concave function for all $\mathbf{u} \in \mathbb{R}^m$ (Theorem 5.18), and all concave functions are continuous on the interior of their domains (see for example Bazaraa and Shetty (1979, p. 82)). It follows from continuity at $\hat{\mathbf{u}}$ that there is some $\hat{\rho} > 0$ such that $\mathbf{u} \geq \mathbf{0}$ and $\|\mathbf{u} - \hat{\mathbf{u}}\| \leq \hat{\rho}$ together imply $v(P_u) \geq \hat{\alpha}$.

Now consider $\{\mathbf{u}^k\}$ generated at Steps 3 and 4 of Algorithm 5E. Each $\mathbf{u}^{k+1}$ is constructed by (5-22) and then projected on the convex set $\{\mathbf{u} \geq \mathbf{0}\}$. Thus, recalling Lemma 5.21, our $\hat{\mathbf{u}} \geq \mathbf{0}$ will have $\|\hat{\mathbf{u}} - \mathbf{u}^{k+1}\|^2 \leq \|\hat{\mathbf{u}} - \mathbf{u}^k - \lambda_k \mathbf{s}^k\|^2$,where $\mathbf{s}^k \triangleq (\mathbf{r} - \mathbf{R}\mathbf{x}^k)/\|\mathbf{r} - \mathbf{R}\mathbf{x}^k\|$. Expanding, then adding and subtracting equal terms,

$$\begin{aligned}\|\mathbf{u} - \mathbf{u}^{k+1}\|^2 &\leq \|\mathbf{u} - \mathbf{u}^k\|^2 - 2\lambda_k \mathbf{s}^k(\hat{\mathbf{u}} - \mathbf{u}^k) + \|\lambda_k \mathbf{s}^k\|^2 \\ &= \|\hat{\mathbf{u}} - \mathbf{u}^k\|^2 + \|\lambda_k \mathbf{s}^k\|^2 \\ &\quad - 2\lambda_k \mathbf{s}^k \left[\hat{\mathbf{u}} - \frac{\hat{\rho}\lambda_k \mathbf{s}^k}{\|\lambda_k \mathbf{s}^k\|} - \mathbf{u}^k + \frac{\hat{\rho}\lambda_k \mathbf{s}^k}{\|\lambda_k \mathbf{s}^k\|} \right] \\ &= \|\hat{\mathbf{u}} - \mathbf{u}^k\|^2 + \|\lambda_k \mathbf{s}^k\|^2 - 2\lambda_k \mathbf{s}^k(\hat{\mathbf{w}} - \mathbf{u}^k) - 2\hat{\rho}\|\lambda_k \mathbf{s}^k\|, \qquad (5\text{-}28)\end{aligned}$$

where $\hat{\mathbf{w}} = \hat{\mathbf{u}} - \hat{\rho}\lambda_k \mathbf{s}^k / \|\lambda_k \mathbf{s}^k\|$. Our definition of $\hat{\mathbf{u}}$ and $\hat{\rho}$ assures $v(P_{\hat{w}}) \geq \hat{\alpha}$. If, by contradiction, we assume $v(P_{u^k}) < \hat{\alpha}$ for all k, then because $\lambda_k > 0$ and $\mathbf{s}^k$ is a subgradient at $\mathbf{u^k}$, $0 \leq 2\lambda_k[v(P_{\hat{w}}) - v(P_{u^k})] \leq 2\lambda_k[\mathbf{s}^k(\hat{\mathbf{w}} - \mathbf{u}^k)]$. Also by definition $\|\mathbf{s}^k\| = 1$, so that $\|\lambda_k \mathbf{s}^k\| = \lambda_k$. Reflecting these observations in (5-28) gives $\|\hat{\mathbf{u}} - \mathbf{u}^{k+1}\|^2 \leq \|\hat{\mathbf{u}} - \mathbf{u}^k\|^2 + (\lambda_k)^2 - 2\hat{\rho}\lambda_k$. Now property (5-24) assures there is some l for which $\lambda_k \leq \hat{\rho}$ for all $k \geq l$. It follows that

$$\|\hat{\mathbf{u}} - \mathbf{u}^{k+1}\|^2 \leq \|\hat{\mathbf{u}} - \mathbf{u}^k\|^2 - \hat{\rho}\lambda_k \qquad \text{for all } k \geq l. \tag{5-29}$$

Summing relations (5-29) for $k = l, l+1, \ldots, l+p$ and eliminating like terms,

$$\|\hat{\mathbf{u}} - \mathbf{u}^{l+p+1}\|^2 \leq \|\hat{\mathbf{u}} - \mathbf{u}^l\|^2 - \hat{\rho} \sum_{k=l}^{l+p} \lambda_k. \tag{5-30}$$

Now the left side of (5-30) is a square, i.e., always nonnegative. But property (5-25) guarantees the right side will be arbitrarily negative as $p \to \infty$. Thus, (5-30) is a contradiction and our assumption that $v(P_{u^k}) < \hat{\alpha}$ for all k is false. There is some finite $\hat{k}$ for which $v(P_{u^{\hat{k}}}) \geq \hat{\alpha}$, and thus $v^{\hat{k}}$. This completes the proof. ■

Dual Ascent

Although we showed, in Example 5.10, that subgradients do not always yield ascent directions, they can be used to precisely characterize ascent directions of any concave function. Recall from Theorem 5.19 that subgradients of the Lagrangean dual function $v(P_u)$ at $\hat{\mathbf{u}}$ are convex combinations of vectors $(\mathbf{r} - \mathbf{R}\hat{\mathbf{x}})$ for $\mathbf{x} \in T(\hat{\mathbf{u}})$, i.e., $\mathbf{x}$ solving $(P_{\hat{u}})$. Ascent directions are exactly those directions creating positive inner products with all subgradients.

Theorem 5.23. Subgradient Characterization of Lagrangean Dual Ascent Directions. *Let (P_u) be the Lagrangean relaxation of the discrete optimization problem (P), where the constraint set, T, is finite and nonempty. Also let $T(\mathbf{u})$ be the set of optimal solutions to (P_u). Then $\mathbf{d}$ is an ascent direction for $v(P_u)$ at a point $\hat{\mathbf{u}}$ if and only if*

$$\mathbf{d}(\mathbf{r} - \mathbf{R}\mathbf{x}) > 0 \qquad \text{for all } \mathbf{x} \in T(\hat{\mathbf{u}}). \tag{5-31}$$

That is, $\mathbf{d}$ is an ascent direction if and only if it creates positive inner product with every subgradient of $v(P_u)$ at $\hat{\mathbf{u}}$.

Proof. The set T is finite. Thus, given any direction, $\mathbf{d}$, and any $\hat{\mathbf{u}}$, there must exist a $\delta > 0$ such that $T(\hat{\mathbf{u}} + \lambda\mathbf{d}) \subseteq T(\hat{\mathbf{u}})$ for all $\lambda \in (0, \delta)$. We have then, via any $\hat{\mathbf{x}} \in T(\hat{\mathbf{u}} + \lambda\mathbf{d})$,

$$\begin{aligned} v(P_{\hat{u}+\lambda\mathbf{d}}) - v(P_{\hat{u}}) &= (\mathbf{c}\hat{\mathbf{x}} + (\hat{\mathbf{u}} + \lambda\mathbf{d})(\mathbf{r} - \mathbf{R}\hat{\mathbf{x}})) - [\mathbf{c}\hat{\mathbf{x}} + \hat{\mathbf{u}}(\mathbf{r} - \mathbf{R}\hat{\mathbf{x}})] \\ &= \lambda\mathbf{d}(\mathbf{r} - \mathbf{R}\hat{\mathbf{x}}). \end{aligned}$$

Certainly, if (5-31) holds, $\mathbf{d}$ will be an ascent direction. Conversely, if $\mathbf{d}$ is an ascent direction, $0 < \mathbf{d}(\mathbf{r} - \mathbf{R}\hat{\mathbf{x}})$ for all $\hat{\mathbf{x}} \in T(\hat{\mathbf{u}} + \lambda\mathbf{d})$. Picking any such $\hat{\mathbf{x}}$, it follows that for any $\mathbf{x} \in T(\hat{\mathbf{u}})$,

$$\begin{aligned} \mathbf{d}(\mathbf{r} - \mathbf{R}\mathbf{x}) &= (\mathbf{c}\mathbf{x} + (\hat{\mathbf{u}} + \lambda\mathbf{d})(\mathbf{r} - \mathbf{R}\mathbf{x})) - (\mathbf{c}\mathbf{x} + \hat{\mathbf{u}}(\mathbf{r} - \mathbf{R}\mathbf{x})) \\ &= (\mathbf{c}\mathbf{x} + (\hat{\mathbf{u}} + \lambda\mathbf{d})(\mathbf{r} - \mathbf{R}\mathbf{x})) - (\mathbf{c}\hat{\mathbf{x}} + \hat{\mathbf{u}}(\mathbf{r} - \mathbf{R}\hat{\mathbf{x}})) \\ &\geq (\mathbf{c}\hat{\mathbf{x}} + (\hat{\mathbf{u}} + \lambda\mathbf{d})(\mathbf{r} - \mathbf{R}\hat{\mathbf{x}})) - (\mathbf{c}\hat{\mathbf{x}} + \hat{\mathbf{u}}(\mathbf{r} - \mathbf{R}\hat{\mathbf{x}})) \\ &= \lambda\mathbf{d}(\mathbf{r} - \mathbf{R}\hat{\mathbf{x}}) > 0. \end{aligned}$$

Finally, recall from Theorem 5.19 that every subgradient of $v(P_u)$ at $\hat{\mathbf{u}}$ can be expressed as a convex combination of terms $(\mathbf{r} - \mathbf{R}\mathbf{x})$ for $\mathbf{x} \in T(\hat{\mathbf{u}})$. Clearly, (5-31) holds for all $(\mathbf{r} - \mathbf{R}\mathbf{x})$ with $\mathbf{x} \in T(\hat{\mathbf{u}})$ if and only if it holds for all convex combinations of such $(\mathbf{r} - \mathbf{R}\mathbf{x})$. Thus, $\mathbf{d}$ is an ascent direction if and only if it makes positive inner product with all subgradients. ■

By employing the characterization (5-31) we can find ascent directions at $\hat{\mathbf{u}}$ instead of merely stepping along subgradients. A precise algorithm is as follows.

Algorithm 5F. Lagrangean Dual Ascent

STEP 0: *Initialization.* Pick any $\mathbf{u}^1 \geq \mathbf{0}$ and set $k \leftarrow 1$.

STEP 1: *Lagrangean Relaxation.* Determine $T(\mathbf{u}^k)$, the set of all optimal solutions to the Lagrangean relaxation (P_{u^k}). If for any $\mathbf{x} \in T(\mathbf{u}^k)$, $(\mathbf{r} - \mathbf{R}\mathbf{x}^k) \leq \mathbf{0}$ and $\mathbf{u}^k(\mathbf{r} - \mathbf{R}\mathbf{x}^k) = 0$, stop; that $\mathbf{x}$ solves the primal problem (P) and $v(P_{u^k}) = v(D_L) = v(P)$.

STEP 2: *Direction Finding.* Solve the optimization problem,

$$\begin{aligned} \max \quad & \alpha \\ \text{s.t.} \quad & \alpha \leq \mathbf{d}(\mathbf{r} - \mathbf{R}\mathbf{x}) \qquad \text{for all } \mathbf{x} \in T(\mathbf{u}^k) \\ & d_i \geq 0 \qquad \text{for all } i \text{ with } u_i^k = 0 \\ & |\mathbf{d}| \leq 1, \end{aligned}$$

where $|\cdot|$ denotes any suitable vector norm. Let $(\alpha^k, \boldsymbol{d}^k)$ be an optimal solution. If $\alpha^k = 0$, stop; $\mathbf{u}^k$ solves the Lagrangean dual and $v(P_{u^k}) = v(D_L)$.

STEP 3: *Ascent Direction Step.* Compute

$$\lambda^k \longleftarrow \sup\{\lambda > 0\colon T(\mathbf{u}^k + \lambda\mathbf{d}^k) \subseteq T(\mathbf{u}^k) \text{ and } \mathbf{u}^k + \lambda\mathbf{d}^k \geq \mathbf{0}\}. \tag{5-32}$$

If $\lambda^k = +\infty$, stop; the primal problem (P) is infeasible. Otherwise, update

$$\mathbf{u}^{k+1} \longleftarrow \mathbf{u}^k + \lambda^k \mathbf{d}^k, \tag{5-33}$$

set $k \leftarrow k + 1$ and return to Step 1. ■

Several details of Algorithm 5F need clarification. Step 2 is clearly just an implementation of the conditions (5-31) in Theorem 5.23. If α can be made positive, $\mathbf{d}$ is an ascent direction; if not, there is none. When there are ascent directions, we might as well maximize the minimum $\mathbf{d}(\mathbf{r} - \mathbf{R}\mathbf{x})$ over unit length vectors, $\mathbf{d}$, in order to steepen the ascent. Any normalization of $\mathbf{d}$ will serve, although some are easier than others to implement. Clearly, the resulting $\mathbf{d}^k$ will differ with the norm.

We must also assume each $\mathbf{u}^k$ remains nonnegative. In Step 2, the nonnegativity constraints on components of $\mathbf{d}^k$, where $u_i^k = 0$, prevent steps to $\mathbf{u} \ngeq \mathbf{0}$.

Finally, consider the stepsize rule (5-32). We can be sure an ascent direction, $\mathbf{d}$, remains so only as long as $T(\mathbf{u}^k + \lambda \mathbf{d}) \subseteq T(\mathbf{u}^k)$. Construction (5-32) steps to the limit of that region, subject to the requirement that $\mathbf{u}^{k+1}$ stay nonnegative. In many applications computing such a λ is so difficult that this last feature makes it impossible to employ Algorithm 5F.

Theorem 5.24. Convergence of Lagrangean Dual Ascent. *Let (P_u) be the Lagrangean relaxation of the discrete optimization problem (P), where the constraint set, T, is nonempty and finite. Then if Algorithm 5F is applied to solve the Lagrangean dual, (D_L), the algorithm stops after finitely many k,*

(i) at Step 1 with $v(P_{u^k}) = v(D_L) = v(P)$; or
(ii) at Step 3 with $v(D_L) = v(P) = +\infty$; or
(iii) at Step 2 with $v(P_{u^k}) = v(D_L)$.

Proof. If Algorithm 5F stops at Step 1, the strong duality conditions of Theorem 5.14 have been fulfilled. It follows that $v(P_{u^k}) = v(D_L) = v(P)$.

If the algorithm stops at Step 3, there is no limit on the improvement in $v(P_u)$ along the direction $\mathbf{u}^k + \lambda \mathbf{d}^k$. Weak duality (Theorem 5.13) implies $v(D_L) \leq v(P)$, and thus $v(P) = +\infty$; the primal is infeasible.

By Theorem 5.23, $\mathbf{u}^k$ can be improved upon in (D_L) only if α^k can be made positive in the direction finding problem of Step 2. Thus if any $\alpha^k = 0$, and we stop at Step 2, $v(P_{u^k}) = v(D_L)$.

We have shown that all three stopping rules perform as claimed. What remains is to show that one of the three stops occurs after finitely many k. Given any specific norm, $|\cdot|$, in Step 2, a proof can be developed by showing that each of the finite number of combinations of elements for

$T(\mathbf{u}^k)$ and zero components of $\mathbf{u}^k$ can arise only once. Exact proofs, however, are technical and norm specific. We shall omit them here and refer the reader to Bazaraa *et al* (1978). ■

Example 5.14. Lagrangean Dual Ascent on Example 5.12. Return to Example 5.12 and pick $\mathbf{u}^1 = (.400, .200)$. To apply Algorithm 5F, we first solve (P_{u^1}) to obtain $T(\mathbf{u}^1) = \{(0, 1)\}$ and $v(P_{u^1}) = 2.000$. For the one element $\mathbf{x} = (0, 1)$, $(\mathbf{r} - \mathbf{R}\mathbf{x}) = (2, -1) \not\leq \mathbf{0}$; so we do not stop at Step 1.

To choose a direction at Step 2, we must specify a normalization $|\cdot|$ on $\mathbf{d}$. We shall apply the most commonly used one

$$-\mathbf{1} \leq \mathbf{d} \leq \mathbf{1}, \tag{5-34}$$

i.e., the maximum component magnitude *sup norm*. Under this norm, the Step 2 problem is the linear program

$$\begin{array}{ll} \max & \alpha \\ \text{s.t.} & \alpha \leq \mathbf{d}(1, -2) \\ & -\mathbf{1} \leq \mathbf{d} \leq \mathbf{1} \end{array}$$

An optimal solution is $\alpha^1 = 3$, $\mathbf{d}^1 = (1, -1)$.

Since $\alpha^1 > 0$, feasible ascent directions exist at $\mathbf{u}^1$. Figure 5.11 shows the limits on a step along the half-line $(.400, .200) + \lambda(1, -1)$. At $\lambda = .200$, the second component drops to zero; at $\lambda = .067$, $T(\mathbf{u}^1)$ changes. Thus, $\lambda^1 \leftarrow \min\{.200, .067\} = .067$. Our new $\mathbf{u}^2 \leftarrow (.467, .133)$.

The relaxation (P_{u^2}) has two alternative optima; $T(\mathbf{u}^2) = \{(0,1)(1, 0)\}$. Both yield $v(P_{u^2}) = 2.200$. The new direction problem is

$$\begin{array}{ll} \max & \alpha \\ \text{s.t.} & \alpha \leq \mathbf{d}(1, -2) \\ & \alpha \leq \mathbf{d}(-2, 1) \\ & -\mathbf{1} \leq \mathbf{d} \leq \mathbf{1}. \end{array} \tag{5-35}$$

An optimal solution is $\alpha^2 = 1$, $\mathbf{d}^2 = (-1, -1)$. Along the half-line $(.467, .133) + \lambda(-1, -1)$, $T(\mathbf{u}^2)$ never changes, however, u_2 drops to zero at $\lambda = .133$. Thus, $\lambda^2 \leftarrow .133$ and $\mathbf{u}^3 \leftarrow (.333, 0.)$.

The relaxation $P(\mathbf{u}^3)$ produces the same $T(\mathbf{u}^3) = T(\mathbf{u}^2) = \{(0, 1), (1, 0)\}$. The direction problem, however, at Step 2 is (5-35) plus the requirement $d_2 \geq 0$. Component u_2 is not allowed to decrease. This linear program has an optimal solution $\alpha^3 = 0$, $\mathbf{d}^3 = (0, 0)$. We conclude $\mathbf{u}^3 = (.333, 0.)$ solves (D_L). ■

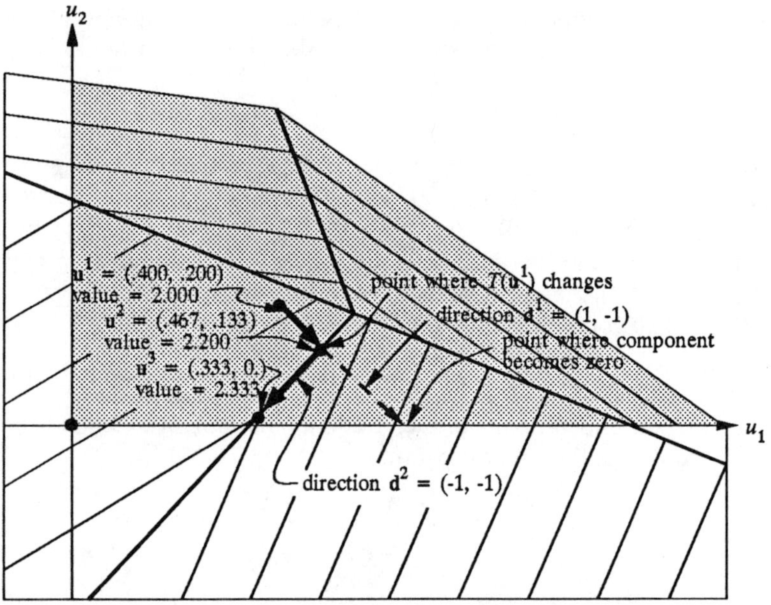

Fig. 5.11. Lagrangean dual ascent on Example 5.12.

Outer Linearization

We have been discussing (D_L) as a search problem over the function $v(P_u)$. We could, however, (at least in principle) write (D_L) as the linear program

$$(OL) \qquad \begin{array}{ll} \max & v \\ \text{s.t.} & v \leq \mathbf{cx} + \mathbf{u}(\mathbf{r} - \mathbf{Rx}) \qquad \text{for all } \mathbf{x} \in T \\ & \mathbf{u} \geq \mathbf{0}. \end{array}$$

The difficulty, of course, is that the cardinality of T is usually enormous. Still, only a few constraints will be tight at optimality. We could thus apply the dual simplex to (OL), generating rows as they are needed. This is precisely the notion of Algorithm 5G.

Algorithm 5G. Outer Lagrangean Linearization

STEP 0: *Initialization* Choose as T_1 any subset of T containing at least one point feasible for the primal problem (P). Also set $k \leftarrow 1$.

STEP 1: *Master Problem* Solve the linear program,

$$(OL_k) \qquad \begin{array}{lll} \max & v & \\ \text{s.t.} & v \leq \mathbf{cx} + \mathbf{u}(\mathbf{r} - \mathbf{Rx}) & \text{for all } \mathbf{x} \in T_k \\ & \mathbf{u} \geq \mathbf{0}. & \end{array}$$

Let $(v^k, \mathbf{u}^k)$ be an optimal solution.

STEP 2: *Lagrangean Relaxation* Solve the Lagrangean relaxation (P_u^k) for an optimal $\mathbf{x}^k \in T$.

STEP 3: *Stopping* Stop if either

$$(\mathbf{r} - \mathbf{Rx}^k) \leq \mathbf{0} \text{ and } \mathbf{u}^k(\mathbf{r} - \mathbf{Rx}^k) = 0 \tag{5-36}$$

or

$$v^k \leq \mathbf{cx}^k + \mathbf{u}^k(\mathbf{r} - \mathbf{Rx}^k). \tag{5-37}$$

In the first case, $\mathbf{x}^k$ solves the primal problem (P), and $v(P_{u^k}) = v(D_L) = v(P)$; in the second $v^k = v(D_L)$. When neither (5-36) nor (5-37) occurs, update

$$T_{k+1} \longleftarrow T_k \cup \{\mathbf{x}^k\}, \tag{5-38}$$

set $k \leftarrow k + 1$, and return to Step 1. ■

Theorem 5.25. Convergence of Outer Lagrangean Linearization. *Let (P_u) be the Lagrangean relaxation of the discrete optimization problem (P), where the constraint set, T, is finite and (P) is feasible. Then, if Algorithm 5G is applied to solve the Lagrangean dual (D_L), the algorithm stops at Step 3 after finitely many of its steps. Furthermore, if stopping is via (5-36), $v(P_{u^k}) = v(D_L) = v(P)$; otherwise, $v^k = v(D_L)$.*

Proof. Our rule for initializing T_1 at Step 0, and the updating scheme (5-38) assumed every T_k contains at least one $\mathbf{x}^f$ feasible in (P). Such an $\mathbf{x}^f \in T$ has $(\mathbf{r} - \mathbf{Rx}^f) \leq \mathbf{0}$. Thus, the feasible set of every (OL_k) is contained in the bounded set $\{(v, \mathbf{u})\colon \mathbf{u} \geq \mathbf{0},\ v \leq \mathbf{cx}^f + \mathbf{u}(\mathbf{r} - \mathbf{Rx}^f)\}$. Noting $\mathbf{u} = \mathbf{0}$ provides a feasible solution for any (OL_k), it follows that every (OL_k) will yield a finite optimum $(v^k, \mathbf{u}^k)$. Furthermore, our assumptions that T is finite and (P) is feasible guarantee each (P_u) has a finite optimum, $\mathbf{x}^k$. We can conclude that Algorithm 5G will iterate until it stops at Step 3.

When stopping is by (5-36), the strong duality contitions of Theorem 5.14 have been established. Clearly, $v(P_{u^k}) = v(D_L) = v(P)$. If (5-36) is never fulfilled, the algorithm either generates $\mathbf{x}^k$ at Step 2 that satisfy

$\mathbf{cx}^k + \mathbf{u}^k(\mathbf{r} - \mathbf{Rx}^k) < v^k = \min_{\mathbf{x}\in T_k} \mathbf{cx} + \mathbf{u}^k(\mathbf{r} - \mathbf{Rx})$ or stops via (5-37). In the first case, the new $\mathbf{x}^k$ cannot belong to T_k. Since every T_k is a subset of the finite set T, stopping will occur after finitely many k. When (5-37) does hold,

$$v(D_L) \le v(OL_k) = v^k \le \mathbf{cx}^k + \mathbf{u}^k(\mathbf{r} - \mathbf{Rx}^k) = v(P_{u^k}) \le v(D_L)$$

and

$$v^k = v(D_L).$$ ■

We have chosen to call Algorithm 5G Outer Lagrangean Linearization, because it can be interpreted as a scheme for producing successively better piecewise-linear approximations to the function, $v(P_u)$. A numerical example will illustrate.

Example 5.15. Outer Lagrangean Linearization on Example 5.12. Return to Example 5.12. Its set, T, contains three feasible points for (P). Let us arbitrarily choose (1, 2) and initialize $T_1 \leftarrow \{(1, 2)\}$. The first master problem for this example would then be

$$\begin{aligned} \max \quad & v \\ \text{s.t.} \quad & v \le 7 - 6u_1 - 9u_2 \\ & u_1, u_2 \ge 0 \end{aligned}$$

with optimal solution $\mathbf{u}^1 = (0, 0)$, $v^1 = 7$. Note that this (OL_1) is bounded because both u_j have nonpositive coefficients in the single constraint.

Continuing at Step 2, the Lagrangean relaxation (P_{u^1}) is solved by $\mathbf{x}^1 = (0, 1)$. The new constraint $v \le \mathbf{cx}^1 + \mathbf{u}(\mathbf{r} - \mathbf{Rx}^1) = 2 + u_1 - 2u_2$ is not satisfied by the old solution $(v^1, \mathbf{u}^1) = (7, (0, 0))$. Also $(\mathbf{r} - \mathbf{Rx}^1) \not\le \mathbf{0}$. Thus neither (5-36) nor (5-37) are satisfied. We revise $T^2 \leftarrow T^1 \cup \{(0, 1)\} = \{(1, 2), (0, 1)\}$ and solve (OL_2)

$$\begin{aligned} \max \quad & v \\ \text{s.t.} \quad & v \le 7 - 6u_1 - 9u_2 \\ & v \le 2 + u_1 - 2u_2 \\ & u_1, u_2 \ge 0. \end{aligned}$$

The unique optimum is $v^2 = 2.714$, $\mathbf{u}^2 = (5/7, 0)$. Continuing with the algorithm, we generate $\mathbf{x}^2 = (1, 0)$, $v^3 = 2.333$, $\mathbf{u}^3 = (1/3, 0)$ and $\mathbf{x}^3 = (0, 1)$ or $(1, 0)$. Either choice of $\mathbf{x}^3$ will fulfill (5-37), and cause us to stop. As claimed, $v^3 = v(D_L) = 2.333$.

Figure 5.12 illustrates the outer linearization interpretation of these

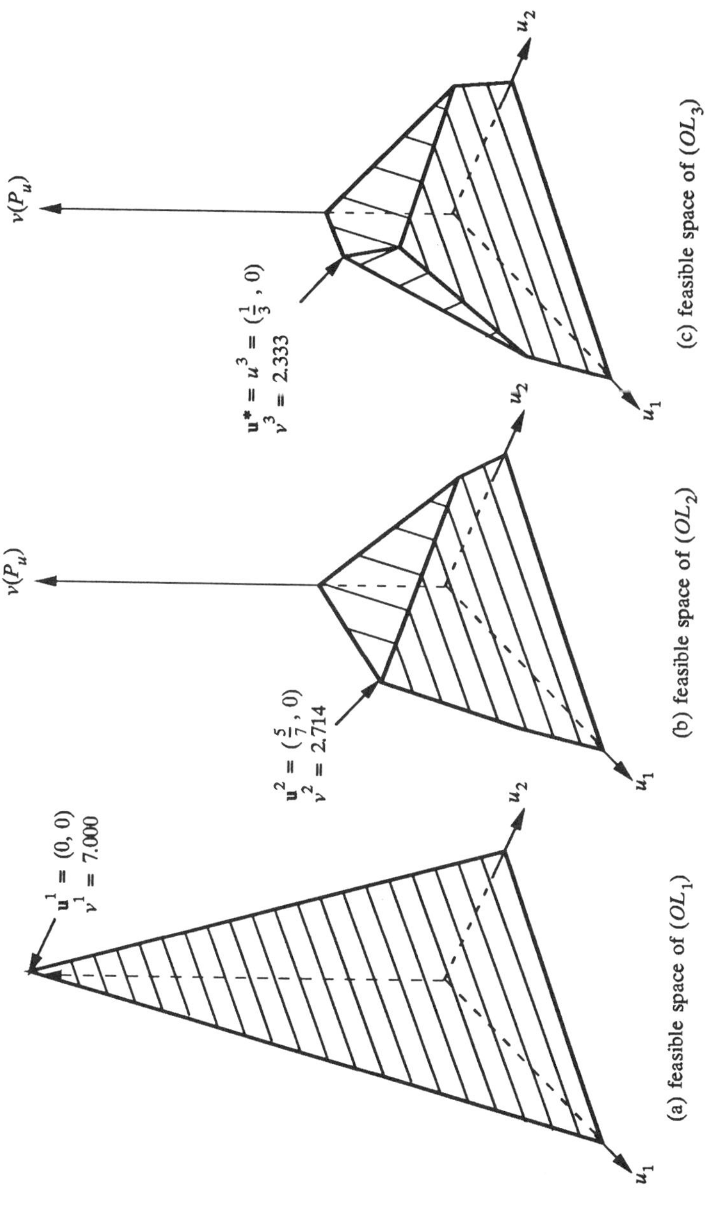

Fig. 5.12. Outer linearization of Lagrangean dual surface for Example 5.12

computations. Each point added to T_k generates a plane cutting off parts of the $(v, \mathbf{u})$ solution space for (OL). As iteration continues, the feasible spaces more and more approximate the $v(P_u)$ surface shown in Figure 5.8. Some of the possible planes, however, were not needed to isolate $v(D_L)$. ∎

5.5.3 Surrogate Duals

The Lagrangean dual is the most familiar dual in mathematical programming, but it is not the only one. In fact, there are many. Only one other, however, the *surrogate dual,* has yet shown much promise computationally. *Surrogate relaxations* of our primal discrete optimization problem (P) are

$$(P^{\mathbf{v}}) \qquad \begin{aligned} \min \quad & \mathbf{cx} \\ \text{s.t.} \quad & \mathbf{v}(\mathbf{r} - \mathbf{Rx}) \leq 0 \\ & \mathbf{x} \in T. \end{aligned}$$

Like its Lagrangean analog, the surrogate relaxation uses a multiplier vector $\mathbf{v} \geq \mathbf{0}$ to roll up the constraints $\mathbf{Rx} \geq \mathbf{r}$ or alternately $(\mathbf{r} - \mathbf{Rx}) \leq \mathbf{0}$. In the surrogate case, however, the resulting single constraint is explicitly enforced in the $(P^{\mathbf{v}})$.

The surrogate dual problem is

$$(D_S) \qquad \begin{aligned} \sup \quad & v(P^{\mathbf{v}}) \\ \text{s.t.} \quad & \mathbf{v} \geq \mathbf{0}. \end{aligned}$$

Weak and strong duality theorems are even simpler than in the Lagrangean case.

Theorem 5.26. Weak Surrogate Duality. *Let (P), $(P^{\mathbf{v}})$ and (D_S) be as defined above. Then for all $\mathbf{v} \geq \mathbf{0}$, $v(P^{\mathbf{v}}) \leq v(P)$. Consequently, $v(D_S) \leq v(P)$.*

Proof. Each $(P^{\mathbf{v}})$ is obviously a relaxation of (P) in the sense of Lemma 5.3, since every feasible solution to (P) is feasible in $(P^{\mathbf{v}})$. It follows that $v(D_S) = \sup_{\mathbf{v} \geq \mathbf{0}} v(P^{\mathbf{v}}) \leq v(P)$. ∎

Theorem 5.27. Strong Surrogate Duality. *Let (P), $(P^{\mathbf{v}})$ and (D_S) be as defined above. Then, if for any $\hat{\mathbf{v}} \geq \mathbf{0}$, there is an $\hat{\mathbf{x}}$ solving $(P^{\hat{\mathbf{v}}})$ and satisfying $\mathbf{R}\hat{\mathbf{x}} \geq \mathbf{r}$, $\hat{\mathbf{x}}$ solves (P) and $v(D_S) = v(P)$.*

Proof. Since $(P^{\mathbf{v}})$ and (P) have the same objective functions, and $(P^{\mathbf{v}})$ is a relaxation of (P), it follows from Lemma 5.4 that $\hat{\mathbf{x}}$ solves (P). Furthermore, weak duality implies $\mathbf{c}\hat{\mathbf{x}} = \nu(P^{\hat{v}}) \leq \nu(D_S) \leq \nu(P) = \mathbf{c}\hat{\mathbf{x}}$ so that $\nu(D_S) = \nu(P)$. ■

Subproblems (P^v) are generally more difficult to solve than their Lagrangean counterparts. The surrogate constraint, $\mathbf{v}(\mathbf{r} - \mathbf{Rx}) \leq 0$, often destroys the neat separability we observed in Examples 5.10 and 5.11. Even the very simple case of Example 5.10, where $T = \{\text{binary } \mathbf{x}\text{-vectors}\}$, yields a (P^v) of the form

$$\begin{aligned} \min \quad & \mathbf{cx} \\ \text{s.t.} \quad & \mathbf{v}(\mathbf{r} - \mathbf{Rx}) \leq 0 \\ & \mathbf{x} \text{ binary.} \end{aligned}$$

Such problems are knapsack problems—relatively easy discrete problems, but discrete ones nonetheless. Regardless, a return to our Example 5.12 will show the extra computation may be justified.

Example 5.16. Surrogate Dual of Example 5.12. Return to the numerical Example 5.12 and consider Figure 5.13. Nonnegative combinations of the two relaxed constraints yield surrogate constraints tangent to $\{\mathbf{x}: \mathbf{Rx} \geq \mathbf{r}\}$. Various multiplier choices can cut off the point $(0, 1) \in T$ ($v_1 = \frac{9}{10}$ $v_2 = \frac{1}{10}$ will do), or the point $(1, 0) \in T$ (consider $v_1 = \frac{1}{8}$ $v_2 = \frac{7}{8}$), but not both. Since $(0, 1)$ is the least cost, (D_S) would prefer to eliminate it. For optimal $v_1^* = \frac{9}{10}$, $v_2^* = \frac{1}{10}$, the surrogate relaxation is

$$(P^{v^*}) \quad \begin{aligned} \min \quad & 3x_1 + 2x_2 \\ \text{s.t.} \quad & (\tfrac{9}{10} \cdot 5 + \tfrac{1}{10} \cdot 2)x_1 + (\tfrac{9}{10} \cdot 2 + \tfrac{1}{10} \cdot 5)x_2 \geq (\tfrac{9}{10} \cdot 3 + \tfrac{1}{10} \cdot 3) \\ & (x_1, x_2) \in \{(0, 1), (0, 2), (1, 0), (1, 1), (1, 2)\}. \end{aligned}$$

The relaxation is solved by $(x_1, x_2) = (1, 0)$ and $\nu(P^{v^*}) = \nu(D_S) = 3$. Referring back to Example 5.12, we observe that $\nu(D_L) = 2.333 < 3 = \nu(D_S) < 4 = \nu(P)$. ■

Example 5.16 has shown the surrogate dual can strictly improve on the Lagrangean. But how frequently does improvement occur? The next theorem relates the strength of the $\nu(D_L)$ and $\nu(D_S)$ bounds.

Theorem 5.28. Surrogate Versus Lagrangean Dual. *Let (P) be the primal discrete optimization problem defined above over a nonempty compact set T. Also let (P_u), (D_L), (P^v) and (D_S) be its Lagrangean and surrogate relaxations and duals. Then $\nu(D_L) \leq \nu(D_S)$. Furthermore, if $\nu(D_L) =$*

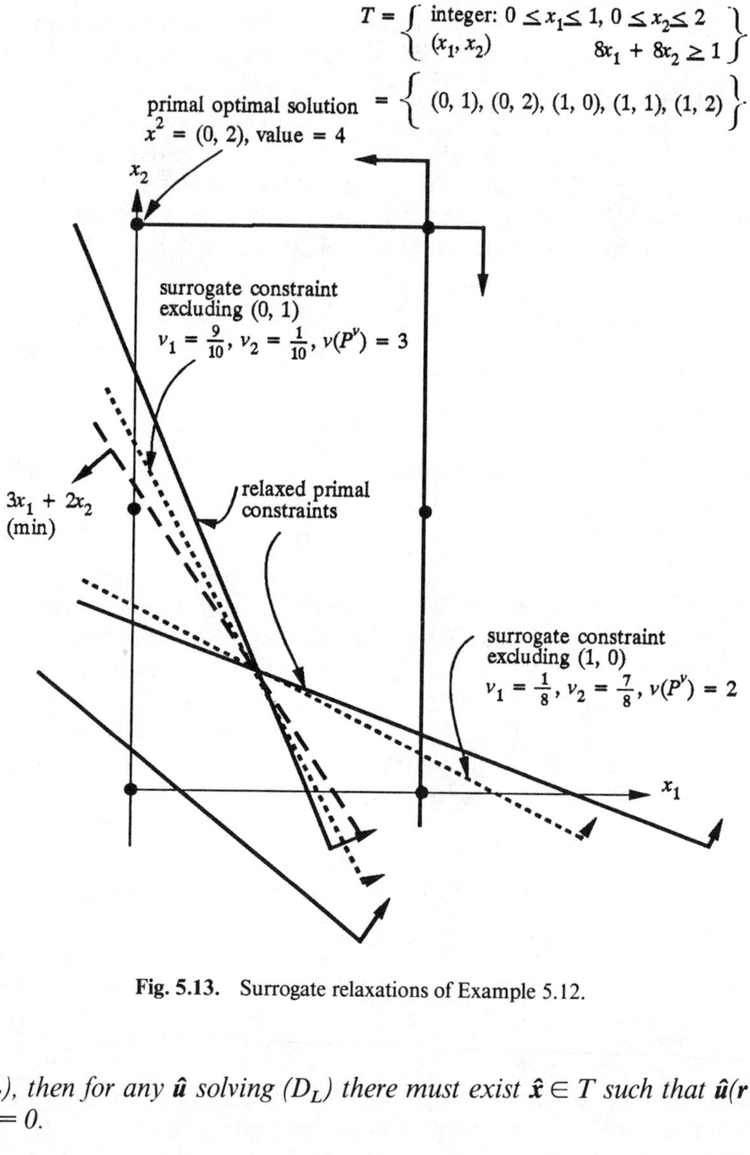

Fig. 5.13. Surrogate relaxations of Example 5.12.

$v(D_S)$, then for any $\hat{\mathbf{u}}$ solving (D_L) there must exist $\hat{\mathbf{x}} \in T$ such that $\hat{\mathbf{u}}(\mathbf{r} - \mathbf{R}\hat{\mathbf{x}}) = 0$.

Proof. First note that for any $\mathbf{w} \geq \mathbf{0}$, $v(P_w) \leq v(P^w)$. This follows since

$$v(P^w) \triangleq v\begin{pmatrix} \min \mathbf{cx} \\ \mathbf{w}(\mathbf{r} - \mathbf{Rx}) \leq 0 \\ \mathbf{x} \in T \end{pmatrix}$$

$$\geq v\begin{pmatrix}\min \mathbf{cx} + \mathbf{w}(\mathbf{r} - \mathbf{Rx}) \\ \mathbf{w}(\mathbf{r} - \mathbf{Rx}) \leq 0 \\ \mathbf{x} \in T\end{pmatrix} \qquad \text{because the nonpositive } \mathbf{w}(\mathbf{r} - \mathbf{Rx}) \text{ is added to the objective}$$

$$\geq v\begin{pmatrix}\min \mathbf{cx} + \mathbf{w}(\mathbf{r} - \mathbf{Rx}) \\ \mathbf{x} \in T\end{pmatrix} \qquad \text{by relaxation}$$

$$\triangleq \quad v(P_w).$$

The immediate consequence is

$$v(D_L) \triangleq \sup_{\mathbf{u} \geq \mathbf{0}} v(P_u) \leq \sup_{\mathbf{v} \geq \mathbf{0}} v(P^v) \triangleq v(D_S). \tag{5-39}$$

Now suppose $\hat{\mathbf{u}} \geq \mathbf{0}$ solves (D_L), and let $\hat{\mathbf{x}}$ solve the surrogate relaxation $(P^{\hat{u}})$. (Our assumptions about T assure there is such an $\hat{\mathbf{x}}$). Feasibility in $(P^{\hat{u}})$ implies $\hat{\mathbf{u}}(\mathbf{r} - \mathbf{R}\hat{\mathbf{x}}) \leq 0$. But since $\hat{\mathbf{x}}$ is also feasible for $(P_{\hat{u}})$,

$$v(P_{\hat{u}}) \leq \mathbf{c}\hat{\mathbf{x}} + \hat{\mathbf{u}}(\mathbf{r} - \mathbf{R}\hat{\mathbf{x}}) \leq \mathbf{c}\hat{\mathbf{x}} = v(P^{\hat{u}}).$$

When $v(P_{\hat{u}}) \triangleq v(D_L) = v(D_S)$, (5-39) yields $\mathbf{c}\hat{\mathbf{x}} = \mathbf{c}\hat{\mathbf{x}} + \hat{\mathbf{u}}(\mathbf{r} - \mathbf{R}\hat{\mathbf{x}})$ and $\hat{\mathbf{u}}(\mathbf{r} - \mathbf{R}\hat{\mathbf{x}}) = 0$. ■

A quick examination of the strong duality Theorems 5.14 and 5.27 will show that the Lagrangean version requires the additional complementarity hypothesis $\mathbf{u}(\mathbf{r} - \mathbf{Rx}) = 0$. Theorem 5.28 shows that complementarity arises again in comparing $v(D_S)$ to $v(D_L)$. It is always the case that $v(D_S) \geq v(D_L)$ and the surrogate dual will provide a superior bound to the Lagrangean unless every solution to (D_L) has a complementary $\mathbf{x} \in T$. Since it seems unlikely that such complementary solutions would always exist, we would expect $v(D_S) > v(D_L)$ in most formulations. Of course, the additional complexity involved in solving (P^v)'s may overcome any advantage.

Theorem 5.28 may seem perplexing to a reader recalling the discussion of surrogate constraints for implicit enumeration in Section 5.2. There, Theorem 5.10 showed that the bound from such a surrogate constraint could never exceed the linear programming relaxation value $v(\overline{P})$. Now Theorem 5.28 suggests that $v(D_S)$ can exceed $v(D_L)$, which we know satisfies $v(D_L) \geq v(\overline{P})$.

The resolution of this dilemma lies in a careful comparison of the surrogate constraints of Bound 5C and the surrogate relaxation of this section. One form of surrogate constraint in 5C ignores cost altogether and the other adds cost to the surrogate constraint in the Lagrangean fashion. In contrast, (P^v) restricts feasibility with surrogate constraints and independently considers cost. This additional independence allows $v(D_S)$ to rise above its implicit enumeration relatives.

Surrogate Search

As with the Lagrangean case, it is a fairly complex matter to find good surrogate multipliers. The similarities, however, between the Lagrangean and surrogate dualizations can be exploited to generalize the Lagrangean methods of Section 5.5.2.

Although we did not always derive them in such terms, Lagrangean search techniques can be viewed as solving the system of inequalities

$$(I_L)\qquad \begin{aligned} \mathbf{cx} + \mathbf{u}(\mathbf{r} - \mathbf{Rx}) &\geq v(D_L) \qquad \text{for every } \mathbf{x} \in T \\ \mathbf{u} &\geq \mathbf{0}. \end{aligned}$$

Outer-linearization techniques work directly with system (I_L), using (P_u) to generate pivot rows as they are needed. Subgradient search solves (P_u) to find a violated constraint of (I_L), and then changes $\mathbf{u}$ in the corresponding $(\mathbf{r} - \mathbf{Rx})$ direction. Dual ascent techniques for lagrangean search use one or more subgradients to compute an (I_L)-infeasibility-reducing direction of change in $\mathbf{u}$ without resorting to the entire inequality system.

Lagrangean relaxations penalize super-optimal $\mathbf{x} \in T$ by the cost term $\mathbf{u}(\mathbf{r} - \mathbf{Rx})$. Surrogate methods deal with such $\mathbf{x} \in T$ by rendering them infeasible in the subproblems $(P^{\mathbf{v}})$. More specifically, they require $\mathbf{v}$ to satisfy

$$(I_S^{\alpha})\qquad \begin{aligned} \mathbf{v}(\mathbf{b} - \mathbf{Ax}) &\geq \epsilon \qquad \text{for all } \mathbf{x} \in T^{\alpha} \\ \mathbf{v} &\geq \mathbf{0} \\ \epsilon &> 0, \end{aligned}$$

where $T^{\alpha} = \{\mathbf{x} \in T\colon \mathbf{cx} \leq \alpha\}$. Any $\mathbf{v}$ satisfying (I_S^{α}) will cut off all $\mathbf{x} \in T^{\alpha}$, i.e., have $v(P^{\mathbf{v}}) > \alpha$. Our next lemma formalizes the idea that the smallest α, not admitting a solution to (I_S^{α}) is $v(D_S)$.

Lemma 5.29. Relation Between System ($\mathbf{I_S}$) and the Surrogate Dual. *Let (P) be the primal discrete optimization problem defined above over a compact set T and let (D_S) and (I_S^{α}) be as above. Also, assume (P) is feasible. Then, (I_S^{α}) has a solution, $(\mathbf{v}, \epsilon)$, if and only if $\alpha < v(D_S)$.*

Proof. First pick α and assume (I_S^{α}) has a solution, $(\mathbf{v}, \epsilon)$. Then no $\mathbf{x} \in T^{\alpha}$ is feasible in $(P^{\mathbf{v}})$ and $\alpha < v(P^{\mathbf{v}}) \leq v(D_S)$.

Conversely, since (P) is assumed feasible, $v(D_S)$ is finite. Noting the definition $v(D_S) \triangleq \sup_{\mathbf{v} \geq \mathbf{0}} v(P^{\mathbf{v}})$, $\hat{\alpha} < v(D_S)$ implies there exist $\hat{\mathbf{v}} \geq \mathbf{0}$ such that $v(P^{\hat{\mathbf{v}}}) > \hat{\alpha}$. It follows that $\hat{\mathbf{v}}(\mathbf{b} - \mathbf{Ax}) > 0$ for all $\mathbf{x} \in T^{\hat{\alpha}}$.

Now, since T is assumed compact, $T^{\hat{\alpha}} \triangleq \{\mathbf{x} \in T\colon \mathbf{cx} \leq \hat{\alpha}\}$ is also compact. Thus, there is some specific $\hat{\mathbf{x}} \in T^{\hat{\alpha}}$ with $0 < \hat{\mathbf{v}}(\mathbf{r} - \mathbf{R}\hat{\mathbf{x}}) = \inf_{\mathbf{x} \in T^{\hat{\alpha}}}$

$\{\hat{\mathbf{v}}(\mathbf{r} - \mathbf{R}\mathbf{x})\}$. Choosing $\hat{\epsilon} = \hat{\mathbf{v}}(\mathbf{b} - \mathbf{A}\hat{\mathbf{x}})$ completes the construction of a solution, $(\hat{\mathbf{v}}, \hat{\epsilon})$, to $(I_S^{\hat{\alpha}})$. ■

There are obviously great similarities between systems (I_L) and (I_S^{α}). Moreover, the implication of Lemma 5.29 is that solving (I_S^{α}) for higher and higher α values can lead to an algorithm for surrogate search. Any α for which the system has a solution is a lower bound on $v(D_S)$. When the system has no solution, $\alpha \geq v(D_S)$.

Such notions invite the adoption of Lagrangean search techniques to the surrogate case via (I_S). A generic algorithm can be stated as follows.

Algorithm 5H. Generic Surrogate Search

STEP 0: *Initialization* Set $k \leftarrow 1$, $\alpha^0 \leftarrow -\infty$ and pick any $\mathbf{v}^1 \geq \mathbf{0}$.

STEP 1: *Surrogate Relaxation* Solve the surrogate relaxation $(P^{\mathbf{v}k})$ for an optimal solution $\mathbf{x}^k$. If $\mathbf{R}\mathbf{x}^k \geq \mathbf{r}$, stop; $\mathbf{x}^k$ solves (P) and $\mathbf{c}\mathbf{x}^k = v(D_S) = v(P)$.

STEP 2: *Incumbent Processing* If $\mathbf{c}\mathbf{x}^k > \alpha^{k-1}$, $(I_S^{\alpha^{k-1}})$ has a solution; advance $\alpha^k \leftarrow \mathbf{c}\mathbf{x}^k$. Otherwise, set $\alpha^k \leftarrow \alpha^{k-1}$.

STEP 3: **v**-*Changing* If by using $\mathbf{x}^k$, it can now be shown that $(I_S^{\alpha^k})$ has no solution, stop; $v(D_S) = \alpha^k$. Otherwise, use $\mathbf{x}^k$ to generate $\mathbf{v}^{k+1}$ as the next element of a sequence converging to a solution to $(I_S^{\alpha^k})$. Then set $k \leftarrow k + 1$ and return to Step 1. ■

The principle behind Algorithm 5H is the solution of (I_S^{α}) for successively higher α. Each time the incumbent solution $\alpha^k > \alpha^{k-1}$ at Step 2, we have shown that $v(D_S) \geq v(P^{\mathbf{v}k}) > \alpha^{k-1}$; by Lemma 5.29, $(I_S^{\alpha^{k-1}})$ has a solution. Stopping at Step 1 is merely an implementation of the strong surrogate duality conditions in Theorem 5.27. If we stop at Step 3 with the conclusion that $(I_S^{\alpha^k})$ has no solution, then by Lemma 5.29 $\alpha^k \geq v(D_S)$. But we generate all α^k as $v(P^{\mathbf{v}l})$ for some $l \leq k$. Thus, α^k equals the surrogate dual value.

To show convergence of Algorithm 5H, we would need to be more specific about Step 3's **v**-changing when stopping does not occur; however, space does not permit a discussion here. We shall merely illustrate the surrogate analog of Outer Lagrangean Linearization (Algorithm 5G) and refer the interested reader to the full treatment in Karwan and Rardin (1980).

Example 5.17. Surrogate Search on Example 5.12. The surrogate analog to outer Lagrangean linearization implements Step 3 of Algorithm 5H as

STEP 3: **v**-*Changing* Add $\mathbf{x}^k$ to a set $X \subseteq T$ containing all $\mathbf{x}^k$ generated so far. Then solve the linear program

$$\begin{aligned} \max \quad & \epsilon \\ \text{s.t.} \quad & \epsilon \leq \mathbf{v}(\mathbf{r} - \mathbf{R}\mathbf{x}) \qquad \mathbf{x} \in X \\ & \mathbf{v} \geq \mathbf{0} \\ & \mathbf{l}\mathbf{v} \leq 1. \end{aligned}$$

Let $(\mathbf{v}^{k+1}, \epsilon^{k+1})$ be an optimal solution. If $\epsilon^{k+1} = 0$, stop; $v(D_S) = \alpha^k$. Otherwise, replace $k \leftarrow k + 1$ and return to Step 1.

The only new element is the constraint, $\mathbf{l}\mathbf{v} \leq 1$. That limit on $\mathbf{v}$ merely keeps ϵ bounded when all $\mathbf{x} \in X$ can simultaneously have $\mathbf{v}(\mathbf{r} - \mathbf{R}\mathbf{x}) > 0$.

Turning to the specific case of Example 5.12, let us begin with $X \leftarrow \varnothing$ $\alpha^0 \leftarrow -\infty$ and $\mathbf{v}^1 \leftarrow (0.5, 0.5)$. The surrogate relaxation (P^{v^1}) would yield $\mathbf{x}^1 = (0, 1)$ with value 2.000. This value produces the new incumbent solution $\alpha^1 \leftarrow 2.000$.

We execute Step 3 next. Adding $\mathbf{x}^1$ to X gives $X = \{(0, 1)\}$. The corresponding **v**-finding linear program is

$$\begin{aligned} \max \quad & \epsilon \\ \text{s.t.} \quad & \epsilon \leq \mathbf{v}(\mathbf{r} - \mathbf{R}\mathbf{x}^1) = v_1 - 2v_2 \\ & v_1 + v_2 \leq 1 \\ & v_1, v_2 \geq 0. \end{aligned}$$

An optimal solution is $\epsilon^2 = 1$, $\mathbf{v}^2 = (1, 0)$.

Since $\epsilon^2 \neq 0$, we return to Step 1. The new relaxation (P^{v^2}) gives $\mathbf{x}^2 = (1, 0)$ and $v(P^{v^2}) = 3.000$. At Step 2 this value provides the new incumbent $\alpha^2 \leftarrow v(P^{v^2}) = 3.000$. Step 3 updates $X \leftarrow X \cup \{(1, 0)\} = \{(0, 1), (1, 0)\}$ and solves

$$\begin{aligned} \max \quad & \epsilon \\ \text{s.t.} \quad & \epsilon \leq v_1 - 2v_2 \\ & \epsilon \leq -2v_1 + v_2 \\ & v_1 + v_2 \leq 1 \\ & v_1, v_2 \geq 0. \end{aligned}$$

An optimal solution is $v_1^3 = 0$, $v_2^3 = 0$, $\epsilon^3 = 0$. Since $\epsilon^3 = 0$, we stop and conclude $v(D_S) = \alpha^2 = 3.000$. ■

5.5.4 *Practical Branch and Bound Considerations*

Although we have presented the essential theory of the use of duality in partial enumeration procedures, space does not permit introduction of a host of detailed questions that must be considered in the implementation of algorithms. If a dual problem, (D), is being used as the principal bound calculation in branch and bound algorithms, there are numerous short cuts that will save significant computational effort. One obvious example is that the maximization of the dual does not always need to be completed. Any time a feasible dual solution $\mathbf{u} \geq \mathbf{0}$ is discovered that yields a relaxation (P_u) or (P^u) with solution value greater than or equal to the branch and bound incumbent solution value v^*, the current candidate problem can be fathomed without further dual maximization. In many other cases, constraints can be dropped before completion without destroying the convergence of the algorithms we have outlined. The reader is referred to Fisher, Northrup and Shapiro (1975), Karwan and Rardin (1977), and Bazaraa and Goode (1979) for additional details.

A more subtle and perplexing consideration is the entire matter of which $\mathbf{u}$ or $\mathbf{v}$ yields the best relaxation. The theory we have presented assumes that the best lower bound is desired; thus, we want to maximize the relaxation's solution value. But suppose we are really using partial enumeration as a nonexact algorithm, i.e., we do not hope to prove the exact optimality of some solution, but seek only a good feasible solution. In a dual-based partial enumeration scheme, one obvious source of feasible solutions derives from solutions to relaxations that happen to be feasible in the main problem. To accelerate the search for a good feasible solution, one might pose the question: How can dual multipliers be found that maximize the quality of such feasible solutions? It remains to be seen whether such a criterion can be implemented.

5.6 Primal Partitioning—Benders Enumeration

The constructive dual techniques of Section 5.5 are designed to deal with discrete optimization problems having special row structure. After some troublesome rows (constraints) are relaxed, the problems have a conveniently solved form.

There are at least as many problems that possess special column structure. After all of a set of troublesome variables are assigned a value, such problems reduce to one with easy to solve structure. Often only a linear program remains.

For an example, return to the uncapacitated warehouse location problems (5-15) through (5-18) in Example 5.11. There, singly subscripted variables decide which supply sources are open and doubly subscripted variables assign flows from supply points to demand points. Once all single-subscript, x_i, are fixed at 0 or 1, a very simple problem remains that is solvable by linear programming. The x_{ij}, which form a substantial majority of the problem variables, can now be optimally chosen by very efficient linear programming techniques.

The partial solution approach to enumeration that we have employed throughout most of this chapter (refer to Section 5.1.2) does fix variables as it creates candidate problems. Still, it does not naturally exploit a complicating variables structure because most candidate problems assign values to only a few of the troublesome variables. All must be fixed to obtain an easy to solve case.

Benders (1962) developed an algorithmic strategy addressed directly to complicating variable problem structures. In Section 5.6.1, we will develop the theory of Benders' partitioning algorithm. In Section 5.6.2, we relate it to the rest of the material in the chapter.

Although more general cases can be treated (see Geoffrion (1972)), we will discuss the following rewritten version of our standard discrete optimization problem $P(S)$ throughout:

$$B(Y) \qquad \begin{aligned} \min \quad & \mathbf{c}^1\mathbf{y} + \mathbf{c}^2\mathbf{z} \\ \text{s.t.} \quad & \mathbf{A}^1\mathbf{y} + \mathbf{A}^2\mathbf{z} \geq \mathbf{b} \\ & \mathbf{y} \in Y, \mathbf{z} \geq \mathbf{0}. \end{aligned}$$

Here the decision vector, $\mathbf{x}$, has been partitioned as $(\mathbf{y}, \mathbf{z})$, where the y_j are the complicating variables. Similarly, $\mathbf{A} = (\mathbf{A}^1, \mathbf{A}^2)$, $\mathbf{c} = (\mathbf{c}^1, \mathbf{c}^2)$ and $S = \{(\mathbf{y}, \mathbf{z}): \mathbf{y} \in Y, \mathbf{z} \geq \mathbf{0}\}$. For fixed $\mathbf{y}$, $B(Y)$ is a linear program over $\mathbf{z}$. To avoid pathological cases we shall also assume throughout that the linear programming relaxation of $B(Y)$ is feasible and bounded in solution value.

5.6.1 *The Primal Partitioning Algorithm*

For fixed $\mathbf{y} \in Y$, the problem $B(Y)$ reduces to the linear program

$$BP(\mathbf{y}) \qquad \begin{aligned} \min \quad & \mathbf{c}^2\mathbf{z} + \mathbf{c}^1\mathbf{y} \\ \text{s.t.} \quad & \mathbf{A}^2\mathbf{z} \geq \mathbf{b} - \mathbf{A}^1\mathbf{y} \\ & \mathbf{z} \geq \mathbf{0}. \end{aligned}$$

Since some $\mathbf{y} \in Y$ solves $B(Y)$, if it has a solution, $B(Y)$ can be rewritten

$$\min_{\mathbf{y} \in Y} v(BP(\mathbf{y})). \tag{5-40}$$

Now each linear program $(BP(\mathbf{y}))$ has a dual

$$BD(\mathbf{y}) \qquad \begin{array}{ll} \max & \mathbf{u}(\mathbf{b} - \mathbf{A}^1\mathbf{y}) + \mathbf{c}^1\mathbf{y} \\ \text{s.t.} & \mathbf{u}\mathbf{A}^2 \leq \mathbf{c}^2 \qquad (5\text{-}41) \\ & \mathbf{u} \geq \mathbf{0}. \qquad (5\text{-}42) \end{array}$$

Furthermore, each $BD(\mathbf{y})$ is feasible because (5-41) and (5-42) are a part of the dual constraints for the linear programming relaxation, $B(\bar{Y})$. Our assumption of a finite optimal value for that relaxation implies its dual is feasible (see Appendix C for a review of linear programming theory).

When $BD(\mathbf{y})$ is feasible, $v(BP(\mathbf{y})) = v(BD(\mathbf{y}))$. Substituting in (5-40), we may write $B(Y)$ as

$$\min_{\mathbf{y} \in Y} v(BD(\mathbf{y})) \tag{5-43}$$

The notion implicit in (5-43) leads immediately to the equivalent Benders' master problem:

$$(BM) \qquad \begin{array}{llll} \min & \beta & & \\ \text{s.t.} & \beta \geq \mathbf{c}^1\mathbf{y} + \mathbf{u}(\mathbf{b} - \mathbf{A}^1\mathbf{y}) & \text{for all } \mathbf{u} \in U & (5\text{-}44) \\ & 0 \geq \mathbf{v}(\mathbf{b} - \mathbf{A}^1\mathbf{y}) & \text{for all } \mathbf{v} \in V & (5\text{-}45) \\ & \mathbf{y} \in Y, & & \end{array}$$

where

$$U = \left\{ \begin{array}{l} \text{extreme points of the set of } \mathbf{u} \text{ defined by} \\ \text{(5-41) and (5-42)} \end{array} \right\} \tag{5-46}$$

and

$$V = \left\{ \begin{array}{l} \text{extreme directions of the set of } \mathbf{u} \text{ defined} \\ \text{by (5-41) and (5-42)} \end{array} \right\} \tag{5-47}$$

Theorem 5.30. Benders' Form of Primal Partitioned Problems. *Suppose $B(Y)$ is the partitionable problem defined above with feasible and bounded-value linear programming relaxation $B(\bar{Y})$ and (BM) as above. Then $B(Y)$ is equivalent to (BM). Specifically, $v(B(Y)) = v(BM)$, and $(\beta^*, \mathbf{y}^*)$ solves (BM), if and only if there is a corresponding $\mathbf{z}^*$ such that $(\mathbf{y}^*, \mathbf{z}^*)$ solves $B(Y)$.*

Proof. In the derivation, we have already argued that $B(Y)$ is equivalent to (5-43), i.e.,

$$v(B(Y)) = v\left(\min_{\mathbf{y}\in Y} v(BD(\mathbf{y}))\right) = v\left(\min_{\mathbf{y}\in Y}\sup_{\mathbf{u}\in D}(\mathbf{c}^1\mathbf{y} + \mathbf{u}(\mathbf{b} - \mathbf{A}^1\mathbf{y}))\right)$$

where $D = \{\mathbf{u}$ satisfying (5-41) and (5-42)$\}$. Thus,

$$v(B(Y)) = v\begin{pmatrix} \min & \beta \\ \text{s.t.} & \beta \geq \mathbf{c}^1\mathbf{y} + \mathbf{u}(\mathbf{b} - \mathbf{A}^1\mathbf{y}) \text{ for all } \mathbf{u} \in D \\ & \mathbf{y} \in Y \end{pmatrix} \tag{5-48}$$

Now, for any $\mathbf{y} \in Y$, we have argued $v(BP(\mathbf{y})) = v(BD(\mathbf{y}))$. But there are two cases. If a given $\mathbf{y}$ leaves $BP(\mathbf{y})$ infeasible, (since $BD(\mathbf{y})$ is always feasible), $BD(\mathbf{y})$ is unbounded, i.e., $v(BD(\mathbf{y})) = +\infty$. When $BP(\mathbf{y})$ is feasible, $BD(\mathbf{y})$ has a finite optimum.

In the minimization of (5-48) we have no interest in the former $\mathbf{y}$, i.e., those yielding unbounded $BD(\mathbf{y})$. *Directions* of the set D are vectors $\mathbf{v}$ such that $\mathbf{v}^0 + \lambda\mathbf{v} \in D$ for some $\mathbf{v}^0 \in D$ and all $\lambda \geq 0$. Clearly, $BD(\mathbf{y})$ will be unbounded exactly when $0 < \mathbf{v}(\mathbf{b} - \mathbf{A}^1\mathbf{y})$ for some direction $\mathbf{v}$ of D. Thus, we may rewrite (5-48) as

$$v(B(Y)) = v\begin{pmatrix} \min \beta \\ \text{s.t. } \beta \geq \mathbf{c}^1\mathbf{y} + \mathbf{u}(\mathbf{b} - \mathbf{A}^1\mathbf{y}) \text{ for all } \mathbf{u} \in D \\ 0 \geq \mathbf{v}(\mathbf{b} - \mathbf{A}^1\mathbf{y}) \text{ for all directions of } D \\ \mathbf{y} \in Y \end{pmatrix}. \tag{5-49}$$

But this is exactly what we wish to prove. The theory of linear programming tells us that the problem in (5-49) is equivalent to one with constraints for only extreme points and extreme directions of D, i.e., equivalent to (BM). We can conclude that $v(B(Y)) = v(BM)$.

It remains to show that $(\beta^*, \mathbf{y}^*)$ solves (BM) if and only if $(\mathbf{y}^*, \mathbf{z}^*)$ solves $B(Y)$. When (BM) has an optimal solution $(\beta^*, \mathbf{y}^*)$, the above demonstrated that $BP(\mathbf{y}^*)$ is feasible and bounded in value. Taking any optimal $\mathbf{z}^*$ for $BP(\mathbf{y}^*)$ gives a $(\mathbf{y}^*, \mathbf{z}^*)$ solving $B(Y)$. Conversely, if (BM) has no solution, there must be no $\mathbf{y} \in Y$ satisfying (5-45) because a β can always be chosen to conform to (5-44). It follows that $BP(\mathbf{y})$ is infeasible for every $\mathbf{y} \in Y$, and $B(Y)$ has no solution. ■

The (BM) formulation is equivalent to $B(Y)$, but it has an enormous number of constraints. Benders' actual algorithm generates constraints only as they are needed by solving subproblems, $BD(\mathbf{y})$.

Algorithm 5I. Benders' Partitioning

STEP 0: *Initialization* Set $\beta^1 \leftarrow -\infty$, the iteration counter $l \leftarrow 1$, an extreme point subset $U^0 \leftarrow \emptyset$, and an extreme direction subset $V^0 \leftarrow \emptyset$. Also, choose any $\mathbf{y}^1 \in Y$.

STEP 1: *Subproblem* Apply some form of the simplex method to $BD(\mathbf{y}^l)$. If $BD(\mathbf{y}^l)$ is unbounded, let $\mathbf{v}^l$ be the extreme direction obtained from the simplex algorithm and go to Step 2. If $BD(\mathbf{y}^l)$ is bounded, let $\mathbf{u}^l$ be an optimal extreme point solution and go to Step 3.

STEP 2: *Extreme Direction Processing* Add $\mathbf{v}^l$ to the extreme direction subset by $V^l \leftarrow V^{l-1} \cup \{\mathbf{v}^l\}$, set $U^l \leftarrow U^{l-1}$ and go to Step 4.

STEP 3: *Extreme Point Processing* If

$$\beta^l \geq \mathbf{c}^1\mathbf{y}^l + \mathbf{u}^l(\mathbf{b} - \mathbf{A}^1\mathbf{y}^l), \tag{5-50}$$

solve $BP(\mathbf{y}^l)$ for $\mathbf{z}^l$ and stop; $(\mathbf{y}^l, \mathbf{z}^l)$ is optimal in $B(Y)$. Otherwise, add $\mathbf{u}^l$ to the extreme point subset by $U^l \leftarrow U^{l-1} \cup \{\mathbf{u}^l\}$, set $V^l \leftarrow V^{l-1}$ and go to Step 4.

STEP 4: *Master Problem* Attempt to solve

$$(BM_l)\quad \begin{array}{lll} \min & \beta & \\ \text{s.t.} & \beta \geq \mathbf{c}^1\mathbf{y} + \mathbf{u}(\mathbf{b} - \mathbf{A}^1\mathbf{y}) & \text{for all } \mathbf{u} \in U^l \\ & 0 \geq \mathbf{v}(\mathbf{b} - \mathbf{A}^1\mathbf{y}) & \text{for all } \mathbf{v} \in V^l \\ & \mathbf{y} \in Y. & \end{array}$$

If (BM_l) is infeasible, stop; $B(Y)$ is infeasible. Otherwise, if (BM_l) is bounded, let $(\beta^{l+1}, \mathbf{y}^{l+1})$ be any optimal solution, and if (BM_l) is unbounded, choose $\beta^{l+1} = \beta^l$ and $\mathbf{y}^{l+1} =$ any $\mathbf{y}$ satisfying $\mathbf{0} \geq \mathbf{v}(\mathbf{b} - \mathbf{A}^1\mathbf{y})$ for all $\mathbf{v} \in V^l$. Then, advance $l \leftarrow l + 1$ and return to Step 1. ■

Theorem 5.31. Finite Convergence of Benders' Partitioning. *Suppose $B(Y)$ is the partitionable problem defined above with feasible, bounded-value linear programming relaxation $B(Y)$. Then, after finitely many of its steps, Algorithm 5I finds an optimal solution to $B(Y)$ or proves that none exists.*

Proof. There are only finitely many extreme points, U, and extreme directions, V, of the feasible space for the dual subproblems $BD(\mathbf{y})$. Thus, Algorithm 5I could only fail to stop finitely if some $\mathbf{u}^l$ or $\mathbf{v}^l$ is duplicated by Step 1. An extreme direction $\mathbf{v}^l$ obtains where $BD(\mathbf{y}^l)$ is unbounded, i.e., when $\mathbf{v}^l(\mathbf{b} - \mathbf{A}^1\mathbf{y}^l) > 0$. Since $\mathbf{y}^l$ solved (BM_{l-1}), we know such a $\mathbf{v}^l$ must be a new extreme direction; $0 \geq \mathbf{v}(\mathbf{b} - \mathbf{A}^1\mathbf{y}^l)$ for all $\mathbf{v} \in V^{l-1}$. Similarly, if Step 1 yields an extreme point $\mathbf{u}^l$,

$$\begin{aligned}
\mathbf{c}^1\mathbf{y}^l + \mathbf{u}^l(\mathbf{b} - \mathbf{A}^1\mathbf{y}^l) &= v(BD(\mathbf{y}^l)) \\
&= v(BP(\mathbf{y}^l)) \text{ by linear programming duality} \\
&\geq v(B(Y)) \text{ because } BP(\mathbf{y}^l) \text{ is a restricted version of } B(Y) \\
&= v(BM) \text{ by Theorem 5.30} \\
&\geq v(BM_{l-1}) \text{ because every } (BM_j) \text{ is a relaxation of } (BM) \\
&= \beta^l \text{ by construction of Step 4} \\
&\geq \mathbf{c}^1\mathbf{y}^l + \mathbf{u}(\mathbf{b} - \mathbf{A}^1\mathbf{y}^l) \text{ because } \mathbf{y}^l \text{ solved} \\
&\quad\ (BM_{l-1}) \text{ for all } \mathbf{u} \in U^{l-1}
\end{aligned}$$

Thus when the stopping rule of Step 3 is not satisfied, i.e., $\mathbf{c}^1\mathbf{y}^l + \mathbf{u}^l(\mathbf{b} - \mathbf{A}^1\mathbf{y}^l) > \beta^l$, $\mathbf{u}^l$ cannot belong to U^{l-1}.

This also proves that when we do stop at Step 3, $\mathbf{y}^l$ is optimal. The stopping criterion (5–50) has $\mathbf{c}^1\mathbf{y}^l + \mathbf{u}^l(\mathbf{b} - \mathbf{A}^1\mathbf{y}^l) \leq \beta^l$. With the above we can conclude $v(BP(\mathbf{y}^l)) = v(B(Y))$.

The only remaining issue is stopping at Step 4 when (BM_l) is infeasible. In such cases, there is no $\mathbf{y} \in Y$ such that $\mathbf{v}(\mathbf{b} - \mathbf{A}^1\mathbf{y}) \leq 0$ for all $\mathbf{v} \in V$, i.e., there is no $\mathbf{y} \in Y$ such that $BD(\mathbf{y})$ is bounded. Clearly, $B(Y)$ is also infeasible. ■

Example 5.18. Benders' Partitioning. Consider the simple mixed 0–1 program

$$\begin{array}{llrl}
 & \min & 42y_1 + 18y_2 + 33y_3 - 8z_1 - 6z_2 + 2z_3 & \\
BX(Y) & \text{s.t.} & 10y_1 + 8y_2 \qquad\quad - 2z_1 - z_2 + z_3 & \geq 4 \\
 & & 5y_1 \qquad + 8y_3 - z_1 - z_2 - z_3 & \geq 3 \\
 & & (y_1, y_2, y_3) \in Y = \{\text{binary 3-vectors}\} & \\
 & & z_1, z_2, z_3 \geq 0. &
\end{array}$$

Dual subproblems of the Benders algorithm have the form

$$\begin{array}{lll}
 & \max & (4 - 10y_1 - 8y_2)u_1 + (3 - 5y_1 - 8y_3)u_2 \\
 & & \qquad + (42y_1 + 18y_2 + 33y_3) \\
BXD(\mathbf{y}) & \text{s.t.} & -2u_1 - u_2 \leq -8 \\
 & & -\ u_1 - u_2 \leq -6 \\
 & & u_1 - u_2 \leq \ 2 \\
 & & u_1, \ u_2 \geq 0.
\end{array}$$

Figure 5.14 plots the feasible space of these duals. The complete set U consists of the extreme points $\{(4, 2), (2, 4), (0, 8)\}$. The two extreme directions give $V = \{(0, 1), (1, 1)\}$. Notice that this dual space is not bounded even though the feasible space for the primal, $BX(Y)$ is bounded.

Algorithm 5I begins with $l \leftarrow 1$, $\beta^1 \leftarrow -\infty$, $U^0 \leftarrow \emptyset$, $V^0 \leftarrow \emptyset$. Let us choose $\mathbf{y}^1 \leftarrow (1, 1, 1)$. Solution of $BXD(\mathbf{y}^1)$ yields the extreme point $\mathbf{u}^1 = (2, 4)$. Adding $U^1 \leftarrow U^0 \cup \{(2, 4)\}$ and $V^1 \leftarrow V^0$, the first master problem is

$$
\begin{array}{lll}
 & \min & \beta \\
(BXM_1) & \text{s.t.} & \beta \geq 42y_1 + 18y_2 + 33y_3 + 2(4 - 10y_1 - 8y_2) \\
 & & \qquad + 4(3 - 5y_1 - 8y_3) \\
 & & y_1, y_2, y_3 = 0 \text{ or } 1.
\end{array}
$$

Notice that it is a discrete optimization problem. Solving by any algorithm, we would obtain the finite optimal solution $\beta^2 = 20$, $\mathbf{y}^2 = (0, 0, 0)$.

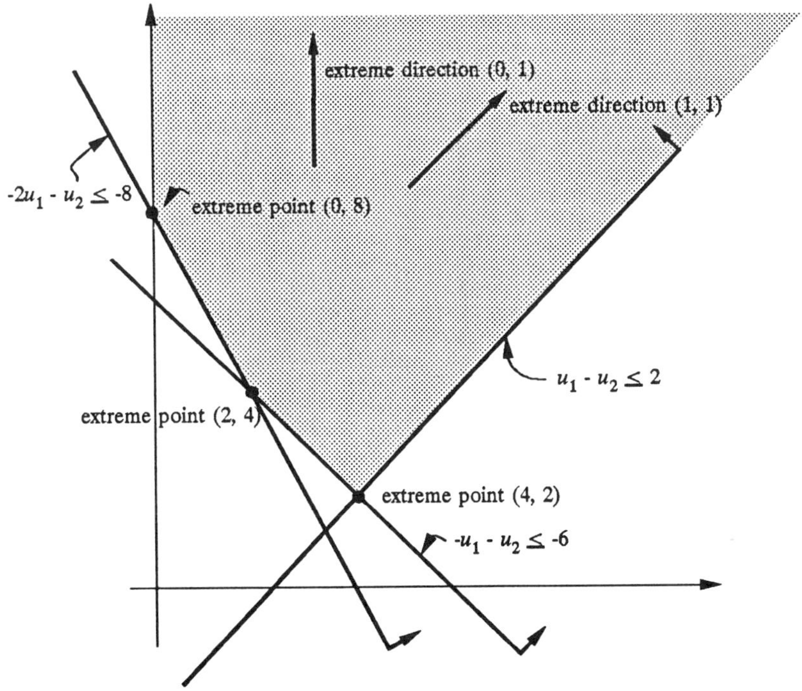

Fig. 5.14. Dual subproblem space of Example 5.18.

Our example is so small that we can see that the primal problem, $BXP(\mathbf{y}^2)$, is infeasible. Accordingly, when Algorithm 5I attempts the dual, $BXD(\mathbf{y}^2)$, an extreme direction results. Actually, either extreme direction might occur, but we shall choose $\mathbf{v}^2 = (0, 1)$. Working subsets become $V^2 = \{(0, 1)\}$, $U^2 = \{(2, 4)\}$. Including the new constraint, the master problem is

$$\begin{array}{lll} & \min & \beta \\ (BXM_2) & \text{s.t.} & \beta \geq 42y_1 + 18y_2 + 33y_3 + 2(4 - 10y_1 - 8y_2) \\ & & \qquad + 4(3 - 5y_1 - 8y_3) \\ & & 0 \geq 3 - 5y_1 - 8y_3 \\ & & y_1, y_2, y_3 = 0 \text{ or } 1. \end{array}$$

Solving again by inspection, $\beta^3 = 21$, $\mathbf{y}^3 = (0, 0, 1)$.

Continuing with the sequence of Algorithm 5I, we obtain $\mathbf{u}^3 = (4, 2)$, $\mathbf{y}^4 = (1, 0, 0)$, $\beta^4 = 22$, $\mathbf{u}^4 = (0, 8)$, $\mathbf{y}^5 = (1, 0, 0)$, $\beta^5 = 26$, $\mathbf{u}^5 = (0, 8)$. At this point the stopping rule of Step 3 is satisfied,

$$26 = \beta^5 \geq \mathbf{c}^1\mathbf{y}^5 + \mathbf{u}^5(\mathbf{b} - \mathbf{A}^1\mathbf{y}^5) = 42 + 8(3 - 5) = 26.$$

We conclude $\mathbf{y}^* = \mathbf{y}^5 = (1, 0, 0)$, $v(BX(Y)) = \beta^5 = 26$. The corresponding $\mathbf{z}^* = (2, 0, 0)$ is derived from $BXP(\mathbf{y}^5)$. ■

As obvious limitation of Algorithm 5I is that it calls for solving discrete optimization problems (BM_l) at Step 4. If more than a few of the huge number of extreme points and extreme directions of $\{\mathbf{u} \geq \mathbf{0}: \mathbf{u}\mathbf{A}^2 \leq \mathbf{c}^2\}$ had to be generated, use of the Benders' technique would be completely impractical.

Fortunately, a careful review of the proof of Theorem 5.31 will show that short cuts are possible. We used the result that (BM_l) are explicitly solved only (i) to assure $\mathbf{u}^l$ and $\mathbf{v}^l$ did not repeat and (ii) to know when to stop. In early stages, when we are certain not to be near stopping, numerous other techniques will allow us to generate a nonrepeating sequence of $\mathbf{u}^l$ and $\mathbf{v}^l$. For example, we could solve $BD(\overline{y}^l)$ where $\overline{y}^l$ is the optimal solution to the linear programming relaxation $(\overline{BM}_{l-1})$. The resulting $\mathbf{u}^l$ or $\mathbf{v}^l$ cannot repeat if $\beta_l < v(B(\overline{Y}))$, the value of the linear programming relaxation of $B(Y)$.

It is the same notion of skipping explicit solution of many (BM_l) that makes the Benders technique well adapted to situations where many similar problems must be solved. A sequence of $\mathbf{y}^l$ saved from one case can be used to generate a starting set of $\mathbf{u}$'s and $\mathbf{v}$'s for a new case, thus, skipping many costly master problem steps.

5.6.2 *Benders Imbedded in Other Enumeration*

The Benders' Partitioning Algorithm 5I is a solution strategy for structured discrete problems that need not have any connection at all with other enumerative techniques. Still, the discrete problems of Step 4 have to be solved in some manner. If one thinks of solving those (BM_l) by enumerative techniques developed earlier in this chapter, a hybrid algorithm emerges. We shall first state and illustrate the technique, and then discuss its strengths and limitations.

Algorithm 5J. Imbedded Benders Enumeration

STEP 0: *Initialization* Establish a candidate list by making the full problem $B(Y)$ its sole entry, and create sets $\tilde{U}$ and $\tilde{V}$ containing, respectively, any known extreme points and extreme directions of $\{\mathbf{u} \geq \mathbf{0}\colon \mathbf{u}\mathbf{A}^2 \leq \mathbf{c}^2\}$. Also set the incumbent solution value $v^* \leftarrow +\infty$.

STEP 1: *Candidate Selection* Choose some member $B(Y_k)$ from the current candidate list.

STEP 2: *Bounding* Compute a lower bound $\tilde{\beta}$ on $v(B(Y_k))$ by solving

$$
\begin{array}{llll}
 & \min & \beta & \\
BM(Y_k) & \text{s.t.} & \beta \geq \mathbf{c}^1\mathbf{y} + \mathbf{u}(\mathbf{b} - \mathbf{A}^1\mathbf{y}) & \text{for all } \mathbf{u} \in \tilde{U} \\
 & & 0 \geq \mathbf{v}(\mathbf{b} - \mathbf{A}^1\mathbf{y}) & \text{for all } \mathbf{v} \in \tilde{V} \\
 & & \mathbf{y} \in Y, &
\end{array}
$$

or some relaxation of it. If $\tilde{\beta} \geq v^*$, go to Step 6 and fathom. Otherwise, go to Step 3.

STEP 3: *Feasible Solutions* Let $\tilde{\mathbf{y}} \in Y_k$ be any feasible solution to $BM(Y_k)$ obtained in the execution of Step 2, and apply some simplex procedure to $BD(\tilde{\mathbf{y}})$ to obtain an extreme point $\tilde{\mathbf{u}}$ or an extreme direction $\tilde{\mathbf{v}}$. If $v(BD(\tilde{\mathbf{y}})) \geq v^*$, go to Step 5 and branch. Otherwise, go to Step 4.

STEP 4: *Incumbent Saving* Save $\tilde{\mathbf{y}}$ from Step 3 as a new incumbent solution by $\mathbf{y}^* \leftarrow \tilde{\mathbf{y}}$, $v^* \leftarrow v(BD(\tilde{\mathbf{y}}))$. If the new $v^* \leq \tilde{\beta}$ (from Step 2), go to Step 6 and fathom. Otherwise, go to Step 5.

STEP 5: *Branching* If a $\tilde{\mathbf{u}} \notin \tilde{U}$ resulted from Step 3, set $\tilde{U} \leftarrow \tilde{U} \cup \{\tilde{\mathbf{u}}\}$. If a $\tilde{\mathbf{v}} \notin \tilde{V}$ was obtained, set $\tilde{V} \leftarrow \tilde{V} \cup \{\tilde{\mathbf{v}}\}$. If the $\tilde{\mathbf{u}}$ or $\tilde{\mathbf{v}}$ from Step 3 was not new, or if it is convenient for some other reason, replace $B(Y_k)$ in the candidate list by one or more further restricted candidate problems, $B(Y_{k1}), B(Y_{k2}), \ldots, B(y_{kp})$. Then return to Step 1.

STEP 6: *Fathoming* Fathom $B(Y_k)$, i.e., delete $B(Y_k)$ from the candidate list. If the candidate list is now empty, stop. Should $v^* = +\infty$, $B(Y)$ is infeasible, and otherwise an optimal solution $(\mathbf{y}^*, \mathbf{z}^*)$ to $B(Y)$ can be completed by solving $BP(\mathbf{y}^*)$. If candidates remain in the candidate list, return to Step 1. ■

Example 5.19. Imbedded Benders Enumeration on Example 5.18. Return to the example $BX(Y)$ of Example 5.18. To begin Algorithm 5J, we need some known subsets $\tilde{U}$ and $\tilde{V}$ of extreme points and extreme directions in $BXD(\mathbf{y})$. Suppose (perhaps by experimenting with $\mathbf{y} = (0, 0, 0)$ and $\mathbf{y} = (1, 1, 1)$), we begin with $\tilde{U} = \{(2, 4)\}$, $\tilde{V} = \{(0, 1)\}$.

After setting $v^* \leftarrow +\infty$, we are ready to begin candidate processing. Initially, the only $BX(Y_k)$ in the candidate list is $BX(Y_1) \triangleq BX(Y)$. Proceeding to Step 2, we must compute a bound on $v(BXM(Y_1))$. Many schemes are possible, but we shall illustrate with a variant of the surrogate with cost bound 5C. Specifically, rows of each $BXM(Y_k)$ will be weighted by surrogate multipliers $1/|\tilde{U}|$ and summed. The resulting one row relaxation is solved by choosing y_j with nonnegative coefficients equal to 0 and y_j with negative coefficients equal to 1.

The first $BXM(Y_k)$ is

$$
\begin{array}{lll}
 & \min & \beta \\
BXM(Y_1) & \text{s.t.} & \beta \geq 2y_1 + 2y_2 + y_3 + 20 \\
 & & 0 \geq -5y_1 - 8y_3 + 3 \\
 & & y_1, y_2, y_3 = 0 \text{ or } 1.
\end{array}
$$

Weighting by $1/|\tilde{U}| = 1$ gives the surrogate

$$\beta \geq -3y_1 + 2y_2 - 7y_3 + 23,$$

which is solved by $\tilde{\beta} = 13$, $\tilde{\mathbf{y}} = (1, 0, 1)$.

Since $\tilde{\beta} \ngeq v^*$, we proceed to Step 3. Solution of $BXD(\tilde{\mathbf{y}})$ yields the extreme point $\tilde{\mathbf{u}} = (4, 2)$ with $v(BXD(\tilde{\mathbf{y}})) = 31$. We proceed to Step 4 to save as an incumbent solution $\mathbf{y}^* \leftarrow \tilde{\mathbf{y}} = (1, 0, 1)$, $v^* \leftarrow v(BXD(\tilde{\mathbf{y}})) = 31$. However, $31 = v^* \nleq \tilde{\beta} = 13$. Thus, we move to Step 5 to branch. There, $\tilde{U} \leftarrow \tilde{U} \cup \{\tilde{\mathbf{u}}\} = \{(2, 4), (4, 2)\}$.

Since our $\tilde{\mathbf{u}}$ was new, we need not partition Y_1. Suppose, however, that we do decide to replace $BX(Y_1)$ in the candidate list by $BX(Y_2)$ and $BX(Y_3)$, where $Y_2 = \{\mathbf{y} \in Y_1 \colon y_2 = 0\}$ and $Y_3 = \{\mathbf{y} \in Y_1 \colon y_2 = 1\}$.

Selection of $BX(Y_2)$ at Step 1 would produce $\tilde{\beta} = 17$, $\tilde{\mathbf{y}} = (1, 0, 0)$, $\tilde{\mathbf{u}} = (0, 8)$, $v^* = 26$, $\mathbf{y}^* = (1, 0, 0)$ and $\tilde{U} = \{(2, 4), (4, 2), (0, 8)\}$ in succeeding steps. If Step 5 did not partition Y_2, reprocessing of candidate $BX(Y_2)$

would give $\tilde{\beta} = 13$, $\tilde{\mathbf{y}} = (1, 0, 1)$, $\tilde{\mathbf{u}} = (4, 2)$. Partitioning at Step 5 would then become manditory because $\tilde{\mathbf{u}} = (4, 2)$ already belongs to $\tilde{U}$.

By continuing in the same way on candidates remaining in the candidate list, we would eventually conclude the incumbent $\mathbf{y}^* = (1, 0, 0)$, $v^* = 26$ is optimal. Solution of $BXP(\mathbf{y}^*)$ completes the optimum with $\mathbf{z}^* = (2, 0, 0)$. ■

We shall not formally prove Algorithm 5J because it so closely resembles methods already treated in Sections 5.2 and 5.6.2. The only important notion to observe is how we can permit reprocessing of the same candidate problem, i.e., how we can fail to partition at Step 5, without risking a loss of finite algorithm convergence. The answer requires refinement in our definition of a candidate problem. Since Algorithm 5J uses $BM(Y_k)$ for bounding, candidates are not completely defined by Y_k. Instead, the restricted space associated with a candidate $BM(Y_k)$ is

$$\left\{\begin{array}{lr} \mathbf{y} \in Y_k\colon v^* \geq \mathbf{c}^1\mathbf{y} + \mathbf{u}(\mathbf{b} - \mathbf{A}^1\mathbf{y}) & \text{for all } \mathbf{u} \in \tilde{U} \\ \text{and } 0 \geq \mathbf{v}(\mathbf{b} - \mathbf{A}^1\mathbf{y}) & \text{for all } \mathbf{v} \in \tilde{V}. \end{array}\right\}$$

This restriction does change as new $\tilde{\mathbf{u}}$ and $\tilde{\mathbf{v}}$ are generated.

More generally, what is the imbedded Benders scheme contributing to Algorithm 5J? One answer is feasible solutions at Step 3. Each time we expose a new extreme point at Step 3, we have generated a feasible solution that might provide a new incumbent optimum. Section 5.4.2 showed the usefulness of good incumbent solutions in any partial enumeration algorithm.

The other area where Benders formulation contributes to Algorithm 5J is in bounding at Step 2. Of course, the degree of that contribution depends on the strength of the bound $\tilde{\beta}$. We can prove two comparisons to the more familiar linear programming relaxation bound $v(B(\overline{Y}_k))$.

Lemma 5.32. Sufficient Subsets $\tilde{\mathbf{U}}$ and $\tilde{\mathbf{V}}$ for Linear Relaxation Benders Bounds. *Let $B(Y)$ be the partitionable problem defined above and $B(Y_k)$ a corresponding candidate problem in Algorithm 5J. Also suppose bounds $\tilde{\beta}$ are obtained at Step 2 via linear programming relaxations as $\tilde{\beta} = v(BM(\overline{Y}_k))$. Then if $\bar{\boldsymbol{u}}$ is an optimal dual solution to the ordinary linear relaxation $B(\overline{Y}_k)$ and $\tilde{U}$ contains $\{\boldsymbol{u}^i\}$ and $\tilde{V}$ contains $\{\boldsymbol{v}^j\}$ such that*

$$\bar{\boldsymbol{u}} = \sum_i \lambda_i \boldsymbol{u}^i + \sum_j \mu_j \boldsymbol{v}^j,$$

where $\mu_j \geq 0$, $\lambda_i \geq 0$, $\Sigma_i \lambda_i = 1$, it must be true that $\tilde{\beta} \geq v(B(\overline{Y}_k))$.

Proof. If $\{\mathbf{u}^i\}$, $\{\mathbf{v}^j\}$, $\{\lambda_i\}$ and $\{\mu_j\}$ exist that satisfy the hypotheses of the lemma, use of the $\{\lambda_i\}$ and $\{\mu_j\}$ as surrogate constraint weights shows

$$\tilde{\beta} = v(BM(\overline{Y}_k)) \geq v \begin{pmatrix} \min & \beta \\ \text{s.t.} & \sum_i \lambda_i \beta \geq \sum_i \lambda_i (\mathbf{c}^1\mathbf{y} + \mathbf{u}^i(\mathbf{b} - \mathbf{A}^1\mathbf{y}) \\ & \qquad + \sum_j \mu_j \mathbf{v}^j(\mathbf{b} - \mathbf{A}^1\mathbf{y}) \\ & y \in \overline{Y}_k \end{pmatrix}$$

$$= v \begin{pmatrix} \min & \beta \\ \text{s.t.} & \beta \geq \mathbf{c}^1\mathbf{y} + \bar{\mathbf{u}}(\mathbf{b} - \mathbf{A}^1\mathbf{y}) \\ & \mathbf{y} \in \overline{Y}_k \end{pmatrix}$$

$$= v \begin{pmatrix} \min & \mathbf{c}^1\mathbf{y} + \bar{\mathbf{u}}(\mathbf{b} - \mathbf{A}^1\mathbf{y}) \\ \text{s.t.} & \mathbf{y} \in \overline{Y}_k \end{pmatrix}$$

$$= v \begin{pmatrix} \min & \mathbf{c}^1\mathbf{y} + \mathbf{c}^2\mathbf{z} + \bar{\mathbf{u}}(\mathbf{b} - \mathbf{A}^1\mathbf{y} - \mathbf{A}^2\mathbf{z}) \\ \text{s.t.} & \mathbf{y} \in \overline{Y}_k,\ \mathbf{z} = \mathbf{0} \end{pmatrix}$$

$$\geq v \begin{pmatrix} \min & \mathbf{c}^1\mathbf{y} + \mathbf{c}^2\mathbf{z} + \bar{\mathbf{u}}(\mathbf{b} - \mathbf{A}^1\mathbf{y} - \mathbf{A}^2\mathbf{z}) \\ \text{s.t.} & \mathbf{y} \in \overline{Y}_k,\ \mathbf{z} \geq \mathbf{0} \end{pmatrix}.$$

But the last problem is the Lagrangean relaxation of the primal problem $B(\overline{Y}_k)$. When the optimal $\bar{\mathbf{u}}$ is used, it has value equal to $v(B(\overline{Y}_k))$ (see Theorem 5.15). ■

Lemma 5.33. Linear Relaxation Benders Bounds. *Let $B(Y)$ be the partitionable problem defined above and $B(\overline{Y}_k)$ a corresponding candidate problem in Algorithm 5J. Then if bounds $\tilde{\beta}$ are obtained from Benders' master problem linear programming relaxations as $\tilde{\beta} = v(BM(\overline{Y}_k))$, the ordinary linear programming relaxation bound $v(B(\overline{Y}_k)) \geq \tilde{\beta}$.*

Proof. Substitution of $\overline{Y}_k$ for Y in Theorem 5.30 shows

$$v(B(\overline{Y}_k)) = v \begin{pmatrix} \min & \beta & \\ \text{s.t.} & \beta \geq \mathbf{c}^1\mathbf{y} + \mathbf{u}(\mathbf{b} - \mathbf{A}^1\mathbf{y}) & \text{for all } \mathbf{u} \in U \\ & 0 \geq \mathbf{v}(\mathbf{b} - \mathbf{A}^1\mathbf{y}) & \text{for all } \mathbf{v} \in V \\ & \mathbf{y} \in \overline{Y}_k & \end{pmatrix}$$

$$\geq \quad v(BM(\overline{Y}_k)),$$

because $\tilde{U} \subseteq U$ and $\tilde{V} \subseteq V$. ■

Lemma 5.32 tells us we can often expect $v(BM(Y_k))$ to provide a useful bound with relatively few $\mathbf{u} \in \tilde{U}$ and $\mathbf{v} \in \tilde{V}$. If we have only a set of $\mathbf{u}^i$ and

$\mathbf{v}^j$ that can represent a $B(\overline{Y}_k)$-optimal dual multiplier, $\bar{\mathbf{u}}$, even the Benders linear programming relaxation, $\overline{BM}(\overline{Y}_k)$, will yield as good a bound as the ordinary linear programming relaxation $B(\overline{Y}_k)$. Of course, if we actually solve the discrete problem $BM(Y_k)$, we would probably do better.

On the other hand, Lemma 5.33 shows that we can never get beyond the ordinary linear programming relaxation bound via Algorithm 5J unless we do more than treat the $BM(Y_k)$ as linear programs. Even with very large sets $\tilde{U}$ and $\tilde{V}$, $v(B(\overline{Y}_k)) \geq v(\overline{BM}(\overline{Y}_k))$. If there are relatively few $\mathbf{y}$ variables and many $\mathbf{z}$ variables, we might find Algorithm 5J convenient even when $v(B(\overline{Y}_k))$ is the best that can be achieved. It seems likely, however, that at least some $BM(Y_k)$ will have to be solved as discrete problems if Algorithm 5J is to have much value in comparison to the more straightforward linear programming based Algorithm 5D.

EXERCISES

5-1. Devise a finite branch and bound procedure for the vertex coloring problem on a given graph $G(V, E)$. Clearly define how subproblems are created, specify a suitable bounding rule or rules and establish finite convergence.

5-2. Consider the degree-constrained spanning tree problem (see Exercise **2-14** of Chapter 2). Repeat **5-1** relative to this problem.

5-3. The branch and bound tree on page 250 shows the enumeration of a 4-variable 0–1 optimization problem by Algorithm 5D. Taking candidate problems (nodes) in the order in which they are numbered, justify at each step why the algorithm branched and/or fathomed and/or updated the incumbent solution, $\mathbf{x}^*$ as is indicated.

5-4. After a number of variables have been fixed in the implicit enumeration of a certain 0–1 optimization problem, variables 4, 5 and 9 remain free, the incumbent solution value is $v^* = 15$, and the current candidate is

$$\begin{aligned} \min \quad & 4x_4 + 8x_5 + 10x_9 + 7 \\ \text{s.t.} \quad & 3x_4 \qquad\quad + 2x_9 + 5 \geq 9 \\ & 2x_4 + 4x_5 + x_9 + 6 \geq 10 \\ & 5x_5 + 4x_9 + 6 \geq 10 \\ & x_4, x_5, x_9 \text{ binary} \end{aligned}$$

(a) Compute implicit enumeration bounds β_1 through β_3 for this candidate, explaining which (if any) of the bounds would result in fathoming.

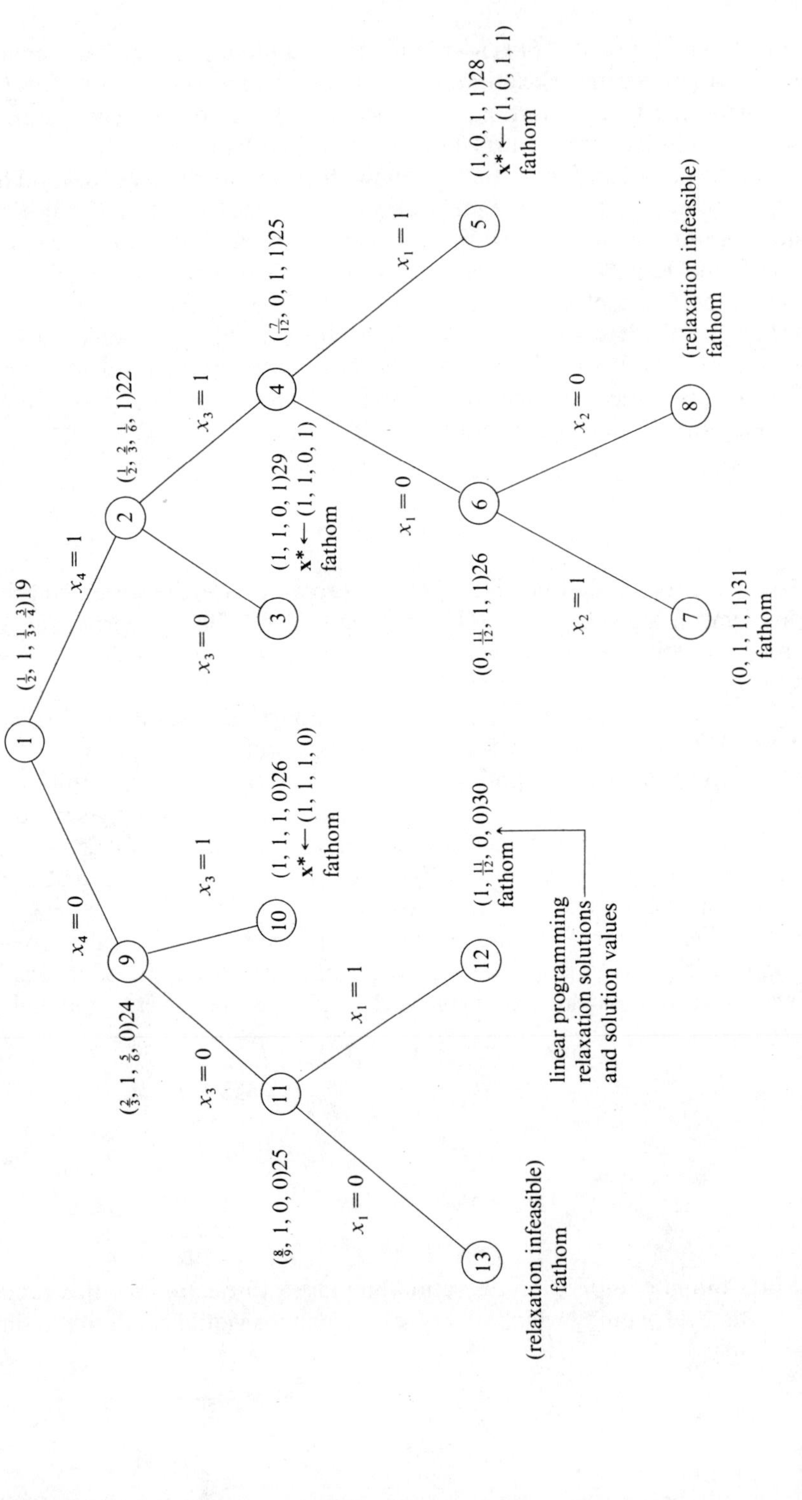
1
$(\frac{1}{2}, 1, \frac{1}{3}, \frac{3}{4})19$
$x_4 = 1$
2
$(\frac{1}{2}, \frac{2}{3}, \frac{1}{6}, 1)22$
$x_3 = 0$
3
(1, 1, 0, 1)29
$\mathbf{x}^* \leftarrow (1, 1, 0, 1)$
fathom
$x_3 = 1$
4
$(\frac{7}{12}, 0, 1, 1)25$
$x_1 = 1$
5
(1, 0, 1, 1)28
$\mathbf{x}^* \leftarrow (1, 0, 1, 1)$
fathom
$x_1 = 0$
6
$(0, \frac{11}{12}, 1, 1)26$
$x_2 = 1$
7
(0, 1, 1, 1)31
fathom
$x_2 = 0$
8
(relaxation infeasible)
fathom
$x_4 = 0$
9
$(\frac{2}{3}, 1, \frac{5}{6}, 0)24$
$x_3 = 1$
10
(1, 1, 1, 0)26
$\mathbf{x}^* \leftarrow (1, 1, 1, 0)$
fathom
$x_3 = 0$
11
$(\frac{8}{9}, 1, 0, 0)25$
$x_1 = 1$
12
$(1, \frac{11}{12}, 0, 0)30$
fathom
$x_1 = 0$
13
(relaxation infeasible)
fathom
linear programming
relaxation solutions
and solution values

(b) Compute implicit enumeration bounds β_4 and β_5 with weights $\mathbf{u} = (0, 1/2, 1)$, and β_6, explaining which (if any) of the bounds would result in fathoming.

5-5. Consider the binary knapsack problem

$$(KP) \qquad \begin{aligned} \min \quad & \sum_{j=1}^{n} c_j x_j \\ \text{s.t.} \quad & \sum_{j=1}^{n} a_j x_j \geq b \\ & \mathbf{x} \text{ binary,} \end{aligned}$$

where all a_j and c_j are positive integers and subscripts are arranged so that $c_j/a_j \leq c_{j+1}/a_{j+1}$ for all $j = 1, 2, \ldots, n-1$.

(a) Show that linear programming relaxations of this problem can be solved recursively for optimum $\bar{\mathbf{x}}$ by

$$\bar{x}_1 \leftarrow \min\{1, b/a_1\}$$

$$\bar{x}_j \leftarrow \min\left\{1, \left(b - \sum_{i=1}^{j-1} a_i \bar{x}_i\right) \Big/ a_j\right\} \qquad j = 2, 3, \ldots, n$$

(b) Develop a calculation based on the results of (a) for establishing an LP-optimal dual solution, and for computing optimal simplex coefficients (see Appendix C) $\bar{a}_j$ and $\bar{c}_j$.

(c) Draw on (b) to develop concise calculations of post-optimality penalties of Theorems 5.11–5.12.

(d) Show that if $\bar{\mathbf{x}}$ is as computed in (a),

$$\tilde{x}_j \leftarrow \lceil \bar{x}_j \rceil \qquad j = 1, 2, \ldots, n$$

gives a feasible solution to problem (KP).

(e) Combine results of (a)–(d) to specify a version of Algorithm 5D specialized to (*KP*) that employs the strongest bounds available from (a) and (c), fixes the value of free variables when implied by (c), and obtains feasible solutions via (d).

5-6. Consider the knapsack problem instance

$$\begin{aligned} \min \quad & 9x_1 + 4x_2 + 4x_3 + 11x_4 + 10x_5 \\ \text{s.t.} \quad & 3x_1 + 3x_2 + 4x_3 + 10x_4 + 12x_5 \geq 5 \\ & 0 \leq x_1, x_2, x_3, x_4, x_5 \leq 1 \\ & x_1, x_2, x_3, x_4, x_5 \text{ integer} \end{aligned}$$

The data for this problem has been arranged so that linear programming relations can be solved by simply using variables in reverse subscript order until feasibility is achieved (see **5-5a**). Solve (*KP*) via standard linear programming branch and bound Algorithm 5D, computing bounds via the above scheme. Do not round. Accept feasible solutions only when they arise naturally in the relaxations. Also, use a depth first enumeration sequence with branches fixed at 1 preferred over those fixed at 0.

5-7. Suppose *a priori* bounds were stored with each candidate in **5-6**. The bounds will merely equal the computed linear programming bound of the parent of the new candidate.

(a) Show the bounds that would be stored in this way as the enumeration of **5-6** continued.

(b) At what point in **5-6** could we have terminated the algorithm with certainty that our incumbent solution value exceeds the true $v(KP)$ by at most 20%? Why?

5-8. Using linear programming based relaxations (see **5-5a**) solve the following knapsack problem by branch and bound.

$$\begin{array}{ll} \min & 7x_1 + 15x_2 + 16x_3 + 13x_4 + 14x_5 \\ \text{s.t.} & 3x_1 + 10x_2 + 9x_3 + 8x_4 + 7x_5 \geq 1 \\ & x_1, x_2, x_3, x_4, x_5 \text{ binary.} \end{array}$$

Do not round solutions. Use a best bound enumeration sequence with a priori bounds equal to those of the parent problem.

5-9. Repeat **5-8** employing penalty calculations first, based on Theorem 5.11 followed by those of Theorem 5.12.

5-10. Suppose we convert 3-SAT (see Chapter 2) to an equivalent integer program. Would the linear programming relaxation of subproblems provide very strong (or clever) lower bounds? Discuss.

5-11. Consider solving the pure 0–1 optimization problem

$$\begin{array}{lrcl} \min & 2x_1 + 3x_2 + 11x_3 + 11x_4 + 9x_5 & & \\ \text{s.t.} & 2x_1 + 10x_3 + 5x_4 + 4x_5 & \geq & 7 \\ & 10x_2 + 10x_3 + x_5 & \geq & 9 \\ & 4x_1 + 10x_3 + 15x_4 + 3x_5 & \geq & 16 \\ & 15x_3 + 5x_4 + 3x_5 & \geq & 7 \\ & x_1, x_2, x_3, x_4, x_5 \text{ binary} & & \end{array}$$

by Algorithm 5D, rounding up linear programming solutions $\bar{x}_j > 0$ to 1 to produce feasible solutions, branching on the fractional variable with lowest subscript, breaking ties by preferring $x_j = 1$ over $x_j = 0$, and employing the bound of the parent problem as *a priori* bound $\sigma(S_k)$ of any candidate.

(a) Solve the problem using depth first enumeration. Show your branch and bound tree.
(b) Solve the problem using best bound enumeration. Show your branch and bound tree.
(c) Produce convergence plots like Figure 5.5 for your results of (a) and (b).
(d) Compare (a) and (b) experience in terms of the four measures of performance discussed in Section 5.4.1.
(e) Show when the enumerations of (a) and (b) could have been terminated had we been willing to accept a solution provably no more than 15% of optimal.
(f) For the last 6 candidates created in (a), estimate their optimal solution values at creation using pseudocost schemes (5-9)-(5-10).

5-12. Consider the usual pure $0-1$ problem

$$\begin{array}{ll} \min & \mathbf{cx} \\ \text{s.t.} & \mathbf{Ax} \geq \mathbf{b} \\ & \mathbf{x} \text{ binary} \end{array}$$

with decision variables $\{x_j: j = 1, 2, \ldots, n\}$ grouped into nonoverlapping intervals $J_1 = \{1, 2, \ldots, n_1\}, J_2 = \{n_1 + 1, \ldots, n_2\}, \ldots J_p = \{n_{p-1} + 1, \ldots, n\}$ of $\{1, 2, \ldots, n\}$. For each of the following *special ordered set constraints* defined in terms of these J_k state and show finite convergence of the version of Algorithm 5A obtained when branching is accomplished by partial solution if $|J_k| = 1$ for all k, and otherwise by the rule shown.

(a) Constraints (Type I): $\Sigma_{j \in J_k} x_i \leq 1$ for all k
Branching: For $J_k \triangleq \{p, p+1, \ldots, q\}$ chose r such that $p \leq r \leq q$ and create two new candidates with $\Sigma_{j=p}^{r} x_j = 0$ and $\Sigma_{j=r+1}^{q} x_j = 0$ respectively.
(b) Constraints (Type II): $\Sigma_{j \in J_k} x_j \leq 2$ for all k

$$x_i = x_j = 1 \text{ for any } i, j \in J_k, j > i \text{ implies } j = i + 1$$

Branching: For $J_k \triangleq \{p, p+1, \ldots, q\}$ choose $p \leq r \leq q$ and create two new candidates with $\Sigma_{j=p}^{r-1} x_j = 0$ and $\Sigma_{j=r+1}^{q} x_j = 0$ respectively.

5-13. Ellwein (1974) proposes a flexible form of depth first enumeration for binary integer programs that proceeds deeper in the search tree by fixing at 1 free variables drawn from a restricted set, and backtracks upon fathoming by switching to 0 the value of some variable fixed at 1 and freeing variables fixed at 0 with greater *level*. *Enumeration constraints* $\mathbf{Ex} \leq \mathbf{p}$ are generated upon backtracking to avoid duplication of candidates. Define $c \triangleq$ the level of the current candidate, $J_0^k \triangleq \{j\colon x_j = 0$ in the hth candidate explored$\}$, $J_1^k \triangleq \{j\colon x_j = 1$ in the hth candidate explored$\}$, $J_F^k \triangleq \{j\colon x_j$ is free in the hth candidate explored$\}$, $l_j^k \triangleq$ level of variable j in candidate k. Initialize $J_0^1 \leftarrow J_1^1 \leftarrow \varnothing$, $J_F^1 \leftarrow \{\text{all } j\}$, $l_j^1 \leftarrow 0$ for all j.

Then forward branching chooses $\bar{j} \in \{j \in J_F^k\colon \Sigma_{t \in J_1^k} e_{it} + e_{ij} \leq p_i \text{ for all } i\}$ and updating $l_c^{k+1} \leftarrow l_c^k + 1$, $J_F^{k+1} \leftarrow J_F^k \backslash \bar{j}$, $J_0^{k+1} \leftarrow J_0^k$, $J_1^{k+1} \leftarrow J_1^k \cup \bar{j}$, $l_j^{k+1} \leftarrow l_c^{k+1}$ if $j = \bar{j}$ and $\leftarrow l_j^k$ otherwise. Backtracking chooses $\bar{j} \in J_1^k$, updates $l_c^{k+1} \leftarrow l_c^k - 1$, $J_F^{k+1} \leftarrow J_F^k \cup \{j \in J_0^k\colon l_j^k \geq l_{\bar{j}}^k\}$, $J_0^{k+1} \leftarrow \{j \in J_0^k\colon l_j^k < l_{\bar{j}}^k\} \cup \bar{j}$, $J_1^{k+1} \leftarrow J_1^k \backslash \bar{j}$, $l_j^{k+1} \leftarrow l_c^{k+1}$ if $j = \bar{j}$ or $l_j^k > l_c^{k+1}$ and $\leftarrow l_j^k$ otherwise. Also upon backtracking additional enumeration constraints, i, are created for each $j^* \in J_0^k$ with $l_{\bar{j}}^k \leq l_{j^*}^k < l_c^k$ by $e_{ij} \leftarrow 1$ if $j = j^*$ or $j \in J_1^k \cap \{j\colon l_j^k \leq l_{j^*}^k\}$ and $\leftarrow 0$ otherwise, $p_i \leftarrow \Sigma_j e_{ij} - 1$.

(a) Illustrate these branching processes by showing all values needed to do the first 6 candidates of depth first enumeration in Figure 5.4, followed by immediate backtracking to $x_2 = 0$. Assume the only variables in the problem are those shown in the figure.

(b) Establish that this Ellwein enumeration is a finitely convergent version of Algorithm 5A.

5-14. Construct a family of specific numerical examples of the form required for Chvátal's Theorem 5.6 and show Algorithm 5D requires exponentially many candidate problems to solve the examples.

***5-15.** Refer to the enumerations of Figure 5.4.

(a) Write a binary integer program $P(S)$ that realizes the σ and β bounds shown. β bounds should be linear programming relaxations and σ bounds the result of post-optimality penalties as in Section 5.3.2.

(b) Develop conditions on σ and β values in such an enumeration tree that render them realizable as such linear programming-based bounds.

5-16. Consider the 0-1 mixed-integer program

$$(P) \qquad \begin{array}{ll} \min & -3x_1 - 2x_2 - 3x_3 + 2x_4 \\ \text{s.t.} & x_1 + x_2 + x_3 + x_4 \geq 2 \\ & 5x_1 + 3x_2 + 4x_3 + 10x_4 \leq 10 \\ & x_1, x_2, x_3 \text{ binary } x_4 \geq 0 \end{array}$$

(a) Which of the following problems are relaxations of (P) in the sense that their solution values lower bound $v(P)$? Explain why.

$$(P_1)\quad \begin{aligned} \min \quad & -3x_3 + 2x_4 \\ \text{s.t.} \quad & x_1 + x_2 + x_3 + x_4 \geq 2 \\ & 5x_1 + 3x_2 + 4x_3 + 10x_4 \leq 10 \\ & 0 \leq x_1, x_2, x_3, \leq 1, x_4 \geq 0 \end{aligned}$$

$$(P_2)\quad \begin{aligned} \min \quad & -3x_1 - 2x_2 - 3x_3 + 2x_4 \\ \text{s.t.} \quad & x_1 + x_2 + x_3 + x_4 \geq 2 \\ & x_1, x_2, x_3, x_4 \geq 0 \end{aligned}$$

$$(P_3)\quad \begin{aligned} \min \quad & -3x_1 - 2x_2 - 3x_3 + 2x_4 \\ & \quad + u(2 - x_1 - x_2 - x_3 - x_4) \\ \text{s.t.} \quad & 5x_1 + 3x_2 + 4x_3 + 10x_4 \leq 10 \\ & x_1, x_2, x_3 \text{ binary } x_4 \geq 0 \end{aligned}$$

where u is a nonnegative parameter

$$(P_4)\quad \begin{aligned} \min \quad & -3x_1 - 2x_2 - 3x_3 + 2x_4 \\ & + v(5x_1 + 3x_2 + 4x_3 + 10x_4 - 10) \\ \text{s.t.} \quad & x_1 + x_2 + x_3 + x_4 \geq 2 \\ & x_1, x_2, x_3 \text{ binary } x_4 \geq 0 \end{aligned}$$

where v is a nonnegative parameter

$$(P_5)\quad \begin{aligned} \min \quad & -3x_1 - 2x_2 - 3x_3 + 2x_4 \\ \text{s.t.} \quad & 5x_1 + 3x_2 + 4x_3 + 10x_4 \leq 10 \\ & x_1, x_2, x_3 \text{ binary } x_4 \geq 0 \end{aligned}$$

(b) The most standard relaxation is, of course, the linear programming one $(\bar{P})$. Considering only those of the above that form valid relaxations, show which (P_k) we can be certain are dominated by $(\bar{P})$, i.e., for which valid relaxations (P_k) is $v(P_k) \leq v(\bar{P})$?

(c) Suppose $\mathbf{x}^k$ is optimal in (P_k) and $\mathbf{x}^k$ is feasible in (P). Again considering only those of the above that form valid relaxations, show which (P_k) permit us to conclude that such an $\mathbf{x}^k$ solves (P). Explain.

5-17. The fixed charge network flow problem *(FN)* on a directed graph $G(V, A)$ can be formulated

$$\begin{aligned} \min \quad & \mathbf{vx} + \mathbf{fy} \\ \text{s.t.} \quad & \mathbf{Ex} = \mathbf{b} \\ & \mathbf{x} \geq \mathbf{0} \\ & x_j \leq u_j y_j \text{ for all } j \in A \\ & \mathbf{y} \text{ binary}, \end{aligned}$$

where $\mathbf{v}$ is a vector of variable (per unit) costs of flows $\mathbf{x}$ on arcs $j \in A$, $\mathbf{f}$ is a corresponding vector of positive fixed costs on arcs, $\mathbf{E}$ is the vertex-arc incidence matrix of the graph, $\mathbf{u}$ is a vector of capacities on arcs $j \in A$, and $\mathbf{b}$ is the vector of net supplies at vertices $i \in V$.

(a) Show that the linear programming relaxation of this model can be solved by solving a single minimum cost network flow problem over the $\mathbf{x}$ variables.

(b) Let $S \triangleq \{i \in V\colon b_i > 0\}$, $T \triangleq \{i \in V\colon b_i < 0\}$, e^i the ith row of $\mathbf{E}$ and define new variables $x_j[s, t] \triangleq$ the flow through arc j originating at sources s and bound for sink t. Show that the problem *(FN)* can be equivalently formulated in terms of these new variables as

$$\begin{aligned} \min \quad & \mathbf{v}\left(\sum_{s \in S} \sum_{t \in T} \mathbf{x}[s, t]\right) + \mathbf{fy} \\ \text{s.t.} \quad & \mathbf{e}^i\mathbf{x}[s, t] = 0 \qquad \text{for all } s \in S, t \in T, i \neq s, t \\ & \sum_{t \in T} \mathbf{e}^s\mathbf{x}[s, t] = b_s \qquad \text{for all } s \in S \\ & \sum_{s \in S} \mathbf{e}^t\mathbf{x}[s, t] = b_t \qquad \text{for all } t \in T \\ & \mathbf{x}[s, t] \geq \mathbf{0} \qquad \text{for all } s \in S, t \in T \\ & \sum_{s \in S} \sum_{t \in T} x_j[s, t] \leq u_j y_j \qquad \text{for all } j \in A \\ & \sum_{s \in S} x_j[s, t] \leq -b_t y_j \qquad \text{for all } j \in A, t \in T \\ & \sum_{t \in T} x_j[s, t] \leq b_s y_j \qquad \text{for all } i \in A, s \in S \\ & \mathbf{y} \text{ binary}. \end{aligned}$$

(c) Prove that the linear programming relaxation of the form in (b) will always yield at least as high a value as that of the *(FN)* form.

(d) Illustrate that the bound relaxation of (c) can be strict by solving both relaxations for the instance

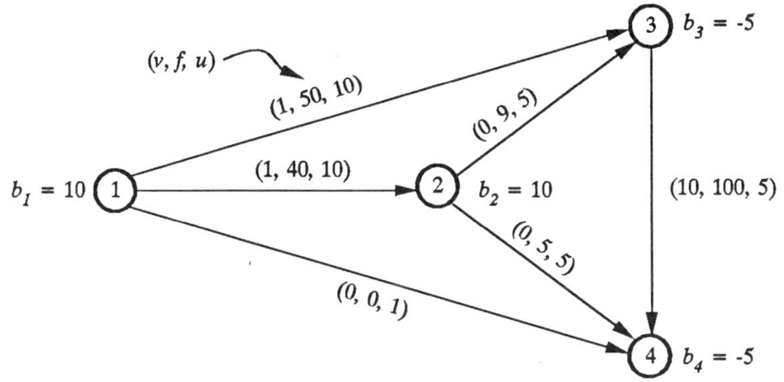

5-18. Consider and branch and bound algorithm satisfying the hypotheses of Theorem 5.1 and define sequences

$\{v_t\} \triangleq$ sequence of incumbent solution values upon the tth execution of Algorithm 5A, Step 1.

$\{\sigma_t\} \triangleq$ sequence of the least bound among candidates stored at (the beginning of) the tth execution of Algorithm 5A, Step 1.

(a) Show that $\{v_t\}$ is monotone nonincreasing with $v_t \geq v(P(S))$ for all t.
(b) Show that $\{\sigma_t\}$ is monotone nondecreasing with $\sigma_t \leq v(P(S))$ for all t.

5-19. Figure 5.4 exhibits results of the first 10 candidate problems under two enumeration sequences. Throughout, the incumbent solution value remains at $v^* = 90$. Suppose now that we *terminate* the branch and bound after those 10 candidates.

(a) If only the information in the depth first tree were available, what could we compute as an upper limit on the percent by which v^* exceeds $v(P(S))$?
(b) What would be your response if we had available only the information in the best bound tree?

5-20. The bounded (not necessarily 0–1) mixed integer linear programming problem has the form

$$\begin{array}{lll} & \min & \mathbf{cx} \\ P(S) & \text{s.t.} & \mathbf{Ax} \geq \mathbf{b} \\ & & \mathbf{x} \in S \triangleq \{\mathbf{x} \geq \mathbf{0}: x_j \leq u_j \text{ and integer for } j \in I\}. \end{array}$$

Algorithm 5D can be adapted for this case by replacing Step 5 by

STEP 5′: *Branching.* Select some component x_p with $p \in I$ having $\bar{x}_p$ noninteger. Replace $P(S_k)$ in the candidate list with two new problems, one characterized by $S_k \cap \{\mathbf{x}\colon x_p \le \lfloor \bar{x}_p \rfloor\}$ and the other by $S_k \cap \{\mathbf{x}\colon x_p \ge \lfloor \bar{x}_p \rfloor + 1\}$. Then return to Step 1.

(a) Construct an instance of P(S) and illustrate the application of this variation of Algorithm 5D.
(b) Apply Theorem 5.1 to prove the procedure will converge to an optimal solution in finitely many steps.
(c) Show how the up and down penalties of Theorem 5.11 can be adapted to this case. Specifically, derive all equations and illustrate with your example of (a).

5-21. Warehouse location Example 5.11 dualizes constraints (5-15) and enforces (5-16), (5-17), and (5-18).

(a) Show the form of an alternative scheme that dualizes (5-16) and retains (5-15), (5-17) and (5-18).
(b) Using the constraints of (UWX) and dual multipliers all equal to 3, solve the implied Lagrangean relaxation.
(c) Can $v(D_L)$ exceed $v(\bar{P})$ for this revised formulation, i.e., does it have Geoffrion's integrality property?

5-22. Return to the uncapacitated warehouse location problems of Example 5.11 and consider replacing (5-16) by $x_{ij} \le d_j x_i$ for all i and j.

(a) Show that the value of the linear programming relaxation for this new (larger) formulation is always at least as high as that of the one with (5-16).
(b) Give a numerical example demonstrating that the new formulation can have a strictly higher linear programming relaxation value than the one with (5-16).
*(c) Show that the value of the Lagrangean dual developed in Example 5.11 has value equal to the value of the linear programming relaxation of the new formulation.

5-23. The development of Section 5.5.1 assumes dualized constraints in Lagrangean relaxations are inequalities. Suppose instead we have equalities $\mathbf{Rx} = \mathbf{r}$.

(a) Show that for any $\mathbf{u}$ (unrestricted in sign) Lagrangean problem (P_u) is a valid relaxation of such a (P).
(b) Show that if $\hat{\mathbf{x}}$ solves some such (P_u) and $\mathbf{R}\hat{\mathbf{x}} = \mathbf{r}$, then $\hat{\mathbf{x}}$ solves (P).

5-24. Held and Karp (1970) propose a Lagrangean relaxation of the traveling salesman problem on a graph $G(V, E)$ based on the formulation

$$\begin{aligned} \min \quad & \sum_i \sum_{j>i} c_{ij} x_{ij} \\ \text{s.t.} \quad & \sum_{i<k} x_{ik} + \sum_{j>k} x_{kj} = 2 \text{ for every } k \in V \\ & \mathbf{x} \in T, \end{aligned}$$

where T is the set of index vectors of *spanning 1-trees* of G, i.e., spanning subgraphs with exactly $|V|$ edges and 1 cycle.

(a) Establish that this is a correct formulation of the traveling salesman problem.

(b) Show that dualizing the first (equality) set of constraints in this formulation leaves a Lagrangean relaxation that can be solved in polynomial time. (Hint: adapt Algorithm 3A)

(c) Express the constraints $\mathbf{x} \in T$ as ones of polynomially many linear constraints on polynomially many binary variables. With respect to your formulation do Lagrangean relaxations (P_u) have Geoffrion's integrality property? Explain.

★(d) Suppose now that $v(D_L)$ of this formulation is used as the β-bound of an implementation of Algorithm 5A. Propose, justify and prove finite convergence of a full statement of the algorithm, including, in particular, an appropriate branching Step 5.

5-25. The generalized assignment problem is

$$(GA) \qquad \begin{aligned} \min \quad & \sum_i \sum_j c_{ij} x_{ij} & & \\ \text{s.t.} \quad & \sum_i x_{ij} \geq 1 & \text{for all } j \quad & (1) \\ & \sum_j a_j x_{ij} \leq s_i & \text{for all } i \quad & (2) \\ & x_{ij} \text{ binary} & \text{for all } i \text{ and } j \quad & (3) \end{aligned}$$

where a_j is the positive integer size of object j and s_i is the positive integer capacity of location i.

(a) Show that the Lagrangean dual formulation with (2) dualized and both (1) and (3) forming T cannot improve of the bound from the linear programming relaxation $(\overline{GA})$.

(b) Show that the Lagrangean dual formulation with (1) dualized and both (2) and (3) forming T has Lagrangean relaxations (GA_u) solvable via a sequence of knapsack problems, one for each i. What does this result tell you about $v(D_L)$ for the formulation vis-a-vis $v(\overline{GA})$?

(c) State the surrogate relaxation corresponding to the Lagrangean structure of part (a). Can it be solved in integers by linear programming? Why?

5-26. Return to exercise **5-12** and consider a Lagrangean relaxation of the problem in **5-12(a)** that dualizes main constraints $\mathbf{Ax} \geq \mathbf{b}$ and retains special constraints $\Sigma_{j \in J_k} x_j \leq 1$ and $\mathbf{x}$ binary. Show that this dualization strategy has Geoffrion's integrality property.

5-27. Return to the equality dualization of **5-23**. Suppose dual multipliers $\mathbf{u}$ are chosen by the computation $\mathbf{u}^1 \leftarrow 0$, $\mathbf{u}^{k+1} \leftarrow \mathbf{u}^k + \lambda_k(\mathbf{r} - \mathbf{Rx}^k)$ for $k = 1, 2, \ldots$, where $\lambda_k \leftarrow 1/(k\|\mathbf{r} - \mathbf{Rx}^k\|)$ and $\mathbf{x}^k$ solves $(P_{\mathbf{u}^k})$. Defining $v^k = v(P_{\mathbf{u}^k})$, show that this computation will either yield a finite $\bar{k}$ with $v^{\bar{k}} = v(P)$, or generate a sequence $\{v^k\}$ with $\lim_{k \to \infty} v^k = v(D_L)$.

5-28. Consider (UW) of Example 5.11. Design an appropriate λ_k (step-size) rule and apply the subgradient search method of Algorithm 5E to execute through 5 changes in the duals u_1 and u_2 on data (UWX). Begin with $u_1 = u_2 = 0$. Indicate which subgradient directions used are ascent directions.

5-29. Apply Algorithm 5F to Example 5.11 with the same starting point as in **5-28.** Proceed until you have changed (u_1, u_2) twice.

5-30. Repeat **5-29** with Algorithm 5G.

5-31. Return to the dualization strategy of **5-25(b).**

(a) Characterize the set of subgradients to $v(P_u)$ at a given point $\mathbf{u}$.

(b) Detail a specialized version of subgradient Algorithm 5E for this model and prove its correctness.

*(c) Repeat (b) for Dual Ascent Algorithm 5F.

(d) Repeat (b) for Outer Linearization Algorithm 5G.

5-32. Show that the closest (Euclidean distance) nonnegative m-vector to a given $\hat{\mathbf{u}} \in \mathbb{R}^m$ is given by $\hat{u}_i \leftarrow 0$ if $\hat{u}_i < 0$ and $\leftarrow \hat{u}_i$ otherwise.

***5-33.** Guignard-Speilberg (1984) develops the notion of a *Lagrangean decomposition* of a problem

$$(LDP)\qquad \begin{aligned} \min\quad & \mathbf{cx} \\ \text{s.t.}\quad & \mathbf{Rx} \geq \mathbf{r} \\ & \mathbf{Hx} \geq \mathbf{h} \\ & \mathbf{x} \text{ binary} \end{aligned}$$

as the form obtained by duplicating variables $\mathbf{x}$ as $\mathbf{y}$ and dualizing a constraint $\mathbf{y} - \mathbf{x} = \mathbf{0}$ with unrestricted dual multiplier $\mathbf{w}$ to obtain the decomposable form

$$(LDP[\mathbf{w}]) \qquad \begin{array}{ll} \min & (\mathbf{c} - \mathbf{w})\mathbf{x} + \mathbf{w}\mathbf{y} \\ \text{s.t.} & \mathbf{R}\mathbf{x} \geq \mathbf{r} \\ & \mathbf{H}\mathbf{y} \geq \mathbf{h} \\ & \mathbf{x}, \mathbf{y} \text{ binary.} \end{array}$$

(a) Show that this Lagrangean decomposition is a relaxation of the original (LDP) for any multipliers $\mathbf{w}$.
(b) Show that any such decomposition can be solved by solving separate problems in $\mathbf{x}$ and $\mathbf{y}$.
(c) Show that the relaxation solution value obtained with the best choice of $\mathbf{w}$ produces a bound on $v(LDP)$ that is at least as large as the value of the corresponding Lagrangean dual (with $\mathbf{R}\mathbf{x} \geq \mathbf{r}$ dualized).
(d) Produce a numerical example showing that the bound relationship of (c) can be strict.
(e) Prove or give a counterexample for the result analogous to (c) comparing the best Lagrangean decomposition and the surrogate dual.
(f) Show that the value of the Lagrangean decomposition $LDP[\mathbf{w}]$ above is a concave function of $\mathbf{w}$.
(g) Develop a modified version of Algorithm 5E suitable to search the function of (f) for good multipliers $\mathbf{w}$.
*(h) Repeat (g) for Algorithm 5F.
(i) Repeat (g) for Algorithm 5G.

***5-34.** Suppose that as in Section 5.5 problem (P) is viewed as having a tractable set of constraints $\mathbf{x} \in T$ and a complicating system $\mathbf{R}\mathbf{x} \geq \mathbf{r}$. Martin (1987) investigates cases where a *variable redefinition* is known to new variables $\mathbf{z}$ and linear constraints $\mathbf{D}\mathbf{z} \geq \mathbf{d}$, $\mathbf{z} \geq \mathbf{0}$ along with a transformation $\mathbf{x} = \mathbf{M}\mathbf{z}$ such that if $\mathbf{z}^*$ solves $\min\{\mathbf{c}\mathbf{M}\mathbf{z}\text{: } \mathbf{D}\mathbf{z} \geq \mathbf{d},\ \mathbf{z} \geq \mathbf{0}\}$ then $\mathbf{x}^* \triangleq \mathbf{M}\mathbf{z}^*$ solves $\min\{\mathbf{c}\mathbf{x}\text{: } \mathbf{x} \in T\}$. Constraints $\mathbf{x} \in T$ are replaced by the system $\mathbf{D}\mathbf{z} \geq \mathbf{d}$, $\mathbf{z} \geq \mathbf{0}$, $\mathbf{R}\mathbf{M}\mathbf{z} \geq \mathbf{r}$.

(a) Show how such a variable redefinition yields a linear program producing the same lower bound as the Lagrangean relaxation (P_u) for any $\mathbf{u} \geq \mathbf{0}$.
(b) Show how such a variable redefinition can be exploited to obtain a linear program with value equal to the value of the Lagrangean dual $v(D_L)$ if constraints $\mathbf{R}\mathbf{M}\mathbf{z} \geq \mathbf{r}$ are dualized.
(c) Is existence of such a variable redefinition the same thing as Geoffrion's integrality property? Why or why not?

(d) Illustrate variable redefinition with the stronger form of the fixed charge network problem in **5-17**, assuming non-network side constraints $\mathbf{Qx} \geq \mathbf{q}$ are part of the (FN) formulation.

5-35. The search strategy of surrogate Algorithm 5H is to show system (I_S^α) has no solution by solving surrogate relaxation $(P^{\mathbf{v}})$.

(a) Show that it would be equally satisfactory to check whether

$$v\begin{pmatrix} \min & \mathbf{v}(\mathbf{r}-\mathbf{Rx}) \\ \text{s.t.} & \mathbf{cx} \leq \alpha \\ & \mathbf{x} \in T \end{pmatrix} > 0.$$

(b) Describe a version of Algorithm 5H that computes $v(D_S)$ by solving a sequence of problems in the form of (a).

5-36. Consider the mixed 0–1 optimization problem

$$\begin{aligned} \min \quad & 11y_1 + 18y_2 + 13y_3 - 12z_1 - 8z_2 - 4z_3 + 12z_4 \\ \text{s.t.} \quad & 5y_1 + 2y_2 + y_3 - 3z_1 - z_2 + z_3 + 3z_4 \geq 1 \\ & -3y_1 + 9y_2 + 3y_3 - 2z_1 - 2z_2 - 4z_3 - 4z_4 \geq 0 \\ & y_1, y_2, y_3 \text{ binary} \\ & z_1, z_2, z_3 \geq 0 \end{aligned}$$

(a) State the Benders dual problem $(BD(\mathbf{y}))$ associated with this example, and graph its feasible region as in Figure 5.14.
(b) State the full Benders master problem (BM) for the example.
(c) Solve by Benders Decomposition Algorithm 5I beginning with $\mathbf{y}^1 \leftarrow \mathbf{0}$, and solving any continuous problems graphically, and 0–1 problems by inspection.

5-37. Consider the uncapacitated warehouse location problems of Example 5.11.

(a) Show that the problems can be viewed as ones suitable for Benders Algorithm 5I with $\mathbf{y} \triangleq$ {single subscripted x's}, $\mathbf{z} \triangleq$ {doubly subscripted x's}.
(b) Derive the corresponding Benders dual problem $(BD(\mathbf{y})$ and illustrate for the numerical (UWX) of Example 5.11.
(c) The constraint $\Sigma_i x_i \geq 1$ is obviously valid for problems (UW). Why? Show that restricting Benders' master problem solutions to $\{x_i\}$ satisfying this constraint will assure no extreme-direction constraint is ever generated by Algorithm 5I.

(d) Show the form of the implied (BM) with all extreme-point constraints present, and illustrate for the numerical data of (UWX).

5-38. Let $\{\beta^l\}$ be the sequence of (BM_{l-1}) solution values generated by Benders' Algorithm 5I and $\{\delta^l\}$ be the sequence derived from the same calculations by $\delta^l \triangleq \min\{v(BD(\mathbf{y}^k)): k = 1, 2, \ldots, l\}$.

(a) Show that $\{\beta^l\}$ is a monotone nondecreasing sequence of lower bounds on problem solution value $v(B(Y))$.
(b) Show that $\{\delta^l\}$ is a monotone nonincreasing sequence of upper bounds on problem solution value $v(B(Y))$.
(c) Produce a numerical example showing that the dual value sequence $\{v(BD(\mathbf{y}^k))\}$ need not be monotone.

***5-39.** Return to the fixed charge network flow problem formulation (FN) of **5-17**.

(a) Show the problem can be viewed as suitable for Benders Algorithm 5I with $\mathbf{z} \triangleq \mathbf{x}$.
(b) Derive the corresponding Benders' dual problem $(BD(\mathbf{y}))$.
(c) Show (in terms of network structure) the form of the extreme-point and extreme-direction set for this $(BD(\mathbf{y}))$.
(d) Show the full Benders' master problem (BM) for this formulation.

6

Nonpolynomial Algorithms—Polyhedral Description

In the previous chapter, we investigated partial enumeration algorithms for exact solution of discrete optimization problems that are apparently beyond the range of well-bounded, polynomial-time procedures. Those algorithms adopt a divide and conquer strategy; they deal with difficult problems by systematically enumerating a number of more tractable cases.

Now we turn to the principal (known) alternative to enumeration—*polyhedral description.* Instead of creating numerous cases, polyhedral description (at least in its purest form) proceeds by better and better defining the given problem until it becomes tractable. Of course, the two approaches can be creatively combined (as we mention in Section 6.1.2).

6.1 Fundamentals of Polyhedral Description

In introducing polyhedral methods, we shall address discrete problems in the general form

$$(P) \qquad \begin{aligned} \min \quad & \mathbf{cx} \\ \text{s.t.} \quad & \mathbf{Ax} \geq \mathbf{b} \\ & \mathbf{x} \in S, \end{aligned}$$

where $S \subseteq \{\mathbf{x} \in \mathbb{R}^n: \mathbf{x} \geq \mathbf{0}\}$ imposes some (as yet unspecified) discrete constraints on the decision vector, $\mathbf{x}$. One familiar example is

$$S = \{\mathbf{x} \geq \mathbf{0}: x_j = 0 \text{ or } 1 \quad \text{for all } j \in I\},$$

where $I \subseteq \{1, 2, \ldots, n\}$. Note that nonnegativity constraints are explicitly included. A number of the results to follow would have to be significantly restated without $\mathbf{x} \geq \mathbf{0}$. Also, we shall usually assume the coefficients $\mathbf{A}$, $\mathbf{b}$ and $\mathbf{c}$ are integers. Such an assumption is no loss of practical generality because the coefficients must be at least rational to be processed by digital computers, and a problem in rationals can easily be converted to one in integers by multiplying by suitable integers. Still, we shall see in Section 6.1.3 that the rationality hypothesis is necessary to many results.

6.1.1 The Relaxation Strategy

Like the enumerative procedures of Chapter 5, polyhedral methods depend centrally on problem relaxations (see Section 5.1.2). When their objective functions match, a problem $(\tilde{P})$ is a relaxation of another problem (P) if every feasible solution to (P) is feasible in $(\tilde{P})$. Enumerative algorithms subdivide cases until some available relaxation gives enough information to resolve each one. Polyhedral algorithms, continue refining the relaxation of a single case.

The purest form of polyhedral description algorithm is termed *cutting*. Each cutting iteration begins by solving the current relaxation. If the optimal solution obtained is also feasible in the discrete problem of interest, we stop. When objective functions match, we know (Lemma 5.4) an optimal solution to a relaxation that is feasible for the original problem is optimal for the latter.

The cutting name derives from how we deal with the event that the relaxation optimum is not feasible in the original problem. Cutting algorithms proceed by generating new constraints for the relaxation. These new constraints must cut off (i.e., render infeasible) the current relaxation optimum, yet eliminate no solutions to the original problem.

We shall be interested only in adding linear constraints to relaxations. More precisely, we study *valid inequalities*, $(\mathbf{g}, g_0)$, i.e., those that satisfy

$$\mathbf{gx} \geq g_0 \text{ for all } \mathbf{x} \in F,$$

where $F \triangleq \{\mathbf{x} \in S: \mathbf{Ax} \geq \mathbf{b}\}$ is the set of feasible solutions to (P). The valid inequalities are also termed *cutting planes* because of their form and cutting role.

Recall that the optimal value function $v(\,\cdot\,)$ is defined as

$$v(\cdot) \triangleq \begin{cases} \text{the value of an optimal solution to problem } (\cdot)\text{, if one exists,} \\ +\infty \text{ if } (\cdot) \text{ is a minimizing and infeasible problem or a maximizing and unbounded one.} \\ -\infty \text{ if } (\cdot) \text{ is a minimizing and unbounded problem or a maximizing and infeasible one.} \end{cases}$$

The relaxation algorithm is as follows.

Algorithm 6A. Relaxation Cutting. *Let F and (P) be as above with $v(P) > -\infty$.*

STEP 0: *Initial Relaxation.* Set $k \leftarrow 0$ and pick as (P^0) any relaxation of (P) in the form

$$\begin{aligned} \min \quad & \mathbf{cx} \\ \text{s.t.} \quad & \mathbf{x} \in T, \end{aligned}$$

where $T \supseteq F$ and $v(P^0) > -\infty$.

STEP 1: *Relaxation.* Attempt to solve relaxation (P^k). If $v(P^k) = +\infty$, stop; (P^k) and thus (P) is infeasible. Otherwise, let $\mathbf{x}^k$ be an optimal solution and go to Step 2.

STEP 2: *Optimality Check.* If $\mathbf{x}^k \in F$, stop; $\mathbf{x}^k$ solves (P) because it solves (P^k) and is feasible in (P). Otherwise, proceed to Step 3.

STEP 3: *Cutting.* Generate a valid inequality $(\mathbf{g}^{k+1}, g_0^{k+1})$ such that

$$\text{(i)} \quad \mathbf{g}^{k+1}\mathbf{x} \geq g_0^{k+1} \quad \text{for all} \quad \mathbf{x} \in F \tag{6-1}$$

and

$$\text{(ii)} \quad \mathbf{g}^{k+1}\mathbf{x}^k < g_0^{k+1}. \tag{6-2}$$

Then, define (P^{k+1}) as (P^k) with the added constraint $\mathbf{g}^{k+1}\mathbf{x} \geq g_0^{k+1}$, replace $k \leftarrow k + 1$, and return to Step 1. ■

Example 6.1. Elementary Cutting. To illustrate Algorithm 6A, consider the problem

$$(IP) \qquad \begin{aligned} \min \quad & 3x_1 + 2x_2 \\ \text{s.t.} \quad & 6x_1 - 4x_2 \geq -1 \\ & 2x_1 + 7x_2 \geq 4 \\ & (x_1, x_2) \in Z \triangleq \{\mathbf{x} \geq \mathbf{0}\text{: } \mathbf{x} \text{ integer}\}. \end{aligned}$$

Figure 6.1a graphs the problem and demonstrates $\mathbf{x}^* = (1, 1)$ is optimal.

To apply Algorithm 6A effectively, we require a relaxation of (IP) that will remain tractable as new inequalities are added. An obvious choice is the linear programming relaxation, obtained when Z is expanded to $\bar{Z} = \{\mathbf{x} \geq \mathbf{0}\}$.

Algorithm 6A begins at Steps 0 and 1 by solving the linear programming relaxation (IP^0). Figure 6.1a shows that $\mathbf{x}^0 = (9/50, 13/25)$ is optimal.

Since $\mathbf{x}^0 \notin Z$ we must proceed to Step 3 and generate an inequality valid for all

$$(x_1, x_2) \in F \triangleq \left\{ \begin{matrix} (x_1, x_2) \geq \mathbf{0}\colon 6x_1 - 4x_2 \geq -1,\ 2x_1 + 7x_2 \geq 4 \\ x_1 \text{ integer, } x_2 \text{ integer} \end{matrix} \right\}$$

and cutting off $\mathbf{x}^0$. The latter is required to prevent $\mathbf{x}^0$ from reappearing as the solution to our next relaxation.

The bulk of the chapter will be devoted to schemes for finding inequalities of Step 3. Here we employ a simple graphical idea. When some component x_j^k of a relaxation solution $\mathbf{x}^k$ is not integer, $\mathbf{x}^k$ lies in the interior of a cylinder set $\{\mathbf{x} \in \mathbb{R}^2\colon \lfloor x_j^k \rfloor \leq x_j \leq \lceil x_j^k \rceil\}$ containing no feasible interior solutions. A cutting plane passing through the points where active constraints of the relaxation (IP^k) intersect the boundary of this cylinder cuts off only interior points of the cylinder. Thus, such a cutting plane is valid and we still see later that its coefficients are readily obtained algebraically.

Figure 6.1b shows the required cylinder set around $\mathbf{x}^0$. The active constraints at $\mathbf{x}^0$ intersect cylinder boundaries at $(\frac{1}{2}, 1)$ and $(0, 2)$. A plane through those points yields the valid inequality $2x_1 + 3x_2 \geq 4$; $(\mathbf{g}^1, g_0^1) = (2,3,4)$. With this new constraint relaxation (IP^1) solves at $\mathbf{x}^1 = (\frac{1}{2}, 1)$. Again, we must generate a cutting plane. Figure 6.1c shows the cylinder set that yields the valid inequality $(\mathbf{g}^2, g_0^2) = (1, 0, 1)$.

Processing in this way we obtain $\mathbf{x}^2 = (1, \frac{4}{7})$, generate the inequality $(\mathbf{g}^3, g_0^3) = (1, 1, 2)$, and solve (IP^3) at $\mathbf{x}^3 = (1, 1)$. Now that $\mathbf{x}^3 \in F$, we can stop. If $\mathbf{x}^3$ solves (IP^3), a relaxation of (IP), and $\mathbf{x}^3$ is also feasible for (IP), it must be optimal for (IP). ■

6.1.2 *Cutting vs. Other Strategies*

Polyhedral methods were originally developed in the context of the cutting strategy; and we shall adhere to that convention in algorithms of this chapter. Most cutting methods, however, remain untried and underdeveloped computationally. Thus, it is quite possible that alternatives to cutting may yield superior strategies for employing valid inequalities in solving particular discrete problems.

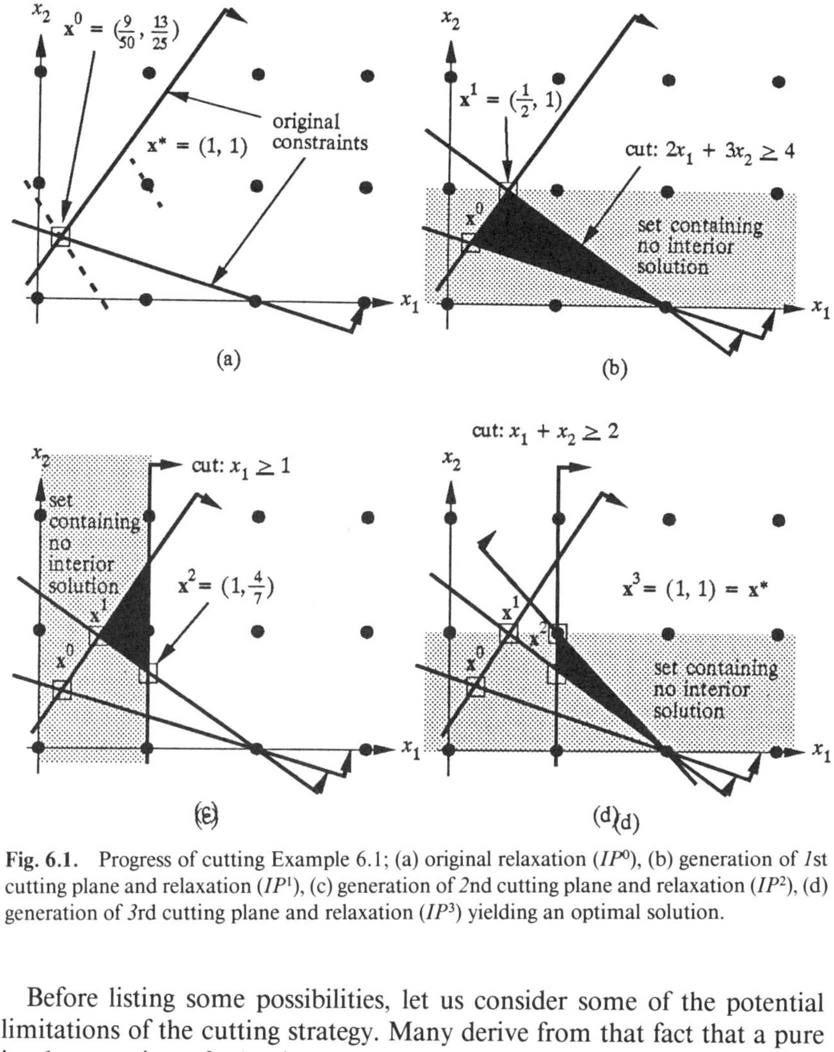

Fig. 6.1. Progress of cutting Example 6.1; (a) original relaxation (IP^0), (b) generation of *1*st cutting plane and relaxation (IP^1), (c) generation of *2*nd cutting plane and relaxation (IP^2), (d) generation of *3*rd cutting plane and relaxation (IP^3) yielding an optimal solution.

Before listing some possibilities, let us consider some of the potential limitations of the cutting strategy. Many derive from that fact that a pure implementation of Algorithm 6A reaches a *feasible* solution for (P) only when it reaches an *optimal* solution. Although we might try some form of rounding, no relaxation solution except the last directly supplies a feasible solution with which we might terminate before reaching provable optimality. To be certain of a good feasible solution, we must carry the algorithm to completion.

If an algorithm must be carried to termination before useful solutions emerge, provable, finite—and hopefully rapid—convergence is essential. We shall see that such convergence is often difficult to demonstrate. At a minimum, it requires that we know a family of valid inequalities that is rich enough to define any point that might be optimal. If potentially optimal points exist that cannot be represented as intersections of a suitable collection of our valid inequalities, convergence is seriously in doubt.

Convergence also requires that we make satisfactorily rapid progress with the algorithm. It is necessary but definitely not sufficient to know that each new cutting plane eliminates the present relaxation optimum. Trivial examples can be derived wherein each new cut eliminates, say, exactly one-half the remaining infeasible portion of the relaxation solution space. Certainly, such a scheme will never achieve optimality in finite time.

Success in a relaxation algorithm will obviously also depend critically on the choice of relaxation employed. Ideally, the relaxations would be both highly tractable and close approximations to the problem of interest, (P). Implementations of Algorithm 6A deal, not with a single relaxation, however, but with a sequence (P^k), $k = 0, 1, 2, \ldots$, formed by imposing more and more valid inequalities. Linear programming relaxations are relatively unhindered by new linear inequalities. But imposition of even one new inequality often destroys the structure of an elegant combinatorial relaxation. Thus, anyone who adopts the cutting strategy is likely to confront a dilemma. If the form of valid inequalities is restricted until the structure of a relaxation is preserved, convergence is threatened; if relaxations are limited to those admitting a full range of linear inequalities, opportunities for creative choice of relaxation are hampered.

What alternatives to relaxation are available? One with apparent promise is dualization. Section 5.5 dealt at length with *Lagrangean relaxations* wherein troublesome constraints are absorbed in the objective function to preserve problem structure. A dualization strategy for cutting algorithms would accomplish the same end by treating cutting planes in the objective. For example, suppose a relaxation

$$(P^0) \qquad \begin{array}{ll} \min & \mathbf{cx} \\ \text{s.t.} & \mathbf{x} \in T \end{array}$$

is available, with T a conveniently structured solution space. As new cutting planes become necessary, a dualization scheme would add them to the objective function with nonnegative multipliers, u_i. Resulting relaxations are

$$(P^k) \qquad \begin{array}{ll} \min & \mathbf{cx} + \sum_{i=1}^{k} u_i(g_0^i - \mathbf{g}^i\mathbf{x}) \\ \text{s.t.} & \mathbf{x} \in T. \end{array}$$

Relaxation structure is preserved, and the methods of Section 5.5.2 offer schemes for finding good multipliers, u_i.

Perhaps the widest range of alternatives to the formal cutting Algorithm 6A derive from the simple idea of combining cutting and enumeration. The tasks of uncovering good feasible solutions and guaranteeing convergence are left to a supervising enumerative algorithm in the form of Section 5.1's Algorithm 5A. Cutting planes are then liberated to assist with a host of subsidiary tasks for which they are more naturally adopted:

- An initial relaxation of the full problem (P) can be improved with valid inequalities so long as it is convenient, without carrying on to optimality. Once progress slows, or structure-preserving cutting planes are no longer available, normal branch and bound can commence. The latter will work more effectively because the improved relaxation will lead to superior bounds.
- Cutting planes can form the basis of range restriction and penalty calculations like those of Section 5.3. Tomlin's penalties of Theorem 5.12 take exactly this tact.
- Cutting can provide an alternative to partial solution in branching at Step 5 of Algorithm 5A. Partial solution further restricts candidate problems by fixing the value of a single decision variable. Cutting permits a wide class of restrictions characterized by the fact that either one or another valid inequality must hold. Such inequalities can involve as many variables as convenient, and take into account known connections between the variables.

We shall not further develop the above algorithmic ideas for alternative use of valid inequalities, because available research results are limited. The small number of cases that have been reported, however, show promise. See, for example, the award winning work of Crowder, Johnson and Padberg (1983).

6.1.3 The Convex Hull of Solutions

Denote by F the set of feasible points to our problem (P), i.e.,

$$F = \{\mathbf{x} \in S\text{: } \mathbf{Ax} \geq \mathbf{b}\}$$

Whatever our strategy for their use, the set of valid inequalities $(\mathbf{g}, g_0)$ satisfying $\mathbf{gx} \geq g_0$ for all $\mathbf{x} \in F$ is our interest in polyhedral description. But all valid inequalities are certainly not equally valuable as cutting planes. For example, each row of the given system $\mathbf{Ax} \geq \mathbf{b}$ defines a trivially valid inequality, but not one of any assistance in obtaining an optimum. Such valid inequalities are satisfied by every $\mathbf{x} \in \{\mathbf{x} \geq \mathbf{0}: \mathbf{Ax} \geq \mathbf{b}\}$.

Is there a particularly desirable subset of valid inequalities? To see that the answer is *yes,* we need to consider the convex hull of solutions to F. Denoted $[F]$, the *convex hull of solutions* to F is the interaction of all convex sets containing F, informally, the smallest such convex set. The importance of $[F]$ for us is that if there is any $\mathbf{x}^* \in F$ that minimizes a linear objective function, $\mathbf{cx}$, over $\mathbf{x} \in F$, then there is such an $\mathbf{x}^*$ that is an extreme point of $[F]$. Also, if $[F]$ is a closed (convex) set, it can be expressed $[F] = \cap_r H^r$, where each of the perhaps infinitely many H^r is a half-space $(\mathbf{x}: \mathbf{h}^r\mathbf{x} \geq h^r{}_0\}$.

Figure 6.2a illustrates one case. Every nonnegative integer (x_1, x_2) satisfying the two specified constraints belongs to F. The convex hull $[F]$ is shaded. Any linear objective function that attains a finite optimal solution in F will be solved by one of the extreme points (0,1) and (2,1).

The convex hull $[F]$ of Figure 6.2a is *polyhedral,* i.e., it can be formed as the intersection of *finitely* many half-spaces H^r. In that example $[F]$,

$$[F] = \{(x_1, x_2): x_1 \geq 0\} \cap \{(x_1, x_2): x_2 \geq 1\} \cap \{(x_1, x_2): -x_1 + 2x_2 \geq 0\}.$$

Figure 6.2b provides an example where $[F]$ is not even a closed set, and thus certainly not polyhedral. The single main constraint is satisfied as an equality when $\sqrt{5} = x_1/x_2$. We know integer (x_1, x_2) can form a rational number x_1/x_2 arbitrarily close to this boundary, but never exactly reach it. Thus, for part (b)

$$[F] = \{(x_1, x_2) \geq \mathbf{0}: -x_1 + \sqrt{5}x_2 > 0\} \cup \{(0, 0)\} \tag{6-3}$$

—neither closed nor polyhedral.

As we hinted earlier, the difficulty in part (b) is irrational coefficients in the $\mathbf{A}$ matrix—a case of little practical interest. Table 6.1 lists most of the forms of F we shall actually encounter. After establishing a useful lemma, we assert $[F]$ is polyhedral for all such forms.

Lemma 6.1. Reduction of Rational Cases to Pure Integer, Equality-Constrained. *Let forms F_7, F_8, F_9 and F_{10} be as defined in Table 6.1. Then any feasible set of the form F_8, F_9 or F_{10} has an equivalent description in form F_7.*

Proof. Consider $F_8 \triangleq \{\mathbf{y}: \mathbf{Ay} \geq \mathbf{b}, \mathbf{y} \geq \mathbf{0}, \mathbf{y} \text{ integer}\}$. Since $\mathbf{A}$ and $\mathbf{b}$ are rational, we may multiply rows by suitable integers to obtain an equivalent $F_8 = \{\mathbf{y}: \hat{\mathbf{A}}\mathbf{y} \geq \hat{\mathbf{b}}, \mathbf{y} \geq \mathbf{0}, \mathbf{y} \text{ integer}\}$ with $\hat{\mathbf{A}}$, $\hat{\mathbf{b}}$ integer. Adding surpluses $\mathbf{s} = \hat{\mathbf{A}}\mathbf{y} - \hat{\mathbf{b}}$ gives the equivalent form $F_8 = \{\mathbf{y},\mathbf{s}: \hat{\mathbf{A}}\mathbf{y} - \mathbf{s} = \hat{\mathbf{b}}, \mathbf{y} \geq \mathbf{0}, \mathbf{s} \geq \mathbf{0},$

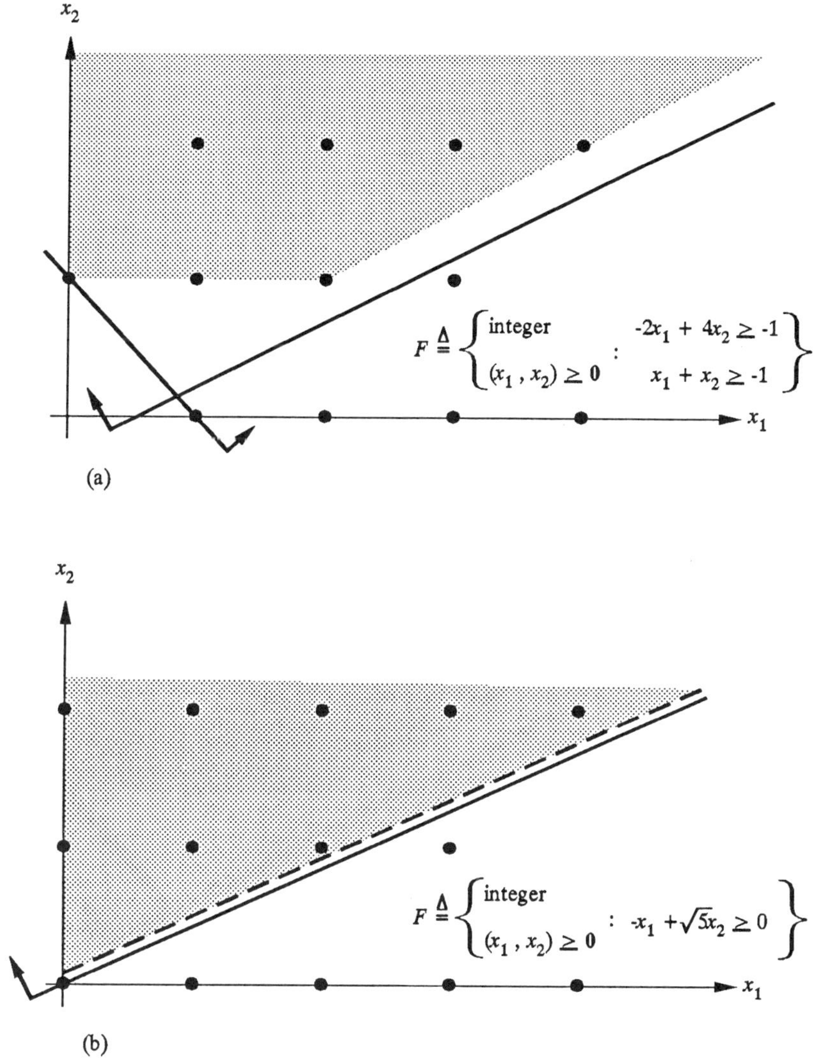

Fig. 6.2. Convex hulls of discrete solutions.

$\mathbf{y}$ integer, $\mathbf{s}$ integer}. Integrality can be enforced on $\mathbf{s}$ because $\hat{\mathbf{A}}\mathbf{y} - \mathbf{b}$ will be integer for all integer $\mathbf{y}$. (Note that we could not have required integer $\mathbf{s}$ if $\mathbf{A}$ contained irrationals.)

F_{10} is easily reduced to F_9 by adding surplus variables in a similar way. Here the surplus variables may be part of the continuous vector, $\mathbf{z}$. F_9 is, in

TABLE 6.1. Discrete Forms with Polyhedral Convex Hulls of Solutions

Form	Feasible Space	Data Characteristics
Bounded Pure Integer	$F_1 = \{\mathbf{y} \in Y: \mathbf{Ay} = \mathbf{b}\}$	Y a bounded set of integer vectors
	$F_2 = \{\mathbf{y} \in Y: \mathbf{Ay} \geq \mathbf{b}\}$	$\mathbf{y}, \mathbf{A}, \mathbf{b}$, arbitrary real
Bounded Integer Part Mixed Integer	$F_3 = \{(\mathbf{y}, \mathbf{z}): \mathbf{y} \in Y, \mathbf{z} \geq \mathbf{0}, \mathbf{A}^1\mathbf{y} + \mathbf{A}^2\mathbf{z} = \mathbf{b}\}$	Y a bounded set of integer vectors
	$F_4 = \{(\mathbf{y}, \mathbf{z}): \mathbf{y} \in Y, \mathbf{z} \geq \mathbf{0}, \mathbf{A}^1\mathbf{y} + \mathbf{A}^2\mathbf{z} \geq \mathbf{b}\}$	$\mathbf{y}, \mathbf{A}^1, \mathbf{A}^2, \mathbf{b}$ arbitrary reals
Disjunctive	$F_5 = \{\mathbf{z}: \mathbf{z} \geq \mathbf{0}, \mathbf{A}^1\mathbf{z} = \mathbf{b}^1 \text{ or } \mathbf{A}^2\mathbf{z} = \mathbf{b}^2 \text{ or} \cdots \text{or } \mathbf{A}^t\mathbf{z} = \mathbf{b}^t\}$	t finite, $\mathbf{A}^1, \ldots, \mathbf{A}^t, \mathbf{b}^0, \mathbf{b}^1, \ldots, \mathbf{b}^t$
	$F_6 = \{\mathbf{z}: \mathbf{z} \geq \mathbf{0}, \mathbf{A}^1\mathbf{z} \geq \mathbf{b}^1 \text{ or } \mathbf{A}^2\mathbf{z} \geq \mathbf{b}^2 \text{ or} \cdots \text{or } \mathbf{A}^t\mathbf{z} \geq \mathbf{b}^t\}$	arbitrary reals
Rational Pure Integer	$F_7 = \{\mathbf{y}: \mathbf{Ay} = \mathbf{b}, \mathbf{y} \geq \mathbf{0}, \mathbf{y} \text{ integer}\}$	$\mathbf{A}$ and $\mathbf{b}$ rational
	$F_8 = \{\mathbf{y}: \mathbf{Ay} \geq \mathbf{b}, \mathbf{y} \geq \mathbf{0}, \mathbf{y} \text{ integer}\}$	
Rational Mixed Integer	$F_9 = \{(\mathbf{y}, \mathbf{z}): \mathbf{y} \geq \mathbf{0} \text{ and integer}, \mathbf{z} \geq \mathbf{0}, \mathbf{A}^1\mathbf{y} + \mathbf{A}^2\mathbf{z} = \mathbf{b}\}$	$\mathbf{A}^1, \mathbf{A}^2, \mathbf{b}$ rational
	$F_{10} = \{(\mathbf{y}, \mathbf{z}): \mathbf{y} \geq \mathbf{0} \text{ and integer}, \mathbf{z} \geq \mathbf{0}, \mathbf{A}^1\mathbf{y} + \mathbf{A}^2\mathbf{z} \geq \mathbf{b}\}$	

turn, reduced to F_7 by a scheme first proposed by Wolsey (1971). As with F_8, multiply rows to obtain $F_9 = \{(\mathbf{y}, \mathbf{z})\colon \mathbf{y} \geq \mathbf{0}$ and integer, $\mathbf{z} \geq \mathbf{0}$, $\hat{\mathbf{A}}^1\mathbf{y} + \hat{\mathbf{A}}^2\mathbf{z} = \hat{\mathbf{b}}\}$ with $\hat{\mathbf{A}}^1$, $\hat{\mathbf{A}}^2$ and $\hat{\mathbf{b}}$ integer. Then to each feasible $\mathbf{y}$, there corresponds a linear programming feasible set $\mathrm{F}_9(\mathbf{y}) \triangleq \{\mathbf{z} \geq \mathbf{0}\colon \hat{\mathbf{A}}^2\mathbf{z} = \hat{\mathbf{b}} - \hat{\mathbf{A}}^1\mathbf{y}\}$. From linear programming theory we know the only $\mathbf{z} \in F_9(\mathbf{y})$ that need be considered are basic solutions of the integer equations $\hat{\mathbf{A}}^2\mathbf{z} = (\hat{\mathbf{b}} - \hat{\mathbf{A}}^1\mathbf{y})$. But the largest denominator of any component z_k of any such basic solution is (by Cramer's rule) the least common multiple, say η, of the absolute values of determinants of the basis matrices of $\hat{\mathbf{A}}^2$. It follows that $\hat{\mathbf{A}}^2\hat{\mathbf{z}} = \eta(\hat{\mathbf{b}} - \hat{\mathbf{A}}^1\mathbf{y})$ has an integer solution $\hat{\mathbf{z}} \triangleq \eta\mathbf{z}$ for every integer vector $\mathbf{y}$. Thus, we can place F_9 in F_7 form by $F_9 = \{(\mathbf{y}, \hat{\mathbf{z}})$ integer: $\mathbf{y} \geq \mathbf{0}$, $\mathbf{z} \geq \mathbf{0}$, $\eta\hat{\mathbf{A}}^1\mathbf{y} + \hat{\mathbf{A}}^2\hat{\mathbf{z}} = \eta\hat{\mathbf{b}}\}$. ■

Theorem 6.2. Polyhedral Convex Hulls. *Let $F_1, F_2, \ldots, F_{10}$ be the feasible forms defined in Table 6.1. Then all corresponding convex hulls $[F_1], [F_2], \ldots, [F_{10}]$ are polyhedral.*

Proof. For $i = 1, 2, \ldots, 6$, $[F_i]$ is a finite union of polyhedral sets, say $S_1, S_2, \ldots, S_s$. With forms F_1 and F_2, each S_j is a singleton set containing one of the finitely many integer vectors in a bounded space. In the case of F_3 and F_4 the S_j are linear programming feasible spaces obtained when $\mathbf{y}$ is fixed at each of its finitely many possible values. Finally, polyhedral sets, S_j, of form F_5 and F_6 are the linear programming feasible spaces

$$\{\mathbf{z}\colon \mathbf{A}^j\mathbf{z} = \mathbf{b}^j\} \text{ or } \{\mathbf{z}\colon \mathbf{A}^j\mathbf{z} \geq \mathbf{b}^j\}$$

A well-known property of any polyhedral set is that it can be equivalently defined in terms of a finite set of extreme points and extreme directions. Thus, in particular, each

$$S_j = \left\{ \sum_{l \in P_j} \lambda_l \mathbf{p}_l^j + \sum_{i \in D_j} \mu_i \mathbf{d}_i^j\colon \sum_{l \in P_j} \lambda_l = 1, \lambda_l \geq 0 \text{ for } l \in P_j, \mu_i \geq 0 \text{ for } i \in D_j \right\},$$

where $\{\mathbf{p}_l^j\colon l \in P_j\}$ and $\{\mathbf{d}_i^j\colon i \in D_j\}$ are the corresponding extreme points and extreme directions respectively. But by definition of $[F]$, every $\mathbf{x} \in [F]$ can be expressed $\mathbf{x} = \Sigma_l\ \alpha_l\mathbf{x}^l$, where $\Sigma_l\ \alpha_l = 1$, all $\alpha_l \geq 0$, and each $\mathbf{x}^l \in F = \cup_{j=1}^{s} S_j$. Thus, we may substitute for each $\mathbf{x}^l$ via an S_j to which it belongs to conclude that $[F]$ is the polyhedral set

$$\left\{ \sum_{j=1}^{s} \left(\sum_{l \in P_j} \lambda_l^j \mathbf{p}_l^j + \sum_{i \in D_j} \mu_i^j \mathbf{d}_i^j \right)\colon \sum_{j=1}^{s} \sum_{l \in P_j} \lambda_l^j = 1, \lambda_l^j \geq 0 \text{ for all } j = 1, \ldots, s, l \in P_j, \mu_i^j \geq 0 \text{ for all } j = 1, \ldots, s, i \in D_j \right\}.$$

Turn now to forms F_7, F_8, F_9 and F_{10}. By Lemma 6.1, all can be reduced to an equivalent F_7. Thus the proof will be complete if we show each rational-equality-constrained, pure integer set $\{\mathbf{y}: \mathbf{Ay} = \mathbf{b}, \mathbf{y} \geq \mathbf{0}, \mathbf{y}$ integer$\}$ has polyhedral convex hull. The proof of this last proposition is lengthy but not difficult. In the interest of brevity we shall omit it here. Interested readers are referred to Meyer (1974). ■

Many technical simplifications accrue when $[F]$ can be treated as polyhedral. For example, $v = \inf\{\mathbf{cx}: \mathbf{x} \in F\}$ may, in general, be finite, and yet no $\mathbf{x} \in F$ has $\mathbf{cx} = v$ because v can only be approached as a limit. The next theorem shows that we need not be concerned about this case when $[F]$ is polyhedral.

Theorem 6.3. Solutions Obtained for Polyhedral Cases. *Let F be as above with polyhedral convex hull [F]. Then for any vector* **c** *the problem*

$$(P) \qquad \begin{array}{ll} \min & \mathbf{cx} \\ \text{s.t.} & \mathbf{x} \in F \end{array}$$

is either infeasible, or unbounded, or obtains a solution $\mathbf{x}^*$ *at an extreme point of [F].*

Proof. Let $H^1, H^2, \ldots, H^r$ be finitely many half spaces $H^j = \{\mathbf{x}: \mathbf{h}^j\mathbf{x} \geq h_0^j\}$ such that $[F] = \cap_{j=1}^r H^i$, and consider the linear program,

$$(\tilde{P}) \qquad \begin{array}{lll} \min & \mathbf{cx} & \\ \text{s.t.} & \mathbf{h}^j\mathbf{x} \geq h_0^j & j = 1, 2, \ldots, r. \end{array}$$

It is well-known that the linear program $(\tilde{P})$ will be infeasible or unbounded, or have finite optimal solution at an extreme point of its feasible space. Also, $(\tilde{P})$ is a relaxation of (P) because $F \subseteq [F]$.

If $(\tilde{P})$ is infeasible, then, since $v(\tilde{P}) \leq v(P)$, (P) is also infeasible. If not, observe that extreme points $\{\mathbf{p}^k: k \in P\}$ and extreme directions $\{\mathbf{d}^l: l \in D\}$ of $[F]$, the feasible space of $(\tilde{P})$, also relate to F. Each $\mathbf{p}^k$ belongs to F, because if not, it would belong to P since it could be expressed as an appropriate convex combination of elements of $F \subseteq [F]$. In the latter case it could not be an extreme point of $[F]$. In a like way, we can argue that if $\mathbf{d}^l$ is an extreme direction of $[F]$, there must be an extreme point $\mathbf{p}^k$ such that $\{\mathbf{x} \in F: \mathbf{x} = \mathbf{p}^k + \lambda\mathbf{d}^l\}$ is nonempty for all $\lambda \geq 0$.

With these observations it is now clear that if $(\tilde{P})$ is solved by an extreme point, $\tilde{\mathbf{p}}$, then $\tilde{\mathbf{p}} \in F$ and so $\tilde{\mathbf{p}}$ solves (P). If $(\tilde{P})$ is unbounded, there must be an extreme direction, $\tilde{\mathbf{d}}$, of $[F]$ with $\mathbf{c}\tilde{\mathbf{d}} < 0$. This same direction will pass through arbitrarily many $\mathbf{x} \in F$, and so (P) too is unbounded. ■

6.1.4 *Classes of Valid Inequalities*

For all cases of Table 6.1 we now know corresponding convex hulls, $[F]$, can be expressed as the intersection of finitely many half-spaces $\{\mathbf{x} \in \mathbb{R}^n\colon \mathbf{h}^r\mathbf{x} \geq h_0^r\}$. Thus, in principle, there is only a finite set of valid inequalities $(\mathbf{h}^r, h_0^r)$ in which we need to be interested to completely settle problems (P) by linear programming.

Unfortunately, we shall rarely know all the inequalities needed to characterize $[F]$. Still, the convex hull provides a standard against which to classify valid inequalities. Recall that the *dimension* of a convex set $X \subseteq \mathbb{R}^n$ (denoted $\dim(X)$, is the maximum number k for which any $\mathbf{x}^0$, $\mathbf{x}^1$, $\mathbf{x}^2, \ldots, \mathbf{x}^k \in X$ have $(\mathbf{x}^1 - \mathbf{x}^0), (\mathbf{x}^2 - \mathbf{x}^0), \ldots, (\mathbf{x}^k - \mathbf{x}^0)$ linearly independent.

- An inequality $(\mathbf{g}, g_0)$ is *facetial*[1] for F if it is valid and $\dim\{\mathbf{x} \in [F]\colon \mathbf{gx} = g_0\} \geq \dim([F]) - 1$. That is, $(\mathbf{g}, g_0)$ is facetial if it is satisfied as an equality by $\mathbf{x}^1, \mathbf{x}^2, \ldots, \mathbf{x}^{\dim[F]}$ points of F for which $(\mathbf{x}^2 - \mathbf{x}^1)$, $(\mathbf{x}^3 - \mathbf{x}^1), \ldots, (\mathbf{x}^{\dim[F]} - \mathbf{x}^1)$ are linearly independent.
- An inequality is *supporting*[2] for F if it is valid and satisfied as an equality by at least one $\mathbf{x} \in F$.
- An inequality is *proper* for F if it is valid and satisfied as a strict inequality by at least one $\mathbf{x} \in F$.

Figure 6.3 illustrates these definitions. In part (a), $[F]$ is *full dimensional* $(\dim([F]) = \dim(\mathbb{R}^2) = 2)$. The indicated supporting inequality $(-x_2 \geq -1)$ is satisfied as an equality by (0,1) of F, and as a strict inequality by all other $\mathbf{x} \in F$. Thus, it is valid, supporting and proper, but not facetial. Facetial inequalities are those that intersect the feasible space in a maximum dimensional face or *facet*. In the example of part (a), there are three facetial inequalities: $x_1 \geq 0$, $x_2 \geq 0$, and $-x_1 - x_2 \geq -1$.

Part (b) of Figure 6.3 shows the more complex situation arising when $[F]$ is less than full dimensional. Here facetial inequalities must support F at only $\dim([F]) = 1$ point. (Since zero does not belong to $[F]$, two points can be linearly independent, but only one difference $(\mathbf{x}^1 - \mathbf{x}^0)$. Thus, there are infinite sets of facetial inequalities supporting at each of the two extreme points, (1,1) and (2,0). Of course, any pair drawn from the two infinite sets would suffice for defining $[F]$, and we shall term *equivalent* valid inequalities that are satisfied as equalities by exactly the same points

[1] Many authors are willing to consider an inequality facetial only if it is proper, i.e., if it is not satisfied by all $\mathbf{x}$ in F. To avoid the constant need to treat separate cases for proper and improper facetial inequalities, we choose to term all facetial.

[2] Some authors use the term *minimal* for this class of inequalities. We prefer to reserve the latter term for the more restricted notion of Section 6.3.

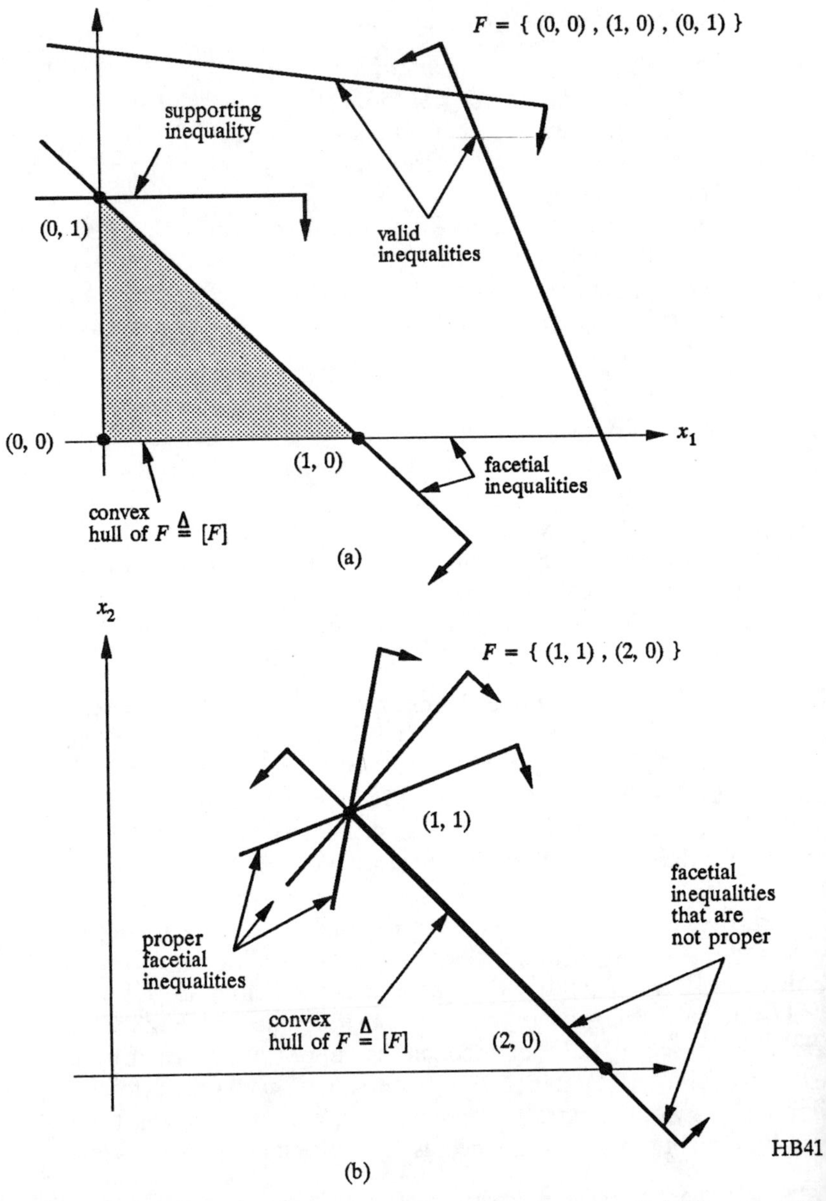

Fig. 6.3. Examples of types of valid inequalities; (a) $[F]$ full dimension, (b) $[F]$ less than full dimension.

of $[F]$. The other two facetial inequalities for the F of part (b) are $x_1 + x_2 \geq 2$ and $-x_1 - x_2 \geq -2$. Neither is proper since both are satisfied by every element of F.

In terms of these definitions, we see that the facetial inequalities are those most desirable ones that define the convex hull of feasible solutions to a discrete problem. Still, it is not necessary to have all—or indeed any—facetial inequalities to produce a relaxation that yields an optimum feasible in (P). For example, the optimum for Example 6.1 occurred at the intersection of the valid inequalities $x_1 \geq 1$ and $x_1 + x_2 \geq 2$. The second is facetial, but the first is merely supporting.

6.1.5 Complexity of Cutting Algorithms

With some nontrivial proofs, we will establish in the later sections of this chapter the *finiteness* of certain implementations of cutting Algorithm 6A. It is customary to proceed as we have, presenting such schemes in a treatise on worse than polynomial-time algorithms. Still, it is interesting to note that we know of no class of problems analogous to those of Chvátal presented in Section 5.1.5. That is, we know of no family of problems provably forcing a convergent implementation of Algorithm 6A to take a number of steps exponential in the length of problem input.

Why then is cutting viewed as probably exponential? There is a variety of circumstantial evidence. First, of course, if the known finite cutting algorithms did not at least sometimes demand more than polynomially many steps on *NP*-Hard problems, they would establish that $P = NP$—a most unlikely result. Second, we shall show in Section 6.3 that available results show it is probably not even possible to construct some facetial inequalities for *NP*-Hard problems in polynomial-time. If some possibly needed cuts cannot be even constructed in polynomial time, no implementation of Algorithm 6A could terminate after only polynomial-time computation. Finally, problem families are known (see Jeroslow (1971)) that force a convergent form of Algorithm 6A to take arbitrarily many steps. Unfortunately, the length of the coefficient data in these problems also grows rapidly. Thus, such families apparently do not seem to show computation can grow as an exponential function of input size.

6.2 Gomory's Cutting Algorithm

The notion of cutting algorithms for discrete problems was apparently first employed by Dantzig, Fulkerson, and Johnson (1954), but the earliest general, provable implementations of cutting Algorithm 6A were developed by R. E. Gomory (1958, 1960a, 1960b). Gomory's algorithms have had only limited success computationally, and they can now be derived as special cases of later ideas (some also Gomory's) that we shall present later in this chapter. Gomory's cutting planes, however, are among the milestone achievements of discrete optimization. Thus, we devote the present section to a development of the algorithms that does not depend upon the more general theory to follow. Connections between Gomory's methods and more general theory will be illustrated as the latter is developed in Sections 6.3–6.5.

6.2.1 Pure Integer Case

The first of Gomory's algorithms addresses discrete optimization problems in the pure integer form

$$(IP) \qquad \begin{aligned} \min \quad & \mathbf{cx} \\ \text{s.t.} \quad & \mathbf{Ax} = \mathbf{b} \\ & \mathbf{x} \geq \mathbf{0} \\ & \mathbf{x} \text{ integer,} \end{aligned}$$

where, as usual in our discussion, **A**, **b**, and **c** have integer coefficients. Of course, this form is easily derived from any rational equality and/or inequality problem (P) where all decision variables are integer-constrained as in Lemma 6.1.

Suppose we solve the linear programming relaxation $(\overline{IP})$ of (IP) for an optimal *basic* solution, $\bar{\mathbf{x}}$, and rewrite the problem in terms of the corresponding optimal basis (see the Appendix C)

$$(IP_N) \qquad \begin{aligned} \min \quad & -x_0 \\ \text{s.t.} \quad & x_i = \bar{x}_i - \sum_{j \in N} \bar{a}_{ij} x_j && \text{for } i \geq 0,\ i \notin N \\ & x_j \geq 0 && \text{for all } j \neq 0 \\ & x_j \text{ integer for all } j. \end{aligned}$$

Here N is the nonbasic variable set, and $\bar{a}_{ij}$ is the coefficient of nonbasic variable x_j in the representation of basic variable x_i. Observe that for later notational convenience we have introduced variable x_0 to represent the negative of the objective function value.

If the $(\overline{IP})$ optimal solution, $\bar{\mathbf{x}}$, is integral, it solves (IP), we would stop at Step 2 in Algorithm 6A. If not, we wish to focus on *fractional parts,*

$$\phi(z) \triangleq z - \lfloor z \rfloor \tag{6-4}$$

of various quantities, z, in the formulation (IP_N). Note that $\phi(z) \geq 0$ for all z. For example, $\phi(6\frac{7}{10}) = \frac{7}{10}$; $\phi(-6\frac{7}{10}) = \frac{3}{10}$. Nonbasic components of $\bar{\mathbf{x}}$ assume by definition, the integer value zero. Thus, if $\bar{\mathbf{x}}$ is not integral there must be some basic x_r with $\phi(\bar{x}_r) > 0$.

Consider the representation of that x_r in (IP_N):

$$x_r = \bar{x}_r - \sum_{j \in N} \bar{a}_{rj}(x_j). \tag{6-5}$$

Substituting via definition (6-4) gives

$$x_r = \phi(\bar{x}_r) + \lfloor \bar{x}_r \rfloor - \sum_{j \in N} (\phi(\bar{a}_{rj}) + \lfloor \bar{a}_{rj} \rfloor) x_j,$$

which can be regrouped as

$$\sum_{j \in N} \phi(\bar{a}_{rj})\, x_j - \phi(\bar{x}_r) = \lfloor \bar{x}_r \rfloor - \sum_{j \in N} \lfloor \bar{a}_{rj} \rfloor x_j - x_r. \tag{6-6}$$

We have only to argue that both sides of (6-6) are nonnegative to show validity of Gomory (1958) *fractional cutting planes*

$$\sum_{j \in N} \phi(\bar{a}_{rj}) x_j \geq \phi(\bar{x}_r) \tag{6-7}$$

6.4. Validity of Gomory Fractional Cuts. *Let (IP) and (IP_N) be as above. Then inequality (6-7) is valid for (IP), i.e., it is satisfied by all integer $\mathbf{x} \geq \mathbf{0}$ such that $\mathbf{Ax} = \mathbf{b}$.*

Proof. Since all feasible solutions to (IP) have $\mathbf{x} \geq \mathbf{0}$, and $1 > \phi(z) \geq 0$ for any number z, the left side of (6-6) satisfies

$$\sum_{j \in N} \phi(\bar{a}_{rj}) x_j - \phi(\bar{x}_r) > -1$$

whenever $\mathbf{x}$ is feasible in (IP). But such feasible $\mathbf{x}$ also have integral components x_r and x_j, $j \in N$. It follows that the right side of (6-6) will always be integer for such solutions. We can conclude that the left side must then be at least the integer 0. That is, cut (6-7) is valid for all $\mathbf{x}$ feasible in (IP). ■

A version of Algorithm 6A employing cut (6-7) can be stated as follows:

Algorithm 6B. Gomory Fractional Cutting Planes. Let (IP) be as above with bounded linear programming relaxation solution value, $v(\overline{IP}) > -\infty$.

STEP 0: *Initial Relaxation.* Assign subscripts $j = 1, 2, \ldots, n$ to the variables x_j of (IP), and let x_0 represent the negative of the solution value. Then set $k \leftarrow 0$ and pick as relaxation (IP^0) the linear programming relaxation ($\overline{IP}$).

STEP 1: *Relaxation.* Attempt to solve relaxation (IP^k). If $v(IP^k) = +\infty$, stop; (IP^k) and thus, (IP) is infeasible. Otherwise, let $\mathbf{x}^k$ be an optimal solution, $-x_0^k$ its solution value, and go to Step 2.

STEP 2: *Optimality Check.* If $\mathbf{x}^k$ is integer, stop; $\mathbf{x}^k$ solves (IP) because it solves (IP^k) and is feasible in (IP). Otherwise, proceed to Step 3.

STEP 3: *Cutting.* Pick any subscript $0 \leq r \leq n$ with x_r^k not integer, and represent x_r in terms of an optimal nonbasic set, N^k for (IP^k) as

$$x_r = \bar{x}_r^k - \sum_{j \in N^k} \bar{a}_{rj}^k x_j \tag{6-8}$$

Next form linear programming relaxation (IP^{k+1}) by adding surplus variable x_{n+k+1} and constraints

$$\sum_{j \in N^k} \phi(\bar{a}_{rj}^k)\, x_j - x_{n+k+1} = \phi(\bar{x}_r^k) \tag{6-9}$$

$$x_{n+k+1} \geq 0 \tag{6-10}$$

to (IP^k). If desired, also delete in (IP^{k+1}) any previously added cutting planes (6-9) with associated surplus variables basic in the (IP^k) optimum. Finally, replace $k \leftarrow k + 1$, and return to Step 1. ■

Example 6.2. Gomory Fractional Cutting Planes. To illustrate Algorithm 6B consider the pure-integer program

$$\begin{aligned} \min \quad & 7x_1 + x_2 \\ \text{s.t.} \quad & x_1 - 3x_2 \geq -1 \\ & 3x_1 + 2x_2 \geq 4 \\ & x_1, x_2 \geq 0 \text{ and integer.} \end{aligned}$$

Adding surplus variables to obtain the form (IP) gives

$$(IPX) \qquad \begin{aligned} \min \quad & 7x_1 + x_2 \\ \text{s.t.} \quad & x_1 - 3x_2 - x_3 = -1 && (6\text{-}11) \\ & 3x_1 + 2x_2 - x_4 = 4 && (6\text{-}12) \\ & x_1, x_2, x_3, x_4 \geq 0 \text{ and integer.} \end{aligned}$$

Here $n = 4$ and our first relaxation (IPX^0) is the linear programming relaxation $(\overline{IPX})$.

As illustrated in Figure 6.4, (IPX^0) solves at $\mathbf{x}^0 = (\frac{10}{11}, \frac{7}{11}, 0, 0)$ with $x_0^0 = -7$. Nonbasic variables in the (IPX^0) optimum are x_3 and x_4, yield-

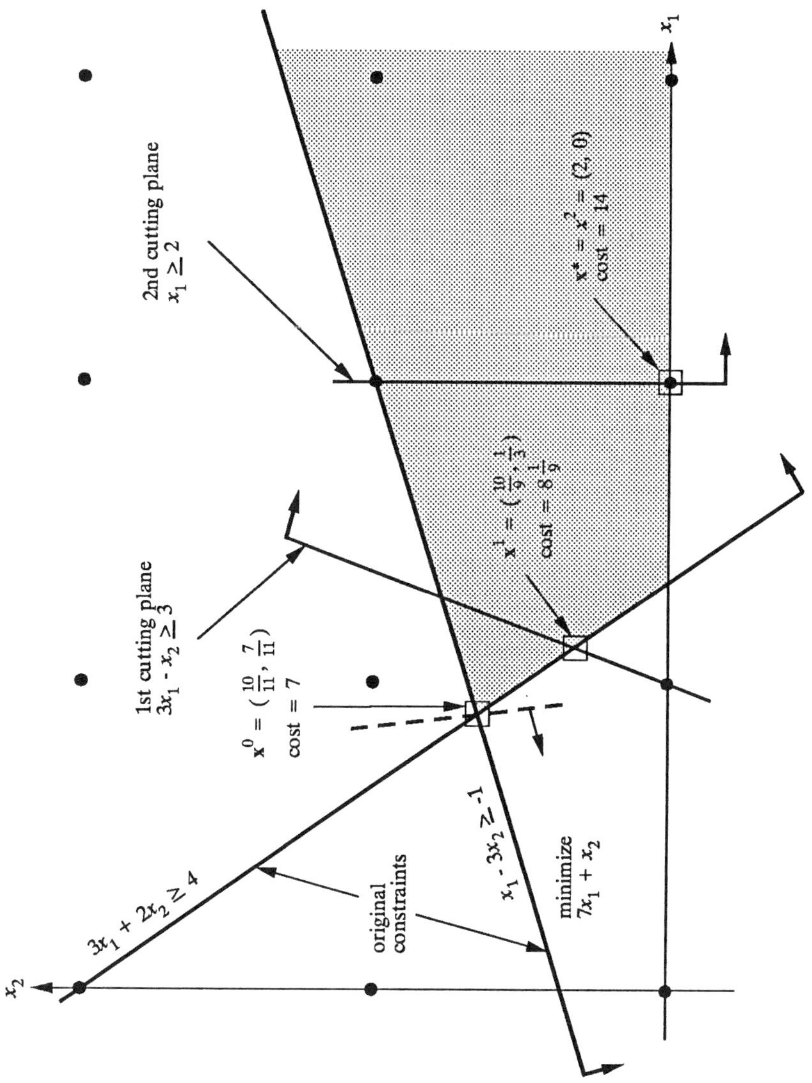

Fig. 6.4. Gomory fractional cutting Example 6.2.

ing the representations

$$x_0 = -7 - (1x_3 + 2x_4)$$
$$x_1 = \tfrac{10}{11} \quad - (-\tfrac{2}{11}x_3 - \tfrac{3}{11}x_4)$$
$$x_2 = \tfrac{7}{11} \quad - (\tfrac{3}{11}x_3 - \tfrac{1}{11}x_4).$$

Since both x_1^0 and x_2^0 are fractional, we must generate a cutting plane. Either $r = 1$ or $r = 2$ could serve as the generating x_r. We shall choose $r = 1$. The corresponding Gomory cut is $\phi(-\tfrac{2}{11})x_3 + \phi(-\tfrac{3}{11})x_4 \geq \phi(\tfrac{10}{11})$ or including the new surplus variable, x_5,

$$(\tfrac{9}{11})x_3 + (\tfrac{8}{11})x_4 - x_5 = \tfrac{10}{11} \tag{6-13}$$

Although it is not necessary for computations, we can restate this new constraint in terms of original variables x_1 and x_2 by substituting for x_3 and x_4 via equations (6-11)–(6-12). The result is

$$(\tfrac{9}{11})(x_1 - 3x_2 + 1) + (\tfrac{8}{11})(3x_1 + 2x_2 - 4) \geq \tfrac{10}{11}$$

or the $3x_1 - x_2 \geq 3$ shown in Figure 6.4. Note that all (*IPX*) feasible solutions do indeed satisfy the cutting plane, i.e., it is a valid inequality.

Adding variable $x_5 \geq 0$ and equation (6-13) to (IPX^0) yields the next relaxation, (IPX^1). Figure 6.4 shows that an optimum for (IPX^1) is $\mathbf{x}^1 = (\tfrac{10}{9}, \tfrac{1}{3}, \tfrac{10}{9}, 0, 0)$, $x_0^1 = -8\tfrac{1}{9}$. The nonbasic variable representation of any of the fractional variables could produce our next cutting plane. We shall pick the cost now $r = 0$, i.e., use the representation

$$x_0^1 = -8\tfrac{1}{9} - ((\tfrac{10}{9})x_4 + (\tfrac{11}{9})x_5)$$

The corresponding cut equation is

$$(\tfrac{1}{9})x_4 + (\tfrac{2}{9})x_5 - x_6 = \tfrac{8}{9} \tag{6-14}$$

with new surplus variable, $x_6 \geq 0$. By substituting for x_3, x_4 and x_5 via (6-11)–(6-13), we again reduce this cutting plane to the original variables. Figure 6.4 shows the result is $x_1 \geq 2$.

Including this new valid inequality gives (IPX^2), Figure 6.4 illustrates that an optimal solution is $\mathbf{x}^2 = (2, 0, 3, 2, 3, 0)$, $x_0^2 = -14$. Here all original variables (indeed, all variables) are integer, and we terminate. Problem (*IPX*) is solved by components $\mathbf{x}^2$ corresponding to original variables, i.e., by $\mathbf{x}^* = (2, 0, 3, 2)$. ■

Before turning to a proof of Algorithm 6B, two new elements should be highlighted. First observe that a rather large number of cutting planes may accumulate as any implementation of the relaxation strategy advances. A very substantial calculation and storage burden results. Step 3 of our statement of Algorithm 6B includes a standard technique for alleviating this buildup of cutting planes. When a cutting plane's surplus variable becomes basic, or more intuitively, when the cutting plane is no longer binding, we can discard it without destroying convergence. It is quite possible that a dropped inequality will later be regenerated, but the savings realized outweigh that potential cost.

A second issue is validity of the fractional cut (6-9). The cut is substantially the one addressed in Theorem 6.4, but the nonbasic set N^k may include surplus variables from previous cutting planes. How can we be sure such variables may be treated as integer-restricted (an essential element in the proof of Theorem 6.4)? Relation (6-6) provides an answer. Each new surplus variable is defined to equal the left, and thus the right side of an equation like (6-6). The first nonbasic set, N^0, is all integer-restricted. This implies the first surplus will also take on an integer value. Continuing in this way, the second surplus variable will be integer because all prior variables are, etc.

In spite of the simplicity of its derivation, fractional cut (6-7) has proved an enduring and fundamental result in cutting theory. The key technical achievement of Gomory (1958) was to show finite convergence of a cutting algorithm.

The proof depends on the notion of *lexicographic ordering.* Vector $\mathbf{z}^1$ is said to be lexicographically greater than $\mathbf{z}^2$, denoted $\mathbf{z}^1 \overset{L}{>} \mathbf{z}^2$, if the first nonzero component of $(\mathbf{z}^1 - \mathbf{z}^2)$ is positive. For example, $(1, -10, 2, 3) \overset{L}{>} (0, 0, 0, 0)$, $(2, 1, 3, 5) \overset{L}{>} (2, 1, 0, 10)$.

Lexicographic ordering is at the heart of some proofs of convergence for the simplex method of linear programming. The proof of Algorithm 6B expands those results to (IP).

Theorem 6.5. Finite Convergence of Fractional Cutting Algorithm 6B. *Let $(\overline{IP})$ be as above with bounded linear programming relaxation value $v(\overline{IP}) > -\infty$, and suppose Algorithm 6B is implemented via a dual simplex procedure passing through basic solutions $\mathbf{w}^0, \mathbf{w}^1, \ldots, \mathbf{w}^{t_0} \triangleq \mathbf{x}^0, \mathbf{w}^{t_0+1}, \ldots, \mathbf{w}^{t_1} \triangleq \mathbf{x}^1, \mathbf{w}^{t_1+1}, \ldots$ Then if*

(i) $(w_0^t, w_1^t, \ldots, w_n^t) \overset{L}{>} (w_0^{t+1}, w_1^{t+1}, \ldots, w_n^{t+1})$ *for all t, and*
(ii) the smallest admissible r is chosen at each execution of Step 3 (i.e., $r \leftarrow \min\{j = 0, 1, \ldots, n\colon \phi(x_j^k) > 0\}$*)*

the algorithm terminates after finitely many simplex iterations, t_p, with either $(x_1^p, \ldots, x_n^p)$ optimal in (IP) or the correct conclusion that no feasible solution exists.

Proof. If the Algorithm stops at Step 1 with a conclusion that (IP) is infeasible, or at Step 2 with a claimed optimum, it is clear that the result is correct. All added constraints (6-9)–(6-10) are valid, so that each (IP^k) is indeed a relaxation of (IP).

We want to show that the procedure cannot continue infinitely. The arguement will take components 0, 1, 2, . . . , n of the solution in order.

Consider first the sequence $\{w_0^t\}$. Each element w_0^t is the negative of a solution value, and thus bounded below by $-v(IP)$. Also, lexicographic ordering, (i), assures the series is monotone nonincreasing. Thus, the series has a limit, say w_0^∞, and after finitely many simplex iterations, the sequence will come arbitrarity close to w_0^∞. Pick a relaxation-solving iteration with $\mathbf{x}^q = \mathbf{w}^{t_q}$ optimal in (IP^q) and $w_0^{t_q} - \lfloor w_0^\infty \rfloor < 1$. If $w_0^{t_q} = \lfloor w_0^\infty \rfloor$ turn to component 1 of the solutions. Otherwise, $\phi(w_0^{t_q}) > 0$ and rule (ii) will cause us to generate a constraint (6-9) for $r = 0$. Consider the effect of dual simplex iteration $t_q + 1$. The implied pivot will occur in the new cut row (the only primal infeasible one) and in the column of some nonbasic variable x_h with $\phi(\bar{a}_{0h}^q) > 0$. Component 0 will be updated

$$w_0^{t_q+1} \leftarrow w_0^{t_q} - \bar{a}_{0h}^q \frac{\phi(w_0^{t_q})}{\phi(\bar{a}_{0h}^q)} \tag{6-15}$$

Now $\phi(w_0^{t_q})$ and $\phi(\bar{a}_{0h}^q)$ are both positive, and $\bar{a}_{0h}^q$ is nonzero because it is fractional. It follows that if $\bar{a}_{0h}^q < 0$, (6-15) would yield $w_0^{t_q+1} > w_0^{t_q}$, a violation of lexicographic ordering. Thus, $\bar{a}_{0h}^q > 0$. But then,

$$\bar{a}_{0h}^q \geq \phi(\bar{a}_{0h}^q) \tag{6-16}$$

and (6-15) implies

$$w_0^{t_q} - \phi(w_0^{t_q}) \geq w_0^{t_q+1} \geq \lfloor w_0^\infty \rfloor .$$

Since we chose t_q with $w_0^{t_q} - \phi(w_0^{t_q}) = \lfloor w_0^\infty \rfloor$, we can conclude that $w_0^{t_q+1} = \lfloor w_0^\infty \rfloor$. The (integer) limit of the nonincreasing series $\{w_0^t\}$ has been reached after finitely many simplex iterations.

Once $\{w_0^t\}$ reaches its limit, lexicographic ordering assures that it retains that level in all future iterations. We can direct our attention to component 1 and the sequence $w_1^{t_q+1}, w_1^{t_q+2}, \ldots,$. As before it is monotone nonincreasing. Also, it must be bounded below by 0; if some w_1^t with $t > t_q$ ever becomes negative the sequence will never return to a feasible linear programming relaxation solution having $w_1 \geq 0$. But if $\{w_1^t\}$ is monotone and bounded below, it too has a limit. By exactly the same arguement as above, this (integer) limit will be achieved after finitely many more simplex iterations.

Continuing the same arguement with $w_2, w_3, \ldots, w_n$, all original problem variables will necessarily reach their integer limit values after finitely many simplex pivots. We need only complete the relaxation (IP^p) during which the last limit is achieved to terminate Algorithm 6B in finitely many steps. ■

In order to prove convergence, we have assumed Algorithm 6B can be implemented with properties (i) and (ii). Rule (ii) is, of course, easily achieved, although it may not be the best practical choice. It is a lengthy-to-prove, but well-known fact of linear programming theory that (i) can also be achieved by a suitable version of the dual simplex algorithm. We refer the reader to any standard book on linear programming (e.g., Bazaraa and Jarvis (1977)) for details.

6.2.2 *Mixed Integer Case*

Fractional cutting plane (6-7) of the previous section was proved valid only for pure-integer programs. Gomory (1960b), however, showed how it can be extended to the mixed integer case:

$$
(MIP) \qquad \begin{array}{ll} \min & \mathbf{cx} \\ \text{s.t.} & \mathbf{Ax} = \mathbf{b} \\ & \mathbf{x} \geq \mathbf{0} \\ & x_j \text{ integer for } j = 1, 2, \ldots n_I \end{array}
$$

Here $\mathbf{x}$ is an n-vector with $n_I < n$.

Just as in the pure-integer case, derivation of Gomory's mixed integer cutting plane begins with the representation

$$x_r = \bar{x}_r - \sum_{j \in N_I} \bar{a}_{rj} x_j - \sum_{j \in N_C} \bar{a}_{rj} x_j \tag{6-17}$$

of an integer-restricted variable x_r with fractional linear programming relaxation value, $\bar{x}_r$. Here the nonbasic set N has been partitioned into N_I, the set of integer-restricted nonbasic variables, and N_C, the continuous nonbasics.

Rewriting all integer-restricted variables of (6-17) in terms of their fractional parts (definition (6-4)) and regrouping

$$\sum_{j \in N_I} \phi(\bar{a}_{rj}) x_j + \sum_{j \in N_C} \bar{a}_{rj} x_j - \phi(\bar{x}_r) = \lfloor \bar{x}_r \rfloor - \sum_{j \in N_I} \lfloor \bar{a}_{rj} \rfloor x_j - x_r. \tag{6-18}$$

Recognition that the left side of (6-18) is an integer ≤ -1 or ≥ 0 is the core of the arguement for validity of one version of Gomory's mixed integer cutting plane:

$$\sum_{j \in N_I} \phi(\bar{a}_{rj}) x_j + \sum_{j \in N_C^+} \bar{a}_{rj} x_j + \frac{\phi(\bar{x}_r)}{1 - \phi(\bar{x}_r)} \sum_{j \in N_C^-} (-\bar{a}_{rj}) x_j \geq \phi(\bar{x}_r), \tag{6-19}$$

where

$$N_C^+ \triangleq \{j \in N_C : \bar{a}_{rj} > 0\}, \; N_C^- \triangleq \{j \in N_C : \bar{a}_{rj} < 0\}.$$

Theorem 6.6. Validity of Gomory Mixed Integer Cut (6-19). *Let (MIP) be as above, and (6-17) the representation, in terms of an optimal linear programming relaxation basis, of some integer-restricted* x_r *having* $\phi(\bar{x}_r) > 0$. *Then mixed integer cut (6-19) is valid for (MIP), i.e., it is satisfied by all* $\boldsymbol{x} \geq \mathbf{0}$ *with* $\boldsymbol{Ax} = \boldsymbol{b}$ *and* x_j *integer for* $j = 1, 2, \ldots, n_I$

Proof. All constants and all variables on the right side of (6-18) are integer for $\mathbf{x}$ feasible in (*MIP*). Thus, the left side is also integer.

In particular, it is either ≥ 0 or ≤ -1, i.e., either

$$\sum_{j \in N_I} \phi(\bar{a}_{rj})x_j + \sum_{j \in N_C} \bar{a}_{rj}x_j - \phi(\bar{x}_r) \geq 0 \tag{6-20}$$

or

$$\sum_{j \in N_I} (-\phi(\bar{a}_{rj}))x_j + \sum_{j \in N_C} (-\bar{a}_{rj})x_j + \phi(\bar{x}_r) \geq 1 \tag{6-21}$$

For $\mathbf{x} \geq \mathbf{0}$, terms with negative or zero coefficients on the left of (6-20) and (6-21) cannot contribute to satisfying the inequalities. Thus, we may rewrite them over the more restricted sets $N_C^+ = \{j \in N_C \colon \bar{a}_{rj} > 0\}$ and $N_C^- = \{j \in N_C \colon \bar{a}_{rj} < 0\}$ as

$$\sum_{j \in N_I} \phi(\bar{a}_{rj})x_j + \sum_{j \in N_C^+} \bar{a}_{rj}x_j \geq \phi(\bar{x}_r) \tag{6-22}$$

or

$$\sum_{j \in N_C^-} (-\bar{a}_{rj})x_j \geq 1 - \phi(\bar{x}_r). \tag{6-23}$$

Noting $1 > \phi(\bar{x}_r) > 0$ because $\bar{\mathbf{x}}$ is fractional, the latter is equivalent to

$$\frac{\phi(\bar{x}_r)}{1 - \phi(\bar{x}_r)} \sum_{j \in N_C^-} (-\bar{a}_{rj})x_j \geq \phi(\bar{x}_r) \tag{6-24}$$

We conclude that either the left side of (6-22) or the left side of (6-24) must be at least equal $\phi(\bar{x}_r)$ in any (*MIP*) solution. But then certainly the sum of those left sides will achieve $\phi(\bar{x}_r)$. The left side of inequality (6-19) is exactly that sum. ■

To illustrate the mixed integer cut computation (6-19), suppose an integer-restricted variable x_{12} is represented in terms of an optimal linear programming basis as

$$x_{12} = 3\tfrac{2}{5} - (-2\tfrac{1}{5}x_3 + \tfrac{3}{5}x_7 + 5x_8 + 1\tfrac{1}{5}x_{10} - 4x_{14} + 5\tfrac{1}{5}x_{15} - \tfrac{2}{5}x_{20}). \tag{6-25}$$

Also, suppose variables through $n_I = 13$ are the integer-restricted ones.

Then,

$$N_I = \{3, 7, 8, 10\},\ N_C^+ = \{15\},\ N_C^- = \{14, 20\}.$$

The implied cutting plane (6-19) is

$$\tfrac{4}{5}x_3 + \tfrac{3}{5}x_7 + \tfrac{1}{5}x_{10} + 5\tfrac{1}{5}x_{15} + \tfrac{2}{3}(4x_{14} + \tfrac{2}{5}x_{20}) \geq \tfrac{2}{5} \tag{6-26}$$

Obviously, cut (6-19) reduces exactly to the fractional cut (6-7) when N_C is empty. Also (because we derived it with $\{-1, 0\}$ instead of some other integer pair), the present nonbasic solution, $x_j = 0$ for $j \in N$, is cut off. However, the mixed cut (6-19)—and indeed fractional cut (6-7)—can be strengthened by a little extra thought.

Suppose we substitute $\bar{a}_{rj} = \lceil \bar{a}_{rj} \rceil - (1 - \phi(\bar{a}_{rj}))$ instead of $\bar{a}_{rj} = \lfloor \bar{a}_{rj} \rfloor + \phi(\bar{a}_{rj})$ in deriving (6-18). The analog of (6-18) will then be

$$-\sum_{j \in N_I} (1 - \phi(\bar{a}_{rj}))x_j + \sum_{j \in N_C} \bar{a}_{rj}x_j - \phi(\bar{x}_r) = \lfloor \bar{x}_r \rfloor - \sum_{j \in N_I} \lceil \bar{a}_{rj} \rceil x_j - x_r.$$

By exactly the arguement of Theorem 6.6 we could proceed to demonstrate validity of the inequality

$$\frac{\phi(\bar{x}_r)}{1 - \phi(\bar{x}_r)} \sum_{j \in N_I} (1 - \phi(\bar{a}_{rj}))x_j + \sum_{j \in N_C^+} \bar{a}_{rj}x_j + \frac{\phi(\bar{x}_r)}{1 - \phi(\bar{x}_r)} \sum_{j \in N_C^-} (-\bar{a}_{rj})x_j \geq \phi(\bar{x}_r). \tag{6-27}$$

Comparing (6-27) to (6-19), we see that the only effect of our new derivation is a change of coefficients on $j \in N_I$. It is also clear that nothing would prevent our mixing the derivations, i.e., substituting $\bar{a}_{rj} = \lceil \bar{a}_{rj} \rceil - (1 - \phi(\bar{a}_{rj}))$ for some $j \in N_I$ and $\bar{a}_{rj} = \lfloor \bar{a}_{rj} \rfloor + \phi(\bar{a}_{rj})$ for the remainder. But then a cutting plane picking the best outcome for each integer-restricted variable is also valid. The result is the standard Gomory mixed integer cutting plane

$$\sum_{j \in N_I} \min\left\{\phi(\bar{a}_{rj}), \frac{\phi(\bar{x}_r)}{1 - \phi(\bar{x}_r)}(1 - \phi(\bar{a}_{rj}))\right\} x_j + \sum_{j \in N_C^+} \bar{a}_{rj}\, x_j + \frac{\phi(\bar{x}_r)}{1 - \phi(\bar{x}_r)} \sum_{j \in N_C^-} (-\bar{a}_{rj})x_j \geq \phi(\bar{x}_r). \tag{6-28}$$

Application of this new computation to example (6-25) gives

$$\min\{\tfrac{4}{5}, \tfrac{2}{3}(\tfrac{1}{5})\}x_3 + \min\{\tfrac{3}{5}, \tfrac{2}{3}(\tfrac{2}{5})\}x_7 + \min\{\tfrac{1}{5}, \tfrac{2}{3}(\tfrac{4}{5})\}x_{10} + 5\tfrac{1}{5}x_{15} + \tfrac{2}{3}(4x_{14} + \tfrac{2}{5}x_{20}) \geq \tfrac{2}{5}$$

or

$$\tfrac{2}{15}x_3 + \tfrac{4}{15}x_7 + \tfrac{1}{5}x_{10} + 5\tfrac{1}{5}x_{15} + 2\tfrac{2}{3}x_{14} + \tfrac{4}{15}x_{20} \geq \tfrac{2}{5}. \tag{6-29}$$

Notice that coefficients on x_3 and x_7 have been reduced from (6-26). This implies the new inequality is stronger in the sense that fewer $\mathbf{x} \geq \mathbf{0}$ satisfy it.

By employing (6-28) at Step 3 one can obtain an algorithm for (*MIP*) that exactly parallels Algorithm 6B. The only change is that there is no longer any guarantee the surplus variables in cutting planes will be integer-valued. Thus, we must place new surplus variables in the continuous set, N_C, instead of the integer set, N_I, as we develop representations (6-17).

A complication also arises in extending the convergence proof of Theorem 6.5 to the mixed integer setting. If the objective function value $-x_0$ of (*MIP*) can be assumed integer-restricted, an exact analog of Theorem 6.5 proves finite convergence of Gomory's mixed integer algorithms. But since $-x_0 \triangleq \Sigma_j c_j x_j$ potentially includes continuous variables at fractional values, we may not always be able to assume integrality. If integrality is not required, the algorithm may not finitely converge.

6.2.3 *Computational Difficulties*

We have seen that Gomory's Fractional Cutting Algorithm 6B for pure integer programs, and the associated mixed integer cutting algorithm for problems with integer objective function values, can be proved to converge in a finite number of dual simplex iterations. Unfortunately, the very limited computational experience with computer implementations of Gomory's algorithms (e.g., Haldi and Isaacson (1965), Trauth and Woolsey (1968)) has been disappointing. Although finiteness proofs guarantee no particular efficiency, the main difficulty has come, not from the number of iterations, but from numerical errors in computer arithmetic.

The key elements of all the cutting planes in Sections 6.2.1 and 6.2.2 were quantities $\phi(\bar{a}_{rj})$, fractional parts of coefficients in dual simplex tableaux. Notice that such quantities are exactly the least significant portions of corresponding coefficients, $\bar{a}_{rj}$. As simplex arithmetic proceeds these least significant parts of computed values are naturally the most likely to include error. It is possible—indeed likely—that the cut computed from erroneous fractional parts will not even be valid. A hopelessly confused algorithm typically results.

Gomory (1963), Young (1965) and others have proposed an approach to circumventing such roundoff errors: if every entry in each dual simplex tableau is integer, there will be no roundoff errors (although integers may become very large in magnitude). Finitely convergent variants on Algorithm 6B can be devised that retain such integer simplex coefficients. Unfortunately, however, the cutting planes employed by these all-integer algorithms seem weaker than the fractional cuts. Computational testing, although limited, has not been encouraging.

6.3 Minimal Inequalities

Upon the relatively unsatisfactory computational experience with Gomory's early cutting planes, most work on specific cutting algorithms simply stopped. Few of the ideas developed in the ensuing 15–20 years have been implemented or tested in computer code until recently.

Still, theoretical investigation has continued at a steady pace. First Gomory, and later Johnson, Araoz, Jeroslow, Blair, Glover, Balas, Chvátal, Padberg and others have produced lines of research yielding mathematical characterizations of the forms of strong valid inequalities. This section, and the three to follow, present that *cutting plane theory* for problems

$$\begin{aligned} & \min && \mathbf{cx} \\ (P) \qquad & \text{s.t.} && \mathbf{x} \in F, \end{aligned}$$

where F is a form (such as those of Table 6.1) having a polyhedral convex hull $[F]$.

A key notion in much of that theory is that a valid inequality $(\mathbf{g}, g_0)$ is a *weakening* of another $(\hat{\mathbf{g}}, \hat{g}_0)$ if

$$\mathbf{g} \geq \hat{\mathbf{g}},\ g_0 \leq \hat{g}_0. \tag{6-30}$$

That is, $(\mathbf{g}, g_0)$ is a weakening of $(\hat{\mathbf{g}}, \hat{g}_0)$, if $\{\mathbf{x} \in \mathbb{R}^n\colon \mathbf{x} \geq \mathbf{0},\ \mathbf{gx} \geq g_0\} \supseteq \{\mathbf{x} \in \mathbb{R}^n\colon \mathbf{x} \geq \mathbf{0},\ \hat{\mathbf{g}}\mathbf{x} \geq \hat{g}_0\}$.

Valid inequalities are termed *minimal* if they are weakenings of no distinct valid inequality. The concept of minimal inequalities was first introduced by Gomory (1967). We shall see that it provides a conveniently manipulated notion of a complete set of valid inequalities, i.e., of a family of valid inequalities containing all those we might need in order to solve appropriate discrete optimization problems by the relaxation Algorithm 6A. All weakenings of minimal inequalities may be ignored since they are implied by the minimal ones.

Figure 6.5 illustrates. If an inequality like $-x_1 - 3x_2 \geq -6$ in Figure 6.5 is not supporting (see Section 6.1.4), its right-hand-side can be improved (to $-x_1 - 3x_1 \geq -4$ in the example). Thus, it cannot be minimal. On the other hand, some supporting valid inequalities are also not minimal. In Figure 6.5 the supporting inequality $-x_1 \geq -2$ is a weakening of the minimal inequality $-x_1 - x_2 \geq -2$.

Before precisely resolving when inequalities are minimal, we must note two pathologies. First, nonnegative multiples of the $x_j \geq 0$ constraints can

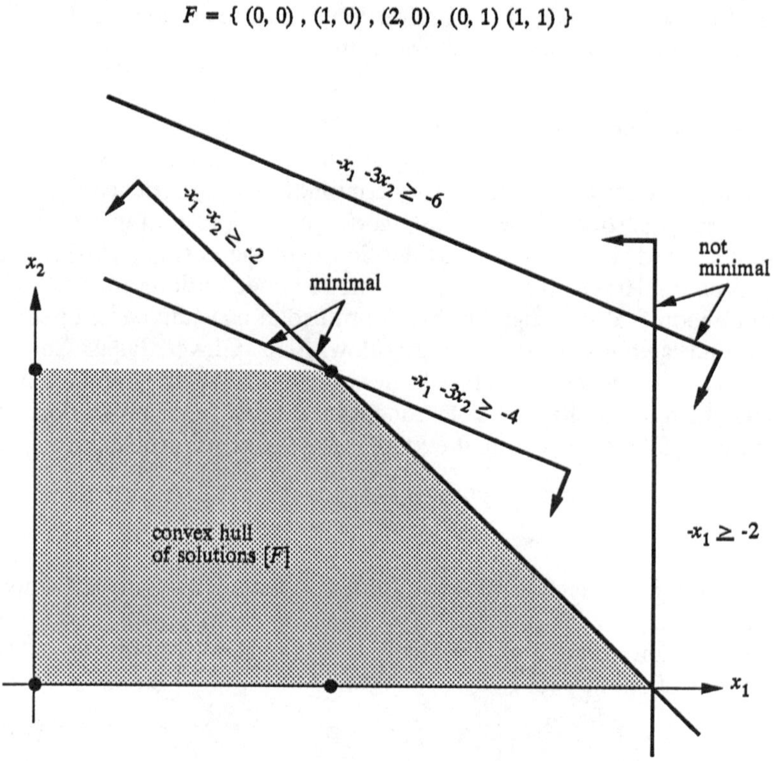

Fig. 6.5. Minimal valid inequalities.

never be minimal even though they can even be facetial. For example, in Figure 6.5, $x_1 \geq 0$ is a facetial—the strongest possible form of valid inequalities. But it can be strengthened in the sense of (6-30) by $\frac{1}{2}x_1 \geq 0, \frac{1}{4}x_1$, etc. We will clearly need to treat the nonnegativities separately in our discussion.

A second problem arises if some $x_j = 0$ for all $\mathbf{x} \in F$. In such a case, the corresponding cutting plane coefficient, g_j, could be chosen arbitrarily without impacting validity. It follows that no valid inequality with finite g_j can be minimal.

To avoid the latter complication we shall often assume that F is *fully positive,* i.e., that

$$\{\mathbf{x} \in F\colon x_j > 0\} \neq \varnothing \qquad \text{for any } j. \tag{6-31}$$

For most practical cases it would not be difficult to check full positivity and eliminate any x_j that are zero in all solutions. Assumption (6-31) is not altogether innocuous, however; we have demanded a nonzero feasible solution to the discrete problem (P) rather than its more easily managed linear programming relaxation $(\bar{P})$. On the other hand, F need not be full dimension to satisfy full positivity. Figure 6.3b provides a counterexample.

6.3.1 Minimal Inequalities Versus Supporting and Facetial

Under the full positivity assumption (6-31), Jeroslow (1978a) has shown how the minimal inequalities relate to the supporting and facetial ones of Section 6.1.4. The next two theorems summarize these relationships.

Theorem 6.7. Minimal vs. Supporting Inequalities. *Let F be as above with polyhedral convex hull $[F]$. Then the minimal inequalities for F are the supporting ones that are satisfied as equalities by at least one strictly positive $\mathbf{x} \in [F]$.*

Proof. If F is empty or fails the full positivity assumption (6-31), the theorem is trivially true. There exists neither minimal inequalities nor strictly positive $\mathbf{x} \in [F]$.

Consider the case where strict positivity does hold and let $(\mathbf{g}, g_0)$ be minimal. First note that $(\mathbf{g}, g_0)$ must be supporting. The fact that $(\mathbf{g}, g_0)$ is valid implies $\{\mathbf{gx}: \mathbf{x} \in F\}$ is bounded below by g_0. Thus, for F with polyhedral $[F]$, Theorem 6.3 assures there is an $\hat{\mathbf{x}} \in F$ such that $\mathbf{g}\hat{\mathbf{x}} = \inf\{\mathbf{gx}: \mathbf{x} \in F\}$. Should $(\mathbf{g}, g_0)$ not support at $\hat{\mathbf{x}}$, g_0 could be increased to $\mathbf{g}\hat{\mathbf{x}}$—a contradiction of minimality.

To show $(\mathbf{g}, g_0)$ supports at a strictly positive point in $[F]$, we shall demonstrate that for each component x_t of $\mathbf{x}$ there is a vector $\mathbf{x}^t \in E \triangleq \{\mathbf{x} \in [F]: \mathbf{gx} = g_0\}$ such that $x_t^t > 0$. It follows from convexity of E and nonnegativity of $\mathbf{x} \in [F] \supset E$ that $(\Sigma_{t=1}^{n} \mathbf{x}^t)/n$ provides the desired strictly positive element of E.

Since $[F]$ is polyhedral, we know every element $\mathbf{x}$ can be represented

$$\mathbf{x} = \sum_{k \in P} \lambda_k \mathbf{p}^k + \sum_{l \in D} \mu_l \mathbf{d}^l, \tag{6-32}$$

where $\{\mathbf{p}^k: k \in P\}$ is a finite list of extreme points of $[F]$, $\{\mathbf{d}^l: l \in D\}$ is a finite list of extreme directions, $\Sigma_{k \in P} \lambda_k = 1$, $\lambda_k \geq 0$ for all $k \in P$, and $\mu_l \geq 0$ for all $l \in D$. Also, $\mathbf{p}^k \geq 0$ for all $k \in P$ and $\mathbf{d}^l \geq 0$ for all $l \in D$ because $[F] \subseteq \{\mathbf{x} \in \mathbb{R}^n: \mathbf{x} \geq \mathbf{0}\}$.

Consider first the $\mathbf{p}^k$. We have by validity of $(\mathbf{g}, g_0)$ that

$$\sum_{j \neq t} g_j p_j^k + g_t p_t^k \geq g_0 \qquad \text{for all } k \in P. \tag{6-33}$$

If $p_t^k > 0$ and $\mathbf{gp}^k = g_0$ for any k, then $\mathbf{x}^t \leftarrow \mathbf{p}^k$ provides the element of E

with $x_t^t > 0$ that we seek. If not, there is an $\epsilon_1 > 0$ such that (6-33) can be strengthened to

$$\sum_{j \neq t} g_j p_j^k + (g_t - \epsilon_1) p_t^k \geq g_0 \qquad \text{for all } k \in P. \tag{6-34}$$

The t-term of the left side is zero when $p_t^k = 0$, and (6-33) must be a strict inequality for each of the finitely many k with $p_t^k > 0$. Similarly, it must be true that

$$\sum_{j \neq t} g_j d_j^l + g_t d_t^l \geq 0 \qquad \text{for all } l \in D. \tag{6-35}$$

Otherwise for $\hat{\mathbf{x}} \in E$, $\hat{\mathbf{x}} + \mathbf{d}^l$, which is by definition within $[F]$, would contradict validity of $(\mathbf{g}, g_0)$ via $\mathbf{g}(\hat{\mathbf{x}} + \mathbf{d}^l) = \mathbf{g}\hat{\mathbf{x}} + \mathbf{g}\mathbf{d}^l < g_0$. If there is any l for which (6-35) is an equality and $d_t^l > 0$, $\mathbf{x}^t \leftarrow \hat{\mathbf{x}} + \mathbf{d}^l$ with $\hat{\mathbf{x}} \in E$ gives the needed point of E having $x_t^t > 0$. If not, there is an $\epsilon_2 > 0$ such that (6-35) may be tightened to

$$\sum_{j \neq t} g_j d_j^l + (g_t - \epsilon_2) d_t^l \geq 0 \qquad \text{for all } l \in D. \tag{6-36}$$

The t-term is zero when $d_t^l = 0$, and (6-35) must be a strict inequality at each of the finitely many l with $d_t^l > 0$.

Now pick any $\mathbf{x} \in [F]$. Replacing ϵ_1 in (6-34) and ϵ_2 in (6-36) by $\epsilon \leftarrow \min\{\epsilon_1, \epsilon_2\}$, and taking a weighted sum of the resulting systems using the nonnegative λ_k and μ_l, of $\mathbf{x}$'s representation (6-32), we have

$$\sum_{j \neq t} g_j \left[\sum_{k \in P} \lambda_k p_j^k + \sum_{l \in D} \mu_l d_j^l \right] + (g_t - \epsilon) \left[\sum_{k \in P} \lambda_k p_t^k + \sum_{l \in D} \mu_l d_t^l \right] \geq g_0. \tag{6-37}$$

But this is a contradiction of $(\mathbf{g}, g_0)$'s minimality since we have shown $(\mathbf{g}, g_0)$ can be strengthened for all $\mathbf{x} \in [F] \supseteq F$ by the now valid inequality (6-37). It follows that there is an $\mathbf{x}^t \in E$ with $x_t^t > 0$, and the forward direction of the proof is complete.

For the converse, pick a valid inequality $(\mathbf{g}, g_0)$ that supports at $\hat{\mathbf{x}} > \mathbf{0}$ with $\hat{\mathbf{x}} \in E \triangleq \{\mathbf{x} \in [F]: \mathbf{g}\mathbf{x} = g_0\}$. Since $\hat{\mathbf{x}} \in [F]$, there must exist finitely many $\mathbf{x}^1, \mathbf{x}^2, \ldots, \mathbf{x}^q \in F$ and $\lambda_1, \lambda_2, \ldots, \lambda_q > 0$ with $\Sigma_{i=1}^q \lambda_i = 1$ such that

$$\hat{\mathbf{x}} = \sum_{i=1}^{q} \lambda_i \mathbf{x}^i. \tag{6-38}$$

Now, validity of $(\mathbf{g}, g_0)$ assumes $\mathbf{g}\mathbf{x}^i \geq g_0$ for each such i. In fact, $\mathbf{g}\mathbf{x}^i = g_0$ for each i, for if not, $\mathbf{g}\hat{\mathbf{x}} = \mathbf{g}\, [\Sigma_{i=1}^q \lambda_i x^i] = \Sigma_{i=1}^q \lambda_i(\mathbf{g}\mathbf{x}^i) > g_0$, which contradicts $\hat{\mathbf{x}} \in E$. It follows that each $\mathbf{x}^i$ of (6-38) belongs to E. But since $\hat{\mathbf{x}} > \mathbf{0}$, (6-38) also implies there that for $t = 1, 2, \ldots, n$ there is at least one $i = 1, 2, \ldots, q$ for which $x_t^i > 0$. Thus, for each $t = 1, \ldots, n$ there is an

$\mathbf{x}^i \in F$ with $\mathbf{g}\mathbf{x}^i = g_0$ and $x_t^i > 0$. Clearly, no coefficient of $(\mathbf{g}, g_0)$ could be reduced while keeping validity for all such $\mathbf{x}^i$. That is, $(\mathbf{g}, g_0)$ is minimal. ■

We have already noted that the nonnegativities need special treatment in the theory of minimal inequalities. But it also true that they are explicity present in our crudest statements of the problem (P). Thus, we shall term *trivial* any inequalities $(\mathbf{g}, g_0)$ *equivalent* to a nonnegative sum $\Sigma_{j=1}^n \lambda_j x_j \geq 0$ of the nonnegativities in the sense that $\{\mathbf{x} \in [F]\colon \mathbf{g}\mathbf{x} = g_0\} = \{\mathbf{x} \in [F]\colon \Sigma_{j=1}^n \lambda_j x_j = 0\}$. If valid inequalities are not trivial, they are *nontrivial.*

Theorem 6.8. Minimality of Nontrivial Facetial Inequalities. *Let F be as above with polyhedral convex hull [F]. Also, assume F satisfies the full-positivity assumption (6-31). Then every nontrivial facetial inequality* $(\mathbf{g}, g_0)$ *of F is a minimal inequality of F.*

Proof. All facetial inequalities are certainly supporting. Thus, the theorem will follow from Theorem 6.7, if each nontrivial facetial inequality $(\mathbf{g}, g_0)$ holds as an equality at some strictly positive $\mathbf{x} \in [F]$.

First consider facetial inequalities that are not proper, i.e., those where $\mathbf{g}\mathbf{x} = g_0$ for all $\mathbf{x} \in [F]$. Under our full positivity assumption, there exist $\mathbf{x}^1, \mathbf{x}^2, \ldots, \mathbf{x}^n \in F$ such that $x_t^t > 0$ for $t = 1, 2, \ldots, n$. Noting $F \subseteq \{\mathbf{x} \in \mathbb{R}^n\colon \mathbf{x} \geq \mathbf{0}\}$, it follows that $(\Sigma_{t=1}^n \mathbf{x}^t)/n$ yields the required positive point of $[F]$.

Now let $(\mathbf{g}, g_0)$ be a proper facetial inequality of F. If there are $\mathbf{x}^1, \mathbf{x}^2, \ldots, \mathbf{x}^n$ in $\{\mathbf{x} \in [F]\colon \mathbf{g}\mathbf{x} = g_0\}$ and $x_t^t > 0$ for all $t = 1, 2, \ldots, n$, the above argument would provide the needed $\mathbf{x} > \mathbf{0}$. If not there is some t such that

$$\{\mathbf{x} \in [F]\colon \mathbf{g}\mathbf{x} = g_0\} \subseteq \{\mathbf{x} \in [F]\colon x_t = 0\} \tag{6-39}$$

But the two sets of (6-39) must also have equal dimension because for $(\mathbf{g}, g_0)$ proper and facetial, $\dim([F]) - 1 = \dim\{\mathbf{x} \in [F]\colon \mathbf{g}\mathbf{x} = g_0\} \leq \dim\{\mathbf{x} \in [F]\colon x_t = 0\} \leq \dim([F]) - 1$. It follows that the two sets, which intersect $[F]$ in a face of the same dimension and which are contained as in (6-39), must be identical. That is, if a strictly positive $\mathbf{x}$ cannot be found in $\{\mathbf{x} \in [F]\colon \mathbf{g}\mathbf{x} = g_0\}$, the facetial inequality $(\mathbf{g}, g_0)$ is trivial. ■

6.3.2 Complexity of Minimal and Facetial Inequalities

Most of the rest of this chapter is devoted to three related characterizations that are comprehensive in the sense that they can construct all minimal valid inequalities. Results presented are at least concise and elegant, and may yet prove the basis of practical algorithms for some *NP*-Hard problems. It seems appropriate to precede presentation of those characterizations, however, by a summary of recent results showing why it is most unlikely any construction could be even nondeterministic polynomial-time in the worst case. That is, it is most unlikely that all potentially required valid inequalities for *NP*-Hard discrete optimization problems can even be verified in polynomial-time.

To discuss these complexity considerations, we follow the recent work of Papadimitriou and Yannakakis (1982) in introducing a new complexity class. Recall from Chapter 2 that class *NP* is the collection of string sets that can be recognized in nondeterministic polynomial-time, or informally, the set of *yes-no* problem instances for which *yes* cases have a polynomial-time proof that might be guessed. Class *CoNP* is the set of complements of sets in *NP*, i.e., the set of problem instances for which the correct answer is *no* (assuming there is no serious difficulty in detecting inputs that are not well-formed problem instances.)

The new complexity class D^p consists of problems (string sets) that are the intersection of one problem in *NP* and one problem in *CoNP*. That is, they are problems that are *yes* only if two properties hold. The first must always have a guessable short proof, and the second must be the negation of a property admitting such a proof. Note the problem's being in D^p says only that it is formed as the intersection of an instance in a member of *NP* and another in a member of *CoNP*. That is qualitatively quite different from saying the problem is itself a member of both *NP* and *CoNP*, i.e., that it belongs to $NP \cap CoNP$.

Our interest here in the class D^p arises from the fact that facetial inequality recognition for most discrete optimization problems belongs to D^p.

Theorem 6.9. Facetial Inequalities in $\mathbf{D^p}$. *Let $\mathscr{F} \triangleq \{F^i\}$ be a family of feasible optimization instances with $\mathscr{F} \in NP$ and each F^i full dimension in $\mathbb{R}^{n_i}$ with polyhedral $[F^i]$. Then $R(\mathscr{F}) \in D^p$ where $R(\mathscr{F})$* is the set of questions of the form: is *$(\mathbf{g}^i, g_0^i)$ a facetial inequality for F^i?*

Proof. Every instance of $R(\mathscr{F})$ consists of a constraint specification F^i and a proposed inequality $(\mathbf{g}^i, g_0^i)$. If $(\mathbf{g}^i, g_0^i)$ is facetial for F^i it must satisfy two properties. First, there must exist $(n_i - 1)$ affinely independent $\mathbf{x} \in F^i \cap \{\mathbf{x} \in \mathbb{R}^{n_i}: \mathbf{g}^i\mathbf{x} = g_0^i\}$. In addition, it must be valid, i.e., there can be no $\hat{\mathbf{x}} \in F^i$ such that $\mathbf{g}^i\hat{\mathbf{x}} < g_0^i$.

When $\mathscr{F} \in NP$ and $[F^i]$ is polyhedral all needed $\mathbf{x}$ with $\mathbf{g}^i\mathbf{x} = g_0^i$ can be polynomially constructed from guessed representations of extreme points and extreme directions. Thus, their existence is a question in *NP*. Similarly, nonexistence of $\hat{\mathbf{x}}$ with $\mathbf{g}^i\hat{\mathbf{x}} < g_0^i$ is the negative of the question: does there exist $\hat{\mathbf{x}} \in F^i$ with $\mathbf{g}^i\hat{\mathbf{x}} < g_0^i$? Under hypotheses of the theorem, the latter $\hat{\mathbf{x}}$ is also derivable from extreme points and extreme directions, so its existence too belongs to *NP*. Thus, its negation is an element of *CoNP*. This completes the proof. ■

In complexity theory, problems (*R*) are termed *complete* for a class when (*R*) belongs to the class and every member of that class polynomially reduces to (*R*). Readers are by now familiar with the infamous *NP*-Complete family to which all members of *NP* reduce. Likewise, there are D^p-Complete problems to which all members of D^p reduce. We first show that exact optimization forms of most *NP*-Hard problems belong to D^p.

Theorem 6.10. Exact Optimization in $\mathbf{D^p}$. *Let $\mathscr{C}^{\leq} \triangleq \{F^i, \mathbf{c}^i, v^i\}$ be a family of threshold optimization problems of the form "Does there exist $\boldsymbol{x} \in F^i$ such that $\mathbf{c}^i\boldsymbol{x} \leq v^i$?", where $\mathbf{c}^i\boldsymbol{x}$ is integer-valued for all $\boldsymbol{x} \in F^i$. Then if $\mathscr{C}^{\leq} \in NP$, the corresponding $\mathscr{C}^{=} \in D^p$ where $\mathscr{C}^{=}$ is the family of exact optimization value problems, "Is $v^i = inf\{\mathbf{c}^i\boldsymbol{x}: \boldsymbol{x} \in F^i\}$?"*

Proof. Affirmative instances in $\mathscr{C}^{=}$ over integer-values and objective functions $\mathbf{c}^i\mathbf{x}$ are those for which (i) there is $\mathbf{x} \in F^i$ with $\mathbf{c}^i\mathbf{x} \leq v^i$, and (ii) there is no $\mathbf{x} \in F^i$ with $\mathbf{c}^i\mathbf{x} \leq v^i - 1$. When the corresponding $\mathscr{C}^{\leq}$ is in *NP* question (i) and the negative of question (ii) are. This establishes $\mathscr{C}^{=} \in D^p$. ■

The problem of interest in this chapter is whether recognition of facetial inequalities for *NP*-Hard optimization problems is D^p-complete. We have already shown such questions usually belong to D^p and that the related exact optimization questions are also in D^p.

General results are not available, but Papadimitriou and Yannakakis (1982) do establish that many exact optimization and some facet recognition problems are D^p-Complete.

Theorem 6.11. $\mathbf{D^p}$-Complete Facet Recognitions. *Let (Clique), (Vertex packing), (Set packing), (Vertex covering) and (Set covering) be as defined in Section 2.3. Then the question of recognizing facetial inequalities for the polytope of feasible solutions to these problems is D^p-Complete.*

Proof. The proof proceeds by showing exact optimization forms of these problems are D^p-Complete and reduce to facetial inequality recognition. See Papadimitriou and Yannakakis (1982) for details. ■

It is easy to see D^p contains both *NP* and *CoNP;* we need only make one of the two conditions for membership trivial. What if any D^p-Complete problem belonged to *NP*? This would imply all members of D^p do, including all members of *CoNP*. That is, if any D^p-Complete problem belongs to *NP* then $NP = CoNP = D^p$. We explained in Section 2.2 why it is most unlikely that even $NP = CoNP$. Thus, we must strongly doubt that any D^p-Complete problem belongs to *NP*.

When facetial inequality proving is D^p-Complete, it follows that facetial inequality proving probably does not belong to *NP*. This means no construction guaranteed to generate all facetial inequalities is likely to be polynomial-time in the worst case. Unless $NP = CoNP = D^p$, there exist facetial inequalities for some discrete optimization problems that simply cannot be validly derived in polynomial-time, even if we allow for inspired guessing in the derivations.

6.4 Disjunctive Characterizations

Having established the notions of valid, minimal, supporting and facetial inequalities, we are now ready to develop some schema by which—at least in principle—all needed valid inequalities can be obtained (although probably not polynomially). One, based on a *disjunctive* description of the feasible space, F, is the object of this section. Sections 6.5 and 6.6 address two others. As usual, we assume throughout that F has a polyhedral convex hull $[F]$.

6.4.1 *The Disjunctive Model*

In general development so far we have remained rather vague about the form of the constraints defining $F = \{\mathbf{x} \in S\colon \mathbf{A}\mathbf{x} \geq \mathbf{b}\}$. Table 6.1 and Theorem 6.2 showed that numerous specific forms yield polyhedral $[F]$.

Balas (1975b) was apparently the first[3] to see that one useful way to describe a discrete feasible space is to write it as a *finite disjunction* of linear programming spaces (F_5 and F_6 of Table 6.1). In mathematical logic, two statements, α that says $\mathbf{x} \in S_1$ and β that says $\mathbf{x} \in S_2$, can be linked in either

- *conjunction* (denoted $\alpha \wedge \beta$) that asserts both α and β must be *true*, i.e., $\mathbf{x} \in S_1 \cap S_2$; or
- *disjunction* (denoted $\alpha \vee \beta$) that asserts either α or β or both must be *true*, i.e., $\mathbf{x} \in S_1 \cup S_2$;

The disjunctive model of discrete problems writes

$$F = \bigcup_{h \in H} \{\mathbf{x} \geq \mathbf{0}\colon \mathbf{A}^h\mathbf{x} \geq \mathbf{b}^h\}, \qquad (6\text{-}40)$$

[3] Section 6.4.5 shows, however, the closely related ideas—some of them Balas'—that predated Balas' original description of the disjunctive approach.

where H is a finite[4] index set and the $\{\mathbf{x} \geq \mathbf{0}\colon \mathbf{A}^h\mathbf{x} \geq \mathbf{b}^h\}$ are finitely many linear programming spaces to which an $\mathbf{x} \in F$ might belong. Some examples will illustrate.

Example 6.3. 0–1 Mixed Integer Programming. One general form that can easily be placed in disjunctive format (6-40) is the mixed 0–1 program,

$$\begin{aligned} \min \quad & \mathbf{cx} \\ \text{s.t.} \quad & \mathbf{Ax} \geq \mathbf{b} \\ & \mathbf{x} \in S = \{\mathbf{x} \geq \mathbf{0}\colon x_j = 0 \text{ or } 1 \text{ for all } j \in I\}. \end{aligned}$$

The problem can be written in terms of conjunctions and disjunctions

$$\begin{aligned} \min \quad & \mathbf{cx} \\ \text{s.t.} \quad & \mathbf{Ax} \geq \mathbf{b} \\ & \mathbf{x} \geq \mathbf{0} \\ & x_j \leq 1 \qquad \text{for all } j \in I \\ & \mathbf{x} \in \bigcap_{j \in I} (\{\mathbf{x}\colon -x_j \geq 0\} \cup \{\mathbf{x}\colon x_j \geq 1\}). \end{aligned} \tag{6-41}$$

Letting, say, I_h^+ be the subset of I containing subscripts with $x_j = 1$ in any of the feasible choices for $\{x_j\colon j \in I\}$, yields the purely disjunctive form

$$\begin{aligned} \min \quad & \mathbf{cx} \\ \text{s.t.} \quad & \boldsymbol{x} \in \bigcup_{h \in H} \{\mathbf{x} \geq \mathbf{0}\colon \mathbf{Ax} \geq \mathbf{b},\ x_j \leq 1 \text{ for all } j \in I, \\ & \qquad x_j \geq 1 \text{ for } j \in I_h^+,\ -x_j \geq 0 \text{ for } j \in I \backslash I_h^+\}, \end{aligned} \tag{6-42}$$

where H enumerates the $|H| = 2^{|I|}$ combinations. ■

Example 6.4. Multiple Choice Constraints. Multiple choice constraints state that exactly one of several 0-1 x_j must equal 1. A general problem format is

$$\begin{aligned} \min \quad & \mathbf{cx} \\ \text{s.t.} \quad & \mathbf{Ax} \geq \mathbf{b} \\ & \mathbf{x} \geq \mathbf{0} \\ & \mathbf{x} \in \bigcap_{k \in K} \left\{\text{integer } \mathbf{x}\colon \sum_{j \in K_k} x_j = 1\right\} \end{aligned} \tag{6-43}$$

[4] Many results of this section are easily extended to infinite H, but useful cases are almost always finite.

where each K_k is a subset of variable indices and K indexes such subsets. Any set {integer $\mathbf{x}$: $\Sigma_{j \in K_k} x_j = 1$} can be expressed

$$\bigcup_{l \in K_k} \left\{ \mathbf{x}: x_l \geq 1, -\sum_{j \in K_k} x_j \geq -1 \right\}. \tag{6-44}$$

Thus, the disjunctive constraint systems $\mathbf{A}^h\mathbf{x} \geq \mathbf{b}^h$, $\mathbf{x} \geq \mathbf{0}$ consist of $\mathbf{A}\mathbf{x} \geq \mathbf{b}$ and one element from each disjunction (6-44). Letting H enumerate the $|H| = \Pi_{k \in K}|K_k|$, such combinations completes the disjunctive form (6-40). ■

Example 6.5. Task Scheduling. Task scheduling problems are concerned with a finite set $T \triangleq \{T_1, T_2, \ldots, T_k\}$ of tasks that must be scheduled on one or more processors or machines. A set $\prec \subseteq T \times T$ specifies required *precedences.* For example, if $(T_i, T_j) \in \prec$, then task T_i must be completed before task T_j can begin. Without loss of generality we may take $\prec$ to be closed under transitivity, i.e., $(T_i, T_j) \in \prec$ and $(T_j, T_k) \in \prec$ imply $(T_i, T_k) \in \prec$. Letting t_i denote the completion time of task T_i, and τ_i the time it will require on the processor, such precedence constraints can be expressed $t_j \geq t_i + \tau_j$.

The combinatorial nature of task scheduling problems arises in resolving the remaining sequencing decisions. In the one machine case, for example, the issue is whether $t_j - t_i \geq \tau_j$ or $t_i - t_j \geq \tau_i$ for all $(T_i, T_j) \in D \triangleq \{(T_i, T_j)$: $(T_i, T_j) \notin \prec$ and $(T_j, T_i) \notin \prec\}$; two tasks cannot be simultaneously processed. Given weights c_j, we can determine a schedule of minimum weighted completion time by solving

$$\min \quad \sum_{T_j \in T} c_j t_j$$

$$\text{s.t.} \quad t_j - t_i \geq \tau_j \quad \text{for all } (T_i, T_j) \in \prec \tag{6-45}$$

$$t_j \geq \tau_j \quad \text{for all } T_j \in T \tag{6-46}$$

$$\mathbf{t} \in \bigcap_{(T_i, T_j) \in D} (\{\mathbf{t}: t_j - t_i \geq \tau_j\} \cup \{\mathbf{t}: t_i - t_j \geq \tau_i\}) \tag{6-47}$$

As with the two previous examples, the disjunctive form (6-40) derives from including in each system $\mathbf{A}^h\mathbf{x} \geq \mathbf{b}^h$ all main constraints of (6-45)–(6-46) and one of the two choices from each of the $|D|$ pairs in (6-47). Here $|H| = 2^{|D|}$. ■

6.4.2 *The Disjunctive Principle*

The advantage of arranging discrete solution spaces like those of Examples 6.3 through 6.5 in disjunctive format is that we can now describe a general cutting plane form valid for any system (6-40).

Certainly if $\mathbf{A}^h\mathbf{x} \geq \boldsymbol{b}^h$, then $(\mathbf{u}^h\mathbf{A}^h)\mathbf{x} \geq \mathbf{u}^h\mathbf{b}^h$ for any $\mathbf{u}^h \geq \mathbf{0}$. Such a collapsing of each $h \in H$ to a single constraint, followed by selection of the worst case among $h \in H$ at each cutting plane coefficient position, yields the following valid inequalities:

$$\sum_{j=1}^{n} \max_{h \in H} \{(\mathbf{u}^h\mathbf{A}^h)_j\}\, x_j \geq \min_{h \in H} \{\mathbf{u}^h\mathbf{b}^h\} \tag{6-48}$$

Here all vectors $\mathbf{u}^1, \mathbf{u}^2, \ldots, \mathbf{u}^{|H|} \geq \mathbf{0}$ and $(\mathbf{u}^h\mathbf{A}^h)_j$ denotes the *jth* component of $(\mathbf{u}^h\mathbf{A}^h)$.

Theorem 6.12. Validity of the Disjunctive Cut Principle. *Let $F \subset \{\boldsymbol{x} \in \mathbb{R}^n: \boldsymbol{x} \geq \boldsymbol{0}\}$ be a feasible set of a finite disjunctive form (6-40). Then for any $\boldsymbol{u}^1, \boldsymbol{u}^2, \ldots, \boldsymbol{u}^{|H|} \geq \boldsymbol{0}$, inequality (6-48) is satisfied by all $\boldsymbol{x} \in F$.*

Proof. If $\hat{\mathbf{x}} \in F$, there must be at least one $\hat{\mathbf{h}} \in H$ such that $\mathbf{A}^{\hat{h}}\hat{\mathbf{x}} \geq \mathbf{b}^{\hat{h}}$. Thus, for any $\mathbf{u}^1, \mathbf{u}^2, \ldots, \mathbf{u}^{|H|} \geq \mathbf{0}$, $(\mathbf{u}^{\hat{h}}\mathbf{A}^{\hat{h}})\hat{\mathbf{x}} \geq \mathbf{u}^{\hat{h}}\mathbf{b}^{\hat{h}}$. It follows—noting $\hat{\mathbf{x}} \geq \mathbf{0}$—that the inequality (6-48) derived from those multipliers, is also satisfied by $\hat{\mathbf{x}}$. Each coefficient $\max_{h \in H}\{(\mathbf{u}^h\mathbf{A}^h)_j\}$ is at least $(\mathbf{u}^{\hat{h}}\mathbf{A}^{\hat{h}})_j$, and the right-hand-side $\min_{h \in H}\{\mathbf{u}^h\mathbf{b}^h\}$ is at most $\mathbf{u}^{\hat{h}}\mathbf{b}^{\hat{h}}$. ■

Example 6.6. Disjunctive Cutting Plane for Task Scheduling. Consider a very simple version of the one processor task scheduling model formulated in Example 6.5. Let $T = \{T_1, T_2\}$ and $\prec = \varnothing$. The implied disjunctive feasible space for completion times t_1 and t_2 is

$$F = \left\{(t_1, t_2): \begin{matrix} t_1 \geq \tau_1, t_2 \geq \tau_1, \\ t_2 - t_1 \geq \tau_2 \end{matrix}\right\} \cup \left\{(t_1, t_2): \begin{matrix} t_1 \geq \tau_1, t_2 \geq \tau_2 \\ t_1 - t_2 \geq \tau_1 \end{matrix}\right\}. \tag{6-49}$$

Figure 6.6 depicts a case with $\tau_1 > \tau_2$. Any (t_1, t_2) in one of the two shaded areas is feasible. In this case the two areas have no common points although such is not required in (6-40).

Any nonnegative choice of $\mathbf{u}^1 = (u_1^1, u_2^1, u_3^1)$ and $\mathbf{u}^2 = (u_1^2, u_2^2, u_3^2)$ to multiply the three constraints in the $|H| = 2$ parts of (6-49) produces a valid inequality. For example $\mathbf{u}^1 = \mathbf{u}^2 = (1, 1, 1)$ gives

$$\max\{1 + 0 - 1, 1 + 0 + 1\}t_1 + \max\{0 + 1 + 1, 0 + 1 - 1\}t_2 \\ \geq \min\{\tau_1 + \tau_2 + \tau_2, \tau_1 + \tau_2 + \tau_1\}$$

or, recalling $\tau_1 > \tau_2$, $2t_1 + 2t_2 \geq \tau_1 + 2\tau_2$. A better choice is $\mathbf{u}^1 = (\tau_1 + \tau_2, 0, \tau_2)$, $\mathbf{u}^2 = (0, \tau_1 + \tau_2, \tau_1)$, which yield the facetial inequality $\tau_1 t_1 + \tau_2 t_2 \geq (\tau_1)^2 + \tau_1\tau_2 + (\tau_2)^2$. ■

Example 6.6 suggests that a clever choice of multipliers $\{\mathbf{u}^h: h \in H\}$ can produce quite strong valid inequalities for disjunctive forms (6-40). In fact, one of the cutting planes we obtained was facetial.

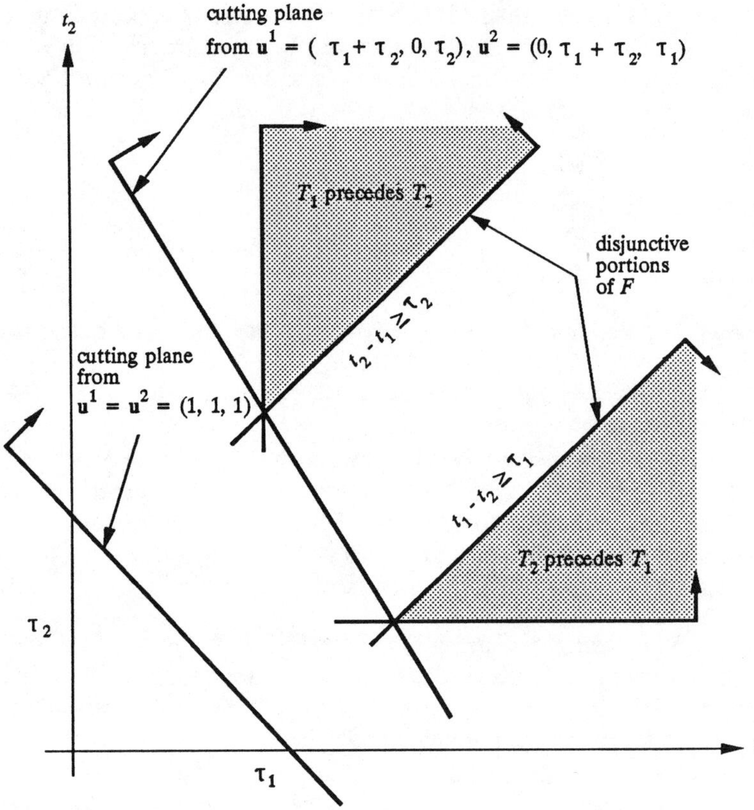

Fig. 6.6. Disjunctive cutting Example 6.6.

Just how rich is the family of cutting planes that can be obtained via the disjunctive cut principle? Jeroslow (1979a) demonstrated that it is very rich indeed. In fact, we show in the next theorem his proofs that the family often includes all the minimal inequalities—the only ones we need in the light of Theorems 6.7 and 6.8.

Theorem 6.13. Sufficiency of the Disjunctive Principle. *Let F be the feasible set of a finite disjunctive form (6-40). Then if for each $h \in H$*

(i) $\{\boldsymbol{x} \geq \boldsymbol{0}: \boldsymbol{A}^h\boldsymbol{x} \geq \boldsymbol{b}^h\}$ *is nonempty; or*

(ii) there is a $\boldsymbol{d}^h$ *such that* $\{\boldsymbol{x} \geq \boldsymbol{0}: \boldsymbol{A}^h\boldsymbol{x} \geq \boldsymbol{d}^h\}$ *is nonempty and bounded; or*

(iii) there is an $i_h \in H$ *with* $\boldsymbol{A}^{i_h} = \boldsymbol{A}^h$ and $\{\boldsymbol{x} \geq \boldsymbol{0}: \boldsymbol{A}^{i_h}\boldsymbol{x} \geq \boldsymbol{b}^{i_h}\}$ *is nonempty*

every minimal valid inequality for F can be computed via the disjunctive principle (6-48) by an appropriate choice of nonnegative $\mathbf{u}^1, \mathbf{u}^2, \ldots, \mathbf{u}^{|H|}$.

Proof. Pick any minimal valid inequality $(\mathbf{g}, g_0)$ for F (if none exist there is nothing to prove), and consider the dual pair of linear programs

$$(P_h) \quad \begin{array}{ll} \min & \mathbf{gx} \\ \text{s.t.} & \mathbf{A}^h\mathbf{x} \geq \mathbf{b}^h \\ & \mathbf{x} \geq \mathbf{0} \end{array} \qquad \text{and} \qquad (D_h) \quad \begin{array}{ll} \max & \mathbf{u}^h\mathbf{b}^h \\ \text{s.t.} & \mathbf{u}^h\mathbf{A}^h \leq \mathbf{g} \\ & \mathbf{u}^h \geq \mathbf{0} \end{array}$$

If condition (i) holds for, say, $h = \bar{h}$, $(P_{\bar{h}})$ is feasible, and since each feasible point belongs to F, validity of $(\mathbf{g}, g_0)$ assures $v(P_{\bar{h}}) \geq g_0$. But then it follows from linear programming duality that the corresponding dual $(D_{\bar{h}})$ is feasible and has optimal $\mathbf{u}^{\bar{h}}$ such that $\mathbf{u}^{\bar{h}}\mathbf{b}^{\bar{h}} = v(D_{\bar{h}}) = v(P_{\bar{h}}) \geq g_0$.

If condition (i) does not hold for $h = \hat{h}$, but condition (ii) does, then $(D_{\hat{h}})$ is still feasible. This is true because for every $\mathbf{g}$, $\mathbf{gx}$ achieves a finite minimum over the nonempty and bounded set $\{\mathbf{x} \geq \mathbf{0}\colon \mathbf{A}^{\hat{h}}\mathbf{x} \geq \mathbf{d}^{\hat{h}}\}$. Moreover, since (i) fails and $(D_{\hat{h}})$ is feasible, $v(D_{\hat{h}}) = +\infty$. In particular, we can find feasible $\mathbf{u}^{\hat{h}}$ with $\mathbf{u}^{\hat{h}}\mathbf{b}^{\hat{h}} \geq g_0$.

If condition (i) fails but (iii) holds for some $h = \tilde{h}$, the situation is similar. Since $(P_{i_{\tilde{h}}})$ is feasible and bounded in solution value by g_0, the dual $(D_{\tilde{h}})$ is feasible; it has the same constraints as $(D_{i_{\tilde{h}}})$. But then when (i) fails, $v(D_{\tilde{h}}) = +\infty$ and feasible $\mathbf{u}^{\tilde{h}}$ can be found with $\mathbf{u}^{\tilde{h}}\mathbf{b}^{\tilde{h}}$ arbitrarily large.

From the above we can conclude that when F satisfies the hypotheses of the theorem, there is for each $h \in H$ a $\mathbf{u}^h \geq \mathbf{0}$ such that $\mathbf{u}^h\mathbf{A}^h \leq \mathbf{g}$ and $\mathbf{u}^h\mathbf{b}^h \geq g_0$. Those $\mathbf{u}^h$ necessarily compute $\mathbf{g}$ and g_0 via (6-48), for if not, $\max_{h \in H}\{(\mathbf{u}^h\mathbf{A}^h)_j\} < g_j$ for some j or $\min_{h \in H}\{\mathbf{u}^h\mathbf{b}^h\} > g_0$. Either would contradict minimality of $(\mathbf{g}, g_0)$. ■

Disjunctions (6-40) do not necessarily have the property that all minimal valid inequalities can be obtained from the disjunctive cut principle. Still, Theorem 6.13, shows that if each system $\mathbf{A}^h\mathbf{x} \geq \mathbf{b}^h$, $\mathbf{x} \geq \mathbf{0}$ is feasible, or can be made feasible and bounded by changing the right-hand-side, or has the same $\mathbf{A}^h$ matrix as an element of the disjunction that is feasible, then all minimal inequalities (and thus, by Theorem 6.8, all nontrivial facetial inequalities for fully positive cases) can be obtained from (6-48). Each of our Examples 6.3–6.5 can be seen to fulfill at least one of these conditions. For example, we could write the disjunctions for the general mixed 0–1 Example 6.3 by placing two constraints $x_j \geq \lambda_j^h$ and $x_j \leq \mu_j^h$ in $\mathbf{A}^h$ for each $j \in I$. Then the lower limit, λ_j^h, and the upper limit, μ_j^h, would vary with h, but all $\mathbf{A}^h$ would be identical. Condition (iii) applies if there are feasible solutions to any part of the disjunction.

To be sure, it is a long step from knowing nonnegative $\mathbf{u}^1, \mathbf{u}^2, \ldots, \mathbf{u}^{|H|}$ exist to showing how to compute needed ones. The proof used $(\mathbf{g}, g_0)$ to show its constructability. Still, the proof of Theorem 6.13 suggests an approach that is at least theoretically available. To see it, assume $\hat{\mathbf{x}} \geq \mathbf{0}$ is an optimal solution for some relaxation of our main problem (P), and that $\hat{\mathbf{x}}$ must be cut off at Step 3 if Relaxation Algorithm 6A is to continue. Then consider the linear program

$$D(\hat{\mathbf{x}}) \qquad \begin{array}{ll} \min & \hat{\mathbf{x}}\mathbf{g} - g_0 \\ \text{s.t.} & \left.\begin{array}{r} \mathbf{u}^h\mathbf{A}^h \leq \mathbf{g} \\ \mathbf{u}^h\mathbf{b}^h \geq g_0 \\ \mathbf{u}^h \geq \mathbf{0} \end{array}\right\} \quad \text{for all } h \in H \\ & \mathbf{1} \geq \mathbf{g} \geq -\mathbf{1} \\ & 1 \geq g_0 \geq -1. \end{array}$$

Note that $\hat{\mathbf{x}}$ is constant in $D(\hat{\mathbf{x}})$; variables are $\mathbf{u}^1, \mathbf{u}^2, \ldots, \mathbf{u}^{|H|}$, $\mathbf{g}$ and g_0. Under conditions of Theorem 6.13, every $(\mathbf{g}, g_0)$ that might cut off $\hat{\mathbf{x}}$ is a feasible solution. An optimum will make $(\mathbf{g}\hat{\mathbf{x}} - g_0)$ as negative as possible, i.e., maximize the cutoff. The last constraint bounds the problem's solution to g_j's in $[-1, +1]$. Without bounds a $(\mathbf{g}\hat{\mathbf{x}} - g_0) < 0$ can be made arbitrarily negative by rescaling $\mathbf{g}$ and g_0.

In the disjunctive setting our goal is to solve discrete optimization problems

$$(P_D) \qquad \begin{array}{ll} \min & \mathbf{c}\mathbf{x} \\ \text{s.t.} & \mathbf{x} \in F \triangleq \bigcup_{h \in H} \{\mathbf{x} \geq \mathbf{0} \colon \mathbf{A}^h\mathbf{x} \geq \mathbf{b}^h\}. \end{array}$$

It is not hard to show that $v(P_D)$ equals the optimal v in the following linear program (sometimes called the *disjunctive dual*), at least under conditions of Theorem 6.13.

$$(D_D) \qquad \begin{array}{ll} \max & v \\ \text{s.t.} & \left.\begin{array}{r} \mathbf{u}^h\mathbf{A}^h \leq \mathbf{c} \\ \mathbf{u}^h\mathbf{b}^h \geq v \\ \mathbf{u}^h \geq \mathbf{0} \end{array}\right\} \quad \text{for all } h \in H \end{array}$$

We need only search for the strongest right-hand-side of an inequality $\mathbf{c}\mathbf{x} \geq v$. Since (D_D) is actually better constrained than any $D(\mathbf{x})$, we see that it would seem ludicrous to actually base an algorithm on repeated solution of problems $D(\mathbf{x})$. Accordingly, we shall not even state such a version of

Algorithm 6A for the general disjunctive case. For completeness, however, we note that one can easily be presented that solves (P) in a finite (but enormous) number of steps.

6.4.3 Cutting Planes for Local Disjunctions

For the reasons just outlined, it seems impractical to take on directly the task of finding strong cutting planes for disjunctive feasible spaces in form (6-40). Instead, we would expect to operate relaxation Algorithm 6A in a more incremental, local manner. Each relaxation that fails to yield an optimum for (P_D) would be seen to induce a local disjunction from which we can deduce a valid inequality to add in forming the next relaxation. Such a procedure is formally stated as follows:

Algorithm 6C. Local Disjunctive Cutting. Let (P_D) be as above with $v(P_D) > -\infty$.

STEP 0: *Initial Relaxation.* Set $k \leftarrow 0$ and pick as (P_D^0) any relaxation of (P_D) in the form

$$\begin{array}{ll} \min & \mathbf{cx} \\ \text{s.t.} & \mathbf{x} \in T, \end{array}$$

where T contains the feasible space, F, of (P_D) and $v(P_D^0) > -\infty$.

STEP 1: *Relaxation.* Attempt to solve (P_D^k). If $v(P_D^k) = +\infty$, stop; (P_D^k) and thus (P_D) is infeasible. Otherwise, let $\mathbf{x}^k$ be an optimal solution and go to Step 2.

STEP 2: *Optimality Check.* If $\mathbf{x}^k \in F$, stop; $\mathbf{x}^k$ solves (P_D) because it solves (P_D^k) and is feasible in (P_D). Otherwise, proceed to Step 3.

STEP 3: *Cutting.* Induce from (P_D^k) and $\mathbf{x}^k$ a relaxed disjunctive feasible space

$$L^k \triangleq \bigcup_{h \in H^k} \{\mathbf{x} \geq \mathbf{0} \colon \mathbf{Q}^h\mathbf{x} \geq \mathbf{q}^h\},$$

where $F \subseteq L^k$ and $\mathbf{x}^k \notin L^k$. Then apply the disjunctive cutting principle (6-48) to find a valid inequality $(\mathbf{g}^{k+1}, g_0^{k+1})$ for L^k (and thus F) such that $\mathbf{g}^{k+1}\mathbf{x}^k < g_0^{k+1}$. Define (P_D^{k+1}) as (P_D^k) with the added constraint $\mathbf{g}^{k+1}\mathbf{x} \geq g_0^{k+1}$, replace $k \leftarrow k+1$, and return to Step 1. ■

Example 6.7. Gomory's Cutting Planes As Local Disjunctive Cutting. One way to illustrate the notion of Algorithm 6C is to see that relaxation algorithms based on the Gomory cutting planes of Section 6.2 can be viewed as a local disjunctive cutting scheme. In particular, consider the weaker mixed integer cutting plane (6-19).

When a mixed integer program relaxation (MIP^k) has failed to yield a relaxation solution $\mathbf{x}^k$ with x_r^k integer on some integer-restricted basic variable x_r, we argued in the proof of Theorem 6.6 that every $\mathbf{x}$ feasible in (MIP) belongs to

$$L^k = \left\{\mathbf{x} \geq \mathbf{0}\colon \sum_{j\in N_I} \phi(\bar{a}_{rj}^k)x_j + \sum_{j\in N_C} \bar{a}_{rj}^k x_j \geq \phi(x_r^k)\right\}$$
$$\cup \left\{\mathbf{x} \geq \mathbf{0}\colon \sum_{j\in N_I} (-\phi(\bar{a}_{rj}^k))x_j + \sum_{j\in N_C} (-\bar{a}_{rj}^k)x_j \geq 1 - \phi(x_r^k)\right\} \quad (6\text{-}50)$$

Observe that the relaxation solution $x_j^k = 0$ for nonbasic x_j with $j \in N_I \cup N_C$. Thus, $\mathbf{x}^k \notin L^k$, and we can employ L^k to generate a valid inequality in Algorithm 6C.

The fractional part $1 > \phi(x_r^k) > 0$, so we may choose $u^1 = 1$, $u^2 = (\phi(x_r^k)/(1 - \phi(x_r^k)))$ as the two nonnegative multipliers in the corresponding cutting plane construction (6-48). The resulting valid inequality is

$$\sum_{j\in N_I} \max\left\{\phi\left((\bar{a}_{rj}^k), \frac{\phi(x_r^k)}{1-\phi(x_r^k)}(-\phi(\bar{a}_{rj}^k))\right)\right\} x_j$$
$$+ \sum_{j\in N_C} \max\left\{\bar{a}_{rj}^k, \frac{\phi(x_r^k)}{1-\phi(x_r^k)}(-\bar{a}_{rj}^k)\right\} x_j$$
$$\geq \min\left\{\phi(x_r^k), \frac{\phi(x_r^k)}{1-\phi(x_r^k)}(1 - \phi(x_r^k))\right\}. \quad (6\text{-}51)$$

Both terms of the right-hand-side minimization equal $\phi(x_r^k)$. We need only sort out the at most one positive choice among each max pair on the left to see that (6-51) is exactly Gomory's mixed integer cutting plane (6-19). ■

If the concept of Algorithm 6C is to be valuable, we would, of course, like to see how it could do more than rederive Gomory's algorithms. The key step in Example 6.7 was the observation that the solution with all nonbasic $x_j = 0$ can be cut off with the disjunction (6-50) over sets defined by single constraints. One direction for broadening the application of Algorithm 6C is to consider a general situation where the relaxation solution $x^k = 0$ and is to be cut off at Step 3 via the *one-row local disjunction.*

$$L \triangleq \sum_{h\in H} \{\mathbf{x} \geq \mathbf{0}\colon \mathbf{q}^h\mathbf{x} \geq q_0^h\} \quad (6\text{-}52)$$

We assume $q_0^h > 0$ for all $h \in H$ so that $\mathbf{x}^k = \mathbf{0}$ does not belong to L. The next theorem provides strong valid inequalities for this case.

Theorem 6.14. Facetial Inequalities for One-Row Disjunctions. *Let L be as in (6-52) with H finite and $q_0^h > 0$, $\{\mathbf{x} \geq \mathbf{0}: \mathbf{q}^h\mathbf{x} \geq q_0^h\}$ nonempty for all $h \in H$. Then if*

$$f_j \longleftarrow \max_{h \in H} \{q_j^h/q_0^h\} \tag{6-53}$$

$$\gamma_h \longleftarrow \min_{\{j:\, q_j^h > 0\}} \{f_j/(q_j^h/q_0^h)\} \tag{6-54}$$

$$g_j \longleftarrow \max_{h \in H} \{(q_j^h/q_0^h)\gamma_h\}, \tag{6-55}$$

the inequalities $\Sigma_j f_j x_j \geq 1$ and $\Sigma_j g_j x_j \geq 1$ are valid for L and render infeasible $\mathbf{x} = \mathbf{0}$. Furthermore, the second is a strengthening of the first because $f_j \geq g_j$ for each component j, and $\Sigma_j g_j x_j \geq 1$ is facetial for L.

Proof. Let $(\mathbf{f}, 1)$ and $(\mathbf{g}, 1)$ denote inequalities constructed in (6-53) and (6-55) respectively. Certainly, if valid, they render infeasible $\mathbf{x} = \mathbf{0}$. To see $(\mathbf{f}, 1)$ is valid for L with $q_0^h > 0$, we need only choose $u^h \leftarrow 1/q_0^h$ for all $h \in H$ and apply the disjunctive cutting principle (6-48). Furthermore, under our hypotheses that each $\{\mathbf{x} \geq \mathbf{0}: \mathbf{q}^h\mathbf{x} \geq q_0^h\}$ is nonempty and $q_0^h > 0$, there must be for each $h \in H$ at least one j with $q_j^h > 0$. But then $f_j \geq q_j^h/q_0^h > 0$, and $\gamma_h \geq 1$. It follows that choosing $u^h \leftarrow \gamma_h/q_0^h$ will demonstrate validity of $(\mathbf{g}, 1)$ via the disjunctive principle. Also, $g_j \leq f_j$ whenever $g_j \leq 0$. But if $g_j > 0$, construction (6-54) assures $g_j = f_j$.

It remains to show that $(\mathbf{g}, 1)$ is facetial for L. Define $J^+ \triangleq \{j: g_j > 0\}$, $J^- \triangleq \{j: g_j < 0\}$ and $J^0 = \{j: g_j = 0$ but $x_j > 0$ for some $\mathbf{x} \in L\}$. We shall proceed by exhibiting $|J^+| + |J^-| + |J^0|$ linearly independent vectors $\mathbf{x}^j$ with $\mathbf{g}\mathbf{x}^j = 1$ and $\mathbf{x}^j \in [L]$, the convex hull of L. This will establish that $\dim\{\mathbf{x} \in [L]: \mathbf{g}\mathbf{x} = 1\} \geq \dim([L]) - 1$ as required for a facetial inequality.

First pick $j \in J^+$ and let $\mathbf{x}^j \leftarrow (0, 0, 1/g_j, 0, \ldots, 0)$, with only its jth component nonzero. Certainly $\mathbf{g}\mathbf{x}^j = 1$. But we have already observed that $g_j = f_j$ for $j \in J^+$. Thus there is an h that defines f_j in (6-53) having $g_j = f_j = q_j^h/q_0^h$. But then $q_j^h(1/g_j) = q_0^h$ and our $\mathbf{x}^j \in \{\mathbf{x} \geq \mathbf{0}: \mathbf{q}^h\mathbf{x} \geq q_0^h\}$, a subset of L and $[L]$.

For $j \in J^0 \cup J^-$, choose h so that

$$g_j = (q_j^h/q_0^h)\gamma_h \tag{6-56}$$

in construction (6-55) and pick i such that $\gamma_h = g_i/(q_i^h/q_0^h)$ in the corresponding (6-54). By definition $q_i^h > 0$, and thus $i \in J^+$ and

$$g_i = f_i = (q_i^h/q_0^h)\gamma_h. \tag{6-57}$$

Now for $j \in J^0$, $g_j = 0$ and so $q_j^h = 0$; all other factors of (6-56) are positive. But then $\mathbf{q}^h\mathbf{d}^j \triangleq 0$ and $\mathbf{g}\mathbf{d}^j = 0$, for $\mathbf{d}^j =$ the jth unit vector. That is, each $\mathbf{d}^j$

is a direction of the set $\{\mathbf{x} \geq \mathbf{0}\colon \mathbf{q}^h\mathbf{x} \geq q_0^h,\ \mathbf{g}\mathbf{x} = 1\}$. Similarly for $j \in J^-$, choosing $\mathbf{d}^j = (0, \ldots, 0, 1/g_i, 0, \ldots, -1/g_j, 0, \ldots, 0)$, with nonzero components only at $d_i^j = 1/g_i$ and $d_j^j = -1/g_j$, also gives a direction of $\{\mathbf{x} \geq \mathbf{0}\colon \mathbf{q}^h\mathbf{x} \geq q_0^h,\ \mathbf{g}\mathbf{x} = 1\}$. This follows because (6-56) and (6-57) yield $\mathbf{q}^h\mathbf{d}^j = q_i^h(q_0^h/\gamma_h q_i^h) + q_j^h(-q_0^h/\gamma_h q_j^h) = 0$ and $\mathbf{g}\mathbf{d}^j$ is obviously 0. Thus, since all the defined $\mathbf{d}^j$ are directions of $\{\mathbf{x} \geq \mathbf{0}\colon \mathbf{q}^h\mathbf{x} \geq q_0^h,\ \mathbf{g}\mathbf{x} = 1\}$, they are also directions of its superset $\{\mathbf{x} \in [L]\colon \mathbf{g}\mathbf{x} = 1\}$. For each $j \in J^0 \cup J^-$ we have only to choose $\mathbf{x}^j \leftarrow \mathbf{x}^{\hat{j}} + \mathbf{d}^j$, with $\hat{j} \in J^+$ and $\mathbf{d}^j$ the above direction, to complete our list of linearly independent vectors in $\{\mathbf{x} \in [L]\colon \mathbf{g}\mathbf{x} = 1\}$. ■

Example 6.8. Cutting Planes from One-Row Disjunctions. Suppose a problem (P) is subject to the condition that either $x_{11} \geq 1$ or $x_{12} \geq 1$. Also, suppose the previous relaxation at Step 1 of Algorithm 6C had solution $x_{11}^k = \frac{3}{5}$ and $x_{12}^k = \frac{2}{5}$. If that relaxation was a linear program solved by some simplex procedure, the variables x_{11} and x_{12} can be represented in terms of nonbasic variables (see (6-8)) as say

$$x_{11} = \frac{3}{5} - \left(-\frac{4}{5}x_1 - \frac{3}{5}x_9 + 5x_{13} + x_{25}\right)$$

$$x_{12} = \frac{2}{5} - \left(-\frac{1}{5}x_1 - \frac{1}{5}x_9 + x_{13} + 2x_{24} + 8x_{25}\right).$$

The implied disjunction violated by $x_1 = x_9 = x_{13} = x_{24} = x_{25} = 0$ is

$$\begin{aligned}\{(x_1, x_9, x_{13}, x_{24}, x_{25}) \in L\} &= \left\{\mathbf{x} \geq \mathbf{0}\colon \frac{4}{5}x_1 + \frac{3}{5}x_9 - 5x_{13} - x_{25} \geq 1 - \frac{3}{5} = \frac{2}{5}\right\} \\ &\quad \cup \left\{\mathbf{x} \geq \mathbf{0}\colon \frac{1}{5}x_1 + \frac{1}{5}x_9 - x_{13} - 2x_{24} - 8x_{25} \geq 1 - \frac{2}{5} = \frac{3}{5}\right\}.\end{aligned}$$

Applying construction (6-53), $f_1 = \max\{\frac{4}{5}/\frac{2}{5}, \frac{1}{5}/\frac{3}{5}\}$, $f_9 = \max\{\frac{3}{5}/\frac{2}{5}, \frac{1}{5}/\frac{3}{5}\}$, $f_{13} = \max\{-5/\frac{2}{5}, -1/\frac{3}{5}\}$, $f_{24} = \max\{0/\frac{2}{5}, -2/\frac{3}{5}\}$, $f_{25} = \max\{-1/\frac{2}{5}, -8/\frac{3}{5}\}$, so that the inequality $\mathbf{f}\mathbf{x} \geq 1$ is

$$2x_1 + \frac{3}{2}x_9 - \frac{5}{3}x_{13} + 0x_{24} - \frac{5}{2}x_{25} \geq 1$$

Notice that—in contrast to the Gomory cutting planes of Section 6.2—inequality (6-53) can have negative components. Improvement to $\Sigma_j g_j x_j \geq 1$ of (6-55) will, in general, strengthen some negative coefficients. Here $\gamma_1 = \min\{2/2, \frac{3}{2}/\frac{3}{2}\} = 1$, $\gamma_2 = \min\{2/\frac{1}{3}, \frac{3}{2}/\frac{1}{3}\} = \frac{9}{2}$. The resulting $\mathbf{g}$-inequality is

$$2x_1 + \frac{3}{2}x_9 - \frac{15}{2}x_{13} + 0x_{24} - \frac{5}{2}x_{25} \geq 1$$

All nonnegative g_j are unchanged, but one negative case improves from $f_{13} = -\frac{5}{3}$ to $g_{13} = -\frac{15}{2}$. ■

6.4.4 *Facial Cases*

The simple logic of Algorithm 6C is appealing as a scheme for using disjunctive cutting planes in relaxation based cutting algorithms. Theorem 6.14 presented a tool for dealing with available local disjunctions, but nothing has been said about how one finds such disjunctions, or whether there is any hope that Algorithm 6C converges to a solution for (P_D).

In general, such issues are quite involved. But Balas (1975b) showed that a wide class of special cases admit a simple approach. Each of the disjunctive models we developed in Section 6.4.1—and indeed most we encounter in discrete optimization—occurs naturally in a *conjunctive normal form*

$$(P_C) \qquad \begin{aligned} \min \quad & \mathbf{cx} \\ \text{s.t.} \quad & \mathbf{A}^0\mathbf{x} \geq \mathbf{b}^0 && \text{(6-58)} \\ & \mathbf{x} \geq \mathbf{0} && \text{(6-59)} \\ & \mathbf{x} \in \bigcap_{i \in I} \left(\bigcup_{h \in H_i} \{\mathbf{x}\colon \mathbf{q}^{ih}\mathbf{x} \geq q_0^{ih}\} \right). && \text{(6-60)} \end{aligned}$$

Here the set I lists a series of one-row disjunctions, at least one element of each of which must hold. For example, in mixed 0–1 programming, I indexes variables subject to 0–1 constraints and each H_i contains two elements corresponding to $x_i = 1$ and $x_i = 0$. The result is

$$\begin{aligned} \min \quad & \mathbf{cx} \\ \text{s.t.} \quad & \mathbf{Ax} \geq \mathbf{b} && \text{(6-61)} \\ & x_i \leq 1 \quad \text{for all } i \in I && \text{(6-62)} \\ & \mathbf{x} \geq \mathbf{0} && \text{(6-63)} \\ & \mathbf{x} \in \bigcap_{i \in I} \left(\{\mathbf{x}\colon x_i \geq 1\} \cup \{\mathbf{x}\colon -x_i \geq 0\} \right). && \text{(6-64)} \end{aligned}$$

Balas termed problems in form (P_C) *facial* if the feasible space for each defining inequality $\mathbf{q}^{ih}\mathbf{x} \geq q_0^{ih}$ intersects $\{\mathbf{x} \geq \mathbf{0}\colon \mathbf{A}^0\mathbf{x} \geq \mathbf{b}^0\}$ in a face of the latter polyhedron. That is, (P_C) is facial if for all $i \in I$, $h \in H_i$

$$\{\mathbf{x} \geq \mathbf{0}: \mathbf{A}^0\mathbf{x} \geq \mathbf{b}^0, \mathbf{q}^{ih}\mathbf{x} \geq q_0^{ih}\} = \{\mathbf{x} \geq \mathbf{0}: \mathbf{A}^0\mathbf{x} \geq \mathbf{b}^0, \mathbf{q}^{ih}\mathbf{x} = q_0^{ih}\}. \quad (6\text{-}65)$$

Many—probably most—discrete optimization problems have facial formulations. The above 0–1 mixed integer case is an obvious example. Thinking of (6-61) and (6-62) as the system $\mathbf{A}^0\mathbf{x} \geq \mathbf{b}^0$, it is clear that for each $i \in I$ $x_i \geq 1$ only if $x_i = 1$, and $-x_i \geq 0$ only if $-x_i = 0$. In a similar straightforward way the multiple choice constraint Example 6.4 can be seen to have a facial form.

Machine scheduling Example 6.5 illustrates a disjunctive formulation that is not facial. The natural format (6-45)–(6-47) certainly does conform to the conjunctive structure (P_C). But it is quite possible that some pair of tasks $(T_i, T_j) \in D$ will not be scheduled adjacently. In that case, both $t_j - t_i > \tau_j$ and $t_i - t_j > \tau_i$; the disjunction (6-47) does not conform to (6-65). It is interesting to note, however, that the machine scheduling problem can be formulated as a 0–1 mixed integer program. In that equivalent formulation it too would be facial.

In our treatment of facial disjunctive problems in the form (P_C), we shall always want to know the corresponding problem in *disjunctive normal form* (6-40) satisfies at least one of the sufficiency conditions of Theorem 6.13. That is, we wish to know all needed valid inequalities can be generated by the disjunctive cut principle (6-48). For this reason, we shall assume throughout this section that

$$F_0 = \{\mathbf{x} \geq \mathbf{0}: \mathbf{A}^0\mathbf{x} \geq \mathbf{b}^0\}$$

is a *nonempty, bounded* set. It follows that we can choose right-hand-sides for other constraints of (P_C) that satisfy condition (ii) of Theorem 6.13.

We have already noted several times that if any optimal solution exists for a problem (P_C), one will occur at an extreme point of the convex hull of

$$F = \left\{\mathbf{x} \in F_0: \mathbf{x} \in \bigcap_{i \in I} \left(\bigcup_{h \in H_i} \{\mathbf{x}: \mathbf{q}^{ih}\mathbf{x} \geq q_0^{ih}\} \right)\right\}$$

the set of solutions to (6-58)–(6-60). The observation from which all results in facial disjunctive programming derive is that all extreme points of $[F]$ will be among those of F_0 when F conforms to (6-65). Facial disjunctions (6-60) eliminate extreme points of F_0, but create no new ones. The next theorem shows some of the implied simplifications.

Theorem 6.15. Intersection Convex Hull for Facial Disjunctive Programs. *Let F_0 be as above, a nonempty, bounded polyhedral set, and*

$$F \triangleq F_0 \cap \left(\bigcap_{i \in I} \left(\bigcup_{h \in H_i} \{x\colon q^{ih}x \geq q_0^{ih}\} \right) \right) \tag{6-66}$$

with all inequalities $q^{ih}x \geq q_0^{ih}$ defining faces if F_0 as in (6-65). Also for any sequencing $i_1, i_2, \ldots, i_{|I|}$ of the elements of I, define

$$F_k \triangleq F_0 \cap \left(\bigcup_{h \in H_{i_1}} \{x\colon q^{i_1h}x \geq q_0^{i_1h}\} \right) \cap \left(\bigcup_{h \in H_{i_2}} \{x\colon q^{i_2h}x \geq q_0^{i_2h}\} \right) \cap \cdots \cdots \cap \left(\bigcup_{h \in H_{i_k}} \{x\colon q^{i_kh}x \geq q_0^{i_kh}\} \right) \tag{6-67}$$

with $F_{|I|} = F$. Then for $k = 1, 2, \ldots, |I|$,

$$[F_k] = \left[[F_{k-1}] \cap \left(\bigcup_{h \in H_{i_k}} \{x\colon q^{i_kh}x \geq q^{i_kh}\} \right) \right] \tag{6-68}$$

where $[\cdot]$ denotes the convex hull of the set $(\cdot)$.

Proof. We have assumed F_0 is polyhedral and bounded, and so all subsets $[F_k]$, are polyhedral and bounded. Thus each $[F_k]$ is merely the set of convex combinations $\{\Sigma_{l \in P_k} \lambda_l \mathbf{p}^l\colon \Sigma_{l \in P_k} \lambda_l = 1, \lambda_l \geq 0 \text{ for all } l \in P_k\}$, where each $\mathbf{p}^l$ is an extreme point of $[F_k]$. But we have already noted that the extreme points of any $[F_k]$ are a subset of those of F_0. Thus, if by (6-65) each $h \in H_{i_k}$ has $\{\mathbf{x} \in F_0\colon \mathbf{q}^{i_kh}\mathbf{x} \geq q_0^{i_kh}\} = \{\mathbf{x} \in F_0\colon \mathbf{q}^{i_kh}\mathbf{x} = q_0^{i_kh}\}$ certainly for subsets $[F_k]$, $\{\mathbf{x} \in [F_k]\colon \mathbf{q}^{i_kh}\mathbf{x} \geq q_0^{i_kh}\} = \{\mathbf{x} \in [F_k]\colon \mathbf{q}^{i_kh}\mathbf{x} = q_0^{i_kh}\}$ (although both could be empty).

Result (6-68) follows from the observations that extreme points of $[F_{k-1}]$ will be those of F_0 that conform to disjunctions $\cup_{h \in H_{i_l}} \{\mathbf{x}\colon \mathbf{q}^{i_lh}\mathbf{x} \geq q^{i_lh}\}$ for $l = 1, 2, \ldots, k-1$. Facial constraints $\mathbf{q}^{i_kh}\mathbf{x} \geq \mathbf{q}^{i_kh}$ will merely select from among those extreme points the ones needed to generate $[F_k]$. ■

At least in principle, Theorem 6.15 provides a recursive method for constructing convex hulls of solutions to bounded, facial disjunctive programs. We first find the convex hull of F_0 intersect one of the disjunctions, then intersect the result with another disjunction, etc. Each of the disjunctions we confront is over a relatively small set, H_i, instead of the usually enormous set H of the full disjunctive normal form (6-40). Two simple examples will illustrate operation of the theorem in the facial case, and its failure in a nonfacial case.

Example 6.9. Construction of Convex Hulls by Successive Intersection. Consider first the facial disjunctive constraints F

$$1 \geq x_1 \geq 0$$

$$1 \geq x_2 \geq 0$$

$$-2x_1 - 2x_2 \geq -3$$

$$\begin{aligned}(x_1, x_2) \in (\{(x_1, x_2)\colon x_1 \geq 1\} \cup \{(x_1, x_2)\colon -x_1 \geq 0\}) \\ \cap (\{(x_1, x_2)\colon x_1 - x_2 \geq 1\} \cup \{(x_1, x_2)\colon -x_1 + x_2 \geq 1\})\end{aligned} \tag{6-69}$$

Figure 6.7 illustrates. Set F_0 is shaded in 6.7a, and each inequality of the disjunction (6-69) is shown. Observe that each does—as required—intersect the F_0 space in a face. For $x_1 \geq 1$, the face is $\{(x_1, x_2)\colon x_1 = 1, 0 \leq x_2 \leq \frac{1}{2}\}$; for $-x_1 + x_2 \geq 1$ the face is the single point (0, 1).

The only points in F_0 feasible in both disjunctions are (0, 1) and (1, 0). Thus, $[F]$ is as in Figure 6.7c. But since the disjunction is facial we may construct $[F]$ by taking disjunctions in order. Let us take them in the order listed. Then Figure 6.7b shows the convex hull corresponding to $[F_1]$ and Figure 6.7c the result of intersecting it with the second disjunction. As required, $[F]$ results. Note that all extreme points of $[F_1]$ and $[F] = [F_2]$ are indeed extreme points of F_0.

Let us now replace (6-69) by the nonfacial disjunction

$$\begin{aligned}(x_1, x_2) \in (\{(x_1, x_2)\colon x_1 \geq 1\} \cup \{(x_1, x_2)\colon -x_1 \geq 0\}) \\ \cap (\{(x_1, x_2)\colon x_1 - x_2 \geq 1\} \cup \{(x_1, x_2)\colon -2x_1 + 2x_2 \geq 1\})\end{aligned} \tag{6-70}$$

Figure 6.8a illustrates that the last inequality fails the definition of facial because $\{\mathbf{x} \in F_0\colon -2x_1 + 2x_2 \geq 1\} \supset \{\mathbf{x} \in F_0\colon -2x_1 + 2x_2 = 1\}$. The remainder of Figure 6.8 shows the construction of Theorem 6.15 also fails. Successive intersection as in (6-67) yields Figure 6.8b, then Figure 6.8c. But $[F]$ is as in Figure 6.8d. ■

Theorem 6.15 describes an elegant construction for the convex hull of solutions to any (bounded) facial disjunctive constraint set, but the effort to actually carry it out would be enormous. Still, it suggests a less global approach that might yield a finitely convergent version of Algorithm 6C. Arranging disjunctions for $i \in I$ in some order $i_1, i_2, \ldots, i_{|I|}$, we could maintain bounded polyhedral approximations, say, $\tilde{F}_1, \tilde{F}_2, \ldots, \tilde{F}_{|I|}$, to the sets $[F_1], [F_2], \ldots, [F_{|I|}]$, respectively. When a current relaxation solution violates a disjunction, i_k, we try to cut it off by generating a valid inequality for $\tilde{F}_{k-1} \cap (\cup_{h \in H_{i_k}} \{\mathbf{x}\colon \mathbf{q}^{i_k h}\mathbf{x} \geq q_0\})$. If the disjunctive cut principle and faciality can be exploited to show such an inequality can be obtained

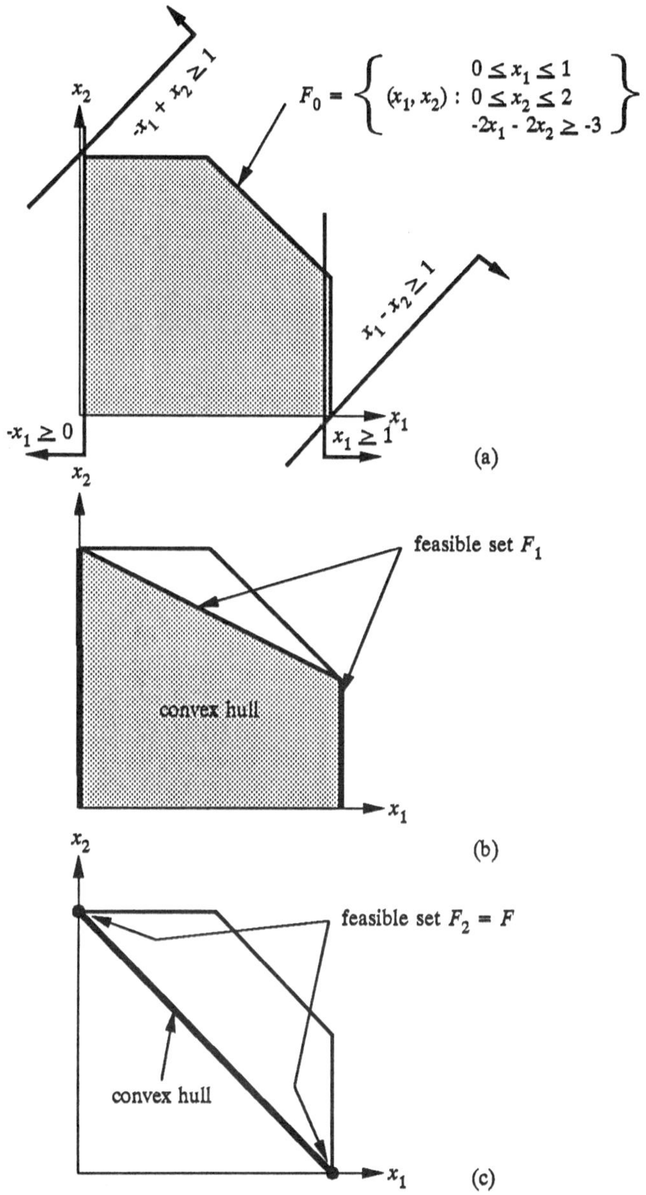

Fig. 6.7. Successive construction of the convex hull of solutions to the facial disjunctive program of Example 6.9; (a) original problem with disjunctive constraints, (b) after intersection with first disjunction, (c) after intersection with second disjunction and convex hull.

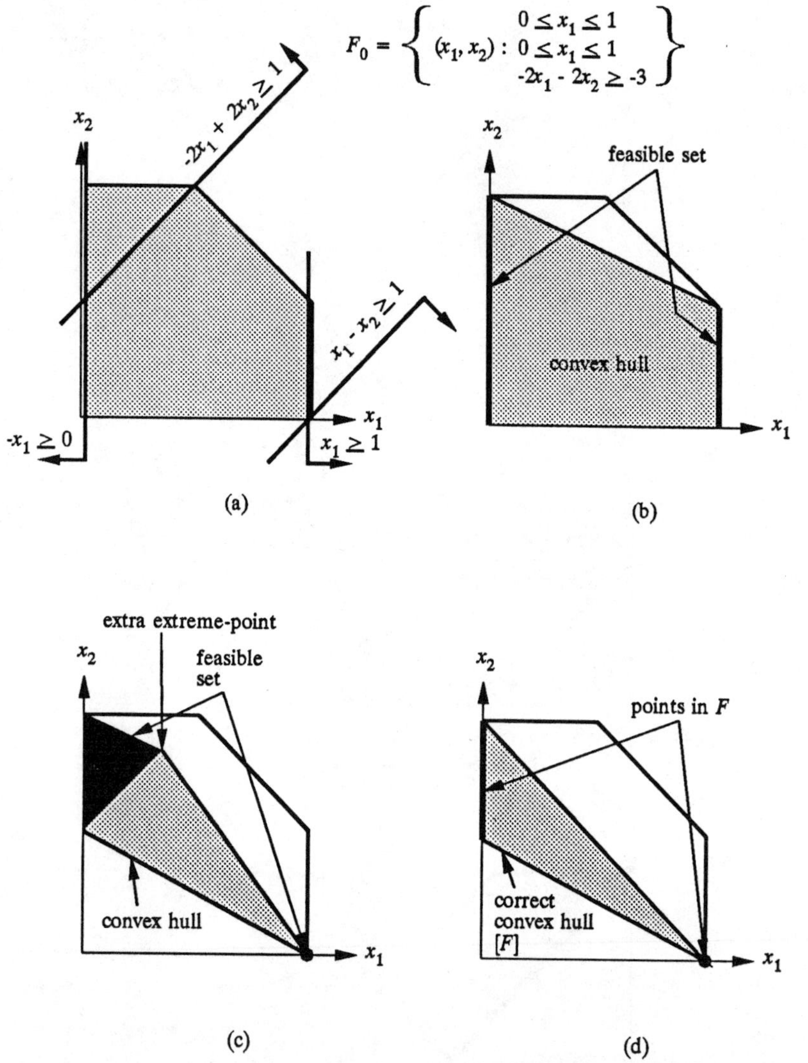

Fig. 6.8. Failure of successive construction of the convex hull of solutions to the nonfacial disjunctive program of Example 6.9; (a) original problem with disjunctive constraints, (b) after intersection with first disjunction, (c) after intersection with second disjunction, (d) correct convex hull.

from among a convenient finite set, the algorithm would provide a practical and convergent approach. We now state, illustrate and prove just such a procedure.

Algorithm 6D. Finitely Convergent Facial Disjunctive Cutting. Let F be as in (6-66), the feasible set for a bounded, facial disjunctive optimization problem (P_C). Also, assume $v(P_C) > -\infty$.

STEP 0: *Initialization.* Set $k \leftarrow 0$, and initialize the current disjunction sequence $i_1, i_2, \ldots, i_s$ as empty with length $s \leftarrow 0$.

STEP 1: *Relaxation.* Attempt to solve

$$(P^k) \qquad \begin{aligned} \min \quad & \mathbf{cx} \\ \text{s.t.} \quad & \mathbf{A}^s\mathbf{x} \geq \mathbf{b}^s \\ & \mathbf{x} \geq \mathbf{0}. \end{aligned}$$

If $v(P^k) = +\infty$, stop; (P^k) and thus, (P_C) is infeasible. Otherwise, let $\mathbf{x}^k$ be an extreme point optimal solution, and go to Step 2.

STEP 2: *Optimality Check.* If $\mathbf{x}^k \in F$, stop; $\mathbf{x}^k$ solves (P_C) because it solves the relaxation (P^k) and is feasible in (P_C). If not, proceed to Step 3.

STEP 3: *Cutting.* If any $i = i_1, i_2, \ldots, i_s$ in the current disjunction sequence has $\mathbf{x}^k \notin \cup_{h \in H_i} \{\mathbf{x}\colon \mathbf{q}^{ih}\mathbf{x} \geq q_0^{ih}\}$, let i_l be the last such i in the sequence. If not, extend $s \leftarrow s+1$, initialize $\mathbf{A}^s \leftarrow \mathbf{A}^{s-1}$ and $\mathbf{b}^s \leftarrow \mathbf{b}^{s-1}$, make i_l any violated disjunction index, and add $i_s \leftarrow i_l$ to the current disjunction sequence. Then choose as $(\mathbf{g}^{k+1}, g_0^{k+1})$ the $(\mathbf{g}, g_0)$ components of any extreme point having $\mathbf{g}^{k+1}\mathbf{x}^k < g_0^k$ of the set defined by

$$\left.\begin{aligned} \mathbf{u}^h\begin{pmatrix}\mathbf{q}^{i_lh} \\ \mathbf{A}^{l-1}\end{pmatrix} &\geq \mathbf{g} \\ \mathbf{u}^h\begin{pmatrix}q_0^{i_lh} \\ \mathbf{b}^{l-1}\end{pmatrix} &\geq g_0 \\ \mathbf{u}^h &\geq \mathbf{0} \end{aligned}\right\} \quad \text{for all } h \in H_{i_l} \tag{6-71}$$

$$-\mathbf{1} \leq \mathbf{g} \leq \mathbf{1} \tag{6-72}$$

$$-1 \leq g_0 \leq 1. \tag{6-73}$$

Then add $\mathbf{g}^{k+1}$ and g_0^{k+1} as the last rows of $\mathbf{A}^r$ and $\mathbf{b}^r$ respectively for $r = l, l+1, \ldots, s$. Finally, advance $k \leftarrow k+1$ and return to Step 1.

■

Example 6.10. Finite Facial Cutting. To illustrate Algorithm 6D, consider the 0–1 problem

$$\begin{aligned}
\min\quad & -3x_1 - x_2 \\
\text{s.t.}\quad & -8x_1 - 2x_2 \geq -9 \\
& -2x_1 + 2x_2 \geq -1 \\
& -x_1 \geq -1 \\
& -x_2 \geq -1 \\
& x_1, x_2 \geq 0
\end{aligned}$$

$$\begin{aligned}
(x_1, x_2) \in (&\{(x_1, x_2)\colon x_1 \geq 1\} \cup \{(x_1, x_2)\colon -x_1 \geq 0\}) \\
\cap (&\{(x_1, x_2)\colon x_2 \geq 1\} \cup \{x_1, x_2)\colon -x_2 \geq 0\})
\end{aligned}$$

Here $\mathbf{A}^0 = \begin{bmatrix} -8 & -2 \\ -2 & +2 \\ -1 & 0 \\ 0 & -1 \end{bmatrix}$ and $\mathbf{b}^0 = \begin{bmatrix} -9 \\ -1 \\ -1 \\ -1 \end{bmatrix}$. The $k = 0$ part of Figure 6.9a shows a first relaxation over $\mathbf{A}^0\mathbf{x} \geq \mathbf{b}^0$, $\mathbf{x} \geq \mathbf{0}$ yields the infeasible optimum $\mathbf{x}^0 = (\frac{7}{8}, 1)$.

The only violated disjunction is $x_1 \geq 1$ or $-x_1 \geq 0$. Thus, we make it the first element of the disjunction sequence via $s \leftarrow 1$, $i_1 \leftarrow 1$, $\mathbf{A}^1 \leftarrow \mathbf{A}^0$, $\mathbf{b}^1 \leftarrow \mathbf{b}^0$. Figure 6.9b shows systems $(\mathbf{A}^1, \mathbf{b}^1)$ as they evolve. A valid inequality is to be generated from an extreme point of the system for $(\mathbf{A}^0, \mathbf{b}^0)$, i.e.,

$$\begin{aligned}
u_0^1 - 8u_1^1 - 2u_2^1 - u_3^1 \qquad\qquad & \leq g_1 \\
-2u_1^1 + 2u_2^1 \qquad - u_4^1 & \leq g_2 \\
u_0^1 - 9u_1^1 - u_2^1 - u_3^1 - u_4^1 & \geq g_0 \\
-u_0^2 - 8u_1^2 - 2u_2^2 - u_3^2 \qquad\qquad & \leq g_1 \\
-2u_1^2 + 2u_2^2 \qquad - u_4^2 & \leq g_2 \\
-9u_1^2 - u_2^2 - u_3^2 - u_4^2 & \geq g_0 \\
u_0^1, u_1^1, u_2^1, u_3^1, u_4^1, u_0^2, u_1^2, u_2^2, u_3^2, u_4^2 & \geq 0 \\
-1 \leq g_0, g_1, g_2 & \leq 1.
\end{aligned}$$

We can obtain an extreme point by trying to cut off $\mathbf{x}^0 = (\frac{7}{8}, 1)$ as much as possible, i.e., minimizing $\frac{7}{8}g_1 + g_2 - g_0$ over the above constraints.

An optimum is $(g_0, \mathbf{g}, u_0^1, \mathbf{u}^1, u_0^2, \mathbf{u}^2) = (-1, -\frac{1}{2}, -1, \frac{7}{2}, \frac{1}{2}, 0, 0, 0, \frac{1}{2}, 0, 0, 0, 1)$ with the implied inequality $-\frac{1}{2}x_1 - x_2 \geq -1$ or integerizing, $(\mathbf{g}^1, g_0^1) = (-1, -2, -2)$. This inequality is added as row 5 of $(\mathbf{A}^1, \mathbf{b}^1)$ in the

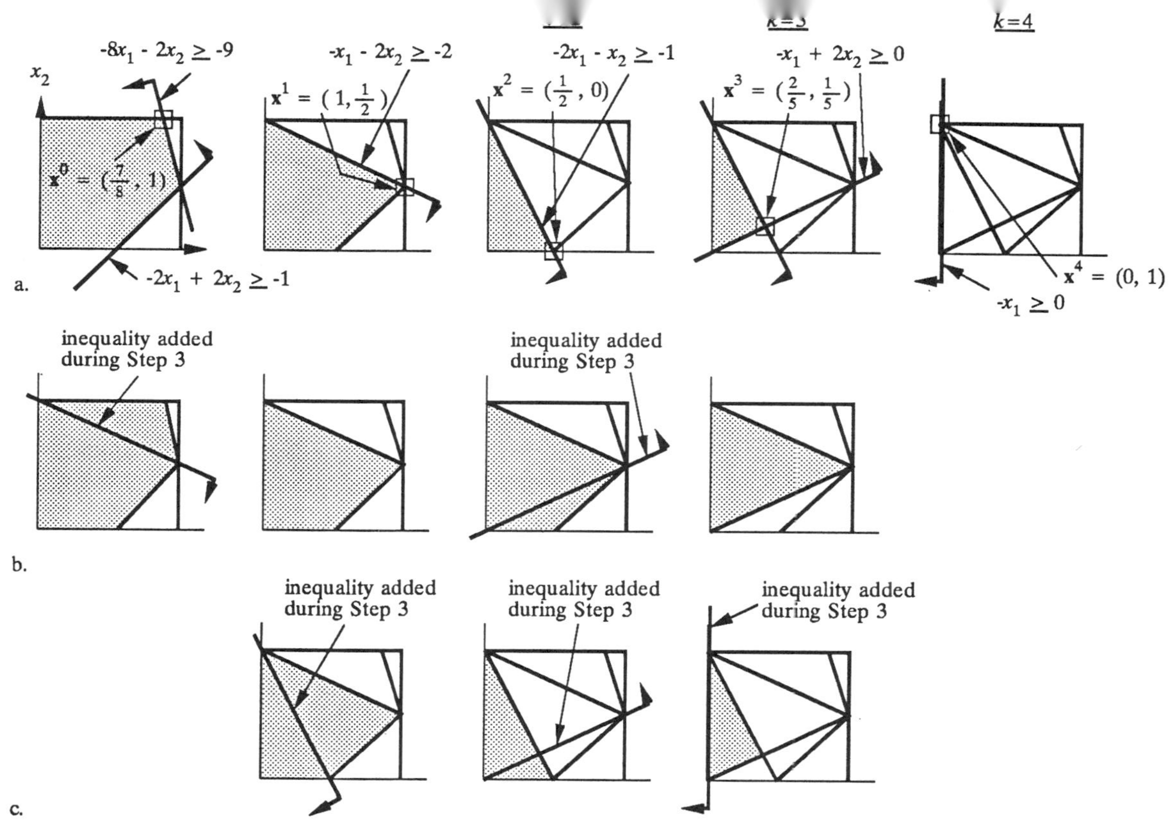

Fig. 6.9. Evolution of algorithm 6D on facial Example 6.10; (a) relaxation feasible spaces ($k = 0$ is $\mathbf{A}^0\mathbf{x} \geq \mathbf{b}^0$), (b) systems $\mathbf{A}^1\mathbf{x} \geq \mathbf{b}^1$ including inequalities from $x_1 \geq 1$ vs. $-x_1 \geq 0$, (c) systems $\mathbf{A}^2\mathbf{x} \geq \mathbf{b}^2$ including inequalities from $x_2 \geq 1$ vs. $-x_2 \geq 0$.

next portion of Figure 6.9. Since $s = 1$, it also forms the $k = 1$ relaxation of part (a).

From the (P^1) relaxation we obtain the infeasible optimum $\mathbf{x}^1 = (1, \frac{1}{2})$. Here only disjunction 2 is violated so we extend $s \leftarrow 2$, $i_2 \leftarrow 2$ and create the first system $(\mathbf{A}^2, \mathbf{b}^2)$ of Figure 6.9c from the current $(\mathbf{A}^1, \mathbf{b}^1)$. Minimizing $g_1 + \frac{1}{2}g_2 - g_0$ over the $(l - 1) = 1$ system (6-71)–(6-73) results in the extreme point $(g_0, \mathbf{g}, u_0^1, \mathbf{u}^1, u_0^2, \mathbf{u}^2) = (-\frac{1}{2}, -1, -\frac{1}{2}, \frac{3}{2}, 0, 0, 0, 0, 1, \frac{3}{2}, 0, 0, 0, 0, 0)$. The implied inequality $(\mathbf{g}^2, g_0^2) = (-2, -1, -1)$ is added to $\mathbf{A}^2\mathbf{x} \geq \mathbf{b}^2$.

Relaxation (P^2) solves at $\mathbf{x}^2 = (\frac{1}{2}, 0)$. Here the only violated disjunction is number 1. Computation over the present (5-77)–(5-79) system for $(\mathbf{A}^0, \mathbf{b}^0)$ produces the new constraint $(\mathbf{g}^3, g_0^3) = (-1, 2, 0)$. It is added to both elements of the current disjunctive sequence.

Since $s = 2$, the last provides relaxation (P^3). As indicated in Figure 6.9a, $\mathbf{x}^3 = (\frac{2}{5}, \frac{1}{5})$. This solution violates both disjunctions, but Algorithm 6D specifies that we must use the last in sequence to obtain a cutting plane. Doing so yields the inequality $-x_1 \geq 0$, and the next relaxation produces $\mathbf{x}^4 = (0, 1)$. Since the point is feasible for (P_C) and optimal for relaxation (P^4) it is optimal for (P_C). ■

Before proceeding to a convergence proof, let us remark on some critical aspects of the solution sequence in Example 6.10. First, note that systems $\mathbf{A}^0\mathbf{x} \geq \mathbf{b}^0$, $\mathbf{A}^1\mathbf{x} \geq \mathbf{b}^1$, etc. from which we generate cutting planes via (6-71)–(6-73) are relatively stable. The same $\mathbf{A}^0\mathbf{x} \geq \mathbf{b}^0$ would be used for all cuts from disjunction 1. Systems $\mathbf{A}^1\mathbf{x} \geq \mathbf{b}^1$ used for cuts on disjunction 2 had only 2 forms.

We shall see below that it is this partial constancy of source systems for new cutting planes that produces finite convergence of Algorithm 6D. But the fact that inequalities are not added to all source systems complicates cut generation. At $k = 3$, for example, the disjunction $x_1 \geq 1$ or $-x_1 \geq 0$ was violated by $\mathbf{x}^3 = (\frac{2}{5}, \frac{1}{5})$. But $\mathbf{x}^3$ could not have been cut off by any valid inequality for $(\mathbf{A}^0, \mathbf{b}^0)$. Point $(\frac{2}{5}, \frac{1}{5})$ belongs to the convex hull $[\{\mathbf{x}: \mathbf{A}^0\mathbf{x} \geq \mathbf{b}^0, \mathbf{x} \geq \mathbf{0}\} \cap (\{\mathbf{x}: x_1 \geq 1\} \cup \{\mathbf{x}: x_1 \geq 0\})]$. In general, the same difficulty could arise with any violated disjunction. But we shall see faciality assures that at least the last violated disjunction can yield a suitable inequality.

Theorem 6.16. Finite Convergence of Algorithm 6D. *Let F be as in (6-66) the feasible set of a bounded, facial disjunctive optimization problem (P_C) with $v(P_C) > -\infty$. Then Algorithm 6D stops after finitely many of its steps with either an $\mathbf{x}^k$ optimal in (P_C) or the correct conclusion that none exists.*

Proof. It is clear that if Algorithm 6D stops at either Step 1 or Step 2, it correctly concludes that (P_C) is infeasible, or solved by $\mathbf{x}^k$ respectively. We have only to show that the algorithm must eventually stop. Consider $i \in I$

in the order $i_1, i_2, \ldots, i_{|I|}$ that they enter the fixed disjunction sequence. Each time we generate a cutting plane from (6-71)–(6-73) with $l = 1$ the corresponding $\mathbf{A}^{l-1}$ and $\mathbf{b}^{l-1}$ are our original system, $\mathbf{A}^0\mathbf{x} \geq \mathbf{b}^0$. Also, all previous cutting planes generated from the same disjunction are part of the relaxation constraints $\mathbf{A}^s\mathbf{x} \geq \mathbf{b}^s$ satisfied by $\mathbf{x}^k$. Thus if any extreme point of (6-71)–(6-73) cuts off $\mathbf{x}^k$, it must be a new one each time. But there are only finitely many. Also, F_0 bounded implies (via Theorem 6.13 (ii)) that every inequality needed to define the convex hull of the disjunction

$$F_1 = \bigcup_{h \in H_{i_1}} \{\mathbf{x} \geq \mathbf{0}\colon \mathbf{A}^0\mathbf{x} \geq \mathbf{b}^0, \mathbf{q}^{i_1 h}\mathbf{x} \geq q_0^{i_1 h}\}$$

will be such an extreme point.

Let $F_2, \ldots, F_{|I|}$ be as in (6-67) of Theorem 6.15. By an identical argument to the above, we can see that any disjunction i_l can lead to only finitely many cutting planes from (6-71)–(6-73) for any fixed $\mathbf{A}^{l-1}$ and $\mathbf{b}^{l-1}$. But we have also proved that only finitely many $(\mathbf{A}^1, \mathbf{b}^1)$ can arise before $[F_1]$ is defined. It follows that each of those could produce only finitely many cutting planes before $[F_2]$ is defined. Proceeding inductively in this way, $[F_{|I|}]$, which under Theorem 6.15 is $[F]$, will eventually result if the algorithm does not stop. But then we know it will stop upon solving the next relaxation; every problem (P_C) achieves an optimum at any extreme point of $[F]$ if any exists.

Note, however, that the argument so far has assumed any infeasible $\mathbf{x}^k$ could be cut off as in Step 3. That is, we have assumed that when we do not stop, $\mathbf{x}^k$ does not belong to the convex hull

$$\left[\bigcup_{h \in H_{i_l}} \left\{\mathbf{x} \geq \mathbf{0}\colon \mathbf{A}^{l-1}\,\mathbf{x} \geq \mathbf{b}^{l-1}, \mathbf{q}^{i_l h}\,\mathbf{x} \geq q_0^{i_l h}\right\}\right], \tag{6-74}$$

the points satisfying all inequalities $(\mathbf{g}, g_0)$ that can result from (6-71)–(6-73).

Now since constraints $\mathbf{q}^{i_l h}\mathbf{x} \geq q_0^{i_l h}$ for $h \in H_{i_l}$ intersect $\{\mathbf{x} \geq \mathbf{0}\colon \mathbf{A}^0\mathbf{x} \geq \mathbf{b}^0\}$ and thus, the subset $\{\mathbf{x} \geq \mathbf{0}\colon \mathbf{A}^{l-1}\mathbf{x} \geq \mathbf{b}^{l-1}\}$ in a face, the convex hull (6-74) is the set generated by extreme points of faces $\{\mathbf{x} \geq \mathbf{0}\colon \mathbf{A}^{l-1}\mathbf{x} \geq \mathbf{b}^{l-1}, \mathbf{q}^{i_l h}\mathbf{x} \geq q_0^{i_l h}\}$. But every row of $\mathbf{A}^l\mathbf{x} \geq \mathbf{b}^l$ not in $\mathbf{A}^{l-1}\mathbf{x} \geq \mathbf{b}^{l-1}$ is a valid inequality for all $\mathbf{x}$ in (6-74). Thus, in particular all indicated extreme points satisfy $\mathbf{A}^l\mathbf{x} \geq \mathbf{b}^l$. However, the $\mathbf{q}^{i_l h}\mathbf{x} \geq q_0^{i_l h}$ also intersect $\tilde{F}_l \triangleq \{\mathbf{x} \geq \mathbf{0}\colon \mathbf{A}^l\mathbf{x} \geq \mathbf{b}^l\}$ in a face. It follows that every extreme point of (6-74) is an extreme point of $\tilde{F}_l$ satisfying at least one inequality $\mathbf{q}^{i_l h}\mathbf{x} \geq q_0^{i_l h}$, $h \in H_{i_l}$.

The proof will be complete if $\mathbf{x}^k$ is an extreme point of $\tilde{F}_l$ because it is not among those satisfying $\mathbf{q}^{i_l h}\mathbf{x} \geq q_0^{i_l h}$ for any $h \in H_{i_l}$, and the fact that it is an extreme point of $\tilde{F}_l$ assures it does not belong to the set (6-74) of convex combinations of such $\tilde{F}_l$ extreme points.

Define $\tilde{F}_r \triangleq \{\mathbf{x} \geq \mathbf{0}: \mathbf{A}^r \mathbf{x} \geq \mathbf{b}^r\}$. Algorithm 6D chooses $\mathbf{x}^k$ as an extreme point of $\tilde{F}_s$. If $l = s$, the proof is complete. If there is an r with $l < r \leq s$, we know $\mathbf{x}^k$ is feasible for disjunction i_r, i.e., there is an $h_r \in H_{i_r}$ with $\mathbf{q}^{i_r h_r}\mathbf{x}^k \geq q_0^{i_r h_r}$. If $\mathbf{x}^k$ is an extreme point of $\tilde{F}_r$, it is also an extreme point of the face $Q_r \triangleq \{\mathbf{x} \in \tilde{F}_r: \mathbf{q}^{i_r h_r}\mathbf{x} \geq q_0^{i_r h_r}\}$. But every inequality of $\mathbf{A}^r\mathbf{x} \geq \mathbf{b}^r$ not present in $\mathbf{A}^{r-1}\mathbf{x} \geq \mathbf{b}^{r-1}$ is a valid inequality for each $\mathbf{x} \in [\tilde{F}_{r-1} \cap (\cup_{h \in H_{i_r}} \{\mathbf{x}: \mathbf{q}^{i_r h}\mathbf{x} \geq q_0^{i_r h}\})]$. Thus, no point of the face $Q_{r-1} \triangleq \{\mathbf{x} \in \tilde{F}_{r-1}: \mathbf{q}^{i_r h_r}\mathbf{x} \geq q_0^{i_r h_r}\}$ could be cut off and $Q_r = Q_{r-1}$. But then $\mathbf{x}^k$, which is extreme in the face Q_{r-1}, must be an extreme point of $\tilde{F}_{r-1}$. Applying this argument inductively it follows that because $\mathbf{x}^k$ is an extreme point of $\tilde{F}_s$ it is one for $\tilde{F}_{s-1}, \tilde{F}_{s-2}, \ldots, \tilde{F}_{l+1}, \tilde{F}_l$. This completes the proof. ■

6.4.5 *Alternative Convexity/Intersection Conceptualization*

Throughout this section, we have followed the disjunctive model (6-40) that at least one of a set of inequality systems must hold. One row models as in (6-52) and Theorem 6.14 have the form

$$L \triangleq \bigcup_{h \in H} \{\mathbf{x} \geq \mathbf{0}: \mathbf{q}^h\mathbf{x} \geq q_0^h\}. \tag{6-75}$$

But the statement "every feasible $\mathbf{x}$ belongs to L" can be equivalently put "no feasible $\mathbf{x}$ lies in the interior of C," where

$$C \triangleq \bigcap_{h \in H} \{\mathbf{x} \geq \mathbf{0}: \mathbf{q}^h\mathbf{x} \geq q_0^h\} \tag{6-76}$$

Historically, cutting planes for logically constrained problems were first developed in this latter setting. In particular, if $q_0^h > 0$ for all $h \in H$, so that a current relaxation solution $\mathbf{x}^k = \mathbf{0}$ is to be cut off, then $\mathbf{q}^h\mathbf{x}^k < q_0^h$ for all $h \in H$ and $\mathbf{x}^k$ lies in the interior of the convex set C. A valid inequality $\hat{\mathbf{f}}\mathbf{x} \geq 1$ can be obtained by fitting $\hat{\mathbf{f}}$ so that it passes through the intersection of each coordinate direction with the boundary of C. That is, we may construct a valid inequality via

$$\hat{f}_j \longleftarrow 1/\sup\{\lambda: \lambda \mathbf{e}^j\mathbf{q}^h \leq q_0^h \text{ for all } h \in H\}, \tag{6-77}$$

where $\mathbf{e}^j$ is the *jth* unit vector and $1/\infty \triangleq 0$. The convex nature of such an interior-infeasible set C and the intersection construction of the implied cutting plane, give this theory the names *convexity cuts* and *intersection cuts.*

Example 6.11. Convexity Cuts of Example 6.1. Our very first cutting example illustrates the above notions. In Example 6.1 we chose as C

cylinder sets around an infeasible relaxation optimum, i.e., $C = \{\mathbf{x} \in \mathbb{R}^2: x_r \geq \lfloor x_r^k \rfloor, -x_r \geq -\lceil x_r^k \rceil\}$. Represent x_r^k in terms of an optimal relaxation basis as in Section 6.2 via $x_r = x_r^k - \Sigma_{j \in N} \bar{a}_{rj} x_j$. Then, $C = \{\mathbf{x}: \Sigma_{j \in N} \bar{a}_{rj} x_j \leq x_r^k - \lfloor x_r^k \rfloor$ and $-\Sigma_{j \in N} \bar{a}_{rj} x_j \leq \lceil x_r^k \rceil - x_r^k, x_j \geq 0$ for all $j \in N\}$. Coefficients $\hat{f}_j$ of (6-77) are easily obtained as $\hat{f}_j \leftarrow 1/\max\{(x_r^k - \lfloor x_r^k \rfloor)/\bar{a}_{rj}, (\lceil x_r^k \rceil - x_r^k)/(-\bar{a}_{rj})\}$.

For example, in relaxation (IP^0) of Example 6.1, basic variables x_1 and x_2 can be represented

$$x_1 = \tfrac{9}{50} - (-\tfrac{7}{50}x_3 - \tfrac{2}{25}x_4)$$

$$x_2 = \tfrac{13}{25} - (\tfrac{1}{25}x_3 - \tfrac{3}{25}x_4),$$

where x_3 and x_4 are surplus variables of the two main constraints. Since both x_1 and x_2 are fractional there are no interior solutions to either $C_1 = \{(x_3, x_4) \geq \mathbf{0}: -\tfrac{7}{50}x_3 - \tfrac{2}{25}x_4 \leq \tfrac{9}{50}, \tfrac{7}{50}x_3 + \tfrac{2}{25}x_4 \leq \tfrac{41}{50}\}$ or $C_2 = \{(x_3, x_4) \geq \mathbf{0}: \tfrac{1}{25}x_3 - \tfrac{3}{25}x_4 \leq \tfrac{13}{25}, -\tfrac{1}{25}x_3 + \tfrac{3}{25}x_4 \leq \tfrac{12}{25}\}$. The inequality, $(\hat{\mathbf{f}}, 1)$, generated from C_2 is $\tfrac{1}{13}x_3 + \tfrac{1}{4}x_4 \geq 1$. ■

It is not hard to see that both inequalities of Theorem 6.14 dominate the simple intersection cut (6-77). Recall that the cut of (6-46) was a strengthening of $(\mathbf{f}, 1)$ in (6-53), where $f_j \leftarrow \max_{h \in H}\{q_j^h/q_0^h\}$. For j with at least one $q_j^h > 0$, (6-77) is equivalent to $\hat{f}_j = 1/\min\{q_0^h/q_j^h: q_h^h > 0\}$, i.e., $\hat{f}_j = f_j$. But when $q_j^h < 0$ for all $h \in H$, $\hat{f}_j = 0$ while $f_j < 0$.

In spite of this dominance of the straightforward intersection cut, improvements can be made to recover results in Theorem 6.14. In fact, Glover (1975b) has shown how one can recover the full generality of the disjunctive cut principle and Theorem 6.13 via the conjunction of statements that polyhedral sets have no interior solutions. He terms the notion *polyhedral annexation* because new polyhedra are annexed to a current set C one by one. We conclude this section by proving and illustrating the key construction.

Theorem 6.17. Polyhedral Annexation. *For $i = 1,2$ let $C_i = \{\mathbf{x} \geq 0: \mathbf{q}_x^{ih} \leq q_0^{ih}$ for all $h \in H^i\}$ and assume no feasible solution $\mathbf{x}$ to a given problem is an interior point of either C_i. Then no feasible solution $\mathbf{x}$ is an interior point of*

$$\hat{C} \longleftarrow \{\mathbf{x} \geq 0: \mathbf{q}^{1h}\mathbf{x} \leq q_0^{1h} \text{ for all } h \in H_1 \backslash \{p\}, \quad (\lambda_h \mathbf{q}^{1p} + \mu_h \mathbf{q}^{2h})\mathbf{x} \leq (\lambda_h q_0^{1p} + \mu_h q_0^{2h}) \text{ for all } h \in H_2\}, \tag{6-78}$$

where p is any element of H_1 and $\{\lambda_h, \mu_h: h \in H_2\}$ are arbitrary nonnegative multipliers.

Proof. Fix $p \in H_1$, $\lambda_h \geq 0$, $\mu_h \geq 0$ for all $h \in H_2$, and assume, by contradiction, there is some feasible interior $\hat{\mathbf{x}} \in \hat{C}$. Then $\mathbf{q}^{1h}\hat{\mathbf{x}} < q_0^{1h}$ for all $h \in H_1 \backslash \{p\}$ and

$$\mathbf{q}^{1p}\hat{\mathbf{x}} \geq q_0^{1p} \tag{6-79}$$

or $\hat{\mathbf{x}}$ is interior for C_1. But $\hat{\mathbf{x}}$ interior in $\hat{C}$ also implies $(\lambda_h q^{1p} + \mu_h q^{2h})\hat{\mathbf{x}} < (\lambda_h q_0^{1p} + \mu_h q_0^{2h})$ for each $h \in H_2$. Rearranging and using (6-79), it follows that $\mu_h(\mathbf{q}^{2h}\hat{\mathbf{x}} - q_0^{2h}) < -\lambda_h(\mathbf{q}^{1p}\hat{\mathbf{x}} - q_0^{1p}) \leq 0$ for all $h \in H_2$. This can only be true if $\lambda_h > 0$ and $\mathbf{q}^{2h}\mathbf{x} < q_0^{2h}$ for all $h \in H_2$. But that would make $\hat{\mathbf{x}}$ a feasible interior point of C_2, which is a contradiction. Thus, no feasible $\hat{\mathbf{x}}$ belong to the interior of $\hat{C}$ and the proof is complete. ■

Example 6.12. Polyhedral Annexation. Return to C_1 and C_2 of Example 6.1 (and Example 6.11). If we pick $p = 2$, $\mu_1 = 0$, $\lambda_1 = 1$, $\lambda_2 = \mu_2 = \frac{1}{2}$, we obtain $\hat{C} = \{(x_3, x_4) \geq \mathbf{0}: -\frac{7}{50}x_3 - \frac{2}{25}x_4 \leq \frac{9}{50}, \frac{1}{25}x_3 - \frac{3}{25}x_4 \leq \frac{13}{25}, \frac{1}{2}(\frac{7}{50} - \frac{1}{25})x_3 + \frac{1}{2}(\frac{2}{25} + \frac{3}{25})x_4 \leq \frac{1}{2}(\frac{41}{50} + \frac{12}{25})\}$. Substituting via $x_3 = 6x_1 - 4x_2 + 1$, $x_4 = 2x_1 + 7x_2 - 4$ we obtain the sets C_1, C_2 and $\hat{C}$ of Figure 6.10. ■

6.5 Subadditive Characterizations

Section 6.4 demonstrated that all minimal valid inequalities (i.e., all the interesting cutting planes) can be generated via the disjunctive cut principle (6-48) when the feasible space is described as a finite disjunction (6-40). Gomory's cutting planes of Section 6.2 derived from the conceptually simpler, although not necessarily superior, scheme of incorporating discreteness through integer restrictions on some or all variables. Can a construct in the latter setting also yield all minimal valid inequalities? In this section we develop an approach.

6.5.1 The Subadditive Principle for Pure Integer Programs

Disjunctive cut generating schemes computed coefficients g_j by taking nonnegative linear combinations of the corresponding constraint columns, $\mathbf{a}^j$, and maximizing or minimizing over elements of the disjunction. Thinking of this process as one of computing g_j as a function, $g(\mathbf{a}^j)$, of its column, we may ask if there are other characterizations of the functions, $g(\cdot)$, that produce valid inequalities.

Gomory (1969) and Gomory and Johnson (1972) described such functions in terms of a subadditive property. A function $g: M \to \mathbb{R}^1$ is *subadditive* on a set M closed under addition if

$$g(\mathbf{m}^1) + g(\mathbf{m}^2) \geq g(\mathbf{m}^1 + \mathbf{m}^2) \qquad \text{for all } \mathbf{m}^1, \mathbf{m}^2 \in M. \tag{6-80}$$

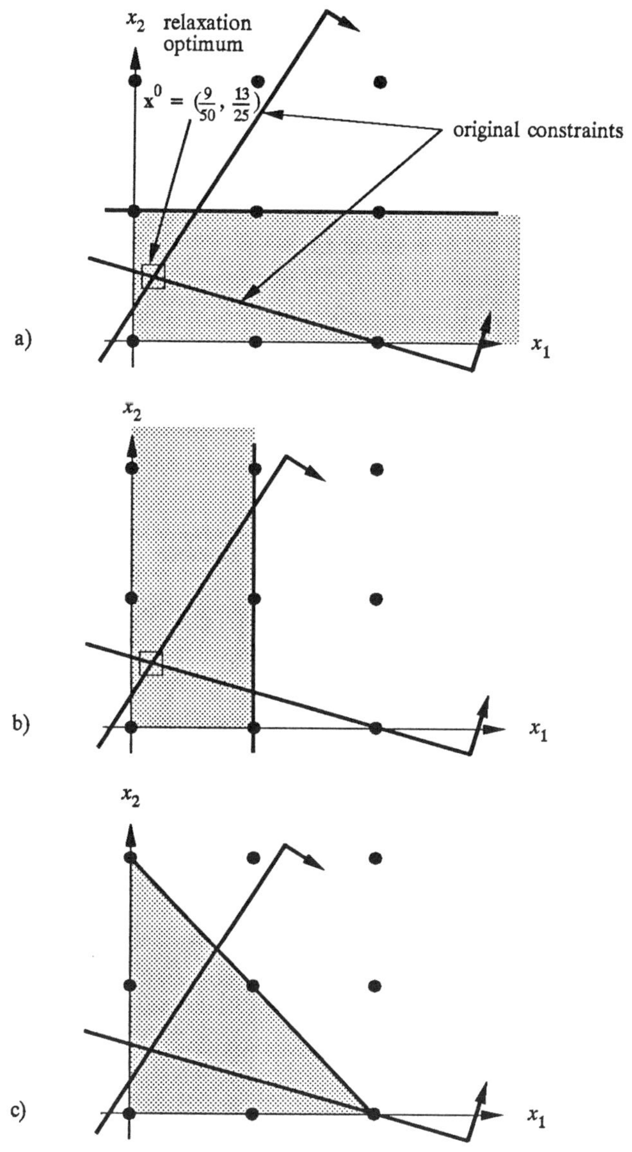

Fig. 6.10. Polyhedral annexation on Example 6.12; (a) infeasible interior set $C_1 = \{(x_1 x_2)$: $0 \le x_1 \le 1\}$, (b) infeasible interior set $C_2 = \{(x_1, x_2)$: $0 \ge x_2 \le 1\}$, (c) annexation result $\hat{C} = \{(x_1, x_2) \le \mathbf{0}$: $x_1 + x_2 \le 2\}$.

Consider the pure integer model

$$(IP) \qquad \begin{aligned} \min \quad & \mathbf{cx} \\ \text{s.t.} \quad & \mathbf{Ax} = \mathbf{b} \\ & \mathbf{x} \geq \mathbf{0} \\ & \mathbf{x} \text{ integer.} \end{aligned}$$

Let F be its feasible space $\{\mathbf{x} \geq \mathbf{0}\text{: } \mathbf{Ax} = \mathbf{b}, \mathbf{x} \text{ integer}\}$, and $\mathbf{a}^j$ the *jth* column of $\mathbf{A}$. As usual, $\mathbf{A}$ and $\mathbf{b}$ will be taken to be rational so that $[F]$ is polyhedral. Taking $M = \{\mathbf{Ax}\text{: } \mathbf{x} \geq \mathbf{0}, \mathbf{x} \text{ integer}\}$ Gomory and Johnson showed subadditivity is almost all that is needed for function g to have the inequality

$$\sum_j g(\mathbf{a}^j)x_j \geq g(\mathbf{b}) \tag{6-81}$$

valid for F.

Theorem 6.18. Subadditive Principle for Pure Integer Programs. *Let F be as above, the feasible space of a rational-equality-constrained, pure integer program (IP), and $M = \{\mathbf{Ax}\text{: } \mathbf{x} \geq \mathbf{0}, \mathbf{x} \text{ integer}\}$. Then if $g\text{: } M \to \mathbb{R}^1$ is subadditive and $g(\mathbf{0}) \leq 0$, inequality (6-81) is valid for all $\mathbf{x} \in F$.*

Proof. If F is empty, every inequality is valid. If not, consider $\hat{\mathbf{x}} \in F$. Noting $\hat{\mathbf{x}} \geq \mathbf{0}$ we shall proceed inductively on the component sum $\Sigma_j \hat{x}_j$ to show

$$\sum_j g(\mathbf{a}^j)\hat{x}_j \geq g\left(\sum_j \mathbf{a}^j \hat{x}_j\right). \tag{6-82}$$

This will establish the theorem because the right side of (6-82) is exactly $g(\mathbf{b})$ when $\hat{\mathbf{x}} \in F$.

Begin with $\Sigma_j \hat{x}_j = 0$, i.e. $\hat{\mathbf{x}} = 0$. Here the left side of (6-82) is zero, and since $g(\mathbf{0}) \leq 0$, the right side is at most zero; (6-82) holds. Now assume (6-82) is true for $\Sigma_j \hat{x}_j = 0, 1, \ldots \sigma$, and consider an $\hat{\mathbf{x}}$ with $\Sigma_j \hat{x}_j = \sigma + 1$. Also let $\hat{x}_{\hat{j}}$ be any positive component. Then

$$\begin{aligned} \sum_j g(\mathbf{a}^j)\hat{x}_j &= \sum_{j \neq \hat{j}} g(\mathbf{a}^j)\hat{x}_j + g(\mathbf{a}^{\hat{j}})(\hat{x}_{\hat{j}} - 1) + g(\mathbf{a}^{\hat{j}}) \\ &\geq g\left(\sum_{j \neq \hat{j}} \mathbf{a}^j \hat{x}_j + \mathbf{a}^{\hat{j}}(\hat{x}_{\hat{j}} - 1)\right) + g(\mathbf{a}^{\hat{j}}) \\ &\geq g\left(\sum_j \mathbf{a}^j \hat{x}_j\right). \end{aligned}$$

The first inequality follows from an inductive hypothesis and the second from subadditivity of g. This establishes (6-82) for $\Sigma_j \hat{x}_j = \sigma + 1$ and completes the proof. ■

The generality of Theorem 6.18 introduces an enormous number of potential schemes for generating valid inequalities because subadditive functions with $g(\mathbf{0}) \leq 0$ are exceedingly common. A very partial list would include the following:

- any *distance metric:* for $\mathbf{a} \in \mathbb{R}^m$, $p > 0$, $g(\mathbf{a}) \triangleq (\Sigma_{i=1}^m (\max\{a_i, -a_i\})^p)^{1/p}$;
- any *modulo arithmetic function:* for $a \in \mathbb{R}^1$ and modulus t, $g(a) \triangleq a(\text{mod } t)$;
- any *linear combination:* for $\mathbf{a} \in \mathbb{R}^m$ and $\mathbf{l}$ an m-vector, $g(\mathbf{a}) \triangleq \mathbf{l}\,\mathbf{a}$;
- the *least integer function:* for $a \in \mathbb{R}^1$, $g(a) \triangleq \lceil a \rceil$;
- the *fractional part function:* for $a \in \mathbb{R}^1$, $\phi(a) \triangleq a - \lfloor a \rfloor$.

It is easy to show that we may also compose conforming functions to obtain another if the outer is monotone. That is, for $a \in M$ $g(a) \triangleq f(h(a))$ is subadditive if h: $M \to \mathbb{R}^1$ is subadditive, and f: $\mathbb{R}^1 \to \mathbb{R}^1$ is subadditive and monotone nondecreasing.

To provide just one numerical illustration return to (6.11)–(6.12) of Example 6.2 in Section 6.2. That system was

$$\begin{aligned} x_1 - 3x_2 - x_3 \qquad &= -1 \\ 3x_1 + 2x_2 \qquad - x_4 &= \;\;\, 4 \\ x_1, x_2, x_3, x_4 \geq 0 \text{ and integer.} \end{aligned}$$

First taking the linear combination $\frac{1}{4}$(row 1) + $\frac{1}{4}$(row 2) and then applying the least integer rounding function, we may deduce

$$\lceil \tfrac{1}{4}(1+3) \rceil x_1 + \lceil \tfrac{1}{4}(-3+2) \rceil x_2 + \lceil \tfrac{1}{4}(-1+0) \rceil x_3 + \lceil \tfrac{1}{4}(0-1) \rceil x_4 \geq \lceil \tfrac{1}{4}(-1+4) \rceil$$

or $x_1 \geq 1$ is a valid inequality. Reference to Figure 6.4 will confirm the veracity of this result.

Another subadditive function of great theoretical interest is the *value function*

$$v(\mathbf{r}) \triangleq v \begin{pmatrix} \min & \mathbf{cx} \\ \text{s.t.} & \mathbf{Ax} = \mathbf{r} \\ & \mathbf{x} \geq \mathbf{0} \\ & x_j \text{ integer for } j \in I \end{pmatrix}. \tag{6-83}$$

We see $v(\mathbf{r})$ is the value of an optimal solution to the indicated integer, or mixed integer problem as a function of the right-hand-side, $\mathbf{r}$. Blair and Jeroslow (1977, 1979) have shown value functions of discrete problems have numerous insightful properties. We shall be interested only in the fact that $v(\mathbf{r})$ conforms to Theorem 6.18.

Lemma 6.19. Subadditivity of Value Functions. *The value function $v(\mathbf{r})$ defined in (6-83) is subadditive at all $\mathbf{r}$ for which the indicated constraints are consistent, i.e., where $v(\mathbf{r}) < +\infty$. Moreover $v(\mathbf{0}) \leq 0$.*

Proof. We first observe that $Q \triangleq \{\mathbf{r}\colon v(\mathbf{r}) < +\infty\}$ is closed under addition, so that the subadditive property is well-defined on Q. This is true because if $\mathbf{x}^1 \in \{\mathbf{x} \geq \mathbf{0}\colon \mathbf{Ax} = \mathbf{r}^1,\ x_j \text{ integer for } j \in I\}$ and $\mathbf{x}^2 \in \{\mathbf{x} \geq \mathbf{0}\colon \mathbf{Ax} = \mathbf{r}^2,\ x_j \text{ integer for } j \in I\}$, then certainly $(\mathbf{x}^1 + \mathbf{x}^2) \in \{\mathbf{x} \geq \mathbf{0}\colon \mathbf{Ax} = \mathbf{r}^1 + \mathbf{r}^2,\ x_j \text{ integer for } j \in I\}$.

Now pick $\mathbf{r}^1, \mathbf{r}^2 \in Q$. Since $v(\mathbf{r}^1) < +\infty$,

$$
\begin{aligned}
v(\mathbf{r}^1) + v(\mathbf{r}^2) &\triangleq v\begin{pmatrix} \min & \mathbf{cx}^1 \\ \text{s.t.} & \mathbf{Ax}^1 = \mathbf{r}^1 \\ & \mathbf{x}^1 \geq \mathbf{0} \\ & x_j^1 \text{ integer for } j \in I \end{pmatrix} \\
&\quad + v\begin{pmatrix} \min & \mathbf{cx}^2 \\ \text{s.t.} & \mathbf{Ax}^2 = \mathbf{r}^2 \\ & \mathbf{x}^2 \geq \mathbf{0} \\ & x_j^2 \text{ integer for } j \in I \end{pmatrix} \\
&\geq v\begin{pmatrix} \min & \mathbf{cx}^1 + \mathbf{cx}^2 \\ \text{s.t.} & \mathbf{Ax}^1 = \mathbf{r}^1 \\ & \mathbf{Ax}^2 = \mathbf{r}^2 \\ & \mathbf{x}^1 \geq \mathbf{0} \\ & \mathbf{x}^2 \geq \mathbf{0} \\ & x_j^1 \text{ integer for } j \in I \\ & x_j^2 \text{ integer for } j \in I \end{pmatrix}.
\end{aligned}
$$

Relaxing by summing pairs of constraints,

$$v(\mathbf{r}^1) + v(\mathbf{r}^2) \geq v\begin{pmatrix} \min & \mathbf{cx}^1 + \mathbf{cx}^2 \\ \text{s.t.} & \mathbf{Ax}^1 + \mathbf{Ax}^2 = \mathbf{r}^1 + \mathbf{r}^2 \\ & \mathbf{x}^1 + \mathbf{x}^2 \geq \mathbf{0} \\ & x_j^1 + x_j^2 \text{integer for } j \in I \end{pmatrix}$$

$$= v\begin{pmatrix} \min & \mathbf{cx} \\ \text{s.t.} & \mathbf{Ax} = \mathbf{r}^1 + \mathbf{r}^2 \\ & \mathbf{x} \geq \mathbf{0} \\ & x_j \text{ integer for } j \in I \end{pmatrix} \triangleq v(\mathbf{r}^1 + \mathbf{r}^2).$$

This establishes subadditivity of $v(\mathbf{r})$. Clearly, since $\mathbf{0} \in \{\mathbf{x} \geq \mathbf{0}\colon \mathbf{Ax} = \mathbf{0},\ x_j$ integer for $j \in I\}$, we may also conclude $v(\mathbf{0}) \leq 0$. ■

Value functions are of limited computational interest because they require us to solve a discrete optimization problem to evaluate them. Jeroslow (1979a) shows, however, that they provide a convenient tool to demonstrate that the subadditive cutting principle (6-81) can generate all interesting valid inequalities for problems (*IP*).

Theorem 6.20. Sufficiency of the Subadditive Principle for Pure Integer Programs. *Let (IP) be as above, a rational-equality-constrained, pure integer program with feasible space F. Then if* $(\boldsymbol{h}, h_0)$ *is a minimal valid inequality for F, there exists a subadditive function* $g\colon M \to \mathbb{R}^1$ *with* $g(0) \leq 0$ *such that*

$$h_j = g(\boldsymbol{a}^j) \qquad \text{for } j \geq 1 \tag{6-84}$$

$$h_0 = g(\boldsymbol{b}), \tag{6-85}$$

where $M = \{\boldsymbol{Ax}\colon \boldsymbol{x} \geq 0,\ \boldsymbol{x} \text{ integer}\}$.

Proof. If there are no minimal inequalities there is nothing to prove. If F admits minimal inequalities, pick one $(\mathbf{h}, h_0)$ and consider the value function

$$v^h(\mathbf{r}) \triangleq v\begin{pmatrix} \min & \mathbf{hx} \\ \text{s.t.} & \mathbf{Ax} = \mathbf{r} \\ & \mathbf{x} \geq \mathbf{0} \\ & \mathbf{x} \text{ integer} \end{pmatrix}. \tag{6-86}$$

We already know from Lemma 6.19 that $v^h(\mathbf{r})$ conforms to Theorem 6.18 so that

$$\sum_j v^h(\mathbf{a}^j)\, x_j \geq v^h(\mathbf{b}) \tag{6-87}$$

is valid for F. We want to show $v^h(\mathbf{a}^j) \leq h_j$ for all $j \geq 1$ and $v^h(\mathbf{b}) \geq h_0$. Then, since none of those inequalities can be strict with $(\mathbf{h}, h_0)$ minimal, $v^h(\mathbf{r})$ provides a $g(\mathbf{r})$ for (6-84) and (6-85) that constructs $(\mathbf{h}, h_0)$.

To show $v^h(\mathbf{a}^j) \leq h_j$, merely consider solutions $\mathbf{x} =$ the jth unit vector in (6-86). Such solutions are feasible in the computation of $v^h(\mathbf{a}^j)$ and have cost h_j. Certainly then $v^h(\mathbf{a}^j) \leq h_j$ for all $j \geq 1$.

For the right-hand-sides, note by definition $v^h(\mathbf{b}) = \inf\{\mathbf{hx}: \mathbf{x} \in F\}$. But then validity of $\mathbf{hx} \geq h_0$ assures $v^h(\mathbf{b}) \geq h_0$. This completes the proof. ■

Theorem 6.20 does establish that the subadditive cutting principle is as sufficient for generating all minimal inequalities for pure integer programs as the disjunctive cutting principle was for disjunctive models. It is important, however, to realize that Theorem 6.20 is no more constructive than the analogous Theorem 6.13. The value function (6-86) used to establish Theorem 6.20 requires an integer program as hard as (*IP*) to be solved in computing each cut coefficient. Moreover, the function (6-86) is parameterized in terms of the cutting plane $(\mathbf{h}, h_0)$ we seek. Thus, we must content ourselves with generating valid inequalities from easier subadditive functions, while knowing such a process is at least theoretically adequate.

One approach (the one taken in Gomory's early development of subadditivity) is to restrict the range of definition of the function $g(\cdot)$ to a finite set of vectors by interpreting the constraints $\mathbf{Ax} = \mathbf{b}$ in modulo arithmetic. Viewed in the right way a finite Abelian group arises. More to the point, there are only finitely many function values $g(k\mathbf{a}^j)$ that need to be established in picking a subadditive $g(\cdot)$. Treating these function values as variables, requirements that they be subadditive with $g(\mathbf{0}) \leq 0$ can be explicitly represented as a linear program. Extreme points of the feasible space for that linear program give a finite list of subadditive functions to consider, with cuts (6-81) being read out as components for $g(\mathbf{a}^1)$, $g(\mathbf{a}^2), \ldots, g(\mathbf{a}^n)$ and $g(\mathbf{b})$. (For further detail on this approach see Jeroslow (1979a)).

6.5.2 *Extensions of the Subadditive Principle to Mixed Integer Programs*

Thus far we have treated the subadditive cutting principle only in the pure integer programming case. Gomory and Johnson (1973a,b) and

Johnson (1974) have also illustrated how results can be extended to the mixed integer case

$$
(MIP) \qquad \begin{aligned} \min \quad & \mathbf{cx} \\ \text{s.t.} \quad & \mathbf{Ax} = \mathbf{b} \\ & \mathbf{x} \geq \mathbf{0} \\ & x_j \text{ integer for } j \in I \end{aligned}
$$

with **A**, **b** rational as usual. We shall see they treat integer restricted columns $\mathbf{a}^j$ with $j \in I$ exactly as in (6-81) of the previous section. The issue now is what to do with $j \notin I$.

The directional derivatives at zero of a function $g(\cdot)$ are defined

$$\breve{g}(\mathbf{d}) \triangleq \limsup_{\lambda \to 0^+} \frac{g(\lambda \mathbf{d})}{\lambda}. \tag{6-88}$$

Intuitively, they are the slopes of the function $g(\cdot)$ as **0** is approached from the direction **d**. The general limsup is necessary in the definition because it is possible that no finite limit exists or that the function varies so dramatically that several different limits can be achieved by different sequences of $\lambda \to 0^+$.

Johnson (1974) showed we need not be concerned about multiple limits if $g(\cdot)$ is subadditive, although $\breve{g}(\mathbf{d})$ may be infinite. More importantly, he showed the appropriate treatment of continuous columns in cutting planes derived from subadditive functions $g(\cdot)$ is to apply the corresponding directional derivative at zero. That is, inequalities are generated as

$$\sum_{j \in I} g(\mathbf{a}^j)x_j + \sum_{j \notin I} \breve{g}(\mathbf{a}^j)x_j \geq g(\mathbf{b}) \tag{6-89}$$

We first establish a lemma on an important property of the derivative (6-88) and then prove validity of (6-89).

Lemma 6.21. Sublinearity of Directional Derivatives of Subadditive Functions at 0. *Let $g: M \to \mathbb{R}^1$ be a subadditive function defined on any $M \subseteq \mathbb{R}^n$ that includes the entire ray $\{\alpha \mathbf{d}: \alpha > 0\}$. Then,*

$$g(\alpha \mathbf{d}) \leq \alpha \breve{g}(\mathbf{d}) \qquad \text{for all } \alpha > 0. \tag{6-90}$$

Proof. Fix $\alpha > 0$, and define a sequence $\{\lambda_k\}_{k=1}^{\infty}$ by $\lambda_k = \alpha/k$. Clearly, $\lambda_k \to 0^+$ as $k \to \infty$. But for each k,

$$g(\alpha \mathbf{d}) \triangleq g(k\lambda_k \mathbf{d}) \leq k g(\lambda_k \mathbf{d}) = (k\lambda_k)\frac{g(\lambda_k \mathbf{d})}{\lambda_k} \triangleq \alpha \frac{g(\lambda_k \mathbf{d})}{\lambda_k}$$

the inequality following from subadditivity because an integer multiple of $(\lambda_k \mathbf{d})$ may be viewed as a finite sum of vectors $(\lambda_k \mathbf{d})$. Thus, for all k

$$g(\alpha \mathbf{d}) \leq \alpha \frac{g(\lambda_k \mathbf{d})}{\lambda_k},$$

and so $g(\alpha \mathbf{d}) \leq \alpha \operatorname{limsup}_{k \to \infty} g(\lambda_k \mathbf{d})/\lambda_k \triangleq \alpha \check{g}(\mathbf{d})$. ■

Theorem 6.22. Subadditive Principle for Mixed Integer Programs. *Let (MIP) be as above, the rational-equality-constrained, mixed integer program with feasible space F and $M = \{\boldsymbol{Ax}\colon \boldsymbol{x} \geq \boldsymbol{0},\ x_j \text{ integer for } j \in I\}$. Then if $g\colon M \to \mathbb{R}^1$ is subadditive and $g(\boldsymbol{0}) \leq 0$, inequality (6-89) is valid for all $\boldsymbol{x} \in F$.*

Proof. If F is empty, every inequality is valid. If not, consider $\hat{\mathbf{x}} \in F$. By Lemma 6.21,

$$\sum_{j \in I} g(\mathbf{a}^j)x_j + \sum_{j \notin I} \check{g}(\mathbf{a}^j)\hat{x}_j \geq \sum_{j \notin I} g(\mathbf{a}^j)\hat{x}_j + \sum_{j \notin I} g(\mathbf{a}^j \hat{x}_j).$$

To complete the proof we need only induct on $\Sigma_{j \in I} \hat{x}_j + |\{j \notin I\}|$ and apply subadditivity as in the proof of Theorem 6.18 to conclude

$$\sum_{j \in I} g(\mathbf{a}^j)\hat{x}_j + \sum_{j \notin I} g(\mathbf{a}^j \hat{x}_j) \geq g\left(\sum_{j \in I} \mathbf{a}^j \hat{x}_j + \sum_{j \notin I} \mathbf{a}^j \hat{x}_j \right) = g(\mathbf{b}).$$ ■

Example 6.13. Gomory Mixed Integer Cutting Planes via the Subadditive Cutting Principle. One way to illustrate computations for inequalities (6-89) is to see that they can reproduce the venerable Gomory mixed integer cutting plane (6-28). The single row mixed integer system from which the latter derived was of the form (6-17)

$$x_r + \sum_{j \in N_I} \bar{a}_{rj} x_j + \sum_{j \in N_C} \bar{a}_{rj} x_j = \bar{x}_r, \tag{6-91}$$

where x_r is an integer-infeasible variable, and N_I and N_C are the integer and continuous nonbasic sets. All variables are nonnegative.

Consider the function $g\colon \mathbb{R}^1 \to \mathbb{R}^1$

$$g(a) = \min\left\{ \frac{a - \lfloor a \rfloor}{\alpha}, \frac{\lceil a \rceil - a}{1 - \alpha} \right\}, \tag{6-92}$$

where α is a constant with $1 > \alpha > 0$. We can show $g(a)$ is subadditive by taking advantage of the fact that value functions are subadditive (Lemma

6.19). It is easy to see

$$g(a) = v\left(\begin{array}{ll} \min & \dfrac{w_1}{\alpha} + \dfrac{w_2}{1-\alpha} \\ \text{s.t.} & (y_1 - y_2) + (w_1 - w_2) = a \\ & y_1, y_2, w_1, w_2 \geq 0 \\ & y_1, y_2 \text{ integer} \end{array}\right).$$

Figure 6.11 illustrates the periodic form of function (6-92). For its domain, $\mathbb{R}^1$, there are only two directions by which $\mathbf{0}$ can be approached, i.e., $d = \pm\delta$ for $\delta > 0$. As illustrated in Figure 6.11 the corresponding directional derivatives are

$$\acute{g}(d) = \frac{1}{\alpha} \qquad \text{for } d \geq 0 \tag{6-93}$$

$$\acute{g}(d) = \frac{-1}{1-\alpha} \qquad \text{for } d < 0 \tag{6-94}$$

Points a_1 and a_2 illustrate the sublinearity property (Lemma 6.21) for these derivatives. One case has $g(a_1) = a_1\acute{g}(a_1)$, and the other $g(a_2) < a_2\acute{g}(a_2)$.

Applying (6-92)–(6-94) to (6-91) with α = fractional part, $\phi(\bar{x}_r)$, we derive the cutting plane (6-89)

$$g(1)x_r + \sum_{j \in N_I} g(\bar{a}_{rj})x_j + \sum_{j \in N_C} \acute{g}(\bar{a}_{rj})x_j \geq g(\bar{x}_r)$$

or

$$\sum_{j \in N_I} \min\left\{\frac{\phi(\bar{a}_{rj})}{\phi(\bar{x}_r)}, \frac{1-\phi(\bar{a}_{rj})}{1-\phi(\bar{x}_r)}\right\} x_j + \sum_{j \in N_C^+} \left(\frac{\bar{a}_{rj}}{\phi(\bar{x}_r)}\right) x_j$$

$$+ \sum_{j \in N_C^-} \left(\frac{-\bar{a}_{rj}}{1-\phi(\bar{x}_r)}\right) x_j \geq \min\left\{\frac{\phi(\bar{x}_r)}{\phi(\bar{x}_r)}, \frac{1-\phi(\bar{x}_r)}{1-\phi(\bar{x}_r)}\right\} = 1,$$

where $N_C^- = \{j \in N_C\colon \bar{a}_{rj} < 0\}$, $N_C^+ = \{j \in N_C\colon \bar{a}_{rj} > 0\}$. This is exactly $1/\phi(\bar{x}_r)$ times the Gomory mixed integer form (6-28). ■

Our proof of the subadditivity of value functions (Lemma 6.19) already encompassed the mixed integer case. Not surprisingly then, we may extend Theorem 6.20 to the mixed integer case using value functions as before, i.e., we can show all interesting valid inequalities for (*MIP*) can be obtained via the subadditive cutting principle (6-89).

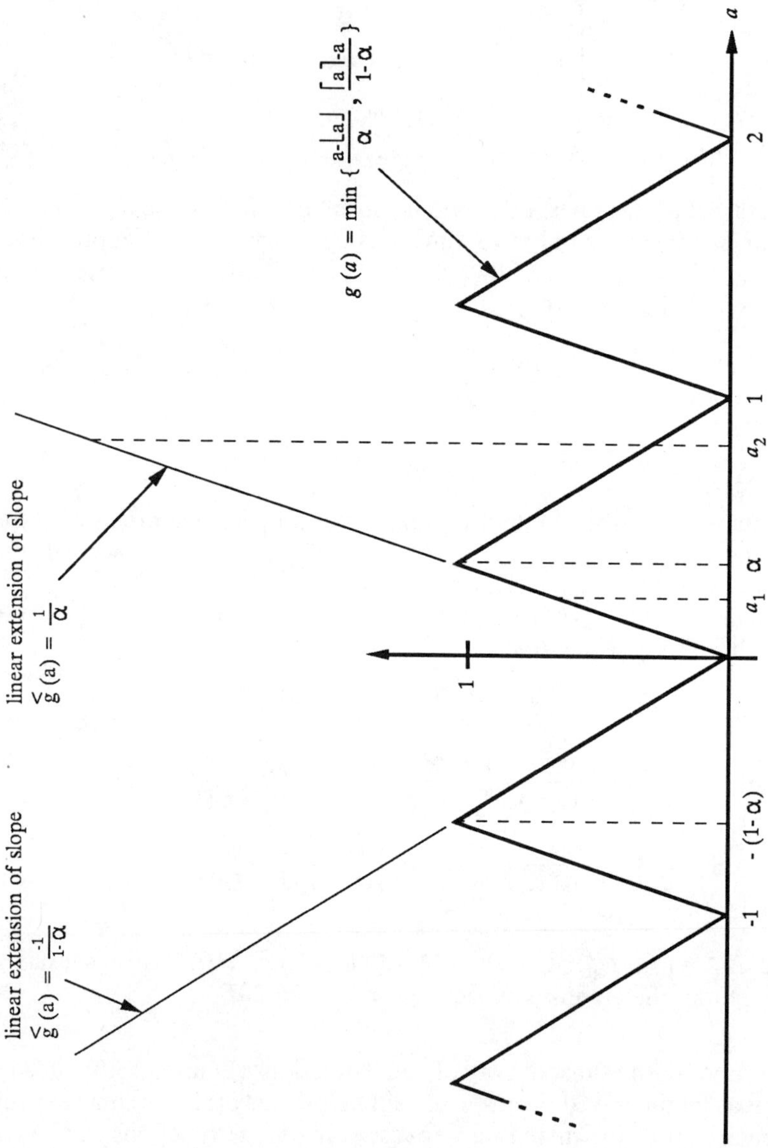

Fig. 6.11. Gomory mixed integer cut function.

Theorem 6.23. Sufficiency of the Subadditive Principle for Mixed-Integer Programming. *Let (MIP) be as above, a rational-equality-constrained, mixed integer program with feasible space F. Then if* $(\mathbf{h}, h_0)$ *is a minimal valid inequality for F, there exist a subadditive function* $g\colon M \to \mathbb{R}^1$ *with* $g(\mathbf{0}) \le 0$ *such that*

$$h_j = g(\mathbf{a}^j) \qquad \text{for all } j \in I \tag{6-95}$$

$$h_j = \check{g}(\mathbf{a}^j) \qquad \text{for all } j \notin I,\ j \ge 1 \tag{6-96}$$

$$h_0 = g(\mathbf{b}) \tag{6-97}$$

Proof. If there are no minimal inequalities there is nothing to prove. If F admits minimal inequalities, pick one $(\mathbf{h}, h_0)$ and consider the value function

$$v^h(r) \triangleq v \begin{pmatrix} \min & \mathbf{hx} \\ \text{s.t.} & \mathbf{Ax} = \mathbf{r} \\ & \mathbf{x} \ge \mathbf{0} \\ & x_j \text{ integer for } j \in I \end{pmatrix}. \tag{6-98}$$

As in the proof of Theorem 6.20, we want to show the subadditive inequality (6-89)

$$\sum_{j \in I} v^h(\mathbf{a}^j)x_j + \sum_{j \notin I} \check{v}^h(\mathbf{a}^j)x_j \ge v^h(\mathbf{b})$$

dominates the minimal $(\mathbf{h}, h_0)$ and thus (6-98) provides the $g(\cdot)$ of (6-95)–(6-97). From the earlier proof $h_j \ge v^h(\mathbf{a}^j)$ for $j \in I$ and $h_0 \le v^h(\mathbf{b})$.

We have only to show $h_j \le \check{v}^h(\mathbf{a}^j)$ for continuous $j \notin I$. Consider solutions to the optimization of (6-98) of the form $\mathbf{x} = \lambda\mathbf{e}^j$, where $\lambda > 0$ and $\mathbf{e}^j$ is the *jth* unit vector, $j \notin I$. Such solutions are feasible in (6-98) for $\mathbf{r} = \lambda\mathbf{a}^j$ and cost λh_j, Thus,

$$v^h(\lambda\mathbf{a}^j) \le \lambda h_j \qquad \text{for all } \lambda > 0,\ j \notin I.$$

Dividing by $\lambda > 0$ and letting $\lambda \to 0^+$

$$\check{v}^h(\mathbf{a}^j) \triangleq \limsup_{\lambda \to 0^+} \left(\frac{v^h(\lambda\mathbf{a}^j)}{\lambda}\right) \le h_j.$$ ■

6.6 Successive Integerized Sum Characterization

The classic (graph) edge covering problem is defined on an undirected graph $G(V, E)$ and seeks a minimum weight subset of edges, at least one of which is incident to each vertex $v \in V$. One formulation is

$$
\begin{array}{llll}
 & \min & \sum_{e \in E} w_e x_e & \\
(WC) & \text{s.t.} & \sum_{e \in E_v} x_e \geq 1 & \text{for every } v \in V \quad (6\text{-}99) \\
 & & 1 \geq x_e \geq 0 & \text{for every } e \in E \quad (6\text{-}100) \\
 & & x_e \text{ integer} & \text{for every } e \in E, \quad (6\text{-}101)
\end{array}
$$

where $\{w_e: e \in E\}$ are weights and $E_v \triangleq \{e \in E$ meeting vertex $v \in V\}$. It is easy to see that the linear program obtained by dropping the integrality requirement (6-101) may have a fractional optimum. Consider, for example, equal weights on the simple graph below:

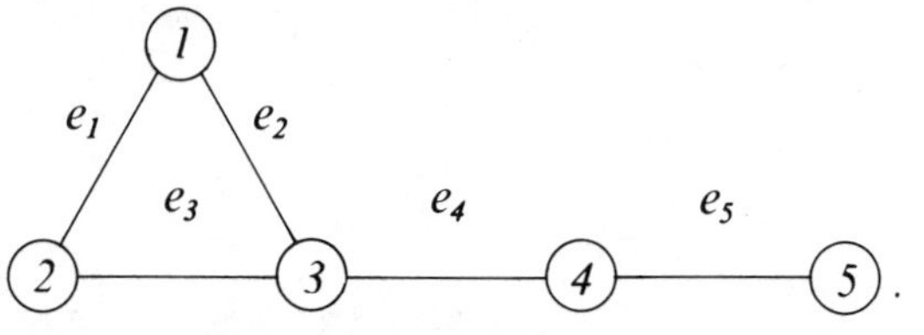

The unique linear programming optimal is $\bar{\mathbf{x}} = (\frac{1}{2}, \frac{1}{2}, \frac{1}{2}, 0, 1)$.

The famous result (derived from Edmonds (1965c)) that has rendered problem (WC) well-solved relates to a set of valid inequalities defined for every odd cardinality vertex subset $S \subseteq V$:

$$\sum_{\substack{e \in \cup E_v \\ v \in S}} x_e \geq \frac{|S| + 1}{2}. \qquad (6\text{-}102)$$

Validity of (6-102) is easily established in two steps. First, sum relations (6-99) for $v \in S$ with (nonnegative) weights of $\frac{1}{2}$. For $S = \{1, 2, 3\}$ in the example above this step gives

$$x_{e_1} + x_{e_2} + x_{e_3} + \tfrac{1}{2} x_{e_4} \geq \tfrac{3}{2}.$$

Then, round up fractional coefficients $\frac{1}{2}$ and $\frac{3}{2}$ to the next integers $\lceil \frac{1}{2} \rceil$ and $\lceil \frac{3}{2} \rceil$, respectively. Here we obtain the (6-102) instance

$$x_{e_1} + x_{e_2} + x_{e_3} + x_{e_4} \geq 2$$

violated by the fractional optimum $\bar{\mathbf{x}}$

It is clear that taking a nonnegative weighted sum of constraints yields a new constraint valid for all $\mathbf{x}$ feasible in (WC). Validity of the round-up operation follows by nonnegativity and the integrality constraint (6-101). Certainly no solutions $\mathbf{x} \geq \mathbf{0}$ are lost if some coefficients on the left-hand-side are weakened to (greater) integers. But then, the left-hand-side is an integer sum of integers, so we may also increase the right-hand-side to an integer.

It can be shown that the linear program over (6-99), (6-100), and all (6-102) has integer extreme points, i.e., (*WC*) can, in principle, be solved by linear programming.

Unfortunately, most problems to which the present chapter is addressed probably do not admit such simple, complete characterizations of their convex hulls (recall Section 6.3.2). Still, the weighted sum and integerize process, for which we have just established validity, is at the heart of derivations of almost all the known facetial inequalities for such problems. In this section, we present Chvátal's (1973) analysis of the approach for bounded, pure integer programs.

6.6.1 *The Chvátal Model*

Suppose we have the bounded pure integer program

$$(BIP) \qquad \begin{aligned} \min \quad & \mathbf{cx} \\ \text{s.t.} \quad & \mathbf{Ax} \geq \mathbf{b} && \text{(6-103)} \\ & \mathbf{d} \geq \mathbf{x} \geq \mathbf{0} && \text{(6-104)} \\ & \mathbf{x} \text{ integer.} && \text{(6-105)} \end{aligned}$$

In general, coefficients of **A, b, c** may take on any real values, but we assume upper bounds are integer. As usual, $\bar{F} \triangleq \{\mathbf{x} \in \mathbb{R}^n$ satisfying (6-103) and (6-104)$\}$, $F \triangleq \{\mathbf{x} \in \bar{F}$ satisfying (6-105)$\}$, and by Theorem 6.2 $[F]$ is polyhedral. Formalization of the above introduction gives the Chvátal construction.

Theorem 6.24. Validity of the Chvátal Construction. *Any inequality* $(\mathbf{g}, g_0)$ *formed as*

$$\begin{aligned} \mathbf{g} &\longleftarrow \lceil \mathbf{uA} - \mathbf{v} + \mathbf{w} \rceil \\ g_0 &\longleftarrow \lceil \mathbf{ub} - \mathbf{vd} \rceil \end{aligned} \qquad \text{(6-106)}$$

with $\mathbf{u} \geq \mathbf{0}$, $\mathbf{v} \geq \mathbf{0}$, $\mathbf{w} \geq \mathbf{0}$ *is valid for F the feasible set in (BIP).*

Proof. Inequalities formed in the indicated way are integerized versions of nonnegative weighted sums of (*BIP*) constraints, with **u, v,** and **w** relating to $\mathbf{Ax} \geq \mathbf{b}$, $-\mathbf{x} \geq -\mathbf{d}$, $\mathbf{x} \geq \mathbf{0}$ respectively. As above, integerizing first the left-hand-side, then the right of the weighted sum establishes validity of $\mathbf{gx} \geq g_0$ for F. ■

Chvátal originally derived the construction of Theorem 6.24 as a generalization of Gomory's fractional cutting plane Algorithm 6B. Before proceeding, we shall briefly illustrate how it recovers fractional cuts (6-7).

Example 6.14. Chvátal Derivation of Gomory's Fractional Cuts. Recall Gomory's fractional cut was derived in Section 6.2 over pure integer, equality-constrained problems with integer data. Taking **A**, **b**, and **c** integer, disregarding upper bounds, and adding surplus variables **y**, we can reach Gomory's format from the above as

$$(IP) \qquad \begin{aligned} \min \quad & \mathbf{cx} \\ \text{s.t.} \quad & \mathbf{Ax} - \mathbf{y} = \mathbf{b} \\ & \mathbf{x} \geq \mathbf{0} \text{ and integer} \\ & \mathbf{y} \geq \mathbf{0} \text{ and integer.} \end{aligned}$$

Take as $\bar{\mathbf{u}}$ an optimal solution to the linear programming dual of the linear relaxation $(\overline{IP})$ of (IP). (Here we assume $(\overline{IP})$ is feasible and bounded in value.) The cost row of the representation (6-15) on which Gomory's cuts are based can be expressed as

$$x_0 = -\bar{\mathbf{u}}\mathbf{b} - (\mathbf{c} - \bar{\mathbf{u}}\mathbf{A})\mathbf{x} - \bar{\mathbf{u}}\mathbf{y}. \tag{6-107}$$

Actually (6-15) would omit terms for basic components x_j and y_i, but we know in such cases $(\mathbf{c} - \bar{\mathbf{u}}\mathbf{A})_j = 0$ and $\bar{u}_i = 0$.

If a fractional cut is generated from cost row (6-107) by Gomory's Algorithm 6B, it will have the form

$$\phi(\mathbf{c} - \bar{\mathbf{u}}\mathbf{A})\mathbf{x} + \phi(\bar{\mathbf{u}})\mathbf{y} \geq \phi(-\bar{\mathbf{u}}\mathbf{b}),$$

where $\phi(\mathbf{p}) \triangleq \mathbf{p} - \lfloor \mathbf{p} \rfloor$. Eliminating surpluses by $\mathbf{y} \triangleq \mathbf{Ax} - \mathbf{b}$ and applying the identity $\phi(\mathbf{p}) = \mathbf{p} + \lceil -\mathbf{p} \rceil$, the cut successively simplifies

$$(\phi(\mathbf{c} - \bar{\mathbf{u}}\mathbf{A}) + \phi(\bar{\mathbf{u}})\mathbf{A})\mathbf{x} \geq \phi(-\bar{\mathbf{u}}\mathbf{b}) + \phi(\bar{\mathbf{u}})\mathbf{b}$$

$$(\mathbf{c} + \lceil \mathbf{uA} - \mathbf{c} \rceil + \lceil -\mathbf{u} \rceil \mathbf{A})\mathbf{x} \geq \lceil \mathbf{ub} \rceil + \lceil -\bar{\mathbf{u}} \rceil \mathbf{b}.$$

Noting $\mathbf{q} + \lceil \mathbf{p} \rceil = \lceil \mathbf{q} + \mathbf{p} \rceil$ for integer $\mathbf{p}$, and again using $\phi(\mathbf{p}) = \mathbf{p} + \lceil -\mathbf{p} \rceil$, we can further simplify

$$(\lceil \bar{\mathbf{u}}\mathbf{A} \rceil + \lceil -\bar{\mathbf{u}} \rceil \mathbf{A})\mathbf{x} \geq \lceil \bar{\mathbf{u}}\mathbf{b} \rceil + \lceil -\bar{\mathbf{u}} \rceil \mathbf{b}$$

$$\lceil (\bar{\mathbf{u}} + \lceil -\bar{\mathbf{u}} \rceil)\mathbf{A} \rceil \mathbf{x} \geq \lceil (\bar{\mathbf{u}} + \lceil -\bar{\mathbf{u}} \rceil)\mathbf{b} \rceil$$

and

$$\lceil \phi(\bar{\mathbf{u}})\mathbf{A} \rceil \mathbf{x} \geq \lceil \phi(\bar{\mathbf{u}})\mathbf{b} \rceil.$$

The last is exactly the cut we would obtain by the construction of Theorem 6.24 using $\mathbf{u} \leftarrow \phi(\bar{\mathbf{u}})$, $\mathbf{v} \leftarrow \mathbf{0}$, $\mathbf{w} \leftarrow \mathbf{0}$. ■

6.6.2 *Sufficiency of Successive Chvátal Construction*

It is obvious that we could apply Chvátal's construction (6-106) successively. That is, generate a valid inequality $\mathbf{g}^1\mathbf{x} \geq g_0^1$, add it to the system $\mathbf{Ax} \geq \mathbf{b}$, and reapply the construction to obtain a new cut $\mathbf{g}^2\mathbf{x} \geq g_0^2$. This time, of course, a nonnegative weight may also be applied to $\mathbf{g}^1\mathbf{x} \geq g_0^1$.

The remarkable fact established by Chvátal (1973) is that finitely many such simple steps can derive every needed inequality for $[F]$. For this reason the minimum number of steps needed has come to be called the *Chvátal rank* of the inequality.

An obvious reason one might doubt that inequalities of finite Chvátal rank are sufficient is that all inequalities produced by (6-106) have integer coefficients, yet we have allowed **A, b** and **c** to even be irrational. It is easy to see, however, that integer inequalities are enough for bounded (*BIP*).

Lemma 6.25. Sufficiency of Integer Inequalities for (BIP). *Let (BIP) be the pure integer program above with bounded solution space F. Then there exists a finite set* $\{\mathbf{g}^k\mathbf{x} \geq g_0^k \colon k \in K\}$ *of integer inequalities* $(\mathbf{g}^k, g_0^k)$ *such that*

$$[F] = \bigcap_{k \in K} \{\mathbf{x}\colon \mathbf{g}^k\mathbf{x} \geq g_0^k\}.$$

Proof. As in the proof of Theorem 6.2, Table 6.1 case F_2, the number of points in F is clearly finite, and each is an integer vector. Any dim($[F]$) or dim($[F]$) + 1 such points determining a proper, respectively improper, facetial inequality can be contained in the hyperplane defined by a rational inequality $\mathbf{gx} \geq g_0$. Multiplying by a suitable integer produces the equivalent integer form $\mathbf{g}^k\mathbf{x} \geq g_0^k$. Also, since $[F]$ is polyhedral (by Theorem 6.2) only finitely many such inequalities are needed to characterize $[F]$. ■

We are now ready to turn to Chvátal's rather technical proof of the sufficiency theorem.

Theorem 6.26. Sufficiency of Finite Chvátal Rank Construction. *Let (BIP) be the feasible pure integer program defined above with bounded solution space F. Then any integer supporting valid inequality for F can be derived by a finite sequence of applications of construction (6-106), each new inequality being added to the linear system,* $\mathbf{Ax} \geq \mathbf{b}$, *from which the next is derived.*

Before proceeding to a proof, let us establish a simplifying lemma. It shows that inequalities can be strengthened by construction (6-106) when they do not even support the feasible space of the current linear programming relaxation.

Lemma 6.27. Improving Inequalities Not Supporting for Linear Relaxations. *Let (BIP) be as in Theorem 6.26, and consider an interim system*

$$\tilde{\mathbf{A}}\mathbf{x} \geq \tilde{\mathbf{b}} \tag{6-108}$$

$$\mathbf{d} \geq \mathbf{x} \geq \mathbf{0} \tag{6-109}$$

$$\mathbf{x} \text{ integer} \tag{6-110}$$

derived from (6-103)–(6-105) by adding valid inequalities in successive application of Chvátal's construction (6-106) to $\mathbf{Ax} \geq \mathbf{b}$. *Then if* $\mathbf{gx} \geq g_0 - 1$ *is an (integer) inequality derivable from (6-108)–(6-110) by the Chvátal construction, and if*

$$\{\mathbf{x}: \mathbf{gx} = g_0 - 1, \tilde{\mathbf{A}}\mathbf{x} \geq \tilde{\mathbf{b}}, \mathbf{d} \geq \mathbf{x} \geq \mathbf{0}\} = \varnothing \tag{6-111}$$

then $\mathbf{gx} \geq g_0$ *can also be derived by (6-106) from (6-108)–(6-110).*

Proof. Consider the linear program

$$\begin{aligned} \min \quad & \mathbf{gx} \\ \text{s.t.} \quad & \tilde{\mathbf{A}}\mathbf{x} \geq \tilde{\mathbf{b}} \\ & \mathbf{d} \geq \mathbf{x} \geq \mathbf{0} \end{aligned}$$

Validity of the $\tilde{\mathbf{A}}\mathbf{x} \geq \tilde{\mathbf{b}}$ system and $\mathbf{gx} \geq g_0 - 1$ assures the problem is feasible and bounded in solution value. Thus, its dual

$$\begin{aligned} \max \quad & \mathbf{u}\tilde{\mathbf{b}} - \mathbf{vd} \\ \text{s.t.} \quad & \mathbf{u}\tilde{\mathbf{A}} - \mathbf{v} + \mathbf{w} = \mathbf{g} \\ & \mathbf{u} \geq \mathbf{0}, \mathbf{v} \geq \mathbf{0}, \mathbf{w} \geq \mathbf{0} \end{aligned}$$

is feasible. Furthermore, this dual has feasible solutions $(\mathbf{u},\mathbf{v},\mathbf{w})$ with values $(\mathbf{ub} - \mathbf{vd})$ as large as $\tilde{\nu} \triangleq \min\{\mathbf{gx}: \tilde{\mathbf{A}}\mathbf{x} \geq \tilde{\mathbf{b}}, \mathbf{d} \geq \mathbf{x} \geq \mathbf{0}\}$. By (6-111), $\tilde{\nu} > g_0 - 1$. Pick a dual feasible solution $(\tilde{\mathbf{u}}, \tilde{\mathbf{v}}, \tilde{\mathbf{w}})$ with $g_0 \geq \tilde{\mathbf{u}}\mathbf{b} - \tilde{\mathbf{v}}\mathbf{d} > g_0 - 1$, and use that solution in construction (6-106). The result will be $\mathbf{g}$ on the left-hand-side, because $(\tilde{\mathbf{u}}, \tilde{\mathbf{v}}, \tilde{\mathbf{w}})$ is dual feasible, and $\lceil \tilde{\mathbf{u}}\tilde{\mathbf{b}} - \tilde{\mathbf{v}}\mathbf{d} \rceil \triangleq g_0$ on the right-hand-side. ■

Lemma 6.27 shows that it is not hard to improve an integer inequality $\mathbf{gx} \geq g_0 - 1$ when (6-111) holds. The core of Chvátal's proof of Theorem 6.26 to follow is a lexicographically-ordered sequence of inequalities $\mathbf{g}^{[s]}\mathbf{x} \geq g_0^{[s]}$, derivable in succession by Chvátal's construction, that will allow us to strengthen a known inequality $\mathbf{hx} \geq h_0$ to $\mathbf{hx} \geq h_0 + 1$ when

(6-111) fails. There is one member of this sequence for each k-vector $\mathbf{s}$, $k \le n$ with integer components, s_j, satisfying $0 \le s_j \le d_j$. The first inequality labeled with the n-vector of zeroes, has

$$\mathbf{g}^{[0, \ldots, 0_n]} \triangleq \mathbf{h} + \sum_{j=1}^{n} \mathbf{e}^j \tag{6-112}$$

$$g_0^{[0, \ldots, 0_n]} \triangleq h_0 + 1 \tag{6-113}$$

where $\mathbf{e}^j$ is the jth unit n-vector. Others follow in an extended notion of lexicographic sequence, treating missing components of vectors with fewer than n components as if they were $d_j + 1$. Specifically, $\mathbf{s} \overset{L}{<} \mathbf{t}$ if the first nonzero component of $(\hat{\mathbf{s}} - \mathbf{t})$ is negative or $\hat{\mathbf{s}} = \mathbf{t}$ and $\hat{\mathbf{s}} \neq \mathbf{s}$ where $\hat{\mathbf{s}}$ is the truncated vector obtained when $\mathbf{s}$ is reduced to the length of $\mathbf{t}$.

Taken in this order, $(\mathbf{g}^{[s]}, g_0^{[s]})$ are derived from their antecedents by the recursive definition

$$\mathbf{g}^{[s_1, s_2, \ldots, s_k]} \triangleq \mathbf{h} + \sum_{\substack{j \le k \\ s_j > 0}} \mathbf{g}^{[s_1, s_2, \ldots, s_j - 1]} + \sum_{\substack{j \le k \\ s_j = 0}} \mathbf{e}^j \tag{6-114}$$

$$g_0^{[s_1, s_2, \ldots, s_k]} \triangleq h_0 + 1 + \sum_{\substack{j \le k \\ s_j > 0}} g_0^{[s_1, s_2, \ldots, s_j - 1]} \tag{6-115}$$

By definition, the last inequality in the sequence, i.e., the one $(\mathbf{g}^{[\,]}, g_0^{[\,]})$ associated with the 0-vector has $\mathbf{g}^{[\,]} \triangleq \mathbf{h}$, $g_0^{[\,]} \triangleq h_0 + 1$ as required. (Readers may wish to study the example in Table 6.2 before passing to the proof.)

Proof of Theorem 6.26. We shall show that for every integer vector $\mathbf{h}$, successive application of Chvátal's construction (6-106) can derive a valid inequality $\mathbf{hx} \ge h_0$ such that either $\mathbf{hx} \ge h_0 + 1$ is not valid for F or it too can be derived by finitely many more Chvátal iterations. Since F is bounded, the latter case can happen only finitely often. Among those integer inequalities $(\mathbf{h}, h_0)$ such that $\mathbf{hx} \ge h_0 + 1$ is not valid are all the supporting integer valid inequalities. Thus, in particular, all of them will derive after finitely many Chvátal iterations.

Pick an integer n-vector, $\mathbf{h}$, and let $h_0 \leftarrow \lceil \bar{v} \rceil$ where $\bar{v} \triangleq \min\{\mathbf{hx}\colon \mathbf{Ax} \ge \mathbf{b}, \mathbf{d} \ge \mathbf{x} \ge \mathbf{0}\}$. Clearly (6-111) holds for $\mathbf{g} \leftarrow \mathbf{h}$, $g_0 \leftarrow h_0$. Thus Lemma 6.27 proves $\mathbf{hx} \ge h_0$ can be generated by successive Chvátal construction.

If $\mathbf{hx} \ge h_0 + 1$ is not valid for F of (*BIP*), there is nothing more to prove. When $\mathbf{hx} \ge h_0 + 1$ is valid, we will show it can be derived by constructing in turn each member of the finite succession of inequalities $\mathbf{g}^{[s]}\mathbf{x} \ge g_0^{[s]}$ defined in (6-112)–(6-115) that ends on $(\mathbf{g}^{[\,]}, g_0^{[\,]}) = (\mathbf{h}, h_0 + 1)$. The proof proceeds in the lexicographic order of those inequalities, showing each can be constructed from its antecedents via (6-106).

We begin at $\mathbf{g}^{[0, \ldots, 0_n]} \triangleq \mathbf{h} + \Sigma_{j=1}^{n} \mathbf{e}^j$, $g_0^{[0, \ldots, 0_n]} \triangleq h_0 + 1$. Inequality $\mathbf{g}^{[0, \ldots, 0]}\mathbf{x} \ge g_0^{[0, \ldots, 0]} - 1$ is merely the sum of (valid) $\mathbf{hx} \ge h_0$ and all nonnegativity constraints of $\bar{F}$. Thus, it can be derived via Chvátal construction from F and $\mathbf{hx} \geq h_0$. Furthermore, for nonzero $\mathbf{x} \in \bar{F}$, at least one inequality $x_j > 0$ holds, and if $\mathbf{x} = \mathbf{0}$ $\mathbf{hx} > h_0$, since $\mathbf{hx} \geq h_0 + 1$ is valid for all integer $\mathbf{x} \in \bar{F}$. Thus, for all $\mathbf{x} \in \bar{F} \cap \{\mathbf{x}\colon \mathbf{hx} \geq h_0\}$, $\mathbf{g}^{[0, \ldots, 0]}\mathbf{x} > g_0^{[0, \ldots, 0]} - 1$. We see that condition (6-111) is fulfilled (with $\mathbf{hx} \geq h_0$ included in $\tilde{\mathbf{A}}\mathbf{x} \geq \tilde{\mathbf{b}}$), so by Lemma 6.27, $\mathbf{g}^{[0, \ldots, 0]}\mathbf{x} \geq g_0^{[0, \ldots, 0]}$ can be derived.

The inductive hypothesis is that $(\mathbf{h}, h_0)$ and all $(\mathbf{g}^{[t]}, g_0^{[t]})$ with $\mathbf{t} \overset{L}{<} \mathbf{s}$ can be constructed. We wish to show, that it follows $(\mathbf{g}^{[s]}, g_0^{[s]})$ can be constructed.

Case 1: **s** *is an n-vector.* When **s** is an n-vector, the argument is very similar to the $(\mathbf{g}^{[0, \ldots, 0]}, g_0^{[0, \ldots, 0]})$ start. Review of definition (6-114)–(6-115) will show $(\mathbf{g}^{[s_1, \ldots, s_n]}, g_0^{[s_1, \ldots, s_n]} - 1)$ is merely the sum of

$$\mathbf{hx} \geq h_0 \tag{6-116}$$

$$\mathbf{g}^{[s_1, \ldots, s_j - 1]}\mathbf{x} \geq g_0^{[s_1, \ldots, s_j - 1]} \qquad \text{for } j \text{ with } s_j > 0 \tag{6-117}$$

$$x_j \geq 0 \qquad \text{for } j \text{ with } s_j = 0 \tag{6-118}$$

By induction, all these inequalities have been derived. If every $\mathbf{x} \in \bar{F}$ that satisfies all these inequalities has at least one of (6-116)–(6-118) strict, the sum gives $\mathbf{g}^{[s_1, \ldots, s_n]}\mathbf{x} > g_0^{[s_1, \ldots, s_n]} - 1$, and Lemma 6.27 shows $(\mathbf{g}^{[s_1, \ldots, s_n]}, g_0^{[s_1, \ldots, s_n]})$ can be derived. We will show that the only possible solution satisfying all of (6-116)–(6-118) as equalities is $\mathbf{s} \triangleq [s_1, \ldots, s_n]$ itself. But **s** is an integer vector, and since $\mathbf{hx} \geq h_0 + 1$ is valid for all integer $\mathbf{x} \in \bar{F}$, $\mathbf{hs} = h_0$ implies $\mathbf{s} \notin \bar{F}$.

We proceed to show **s** is the unique possible equality solution to (6-116)–(6-118) by inducting on its length k. For $k = 0$ there is nothing to prove. Assuming the claim is true for $i = 0, 1, \ldots, k - 1$, consider a possible $\mathbf{s} = [s_1, s_2, \ldots, s_k]$. If (6-116)–(6-118) yields inconsistent equalities there is again nothing to prove. Otherwise, deleting whichever of (6-117) and (6-118) applies for $j = k$, shows by induction that we are left with an equality version of (6-116)–(6-118) having unique solution $[s_1, s_2, \ldots, s_{k-1}]$. If $s_k = 0$, $x_j = 0$ is a part of the full system, and the claim follows. If $s_k > 0$ we may sum all equalities but the $j = k$ entry in (6-116)–(6-118) to obtain

$$\left(\mathbf{h} + \sum_{\substack{s_j > 0 \\ j \leq k-1}} \mathbf{g}^{[s_1, \ldots, s_j - 1]} + \sum_{\substack{s_j = 0 \\ j \leq k-1}} \mathbf{e}^j\right)\mathbf{x} = \left(h_0 + \sum_{\substack{s_j > 0 \\ j \leq k-1}} g_0^{[s_1, \ldots, s_j - 1]}\right),$$

or by definitions (6-114)–(6-115)

$$\mathbf{g}^{[s_1, \ldots, s_{k-1}]}\mathbf{x} = g_0^{[s_1, \ldots, s_{k-1}]} - 1. \tag{6-119}$$

The kth (6-117) form is

$$\mathbf{g}^{[s_1, \ldots, s_k-1]}\mathbf{x} = g_0^{[s_1, \ldots, s_k-1]}.$$

It can be written

$$(\mathbf{g}^{[s_1, \ldots, s_k-2]} + \mathbf{g}^{[s_1, \ldots, s_{k-1}]})\mathbf{x} = g_0^{[s_1, \ldots, s_k-2]} + g_0^{[s_1, \ldots, s_{k-1}]}$$
$$(\mathbf{g}^{[s_1, \ldots, s_k-3]} + 2\mathbf{g}^{[s_1, \ldots, s_{k-1}]})\mathbf{x} = g_0^{[s_1, \ldots, s_k-3]} + 2g_0^{[s_1, \ldots, s_{k-1}]}$$
$$\vdots \qquad\qquad \vdots$$
$$(\mathbf{g}^{[s_1, \ldots, 1]} + (s_k - 1)\mathbf{g}^{[s_1, \ldots, s_{k-1}]})\mathbf{x} = g_0^{[s_1, \ldots, 1]} + (s_k - 1)g_0^{[s_1, \ldots, s_{k-1}]}$$
$$x_k + s_k(\mathbf{g}^{[s_1, \ldots, s_{k-1}]}) = s_k g_0^{[s_1, \ldots, s_{k-1}]} \tag{6-120}$$

Substituting (6-119) in (6-120) gives

$$x_k + s_k(g_0^{[s_1, \ldots, s_{k-1}]} - 1)\mathbf{x} = s_k g_0^{[s_1, \ldots, s_{k-1}]}$$

or

$$x_k = s_k.$$

We may conclude $\mathbf{x} = \mathbf{s}$ is the unique possible equality solution to (6-116)–(6-118) for all k, and Case 1 is complete.

CASE 2: $\mathbf{s}$ *is a k-vector, $k < n$.* In lexicographic order, we come to such cases immediately after

$$\mathbf{g}^{[s_1, \ldots, s_k, d_{k+1}]}\mathbf{x} \geq g_0^{[s_1, \ldots, s_k, d_{k+1}]}. \tag{6-121}$$

By the inductive hypothesis, (6-121) can be derived by the Chvátal construction. Proceeding as with (6-120) above, (6-121) can be rewritten

$$(\mathbf{g}^{[s_1, \ldots, s_k, d_{k+1}-1]} + \mathbf{g}^{[s_1, \ldots, s_k]})\mathbf{x} \geq g_0^{[s_1, \ldots, s_{n-1}, d_{k+1}-1]} + g_0^{[s_1, \ldots, s_k]}$$
$$\vdots \qquad\qquad \vdots$$
$$x_{k+1} + (d_{k+1} + 1)(\mathbf{g}^{[s_1, \ldots, s_k]})\mathbf{x} \geq (d_{k+1} + 1)(g_0^{[s_1, \ldots, s_k]}). \tag{6-122}$$

Also, one of the original $\bar{F}$ constraints is

$$-x_{k+1} \geq -d_{k+1}. \tag{6-123}$$

Applying Chvátal's construction directly with weights $1/(d_{k+1} + 1)$ on (6-122) and (6-123), we derive the inequality

$$\mathbf{g}^{[s_1, \ldots, s_k]}\mathbf{x} \geq \left\lceil g_0^{[s_1, \ldots, s_k]} - \frac{d_{k+1}}{d_{k+1} + 1} \right\rceil = g_0^{[s_1, \ldots, s_k]}.$$

This completes Case 2 and the proof of the theorem. ■

Example 6.15. Chvátal's Construction. To illustrate the derivation of an integer inequality via the sequence of the above proof, return to the Gomory cut Example 6.2 and Figure 6.2. Add bounds to create F as the set of integer (x_1, x_2) satisfying

$$x_1 - 3x_2 \geq -1$$

$$3x_1 + 2x_2 \geq 4$$

$$2 \geq x_1 \geq 0$$

$$1 \geq x_2 \geq 0.$$

We shall derive constraints with $\mathbf{h} \triangleq (1, 0)$. Figure 6.4 shows that the strongest is $x_1 \geq 2$. The start process of the proof of Theorem 6.26 would derive a start of $x_1 \geq \lceil \frac{10}{11} \rceil = 1$. To show more cases, however, we begin at $(\mathbf{h}, h_0) = (1, 0, 0)$ or $x_1 \geq 0$ and derive $x_1 \geq 1$. Repetition of the process would produce $x_1 \geq 2$.

Table 6.2 shows the $x_1 \geq 0$ to $x_1 \geq 1$ development. The sequence begins by summing $\mathbf{hx} \geq h_0$ and both nonnegativities to obtain $\mathbf{g}^{[0,0]}\mathbf{x} \geq g_0^{[0,0]} - 1$ or $2x_1 + x_2 \geq 0$. As in Case 1 of the proof, the only solution satisfying this constraint exactly is $\mathbf{x} = \mathbf{s} = [0, 0]$, and $[0, 0]$ violates the known (but not yet derived) $x_1 \geq 1$. Thus the conditions of Lemma 6.27 are fulfilled, and we can derive $2x_1 + x_2 \geq 1$.

Derivation of the next $\mathbf{g}^{[0,1]}\mathbf{x} \geq g_0^{[0,1]}$ is similar. The following step illustrates Case 2 of the proof of Theorem 6.26. We weight $4x_1 + x_2 \geq 2$ and upper bound $-x_2 \geq -1$ by $1/(d_2 + 1) = \frac{1}{2}$ to derive $2x_1 \geq \lceil \frac{1}{2} \rceil = 1$.

The process then continues as in Case 1 of the proof. Inequality $\mathbf{g}^{[1,0]}\mathbf{x} \geq g_0^{[1,0]} - 1$ is derived by summing $x_1 \geq 0$, $x_2 \geq 0$, $2x_1 \geq 1$ as $3x_1 + x_2 \geq 1$. This time (since $x_1 = 0$ cannot be consistent with $2x_1 = 1$) even $\mathbf{x} \leftarrow \mathbf{s}$ does not satisfy the constraint as an equality. Lemma 6.27 applies to derive $3x_1 + x_2 \geq 2$.

Details of the remainder of the lexicographic sequence are similar. The last entry is $\mathbf{g}^{[\,]}\mathbf{x} \geq g_0^{[\,]}$ or the desired $x_1 \geq 1$. ■

Like its subadditive and disjunctive analogs, Theorem 6.26 does show completeness of a valid inequality construction, but is not in any real sense constructive. Given the inequality $(\mathbf{h}, h_0)$ we wish to construct, the above proof shows it can be obtained by finite application of Chvátal's construction. Nothing tells us how to find an inequality we do not yet know, or even suggests that one we do know can be derived in polynomial time. Thus, like the prior theories, the algorithmic merit of Chvátal's result derives mostly from the easy method it gives for constructing simplified families of inequalities such as the edge covering ones of (6-102).

6.7 Direct Techniques

Sections 6.4–6.6 have presented the available complete theories of valid inequality generation for generic discrete optimization models. The methods give elegant theoretical tools and suggest many actual constructions for relaxed forms. Still, they do not exhaust active research in valid inequalities for discrete optimization.

For a host of specific discrete problems, *direct* investigation of underlying models has produced families of inequalities that are provably facetial. Since these results are inherently problem-specific, they all cannot be presented here. We will illustrate some main ideas in this section by

TABLE 6.2. Chvátal Derivation on Example 6.15

	To derive $\mathbf{hx} \geq h_0 + 1$:		
	$x_1 \geq 1$		
	Original Inequalities		
(*a*)	$x_1 - 3x_2 \geq -1$		
(*b*)	$3x_1 + 2x_2 \geq 4$		
(*c*)	$-x_1 \geq -2$		
(*d*)	$-x_2 \geq -1$		
(*e*)	$x_1 \geq 0$		
(*f*)	$x_2 \geq 0$		
	Partially Derived Inequality $\mathbf{hx} \geq h_0$:		
(*g*)	$x_1 \geq 0$		
	New Inequalities	**Lexicographic s**	**Derivation**
(*h*)	$2x_1 + x_2 \geq 1$	[0, 0]	(*g*) + (*e*) + (*f*), Lemma
(*i*)	$4x_1 + x_2 \geq 2$	[0, 1]	(*h*) + (*g*) + (*e*), Lemma
(*j*)	$2x_1 \geq 1$	[0]	$\frac{1}{d_2+1}((i)+(d))$
(*k*)	$3x_1 + x_2 \geq 2$	[1, 0]	(*j*) + (*g*) + (*f*), Lemma
(*l*)	$6x_1 + x_2 \geq 4$	[1, 1]	(*k*) + (*j*) + (*g*), Lemma
(*m*)	$3x_1 \geq 2$	[1]	$\frac{1}{d_2+1}((l)+(d))$
(*n*)	$4x_1 + x_2 \geq 3$	[2, 0]	(*m*) + (*g*) + (*f*), Lemma
(*o*)	$8x_1 + x_2 \geq 6$	[2, 1]	(*n*) + (*m*) + (*g*), Lemma
(*p*)	$4x_1 \geq 3$	[2]	$\frac{1}{d_2+1}((o)+(d))$
(*q*)	$x_1 \geq 1$	[ ]	$\frac{1}{d_1+1}((p)+(c))$

presenting Balas's (1975a) analysis of facetial inequalities for the knapsack problem.

The 0–1 minimization form of the knapsack problem is

$$(KP) \qquad \begin{aligned} \min \quad & \sum_{j\in N} c_j x_j \\ \text{s.t.} \quad & \sum_{j\in N} a_j x_j \geq b \\ & \left.\begin{aligned} 1 \geq x_j \geq 0 \\ x_j \text{ integer} \end{aligned}\right\} \quad \text{for all } j \in N \end{aligned}$$

with $N=\{1, 2, \ldots, n\}$ and $a_j > 0$ for all $j \in N$; $\Sigma_{j\in N} a_j \geq b > 0$. To discuss facetial inequalities we must first establish the dimension of the feasible space F of (KP).

Lemma 6.28. Dimension of the Knapsack Polytope. *Let (KP) be the above knapsack problem and F the polytope of feasible solutions. Then* $dim(F) = |J_0|$ *where* $J_0 = \{j\colon \Sigma_{k\in N\setminus j} a_k \geq b\}$.

Proof. As defined, J_0 contains all indices of variables that take on the value zero in some feasible solution. Clearly, for any $j \notin J_0$, $x_j = 1$ is valid for F. Thus $\dim(F) \leq |J_0|$. To see equality holds, pick $|J_0|$ vectors $\mathbf{x}^j$ with $x^j_j = 0$ and $\Sigma_{k\in N\setminus j} a_k x^j_k \geq b$. By definition of J_0, all these $\mathbf{x}^j$ belong to F, and by our assumption that $\Sigma_{j\in N} a_j \geq b$, the all ones vector belongs to F. But then differences $\{(\mathbf{x}^j - \mathbf{1})\colon\ j \in J_0\}$ are distinct unit vectors and $\dim(F) = |J_0|$. ■

Lemma 6.28 assures (KP) is full dimension if there is no j for which $x_j = 1$ in every solution. For simplicity in the remainder of this section, we *assume F is full dimension.* Also note that $b > 0$ implies $\mathbf{x} = \mathbf{0}$ is not feasible in (KP). Thus, in proving a valid inequality $(\mathbf{g}, g_0)$ is facetial for (KP), it is sufficient to show there are n linearly independent solutions $\{\mathbf{x}^j\colon j \in N\}$ such that $\mathbf{g}\mathbf{x}^j = g_0$ for all $j \in N$.

A *minimal cover* $C \subseteq N$ is a collection of indices such that there is no feasible solution with all $x_j = 0$ for $j \in C$, i.e.,

$$\sum_{j\in N\setminus C} a_j < b \tag{6-124}$$

yet with any element of C at one, solutions exist;

$$\sum_{j\in N\setminus C\cup k} a_j \geq b \qquad \text{for all } k \in C. \tag{6-125}$$

It follows by definition that some element of C must equal one in every solution. That is, inequality

$$\sum_{j\in C} x_j \geq 1 \tag{6-126}$$

is valid for (*KP*).

Suppose we order the elements of minimal covers C so that $j[1]$ is a $j \in C$ with $a_{j[1]} = \max_{k\in C}\{a_k\}$, $a_{j[2]} = \max_{k\in C\setminus j[1]}\{a_k\}$, etc. Then we may define the *extension,* of a minimal cover C by $\hat{C} \triangleq \{k \in N\setminus C: a_k \geq a_{j[1]}\}$. That is, $\hat{C}$ includes all $k_j \in N\setminus C$ with a_k as large as the largest of a_j among $j \in C$. From this definition we can establish validity of *minimal cover inequalities*

$$\sum_{j\in C\cup\hat{C}} x_j \geq |\hat{C}| + 1 \tag{6-127}$$

Lemma 6.29. Validity of Minimal Cover Inequalities. *Let (KP) be as above, C a minimal cover for (KP) and $\hat{C}$ the extension of C. Then minimal cover inequality (6-127) is valid for (KP).*

Proof. Inequality (6-127) can fail to be valid only if there is a solution $\hat{\mathbf{x}} \in F$ with $\Sigma_{j\in C\cup\hat{C}}\hat{x}_j \leq |\hat{C}|$. The definition of $\hat{C}$ shows that the largest $|\hat{C}|$ coefficients a_j among $\{a_j: j \in C \cup \hat{C}\}$ are the $\{a_j: j \in \hat{C}\}$. But, (6-124) establishes that even with all such $\hat{x}_j = 1$, and also $\hat{x}_j = 1$ for $j \in N\setminus(C \cup \hat{C})$, we cannot have a feasible solution to (*KP*). Thus, (6-127) always holds. ■

Example 6.16. Minimal Knapsack Covers. Before proceeding, let us illustrate the above definitions with an example. Consider the (*KP*)

$$\begin{aligned} \min \quad & 8x_1 + 7x_2 + 2x_3 + 9x_4 + 6x_5 + x_6 \\ \text{s.t.} \quad & 4x_1 + 3x_2 + 3x_3 + 6x_4 + 5x_5 + x_6 \geq 14 \\ & x_1, x_2, x_3, x_4, x_5, x_6 \text{ binary.} \end{aligned}$$

Set $C_1 = \{1, 5\}$ is a minimal cover because $\Sigma_{j\notin C_1} a_j = 13 < b$ and addition of any a_j with $j \in C_1$ brings the total to a feasible value. The extension $\hat{C}_1 \triangleq \{j \notin C_1: a_j \geq \max_{k\in C_1}\{a_k\}\} = \{4\}$. Thus, the full minimal cover inequality is

$$x_1 + x_4 + x_5 \geq 2.$$ ■

Not all minimal cover inequalities (6-127) are facetial. In Example 6.16, for example, the one for $C_1 = \{1, 5\}$ dominates that of $C_2 = \{4, 5\}$. Cardinality $|C_1| = |C_2|$, but $C_1 \cup \hat{C}_1 = C_2 \cup \hat{C}_2 \cup \{4\}$. Thus, $\{\mathbf{x}\colon \Sigma_{j\in C_2\cup\hat{C}_2} x_j \geq |\hat{C}_2| + 1\} \subseteq \{\mathbf{x}\colon \Sigma_{j\in C_1\cup\hat{C}_1} x_j \geq |\hat{C}_1| + 1\}$.

Balas (1975a) identifies exactly which minimal cover inequalities are facetial. The following theorem summarizes his results.

Theorem 6.30. Facetial Minimum Cover Inequalities. *Let (KP) be as above, C a minimum cover with elements j[1], j[2], . . . , j[|C|]* arranged in nonincreasing order of $a_{j[k]}$, and $\hat{C}$ the extension of C. Also, define *$\tilde{C} = N\setminus C$ and assume $i \in \tilde{C}$ are arranged i[1], i[2], . . . , i[|$\tilde{C}$|] in nonincreasing order of $a_{i[l]}$. Then if,*

$$(i) \qquad \sum_{k\in\tilde{C}\setminus i[|\hat{C}|+1]\cup j[1]} a_k \geq b$$

$$(ii) \qquad \sum_{k\in\tilde{C}\setminus i[1]\cup j[1]\cup j[2]} a_k \geq b$$

the corresponding strong minimal covering inequality (6-127) is facetial for (KP).

Proof. We shall show the result directly by exhibiting $n \triangleq |N|$ linearly independent solutions $\mathbf{x}^1, \mathbf{x}^2, \ldots, \mathbf{x}^n$ satisfying (6-127) as an equality. The matrix $\mathbf{X}$ with required solutions as columns has the form

$$
\mathbf{X} \triangleq
\begin{array}{r}
\\ j[1] \\ j[2] \\ \vdots \\ j[|C|] \\ i[1] \\ i[2] \\ \vdots \\ i[|\hat{C}|] \\ i[|\hat{C}|+1] \\ \vdots \\ i[|\tilde{C}|]
\end{array}
\begin{array}{c}
\begin{array}{ccc} C & \hat{C} & \tilde{C}\setminus\hat{C} \end{array} \\
\left[\begin{array}{c|c|c}
 & \mathbf{1} & \mathbf{1} \\ \hline
\mathbf{I} & \mathbf{1} & \mathbf{0} \\ \hline
 & \mathbf{0} & \mathbf{0} \\ \hline
\mathbf{1} & \mathbf{1}-\mathbf{I} & \mathbf{1} \\ \hline
\mathbf{1} & \mathbf{1} & \mathbf{1}-\mathbf{I}
\end{array}\right]
\end{array},
$$

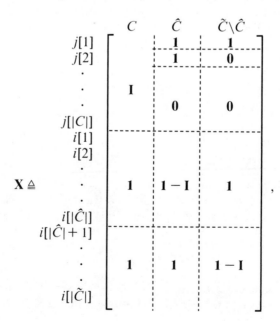

where **0** is a suitable vector or matrix of 0's, **1** is a vector or matrix of 1's, and **I** is an identity matrix. The first $|C|$ solutions follow directly from the definition of a minimal cover. Elements of C are combined in turn with all $i \in \tilde{C}$. This must produce a feasible $\mathbf{x}^k \in F$, and it has exactly $|\hat{C}| + 1$ elements of $C \cup \hat{C}$ at value 1.

The next $|\hat{C}|$ solutions are derived from property (ii). They are obtained by deleting elements of $\hat{C}$ in turn from $\tilde{C}$ and replacing with the two j's of C with greatest $a_{j[1]}$ and $a_{j[2]}$. Each such solution also has exactly $|\hat{C}| + 1$ components of $C \cup \hat{C}$, at value 1. Also, property (ii) implies that $\Sigma_{k \in \hat{C} \setminus i \cup j[1] \cup j[2]} a_k \geq b$ for all $i \in \hat{C}$, because the result holds when the $i[1]$ with greatest $a_{i[1]}$ is replaced.

The final $|\tilde{C} \setminus \hat{C}|$ solutions follow by property (i) of the theorem. Those solutions are constructed by replacing elements of $\tilde{C} \setminus \hat{C}$ in turn by $j[1]$ of C. Property (i) shows that all solutions are feasible because a feasible solution is obtained even when the largest a_i with $i \subset \tilde{C} \setminus \hat{C}$ is substituted by $a_{j[1]}$. Also $x_{j[1]} = 1$ and all $x_k = 1$ with $k \in \hat{C}$ in such solutions, so inequality (6-127) is an equality.

To complete the proof, we need to show matrix **X** is nonsingular. This is easily seen from the fact that diagonal submatrices are identities or complementary identity matrices **I**, and **1-I**. ■

The $C_1 = \{1, 5\}$ of Example 6.16 can easily be seen to fulfill conditions of Theorem 6.30. Here the ordered version of C_1 is $\{5, 1\}$ and ordered $\tilde{C}_1 = \{4, 2, 3, 6\}$. Substitution of $a_{j[1]} = 5$ for $a_{i[2]} = 3$ produces feasibility as in (i), and using $a_{j[1]} + a_{j[2]} = 9$ in lieu of $a_{i[1]} = 6$ demonstrates (ii). The **X** matrix of proof expression (6-128) is

$$\mathbf{X} = \begin{array}{c} \\ 5 \\ 1 \\ 4 \\ 2 \\ 3 \\ 6 \end{array} \begin{array}{c} \begin{array}{cc:c:cccc} 5 & 1 & 4 & 2 & 3 & 6 \end{array} \\ \left[\begin{array}{cc:c:ccc} 1 & 0 & 1 & 1 & 1 & 1 \\ 0 & 1 & 1 & 0 & 0 & 0 \\ \hdashline 1 & 1 & 0 & 1 & 1 & 1 \\ \hdashline 1 & 1 & 1 & 0 & 1 & 1 \\ 1 & 1 & 1 & 1 & 0 & 1 \\ 1 & 1 & 1 & 1 & 1 & 0 \end{array}\right] \end{array}.$$

EXERCISES

6-1. Suppose inequality $(\mathbf{g}, g_0)$ is such that $\mathbf{g} = \lambda_1 \mathbf{g}^1 + \lambda_2 \mathbf{g}^2$, $g_0 = \lambda_1 g_0^1 + \lambda_2 g_0^2$ for given $\lambda_1, \lambda_2 > 0$ and distinct valid inequalities $(\mathbf{g}^1, g_0^1)$, $(\mathbf{g}^2, g_0^2)$ for full dimension discrete set F.

(a) Show that $(\mathbf{g}, g_0)$ is valid for F.
(b) Show that $(\mathbf{g}, g_0)$ cannot be facetial for F.

6-2. Consider a nonempty discrete set F with polyhedral convex hull and a relaxation $\tilde{F}$ such that $\tilde{F} \supseteq F$ and $\tilde{F}$ also has polyhedral convex hull.

(a) Show that $(\mathbf{g}, g_0)$ valid for $\tilde{F}$ implies $(\mathbf{g}, g_0)$ valid for F.
(b) Show that $(\mathbf{g}, g_0)$ supporting for F and $(\mathbf{g}, g_0)$ valid for $\tilde{F}$ imply $(\mathbf{g}, g_0)$ supporting for $\tilde{F}$.
(c) Show that $(\mathbf{g}, g_0)$ facetial for F and $(\mathbf{g}, g_0)$ valid for $\tilde{F}$ imply $(\mathbf{g}, g_0)$ facetial for $\tilde{F}$.

6-3. Let F be a nonempty subset of $\mathbb{R}^n$. Show that F admits an improper valid inequality if and only if $\dim([F]) < n$.

6-4. Let F be a nonempty discrete set with polyhedral convex hull and $(\mathbf{g}, g_0)$ a valid inequality for F. Show that there exists a $\bar{g}_0 \in \mathbb{R}^1$, $\bar{g}_0 \geq g_0$ such that $(\mathbf{g}, \bar{g}_0)$ is supporting for F.

6-5. Consider the rational, mixed-integer, inequality constrained problem

$$
(MI) \qquad
\begin{aligned}
\min \quad & 5x_1 + \tfrac{3}{2}x_2 + x_3 - \tfrac{3}{4}x_4 - 2x_5 \\
\text{s.t.} \quad & \tfrac{2}{3}x_1 - \tfrac{1}{4}x_2 + x_3 + 3x_4 + \tfrac{2}{3}x_5 \leq 1 \\
& -\tfrac{5}{4}x_1 + \tfrac{3}{2}x_2 + 2x_3 + \tfrac{1}{2}x_4 + 6x_5 \leq 4 \\
& x_1, x_2, \geq 0 \text{ and integer; } x_3, x_4, x_5 \geq 0
\end{aligned}
$$

(a) Use the methods of Lemma 6.1 to reduce (MI) to an equivalent integer data, pure integer program with equality main constraints (F_7 of Table 6.1)
(b) Determine bounds on x_1 and x_2 and use them to reduce (MI) to an equivalent disjunctive program over equality constraints (F_5 of Table 6.1).

***6-6.** Consider a composite of parts (a) and (b) of Figure 6.2 in $F \triangleq \{\text{integer}(x_1, x_2) \geq \mathbf{0}\colon x_1 + x_2 \geq 1, \sqrt{5}\, x_2 \geq x_1\}$.

(a) Show that the convex hull $[F]$ is not closed.
(b) Show that $[F]$ has infinitely many facets, i.e., requires infinitely many facetial inequalities to define.
(c) Use the result of (b) to show that for any number k there exists a related set

$$F_k \triangleq \{\text{integer}(x_1, x_2) \geq \mathbf{0}\colon x_1 + x_2 \geq 1, \alpha_k x_2 \geq x_1\},$$

where α_k rational such that $[F_k]$ has at least k facets.

6-7. Solve the (*IP*) of Example 6.1 by Gomory Fractional Cutting Algorithm 6B. Show your progress graphically as in Figure 6.4 and comment on the numerical issues raised in Section 6.2.3.

6-8. Consider the simple pure integer program

$$\begin{aligned} \min \quad & 3x_1 + 9x_2 \\ \text{s.t.} \quad & 37x_1 - 9x_2 \geq 23 \\ & 13x_1 + 2x_2 \geq 36 \\ & x_1, x_2 \geq 0 \text{ and integer.} \end{aligned}$$

Solve by Gomory Fractional Cutting Algorithm 6B. Show your progress graphically as in Figure 6.4 and comment on the numerical issues of Section 6.2.3.

6-9. Consider the equivalent pure integer program form (IP_N) of Section 6.2.1. Show that each of the following *Dantzig cuts* is valid for the corresponding (*IP*) when $\bar{x}_i$ is noninteger for any $i \notin N$:

(a) $\sum_{j \in N} x_j \geq 1$

(b) $\sum_{j \in N_i} x_j \geq 1$ for any i with $\phi(\bar{x}_i) > 0$ and $N_i \triangleq \{j \in N: \bar{a}_{ij} \neq 0\}$

(c) $\sum_{j \in N_i} x_j \geq 1$ for any i with $\phi(\bar{x}_i) > 0$ and $N_i \triangleq \{j \in N: \phi(\bar{a}_{ij}) > 0\}$

6-10. Consider a primal all-integer cutting plane algorithm that interacts with the primal simplex algorithm so that a pivot always occurs in a +1, i.e., simplex representations in the form of (IP_N) of Section 6.2.1 have all $\bar{a}_{ij}$ and $\bar{x}_i$ integer. Young's (1965) algorithm takes this approach.

(a) Show that if normal primal simplex computations lead to a pivot in row r, column p and $\bar{a}_{rp} > 1$, then inequality

$$\sum_{j \in N} \left\lfloor \frac{\bar{a}_{rj}}{\bar{a}_{rp}} \right\rfloor x_j \leq \left\lfloor \frac{\bar{x}_r}{\bar{a}_{rp}} \right\rfloor$$

is valid.

(b) Show that if the inequality of (a) is now added to the simplex representation with slack, s basic, a pivot on the +1 in that row at column p is now consistent with primal simplex rules.

6-11. Show in detail that finite convergence of Gomory Fractional Cutting Algorithm 6B is still guaranteed if rule (ii) in Theorem 6.5 is relaxed by requiring choice of the smallest admissable r only on cut generation epoch numbers l, $2l$, $3l$, . . . for positive integer l.

***6-12.** Consider mixed integer program (*MIP*) of Section 6.2.2 with the added assumption that **cx** is integer for all feasible **x**. Show that application of Algorithm 6B to such an (*MIP*) with mixed-integer cuts (6-19) being used and rules (i) and (ii) of Theorem 6.5 observed, leads to finite solution of any such (*MIP*).

6-13. Theorem 6.9 established that for most discrete optimizations, facetial inequality recognition belongs to class D^p. Under the same hypothesis plus full positivity assumption (6-31), show that minimal valid inequality recognition is also in D^p.

6-14. Explain how membership of a problem in complexity class D^p is qualitatively quite different from membership in $NP \cap CoNP$.

***6-15.** Return to problems (P_D) and (D_D) of Section 6.4.2 and assume the hypotheses of Theorem 6.13 hold.

(a) Write the linear programming dual of disjunctive dual (D_D).
(b) Interpret the result of (a) as the version of (P_D) with $\mathbf{x} \in F$ replaced by $\mathbf{x} \in [F]$.
(c) Show how it follows from (b) that (D_D) provides a gapless dual for (P_D), i.e., $v(P_D) = v(D_D)$.

6-16. Consider a disjunctive program with feasible space $F = F_0 \cap D_1 \cap D_2$, where $F_0 = \{x_1, x_2 \geq \mathbf{0}: -2x_1 + 2x_2 \leq 1,\ 2x_1 + x_2 \leq 2\}$, and for $i = 1, 2$, $D_i = \{x_1, x_2: x_i \leq 0 \text{ or } x_i \geq 1\}$.

(a) Establish that this disjunction is facial.
(b) Illustrate Theorem 6.15 by constructing successive convex hulls and showing them graphically as in Figure 6.7.

6-17. As detailed in Section 6.4.2, task scheduling Example 6.5 is not facial. Show that facial formulation of the same instance can be derived by introducing new 0–1 variables.

6-18. Establish that the multiple-choice constraint model in Example 6.4 is a facial disjunctive program.

6-19. Suppose linear programming relations of multiple choice Example 6.4 are solved to optimality and basic variables then reexpressed in terms of nonbasic ones as in (IP_N) of **6-21.** If for some $k \in K$ at least two x_j with $j \in K_k$ are basic, detail how to apply corresponding disjunction (6-44) as a local disjunction in producing inequalities of Theorem 6.14.

6-20. Establish that each of the following functions is subadditive over the indicated domains and has value 0 at the zero vector.

(a) $g(a) \triangleq \lceil a \rceil$ over $a \in \mathbb{R}^1$
(b) $g(a) \triangleq \phi(a) \triangleq a - \lfloor a \rfloor$ over $a \in \mathbb{R}^1$

(c) $g(a) \triangleq a \bmod t$ over $a \in \mathbb{R}^1$ (for given integer modulus t)
(d) $g(\mathbf{a}) \triangleq \mathbf{la}$ over $\mathbf{a} \in \mathbb{R}^m$ (for given $\mathbf{l} \in \mathbb{R}^m$)
(e) $g(\mathbf{a}) \triangleq \left(\sum_{i=1}^{m} (\max\{a_i, -a_i\})^p \right)^{1/p}$ over $\mathbf{a} \in \mathbb{R}^m$ (fixed integer $p \geq 1$)

★6-21. Consider pure integer program (IP) of Section 6.5.1 and suppose it is reexpressed in terms of the optimal linear programming basis as

$$
\begin{array}{llll}
 & \min & \sum_{j \in N} \bar{c}_j x_j + v(\overline{IP}) & \\
(IP_N) & \text{s.t.} & \mathbf{x}^B = \overline{\mathbf{x}}^B - \sum_{j \in N} \bar{\mathbf{a}}^j x_j & (1) \\
 & & \mathbf{x}^B \geq 0 \text{ and integer} & (2) \\
 & & x_j \geq 0 \text{ and integer for all } j \in N, & (3)
\end{array}
$$

where N is the nonbasic set, $\mathbf{x}^B$ is the vector of basic variables and $\bar{\mathbf{a}}^j$ is the vector of simplex tableau coefficients for variable j. Gomory's (1965) group relaxation of the problem replaces (1) and (2) by

$$\sum_{j \in N} \phi(\bar{\mathbf{a}}^j)\, x_j \overset{\text{mod } 1}{\equiv} \phi(\overline{\mathbf{x}}^B), \tag{4}$$

where ϕ is the fractional part operator (generalized to apply to vectors).

(a) Show that this restatement is a relaxation of form (IP_N) in the sense of Section 5.1.2 and 6.1.1.
(b) State this group relaxation for the linear programming optimal basis of instance (IPX) in Example 6.2.
(c) Show that the modulo 1 sums computable on the left-hand-side of (4), i.e.,

$$G \triangleq \{\phi(\mathbf{a}) : \mathbf{a} = \sum_{j \in N} \phi(\bar{\mathbf{a}}^j) x_j,\ x_j \geq 0 \text{ and integer for all } j \in N\}$$

are finite in number. (Hint: Consider the determinant of the optimal basis and Cramer's rule.)
(d) Illustrate (c) on the instance of (b) by enumerating the 11 members of that G.
(e) Apply (c) to show that the group relaxation can be viewed as a shortest path problem from source $\mathbf{0}$ to sink $\phi(\overline{\mathbf{x}}^B)$ over a directed graph with vertex set G and arcs with cost $\bar{c}_k$ between all $\mathbf{v}^i$, $\mathbf{v}^j \in G$ such that $\mathbf{v}^j = \phi(\mathbf{v}^i + \phi(\bar{\mathbf{a}}^k))$ for some $k \in N$.
(f) State and solve the shortest path problem of (e) for the instance of (b).
(g) State the optimal relaxation solution value obtained and determine the $\mathbf{x}^B$ implied by optimal values for x_3 and x_4.

6-22. The *disjunctive dual* of Section 6.4.2 is formed by seeking the strongest possible valid inequality $\mathbf{cx} \geq v$ that can be derived by the disjunctive cut principle.

(a) Write a similar optimization over the set of subadditive functions with value zero at $\mathbf{0}$ that provides a *subadditive dual* for (*IP*) of Section 6.5.1.

(b) Rewrite the formulation of (a) as a linear program over infinitely many variables $g_{\mathbf{a}}$ defining the value of the needed subadditive function at each $\mathbf{a} \in M \triangleq \{\mathbf{Ax}: \mathbf{x} \geq \mathbf{0}$ and integer$\}$. (Hint: Express subadditivity requirement (6-80) as an inequality in the specified variables.)

(c) Repeat (b) for the problem (*IP*) as equivalently reexpressed in terms of the optimal linear programming relaxation basis (i.e. (IP_N) of **6-21**).

*(d) Show how (c) can be reduced to a finite linear program by considering only the finite set G of the group relaxation in **6-21** (c).

(e) Conclude that this finite linear program provides a gapless dual to the group relaxation and thus a lower bound on $v(IP)$.

(f) Detail the dual of (d) and (e) for the numerical example in **6-21** part (b).

6-23. Let F be the equality-constrained disjunction $\cup_{h\in H}\{\mathbf{x} \geq \mathbf{0}: \mathbf{A}^h\mathbf{x} = \mathbf{b}^h\}$ with H finite.

(a) Show (analogously to Theorem 6.12) that for any multiplier vectors $\mathbf{u}^1, \mathbf{u}^2, \ldots, \mathbf{u}^{|H|}$,

$$\sum_j \left(\max_{h\in H} \{(\mathbf{u}^h\mathbf{A}^h)_j\} \right) x_j \geq \min_{h \in H} \{\mathbf{u}^h\mathbf{b}^h\}.$$

(b) Show (analogously to Theorem 6.13i) that if $\{\mathbf{x} \geq \mathbf{0}: \mathbf{A}^h\mathbf{x} = \mathbf{b}^h\}$ is nonempty for every $h \in H$, every minimal valid inequality for F can be computed as in (a) for an appropriate choice of multiplier vectors.

6-24. Consider a finite disjunctive $\mathbb{R}^n$ set $F \triangleq \cup_{h\in H}\{\mathbf{x} \geq \mathbf{0}: \mathbf{Ax} = \mathbf{b}^h\}$ for integer $\mathbf{A}, \mathbf{b}^1, \mathbf{b}^2, \ldots, \mathbf{b}^{|H|}$, and assume $\{\mathbf{x} \geq \mathbf{0}: \mathbf{Ax} = \mathbf{b}^h\}$ is nonempty for every $h \in H$.

(a) Derive an equivalent mixed-integer programming space of the form $G \triangleq \{(\mathbf{x}, \mathbf{y}) \geq \mathbf{0}: \mathbf{Qx} + \mathbf{Ry} = \mathbf{d}, \mathbf{y}$ integer$\}$ where $G \subseteq \mathbb{R}^{n+|H|}$ and $\mathbf{x} \in F$ if and only if there exist $\mathbf{y}$ such that $(\mathbf{x}, \mathbf{y}) \in G$.

(b) Show that an inequality $\mathbf{gx} \geq g_0$ valid for the disjunctive formulation is also valid for the mixed-integer form of (a).

(c) Express the value function v^g calculations required to prove Theorem 6.23 for set G as optimizations over disjunctions in $\mathbf{x}$.

(d) Write the linear program of the proofs of Theorem 6.13*i* (adjusted to equality constraints as in **6-23**) yielding a disjunctive inequality as strong as $\mathbf{gx} \geq g_0$.

(e) Compare the inequality produced by computations (c) with the one of (d).

6-25. Consider the bounded pure integer program with feasible set

$$3x_1 + 2x_2 \geq 3$$
$$-x_1 + 3x_2 \geq 0$$
$$0 \leq x_1, x_2 \leq 2$$
$$x_1, x_2 \text{ integer}$$

(a) Show graphically the convex hull of solutions and illustrate that $x_1 + x_2 \geq 1$ is a face.

(b) Show a sequence like Table 6.2 of Chvátal inequality constructions that yields $x_1 + x_2 \geq 1$.

6-26. Suppose the (*BIP*) model of Section 6.6.1 is generalized to include equality constraints, i.e.,

$$\begin{aligned} \min \quad & \mathbf{cx} \\ \text{s.t.} \quad & \mathbf{Ax} \geq \mathbf{b} \\ & \mathbf{Rx} = \mathbf{r} \\ & \mathbf{d} \geq \mathbf{x} \geq \mathbf{0} \\ & \mathbf{x} \text{ integer.} \end{aligned}$$

Show that for any $\mathbf{t}$, $\mathbf{u} \geq \mathbf{0}$, $\mathbf{v} \geq \mathbf{0}$, $\mathbf{w} \geq \mathbf{0}$ the inequality

$$\mathbf{g} \longleftarrow \lceil \mathbf{uA} + \mathbf{tR} - \mathbf{v} + \mathbf{w} \rceil$$
$$g_0 \longleftarrow \lceil \mathbf{ub} + \mathbf{tr} - \mathbf{vd} \rceil$$

is valid for the above problem.

***6-27.** Consider the decision version ($BIP^{\leq}$) of (*BIP*) in Section 6.6.1, "Given integer $\mathbf{c}$, $\mathbf{A}$, $\mathbf{b}$, $\mathbf{d}$, and v, does there exist an integer $\mathbf{x}$ such that $\mathbf{Ax} \geq \mathbf{b}$, $\mathbf{d} \geq \mathbf{x} \geq \mathbf{0}$, $\mathbf{cx} \leq v$?" Show that if there is a k such that every facetial inequality for $F \triangleq \{\text{integer } \mathbf{x} \geq \mathbf{0}\text{: } \mathbf{Ax} \geq \mathbf{b},\ \mathbf{d} \geq \mathbf{x} \geq \mathbf{0}\}$ can be derived by at most k applications of Chvátal construction (6-106), then ($BIP^{\leq}$) $\in NP \cap CoNP$.

6-28. The symmetric traveling salesman problem on graph $G(V, E)$ can be formulated

$$\min \quad \sum_{j}\sum_{j>i} c_{ij}x_{ij}$$

$$\text{s.t.} \quad \sum_{i<k} x_{ik} + \sum_{j>k} x_{kj} = 2 \text{ for all } k \in V \tag{1}$$

$$\sum_{i,j\in S} x_{ij} \leq |S| - 1 \text{ for all } S \subset V \tag{2}$$

$$0 \leq x_{ij} \leq 1 \text{ for all } i, j > i \tag{3}$$

$$x_{ij} \text{ integer for all } i, j > i \tag{4}$$

For odd integer k, *comb inequalities* are derived from a *handle* vertex set H and k *teeth* sets $T_1, T_2, \ldots, T_k$ with $T_l \cap T_j = \cap$ for all l and j, $|T_l \cap H| \geq 1$ and $|T_l \backslash H| \geq 1$ for $l = 1, 2, \ldots, k$. The inequality implied is

$$\sum_{i,j\in H} x_{ij} + \sum_{l=1}^{k}\sum_{i,j\in T_l} x_{ij} \leq \sum_{l=1}^{k}(|T_l| - 1) - \left\lfloor \frac{k}{2} \right\rfloor$$

Show that this inequality can be derived by one application of Chvátal construction (6-106) with suitable multipliers on equations (1) for $i \in H$, and inequalities (2) for all T_l, $T_l \cap H$ and $T_l \backslash H$.

6-29. The uncapacitated facilities location problem can be formulated

$$\min \quad \sum_{i\in S}\sum_{j\in T} c_{ij}x_{ij} + \sum_{i\in S} f_i y_i$$

$$\text{s.t.} \quad \sum_{i\in S} x_{ij} = 1 \text{ for every } j \in T \tag{1}$$

$$0 \leq x_{ij} \leq y_i \text{ for every } i \in S \text{ and } j \in T \tag{2}$$

$$y_i \leq 1 \text{ for every } i \in S \tag{3}$$

$$y_i \text{ integer for every } i \in S \tag{4}$$

$$x_{ij} \text{ integer for every } i \in S\, j \in T, \tag{5}$$

where S is a set of candidate source locations and T a list of sinks to be served. For a subset $\{i_1, i_2, i_3\} \subseteq S$ of 3 sources and corresponding symmetric $J_1 = \{j_1, j_2\}$, $J_2 = \{j_2, j_3\}$, $J_3 = \{j_1, j_3\}$ defined by distinct $j_1, j_2, j_3 \in T$ the inequality

$$\sum_{k=1}^{3}\left(\sum_{j\in J_k} x_{i_k j} - y_{i_k}\right) \leq 1$$

is valid. Show that it can be derived by one application of Chvátal (6-106) on appropriate equalities (1) and inequalities (2).

6-30. The matching problem on a graph $G(V, E)$ can be formulated

$$
(MP) \qquad
\begin{aligned}
\min \quad & \sum_{(i,j)\in E} c_{ij}x_{ij} \\
\text{s.t.} \quad & \sum_{i<k} x_{ik} + \sum_{j>k} x_{kj} \leq 1 \text{ for all } k \in V \\
& 0 \leq x_{ij} \leq 1 \text{ for all } (i, j) \in E \\
& x_{ij} \text{ integer for all } (i, j) \in E
\end{aligned}
$$

For any odd cardinality subset $S \subseteq V$, $|S| \geq 3$, the inequality $\Sigma_{i,j\in S}\, x_{ij} \leq \frac{1}{2}(|S| - 1)$ is valid for (MP). Show that this inequality can be derived by one application of Chvátal construction (6-106).

***6-31.** Consider the Chvátal cut construction on a bounded integer programming feasible $\mathbb{R}^n$ set.

$$F = \{\mathbf{x} \geq \mathbf{0}\text{: } \mathbf{Ax} \geq \mathbf{b},\ \mathbf{x} \leq \mathbf{d},\ \text{x integer}\}.$$

(a) Add surplus variables to define an equivalent set $G \subseteq \mathbb{R}^{n+m}$ such that $\mathbf{x} \in F$ if and only if $\mathbf{y} \triangleq \mathbf{Ax} - \mathbf{b}$ implies $(\mathbf{x}, \mathbf{y}) \in G$.

(b) Show that one application of the Chvátal cut construction (6-106) on the original model can be viewed as applying a subadditive function to the columns of the form of (a) as in (6-81).

(c) Show inductively that inequalities derived by a finite sequence of Chvátal constructions beginning from the original form may also be viewed as applying a subadditive function on the form of (a).

7

Nonexact Algorithms

The use of nonexact (heuristic or approximate) procedures, particularly in discrete optimization, is certainly not of recent vintage. What is relatively new is the respectability associated with such procedures. Once, viewed as second class citizens, heuristics now occupy a position of substantially greater prestige. Accompanying and/or responsible for this change in status is a rapidly growing literature that has provided a mathematical underpinning for nonexact methods and produced numerous results and insights that often rival their exact counterparts in elegance, sophistication, and usefulness.

Our aim in this chapter is to present a unified view of nonexact procedures in discrete optimization. Following rather general discussions regarding the nature, use, and evaluation of such approaches, we consider a three-way classification of heuristics:

(i) greedy procedures,
(ii) local improvement schemes,
(iii) truncated exponential procedures.

We conclude the chapter with an examination of some formal limitations of nonexact procedures.

7.1 The Nature of Nonexact Procedures

7.1.1 Definition and Applicability

Stated in elementary terms, a nonexact procedure is one whose application yields a feasible solution that cannot be guaranteed to be optimal. From a slightly more formal perspective, we can define a nonexact algorithm to be one that, when applied to an instance of a particular problem, is guaranteed to find a *candidate solution* that is feasible. This candidate solution may be optimal or it may not be. As we shall see shortly, it may even be arbitrarily distant from an optimal solution.

Naturally, we would expect a nonexact procedure to at least produce feasible candidate solutions. In addition, the rapidity with which such solutions are obtained must be substantial in order to justify the use of any procedure that cannot guarantee optimality. There would be little interest in a nonexact algorithm whose computational demand differed only marginally from that of its optimal counterpart. Clearly then, a minimal requirement of a nonexact procedure would be that it be formally efficient. That is, our interest is confined usually to nonexact algorithms that stop in polynomially-bounded time.

Assuming that we have a choice in the matter, when does one resort to the use of a nonexact procedure? Indeed, under what circumstances would we be justified in opting for the less ambitious objectives of nonexact algorithms? Some authors suggest (rather unyieldingly) that verification of a problem's *NP*-hardness makes using an exact procedure questionable and thus is evidence enough that we must turn to a nonexact algorithm.

We believe the issue is a little more subjective. Certainly, a problem's being *NP*-Hard presents strong evidence that optimization of large instances may be economically unjustifiable or even hopeless. Still, the size at which this large instance phenomenon is encountered varies immensely with model form; methods of Chapters 5 and 6 can be used to solve many problems of practical importance. All *NP*-Hard problems, however, eventually reach sizes where exact methods are impractical and nonexact schemes must be considered.

7.1.2 Difficulties with Nonexact Procedures

Part of the reason for equivocation regarding whether or not to use a nonexact procedure is that we frequently have little or no notion of how well (or poorly) such procedures will perform. Clearly, if we could be certain that a heuristic would always produce solutions within some acceptable distance from optimality and do so efficiently, then the nonexact vs exact dilemma would be easier to resolve. Unfortunately, many nonexact procedures do not possess this sort of formal analysis, and for those that do, it is generally the case that our guarantees in terms of closeness to optimality admit ranges much greater than we would like.

The extent to which solution quality can suffer when using heuristics is easily illustrated. Consider a frequently cited scheme for treating the traveling salesman problem (find a minimum weight cycle in a graph that includes each vertex exactly once) known as the *nearest unvisited city* approach. This simple procedure selects an arbitrary starting vertex (city) and then grows a tour by always moving from a current vertex to one nearest it, selecting only from those not yet chosen. After all cities are included, the tour is closed by returning to the starting point. A simple 4-city instance is given below where entries in the matrix represent intercity travel costs.

	1	2	3	4
1	∞	2	70	50
2	6	∞	3	4
3	8	3	∞	2
4	50	5	6	∞

Starting with, say, city 1 we would generate the tour 1-2-3-4-1 having total cost of 57.

Now, it is easily verified that the optimal tour for the problem above is 1-2-4-3-1 having total cost of 20. For this instance, the tour generated is certainly not a good one. The quality of a solution constructed by this particular heuristic can be influenced, however, by the choice of starting point. As a precautionary measure, let us select each of the remaining three cities as initial ones and repeat the procedure (clearly this does not change the polynomiality of the heuristic). The tours thus generated are 2-3-4-1-2, 3-4-2-1-3, and 4-2-3-1-4 with a total cost of 57, 83, and 66 respectively. Unfortunately, none of these are improvements on the original solution. Even more negatively, if the intercity costs for city pairs (1, 3), (1, 4) and (4, 1) are altered appropriately, the nearest unvisited city approach can be made to produce solutions (for this instance) that are arbitrarily poor!

While the foregoing illustration underscores the limits of deterioration of solution quality relative to nonexact procedures, we might also ask: will nonexact procedures be devious? That is, while heuristic procedures may produce results that are disappointingly poor, will they also produce results that are counterintuitive? This issue is easily addressed by considering another illustration due to Graham (1976).

A fundamental problem in deterministic scheduling is one of assigning a batch of jobs to a set of identical processors so as to minimize the overall completion time of the entire batch. Precedence restrictions governing job processing order must be adhered to by any schedule, and no processor can work on more than one job at a time. This is a formally difficult and much studied problem in scheduling and one which is often treated by nonexact procedures. The list processing algorithm is a popular choice. With such a scheme, we form a list of the jobs based on some rule and then begin selecting jobs from the list for scheduling. When a processor becomes available, the first unscheduled job in the list is found that can be performed subject to precedence. The list is repeatedly scanned in this manner until all jobs have been scheduled.

Consider the directed graph in Figure 7.1 where vertices depict jobs and arcs, precedence relationships. Job duration times are labeled next to each vertex. Suppose there are two processors available at time zero and let our list be given by L as shown. Applying the list processing procedure would produce the schedule shown in (a) and having completion time of 20.

Now, suppose we repeat the scheme on a slightly modified version of this instance. In particular, let us keep everything as before but change only the job duration times by adding one time unit to the previous times. Applying the list processing algorithm to the same list yields the schedule shown in (b) requiring only 19 time units. Curiously, we have taken longer to perform each separate job but less time to process the entire batch!

The two illustrations just presented demonstrate and even more starkly underscore the potential weakness of nonexact procedures. Of course, under special problem structures, phenomena such as these may be avoidable. Even for arbitrary structures their occurrence may be so infrequent that as a practical matter they can be neglected. The point is that their occurrence is possible, and, if we are to say anything formal regarding the performance of nonexact algorithms, we must be able to account for any poor and/or anomalistic behavior they might exhibit.

7.2 Measures of Algorithm Performance

In practical problem solving the notion of performance relative to nonexact algorithms can be somewhat elusive or at the very least subject to interpretation. For example, the quality of computer codes for identical procedures can often exhibit enough variation to introduce distinguishable differences in computational behavior. Such variations, of course, are influenced by factors independent of the specific algorithm and in any event, should not affect an algorithm's formal efficiency classification (see Chapter 2).

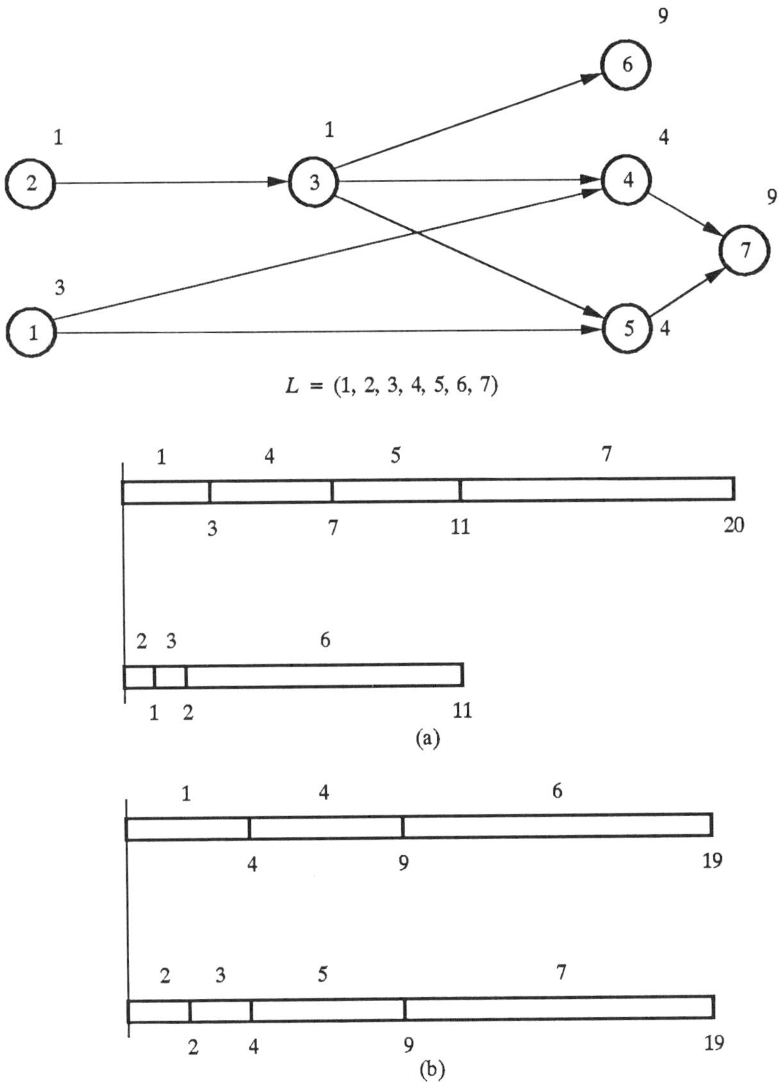

Fig. 7.1. Anomalistic behavior of heuristic procedures.

As a consequence, we shall discount these issues and direct our attention solely in terms of the quality of solutions produced by an algorithm. Specifically, we consider the *performance ratio,*

$$\rho_H(t) \triangleq \begin{cases} \dfrac{v(t)}{v_H(t)} & \text{if } t \text{ is a maximize problem} \\ \\ \dfrac{v_H(t)}{v(t)} & \text{if } t \text{ is a minimize problem} \end{cases}, \quad (7\text{-}1)$$

where $v_H(t)$ is the value of the solution returned by nonexact algorithm H on instance t and $v(t)$ is the value of an optimal solution to instance t.

Observe that this measure already admits a variety of anomalies. For example, adding a constant in the objective changes no solutions but *does* impact the ratio. More crucially, the ratios make sense only when one assumes solution values to be nonnegative. Also, in contrast to the usual situation in optimization, minimization cases are qualitatively different from those of the maximize form.

7.2.1 Empirical Analysis

Prior to the Cook/Karp developments in complexity, it was very common (if indeed not the rule) for nonexact algorithms to be evaluated by purely observational means. That is, procedures were put to various computational tests, statistics were drawn and inferences were made regarding the routine application of the algorithms. This form of analysis can, certainly, provide valuable insights. It is the case, however, that it can also be plagued by problems.

The following scenario would be typical of the empirical analysis suggested. The algorithm designer, having put together a (nonexact) procedure, say A, would then create a set of test problems of varying sizes. Within each size group various replications might be considered, varying in each the problem's data (e.g., coefficients, etc.). Each such sample problem generated would be solved by an exact algorithm in order that a known optimum would be at hand. The problems then would be fed to the nonexact algorithm and a grand mean (performance) ratio would then be calculated and/or the mean ratios within size groups.

Assuming that an overall mean ratio will suffice, we would report the average ratio for A as $\bar{\rho}_A \geq 1$. Such ratios are frequently very close to 1 (e.g., 1.15, 1.09, etc.) and can be useful as a performance estimate of an algorithm.

Unfortunately, such testing contains a built-in contradiction. Normally, performance ratios can be measured exactly only for problems that can be solved optimally. These are precisely the instances for which heuristic solution is not recommended. A good lower bound on the optimum (for a minimization problem) can suffice for a satisfactory estimate of $\bar{\rho}$, but this too suggests there is a branch and bound scheme where employing the available bound would yield good exact results.

A second indictment of empirical testing arises in conjunction with the specific structures of test problem instances. That is, can we be confident that the test problems are suitably difficult? For a few model forms there are bench-mark problems—ones specially constructed and universally accepted as strong tests for an algorithm. Such problems, however, are frequently not readily available or certainly not very abundant.

In the end, one is often left with personally constructed instances where a host of subtle issues may arise. For example, a given algorithm may perform very well when coefficients possess a particular characteristic, and yet perform dreadfully when such a characteristic is not present. If the computational testing is performed only on instances from the former more desirable class, any statement regarding general solution quality (i.e., average ratio $\bar{\rho}$) must be taken with a heavy dose of skepticism.

The provable approaches to heuristic performance analysis that we introduce in the next two subsections at least offer a mathematical surety empirical tests cannot match. The reader should be warned, however, that they too are frought with problems and difficulties.

7.2.2 Performance Guarantees

One of the first mathematical developments for nonexact algorithms appeared in the mid-1960s and was due to Graham (1966) who demonstrated that under fairly mild conditions, various heuristic procedures for machine scheduling not only could produce solutions of poor quality, but ones with anomalous characteristics as well (see the discussion regarding Figure 7.1). Moreover, Graham was able to establish realizable upper bounds on the ratio of solution values produced by a class of algorithms to optimal values given any instance of the particular class of scheduling problems. These worst case bounds provide a warning regarding a solution's potential departure from the optimality but at the same time confine, in a formal way, the limits of such departure.

Graham's contributions stood somewhat alone for a few years and it was not until later that similar developments were advanced in other areas. Since that time, research in the field has been quite productive, giving rise to numerous and varied results.

To develop a formal framework, consider a heuristic algorithm A and let $\rho_A(t)$ be as defined in (7-1) (i.e., $v(t)/v_A(t)$ for a maximize and $v_A(t)/v(t)$ for a minimize problem). A *guaranteed absolute performance ratio*, ρ_A satisfies

$$\rho_A \triangleq \inf\{\alpha\colon \alpha \geq \rho_A(t) \text{ for every instance } t\}. \tag{7-2}$$

Graham's work showed that a specific nonexact algorithm for parallel processor scheduling with m independent processors guaranteed the performance ratio $\rho_A = 2\text{-}1/m$. A much simpler ratio to derive is available for the nonnegative integer knapsack problem.

Example 7.1. Guaranteed Performance Ratio for Nonnegative Integer Knapsack Problems. The stated problem can be given as

$$(IKP) \qquad \begin{aligned} \max \quad & \Sigma_j\, c_j x_j \\ \text{s.t.} \quad & \Sigma_j\, a_j x_j \le b \\ & x_j \ge 0 \text{ and integer for all } j, \end{aligned}$$

where the c_j are positive weights and all a_j and b are positive integers. We also assume (to avoid triviality) that $b/a_j \ge 1$ for all j.

Now, it is easy to show the linear programming relaxation $(\overline{IKP})$ of (IKP) has value bc_{j*}/a_{j*}, where j^* is such that $c_{j*}/a_{j*} = \max_j\{c_j/a_j\}$. This value provides an upper bound on $v(IKP)$.

Now, the heuristic solution inspired by this linear programming result is to choose

$$x_j = \begin{cases} \left\lfloor \dfrac{b}{a_J} \right\rfloor & j = j^* \\ 0 & \text{otherwise} \end{cases}.$$

The implied performance ratio is at most

$$\frac{c_{j*}\dfrac{b}{a_{j*}}}{c_{j*}\left\lfloor \dfrac{b}{a_{j*}} \right\rfloor} = \frac{\dfrac{b}{a_{j*}}}{\left\lfloor \dfrac{b}{a_{j*}} \right\rfloor}.$$

We have assumed $b \ge a_{j*}$ so that the denominator is at least 1. Also, the numerator and denominator cannot differ by more than 1. From these facts, it is easy to see that the ratio will be worst when $\lfloor b/a_{j*} \rfloor = 1$ and b/a_{j*} is very close to 2. Hence, we can conclude that $\rho_A = 2$ is a valid absolute performance guarantee for the indicated heuristic and (IKP). ■

Complementing the notion of an absolute performance result is that of an *asymptotic performance guarantee.* As before, let our algorithm be A and denote the asymptotic ratio by $\hat{\rho}_A$. We have

$$\hat{\rho}_A \triangleq \inf \left\{ \alpha: \begin{array}{l} \text{for some strictly positive integer } k, \\ \hat{\rho}_A(t) \le \alpha \text{ for every instance } t \text{ where } v(t) \ge k \end{array} \right\}.$$

To illustrate, we might have a minimize problem with $v_A(t)/v(t) \leq 5/4 + 3/v(t)$; where $\hat{\rho}_A$ is 5/4. For arbitrarily large instances the term $3/v(t)$ becomes negligible. Instances of smaller size, however, might exist with $\rho_A(t) > 5/4$ revealing that ρ_A and $\hat{\rho}_A$ need not be equivalent.

7.2.3 *Approximation Schemes*

We could view the performance ratios defined above in terms of some constant perturbation from 1. If this perturbation is given by $\epsilon > 0$, then we might refer to an algorithm possessing finite worst case performance bound as an *ϵ-approximate algorithm.* That is, for A applied to an instance of a given problem we have that

$$\rho_A(t) \leq \rho = 1 + \epsilon$$

for every instance t.

Above, the value of ϵ is fixed. Suppose, however, that as part of the input to our algorithm we specify not only the instance t, but also a value of ϵ. In other words, ϵ becomes something of a desired accuracy or accuracy requirement. Applying A on instance t augmented with requirement $\epsilon > 0$ would then produce solution value $v_A(t, \epsilon)$ whereupon

$$1 + \epsilon \geq \rho_A(t, \epsilon) \triangleq \begin{cases} \dfrac{v_A(t, \epsilon)}{v(t)} & \text{for minimize problems} \\ \\ \dfrac{v(t)}{v_A(t, \epsilon)} & \text{for maximize problems} \end{cases}.$$

Such a procedure, which we can relativize as $A(\epsilon)$, is referred to by Garey and Johnson (1979) as an *approximation scheme.* This choice of terms signals only that A provides a spectrum of ϵ-approximate procedures for a given problem, one for each fixed value of $\epsilon > 0$.

Now, if $A(\epsilon)$ is a polynomial time procedure for each fixed $\epsilon > 0$, then we can say that the algorithm A is a *polynomial time approximation scheme.* If the order of computation of algorithm A itself is bounded by a polynomial function of the length of an instance t and $1/\epsilon$ then A is said to be a *fully polynomial time approximation scheme.* Short of establishing the equivalence of P and NP, the construction of fully polynomial time approximation procedures is about as much as we can hope for in negotiating many discrete optimization problems.

Because they do approach the satisfaction provided by polynomial time exact algorithms, polynomial approximation schemes of all types tend to be rare and technically demanding. We shall simply illustrate their flavor with a polynomial time approximation scheme for the familiar knapsack problem by Sahni (1975) (see Ibarra and Kim (1975) for a fully polynomial scheme for the same problem).

Example 7.2. Polynomial Approximation of the Knapsack Problem. Specifically, let us consider the 0–1 version of (*IKP*) and assume, without loss of generality, that variables x_j are indexed in nonincreasing c_j/a_j ratio. We may define the following heuristic: Let $I \subseteq \{1, 2, \ldots, n\}$ and for a fixed integer $k \geq 0$, generate all subsets for which $|I| \leq k$ and $\Sigma_{i \in I} a_i \leq b$. Then relative to the reduced problem defined by x_i, $i \in I$, fill the resultant knapsack (with capacity $b - \Sigma_{i\in I} a_i$) in the stated c_i/a_i-order rejecting a choice only if there is insufficient remaining capacity. Letting the value of this heuristically constructed solution to the reduced problem be $v_r(\tilde{I})$, then the corresponding overall solution (for given I) will have value $\Sigma_{i\in I} c_i + v_r(\tilde{I})$. Here $\tilde{I}$ is the complement of I. Thus, any suitable subset, I, leading to a maximum value defines the solution (for given k) by the stated approximation algorithm.

In order to demonstrate, consider the following instance:

$$\begin{aligned} \max \quad & 3x_1 + 2x_2 + 7x_3 + 4x_4 + 5x_5 \\ \text{s.t.} \quad & x_1 + x_2 + 5x_3 + 3x_4 + 9x_5 \leq 17 \\ & 0 \leq x_j \leq 1 \text{ and integer for all } j. \end{aligned}$$

If $k = 0$, then $I = \emptyset$ and we define the reduced knapsack instance given by $\tilde{I} = \{1, 2, 3, 4, 5\}$. In this case, we would generate the heuristic solution as $\mathbf{x} = (1, 1, 1, 1, 0)$ with value 16. If $k = 1$, we see that the singletons $\{1\}$, $\{2\}$, $\{3\}$, and $\{4\}$ produce the same solution as that when $I = \emptyset$. Letting $I = \{5\}$, we see that the 4-variable reduced problem (with capacity 8) sets $x_1 = x_2 = x_3 = 1$ and $x_4 = 0$. Thus, $v_r(\{1, 2, 3, 4\}) = 12$ and the overall solution $(1, 1, 1, 0, 1)$ results with value $c_5 + v_r(\{1, 2, 3, 4\}) = 17$. We leave it to the reader to perform the calculations for $k \geq 2$, while noting that the stated solution for $I = \{5\}$ is optimal.

In order to see that this procedure constitutes a polynomial time approximation scheme, observe that for given k, the effort required to consider the subsets I is $\Sigma_{i=0}^{k} \binom{n}{i} \leq \Sigma_{i=0}^{k} n^i$, i.e., $O(n^k)$. Moreover, for a given I, the time required by the heuristic applied to the reduced knapsack problem (given by $\tilde{I}$) is $O(n)$ and so the overall procedure has complexity $O(n^{k+1})$. Now, if we denote our heuristic by $A(k)$, then Sahni (1975) has shown that the stated procedure will produce solutions satisfying $\rho_{A(k)} \leq 1 + 1/k$. We will not give the details of the proof but rather simply indicate the key ideas. Essentially, one can argue that the difference $v(t) - v_{A(k)}(t)$ is bounded from above by $\bar{c}$, which has magnitude no greater than the $(k + 1)$-*st* largest c_j. This, along with the inequality $\bar{c} \leq v(t)/k + 1$, yields the bound on $\rho_{A(k)}$. Thus, for a desired accuracy ϵ, we simply let k be the smallest value no less than ϵ^{-1}. Observe that the resultant algorithm is not a fully polynomial time approximation scheme since ϵ^{-1} appears in the exponent. ■

7.2.4 *Expected Performance Analysis*

The worst case perspective is often criticized as the pessimistic or catastrophic view of nonexact algorithm behavior. It is frequently pointed out that the more meaningful measure of algorithm performance is that which reflects what might be expected to happen in a typical problem encounter. Usually accompanying such sentiment is evidence suggesting that, while a particular procedure in the worst case can be very poor, it will generally produce solutions that on average are only a few percent (i.e., 5%, 6%, etc.) away from optimality.

This sort of typical behavior perspective is an important one but one of which to be wary. Certainly, if we possessed a crystal ball indicating that instances forcing an algorithm to achieve its worst case performance were truly pathological, then we might be legitimately less interested in worst case issues. Indeed, such pathologies could be considered, as in a statistical sense, to be outliers and dismissed accordingly. We could then concentrate on algorithm behavior relative to the so-called real instances of our problem.

First of all, note that any sort of analysis of these real problem instances will not reveal how an algorithm will perform on a specific instance. At least with guaranteed performance analysis, we have an element of certainty in terms of a bound on such performance. Another nontrivial issue is what constitutes one of these real problem instances and what would an analysis of a set of such instances provide? What would we expect it to provide?

Sometimes, the mode of evaluation sketched in the earlier section on empirical analysis is termed *average-case analysis.* Performed as described above, such studies do not, in fact, provide average-case results. To do so implies that we know a probability distribution characterizing problem instances.

The structure of a true average or expected case analysis of a nonexact algorithm is simple to state. The components are:

(i) the algorithm,
(ii) a measure reflecting the (successful) performance of the algorithm, such as solving a problem within a specified performance, and
(iii) a sample space of input instances for the algorithm.

Of these three constituent components, the third is clearly the greatest threat to validity of analysis. Certainly, in setting up a sample space we must make allowances for instances of arbitrarily large size. For example, if our algorithm is for solving traveling salesman problems, we must be capable of accepting instances with an arbitrarily large number of cities or vertices. This would seem to deny one formal requirement for a proper characterization of the distribution of instance sizes. Fortunately, we can skirt this issue by constructing a sample space in two stages. As suggested in Karp (1976), we first define a sample space S_n for each n, from which instances of size n are drawn. After this, we can construct an overall sample space $\zeta = S_1 \times S_2 \times \ldots \times S_n \times \ldots$. That is, ζ is simply the Cartesian product of the spaces S_k, $k = 1, 2, \ldots$. Letting x_n denote an element drawn from S_n, then an element of ζ is $\mathbf{x} = x_1, x_2, \ldots, x_n, \ldots$. We would say that our algorithm almost surely is successful if, when $\mathbf{x}$ is drawn from ζ, the number of x_n on which the algorithm does not meet its performance measure is finite with probability 1.

Even when a suitable sample space can be designed, the analysis leading to successful expected case results can be exceedingly tedious. As some form of testimony to this phenomenon, the reader may want to compare the spartan list of the latter results in the literature versus those for guaranteed performance.

There are, of course, some success stories. One notable probabilistic model was developed by Karp (1977) relative to the Euclidean distance traveling salesman problem. Essentially, the algorithm partitions the plane into a set of regions within each of which a small traveling salesman problem is solved. These mini problem solutions are then patched together in a particular way forming an overall tour through the original points. If each of the small traveling salesman problems contain no more than l points, Karp was able to show that for points distributed randomly in the plane

$$\rho \leq 1 + O(l^{-1/2})$$

as p, the total number of vertices, becomes large.

It is interesting to compare the above result to its known worst case analog. The Euclidean traveling salesman problem is but a special case of the metric problem (symmetric and nonnegative edge weights obeying the triangle inequality) in which case its solution by an algorithm due to Christofides (1976) can be, in the worst case, 50% greater than optimal. Moreover, this $\rho = 3/2$ result is presently the best known.

7.3 Greedy Procedures

Having examined available ways for mathematically evaluating nonexact algorithms, we are now ready to discuss characteristics of typical forms such algorithms may take. We classify the forms into three main strategies —greedy, local improvement and truncated exponential.

7.3.1 General Concepts

In a most general sense a *greedy* procedure is one in which the decision-maker selects, at each stage of the process, an alternative that is best among the feasible alternatives without regard to the impact the choice may have on subsequent decisions. Normally, best simply implies most favorable with respect to the objective function. The nearest unvisited city heuristic for the traveling salesman problem mentioned earlier in this chapter offers one example of a greedy procedure. Another interesting one arises in change-making.

Example 7.3. Greedy Solution of Change-Making Knapsacks. Consider the simple knapsack problem of minimizing the number of coins required to make change totaling κ in value. It is obvious that the greedy choice, i.e., the coin making the greatest progress toward a specified amount of change, is the largest that will fit. Taking U.S. coins as an example, we could start this greedy idea by setting the number of fifty-cent pieces, say x_{50}, as $\lfloor \kappa/50 \rfloor$ (assume kappa is expressed in cents). Continuing, the number of quarters would be $x_{25} = \lfloor (\kappa - 50x_{50})/25 \rfloor$, the number of dimes, $x_{10} = \lfloor (\kappa - 50x_{50} - 25x_{25})/10 \rfloor$, the number of nickels, $x_5 = \lfloor (\kappa - 50x_{50} - 25x_{25} - 10x_{10})/5 \rfloor$, and the number of pennies, $x_1 = \kappa - 50x_{50} - 25x_{25} - 10x_{10} - 5x_5$. Hence, for $\kappa = 90$ the procedure would imply 1 fifty-cent piece, 1 quarter, 1 dime and 1 nickel.

The surprising thing about this greedy heuristic is that it always gives an optimal solution for U.S. denomination coins (see Magazine, et al. (1975)) for a proof and conditions for related problems). Still, it would be easy to invent coinage systems where the heuristic fails. Consider, for example, adding a twenty-cent piece to U.S. coinage. For $\kappa = 40$ the greedy algorithm would suggest a quarter, a dime and a nickel when two of the new coins would suffice. ■

There are not many general conclusions one can draw from such examples. We know from the earlier discussion of the nearest unvisited city algorithm for traveling salesman problems that greedy heuristics can frequently produce arbitrarily bad solutions. Still, in special circumstances, we have seen thc solutions may even be optimal.

7.3.2 Performance Guarantees for Independence Systems

Chapter 3 showed how matroid theory has provided a formal underlying structure for a host of polynomially-solved problems in discrete optimization. We also demonstrated that the problem of finding the maximum weight independent set of a matroid could be solved optimally be selecting elements in order by weight.

Independence systems may be viewed as relaxations of matroids. They, too, are posed over a nonempty finite set G, and a collection $\mathscr{I}$ of *independent* subsets $I \subseteq G$, closed under inclusion, i.e.,

$$I_1 \subseteq I_2,\ I_2 \in \mathscr{I} \text{ imply } I_1 \in \mathscr{I}. \tag{7-3}$$

With nonnegative weights w_e, the natural optimization problem for an independence system is the maximize form

$$\max\left\{\sum_{e \in I} w_e\colon I \in \mathscr{I}\right\}. \tag{7-4}$$

If $\mathscr{I}$ also satisfied the axiom that all maximal independent subsets of every $S \subseteq G$ have equal cardinality, the independence system would be a matroid, and (7-4) would be exactly the search for a maximum weight independent set. We know the pure greedy algorithm of selecting elements in nonincreasing weight order yields optimal solutions.

Following Korte and Hausmann (1978) and Hausmann, Korte and Jenkyns (1980), we can show how the same greedy procedure yields at least a guaranteed performance ratio for any independence system. To do so, define for any $S \subseteq G$ *lower* and *upper rank functions*

$$\text{lr}(S) \triangleq \min\{|I|\colon I \text{ is a maximal independent subset of } S\}$$

$$\text{ur}(S) \triangleq \max\{|I|\colon I \text{ is a maximal independent subset of } S\}.$$

These functions taken together define the *rank quotient*

$$\max_{S \subseteq G} \frac{\text{ur}(S)}{\text{lr}(S)}. \tag{7-5}$$

For matroids, lower and upper rank are equal and the rank quotient is 1. For more general independence systems we have the following result.

Theorem 7.1. Performance Guarantee for Greedy Maximization. *Let $(G, \mathscr{I})$ be an independence system, I^* an optimum in (7-4), and I_g the feasible solution to (7-4) obtained by sequentially placing in I_g the $e \in G$ defined by $w_g = \max\{w_e\colon I_g \cup e \in \mathscr{I}\}$. Then the rank quotient of $(G, \mathscr{I})$ yields a performance guarantee on this greedy procedure, i.e.,*

$$1 \le \frac{\sum_{e \in I^*} w_e}{\sum_{e \in I_g} w_e} \le \max_{S \subseteq G} \frac{\text{ur}(S)}{\text{lr}(S)}. \tag{7-6}$$

Proof. Order G as $\{e[1], e[2], \ldots, e[n]\}$ with $w_{e[i]} \geq w_{e[j]}$ for $i<j$. Then define sets $G_i \triangleq \{e[1], e[2], \ldots, e[i]\}$. In terms of this notation (taking $w_{e[n+1]} \triangleq 0$) the values of the two solutions I^* and I_g are

$$\sum_{e \in I^*} w_e = \sum_{i=1}^{n} |I^* \cap G_i|(w_{e[i]} - w_{e[i+1]}) \tag{7-7}$$

$$\sum_{e \in I_g} w_e = \sum_{i=1}^{n} |I_g \cap G_i|(w_{e[i]} - w_{e[i+1]}). \tag{7-8}$$

Now, since $I^* \in \mathscr{I}$, it must be true that $|I^* \cap G_i| \leq \mathrm{ur}(G_i)$. Similarly, the greedy construction of I_g assures that $I_g \cap G_i$ is a maximal independent subset of G_i, so that $|I_g \cap G_i| \geq \mathrm{lr}(G_i)$. Combining these observations,

$$\frac{|I^* \cap G_i|}{|I_g \cap G_i|} \leq \frac{\mathrm{ur}(G_i)}{|I_g \cap G_i|} \leq \frac{\mathrm{ur}(G_i)}{\mathrm{lr}(G_i)} \leq \max_{S \subset G} \frac{\mathrm{ur}(S)}{\mathrm{lr}(S)}.$$

Substituting in (7-7) and (7-8) gives

$$\sum_{e \in I^*} w_e \leq \sum_{i=1}^{n} \left(\max_{S \subset G} \frac{\mathrm{ur}(S)}{\mathrm{lr}(S)} \right) |I_g \cap G_i|(w_{e[i]} - w_{e[i+1]})$$

$$\leq \left(\max_{S \subset G} \frac{\mathrm{ur}(S)}{\mathrm{lr}(S)} \right) \sum_{e \in I_g} w_e$$

and the theorem follows. ■

It is easy to pick weights that show the guarantee of (7-6) to be the best possible. Given any independence system $(G, \mathscr{I})$, pick S_0 so that

$$\frac{\mathrm{ur}(S_0)}{\mathrm{lr}(S_0)} = \max_{S \subseteq G} \frac{\mathrm{ur}(S)}{\mathrm{lr}(S)} \tag{7-9}$$

and let I_l and I_u be maximal independent subsets of S_0 that establish $\mathrm{lr}(S_0)$ and $\mathrm{ur}(S_0)$, respectively. Consider weights

$$w_e = \begin{cases} 1 \text{ if } e \in S_0 \\ 0 \text{ if } e \notin S_0 \end{cases}. \tag{7-10}$$

If elements of G are ordered so that those in I_l come first, followed by remaining members of S_0, followed by the rest of G, the greedy algorithm will yield exactly $I_g = I_l$ at value $|I_l|$. But since (7-10) gives weight only to members of S_0, we know I_u is a maximum weight solution to this maximum weight independent set problem. It follows that the ratio of the optimal I_u to the greedy I_l is exactly the rank quotient of (7-9).

Hausmann, et al. (1980) present a variety of applications of Theorem 7.1 to specific discrete problems, including a maximize version of the traveling salesman problem. We shall illustrate three of their results.

Example 7.4. Guaranteed Greedy Performance on Weighted Matching. Recall that a matching on a graph $G(V, E)$ is a subset $M \subseteq E$ with at most one member of M incident to any vertex in V. While exact, polynomial-time algorithms for the problem of finding a maximum total weight matching exist, Theorem 7.1 shows that the greedy heuristic will achieve at least the rank quotient of the independence system defined by $\mathcal{I} \triangleq$ {matchings M of G}.

Suppose, to avoid uninteresting cases, that our graph G is connected and $|V| \geq 4$. Then we can prove the rank quotient for this problem and thus (by Theorem 7.1) the guaranteed performance ratio of the greedy heuristic is 2.

Pick an arbitrary set $S \subseteq E$ and two maximal matchings M_1, $M_2 \subseteq S$ with $|M_1| \leq |M_2|$. We will demonstrate that $|M_2|/|M_1| \leq 2$, which implies, since S, M_1, and M_2 are arbitrary, that the rank quotient is at most 2. A path of three edges shows that the value 2 can actually happen.

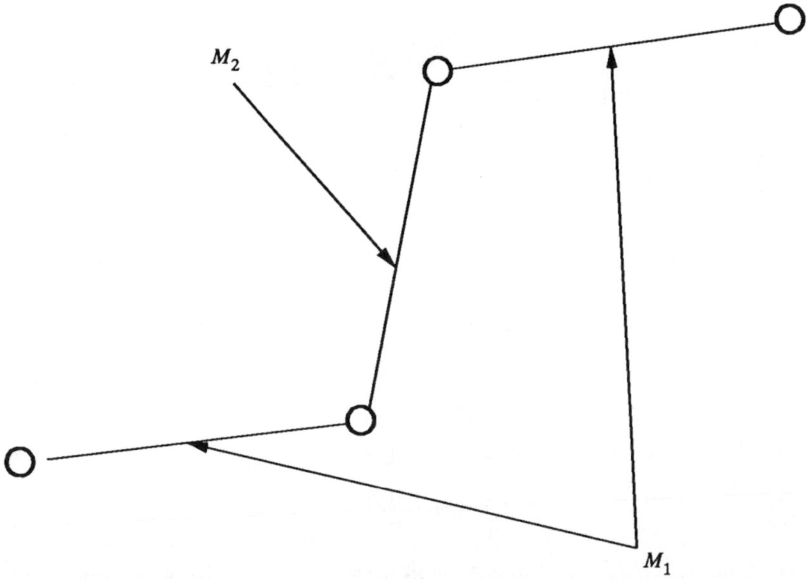

Neither of M_1 nor M_2 can contain the other because both are maximal. Pick $e \in M_1 \backslash M_2$. Since M_2 is maximal $M_2 \cup \{e\}$ is not a matching, and there are M_2 edges meeting one or both of the vertices of edge e. Then one or two such edges belong to $M_2 \backslash M_1$ because they are not consistent with the M_1 matching that includes e. Furthermore, every $M_2 \backslash M_1$ edge conflicts in this way with some $e \in M_1 \backslash M_2$. Repeating the argument for each $e \in M_1 \backslash M_2$ we see that $|M_2 \backslash M_1| \leq 2|M_1 \backslash M_2|$. It then follows that

$$|M_2| = |M_2 \backslash M_1| + |M_1 \cap M_2| \leq 2\,|M_1 \backslash M_2| + |M_1 \cap M_2| \leq 2\,|M_1| \quad \blacksquare$$

Example 7.5. Greedy on k-Matroid Intersection. The k-matroid intersection problem ($k \geq 3$) is the NP-Complete generalization of the 2-matroid intersection problem for which algorithms were presented in Chapter 3. Given a single set G, weights $\{w_e: e \in G\}$, and k matroids $(G, \mathscr{I}_1)$, $(G, \mathscr{I}_2)$, . . . , $(G, \mathscr{I}_k)$, we seek a maximum total weight subset of G that is independent in all matroids. That is, we seek a maximum weight solution for the independence system $(G, \mathscr{I})$, where

$$\mathscr{I} \triangleq \bigcap_{i=1}^{k} \mathscr{I}_i.$$

For this independence system, we can show that the rank quotient is k, so that the greedy heuristic will achieve a performance ratio of at least k. That this is so follows from two results. The first states that given an independence system $(G, \mathscr{I})$, if for any $A \in \mathscr{I}$ and $e \in G$, $A \cup \{e\}$ has no more than k circuits, then the rank quotient of $(G, \mathscr{I})$ is no smaller than k. This is established in Hausmann, et al. (1980). Secondly, for a matroid $(G, \mathscr{I}_i)$, the union of an independent set $A \in \mathscr{I}_i$ and $\{e\}$ contains at most a single circuit. Combining these results establishes the bound of the example. ■

The last example is a bad one showing the rank quotient can be arbitrarily poor on some problems.

Example 7.6. Greedy Is Arbitrarily Poor on Vertex Packing. The vertex packing problem is complementary to matching on a graph $G(V, E)$ in that a packing J is a collection of vertices that meet each edge at most once. That is, packings are subsets $J \subseteq V$ that contain no adjacent vertices.

It is easy to see that $\mathscr{I} \triangleq \{\text{packings } J\}$ is an independence system, and we might consider approaching the maximum total weight vertex packing problem via the greedy algorithm. Unfortunately, the performance guarantee of Theorem 7.1 deteriorates as the size of the given graph increases.

To see this simply, consider a *star* graph on p vertices, i.e., the graph $K_{1,p-1}$. Let v_0 be the central vertex of the star. Then $\{v_0\}$ is one maximal vertex packing and $V \backslash v_0$ is another. The rank quotient is then $(p-1)/1$, which grows arbitrarily large with G. We know from earlier discussion that weights can be chosen to make the greedy algorithm actually achieve this poor performance. ■

To this point our development has pertained to maximize problems over independence systems. What can be said about minimize models?

First, observe that a simple minimize problem with nonnegative weights would be trivial; the empty set is an obvious optimal solution. To form a more interesting model, we must constrain the feasible space to the maximal members of $\mathscr{I}$. That is, we consider the problem

$$\min \left\{ \sum_{e \in I} w_e : I \text{ is a maximal member of } \mathscr{I} \right\}. \tag{7-11}$$

If we add the further assumption that all maximal members of $\mathscr{I}$ have the same cardinality (this is not the same as assuming $(G, \mathscr{I})$ is a matroid), we obtain following consequence of Theorem 7.1.

Theorem 7.2. Performance Guarantee for Greedy Minimization. *Let $(G, \mathscr{I})$ be an independence system for which all maximal members of $\mathscr{I}$ have cardinality p, I^* an optimum in (7-11), I_g be a feasible solution to (7-11) obtained by sequentially placing in I_g the $e \in G$ defined by $w_g = \min\{w_e: I_g \cup e \in \mathscr{I}\}$, and w_{max} be the largest w_e. Then,*

$$\sum_{e \in I_g} w_e \leq \frac{1}{q} \sum_{e \in I^*} w_e + \left(1 - \frac{1}{q}\right) p w_{\max}, \tag{7-12}$$

where q is the rank quotient of $(G, \mathscr{I})$.

Proof. Since all feasible solutions to (7-11) are assumed to have the same cardinality, p, our minimization can be converted to an equivalent maximization by employing weights $\overline{w}_e \triangleq w_{\max} - w_e$. Every solution in this maximization form will have objective function value $pw_{\max}$ minus its cost in the minimization. Also, the greedy algorithm for the maximization over $\overline{w}_e$ will select exactly the same sequence of elements e as the corresponding minimization over w_e.

Applying Theorem 7.1, to the maximization, we know

$$q \geq \frac{\sum_{e \in I^*} \overline{w}_e}{\sum_{e \in I_g} \overline{w}_e} \triangleq \frac{pw_{\max} - \sum_{e \in I^*} w_e}{pw_{\max} - \sum_{e \in I_g} w_e}. \tag{7-13}$$

Rearrangement of (7-13) yields (7-12). ■

We demonstrated above that the maximization guarantee of (7-13) can hold as an equality. It is not difficult to adapt the same approach to show that the bound of Theorem 7.2 is the best possible.

Comparison of Theorems 7.1 and 7.2 also shows the dissatisfying lack of transitivity in performance guarantees. The maximization result gave a constant performance ratio for all problem instances. Even under the assumption that all maximal members of $\mathscr{I}$ have equal cardinality, the corresponding minimization guarantee is not even a ratio. Instead, it bounds the greedy's performance by a weighted sum of the true optimum and the data-dependent upper bound $pw_{\max}$.

7.4 Local Improvement

In this section, we examine nonexact procedures based on the search notion that forms the core of most continuous optimization. Search proceeds by sequential improvement of problem solutions, advancing at each step from a current solution to an objective-function superior neighbor. We term the discrete optimization adaptation of search *local improvement;* however, terms such as local optimization or neighborhood search are equally suitable.

7.4.1 General Strategy

Consider the general (minimize) discrete optimization form

$$\min \{\mathbf{cx}: \mathbf{x} \in F\}, \tag{7-14}$$

where the feasible set F is defined by

$$F \triangleq \{\mathbf{x} \geq \mathbf{0}: \mathbf{Ax} \geq \mathbf{b}, x_j = 0 \text{ or } 1 \text{ for all } j \in I\}. \tag{7-15}$$

Local improvement algorithms proceed from a *current solution*, $\mathbf{x}^k \in F$. We attempt to improve $\mathbf{x}^k$ by looking for a superior solution in an appropriate *neighborhood* $N(\mathbf{x}^k)$ around $\mathbf{x}^k$. A formal statement is as follows:

Algorithm 7A. Local Improvement

STEP 0: *Initialization.* Choose any initial point $\mathbf{x}^1 \in F$, and let iteration counter $k \leftarrow 1$.

STEP 1: *Improvement.* Consider the local improvement set

$$\{\mathbf{x} \in F \cap N(\mathbf{x}^k): \mathbf{cx} < \mathbf{cx}^k\}, \tag{7-16}$$

where $N(\mathbf{x}^k)$ is a neighborhood around $\mathbf{x}^k$. If the set is empty, stop; $\mathbf{x}^k$ is a local minimum for (7-14) with respect to the neighborhood $N(\mathbf{x})$. Otherwise, choose a member of the set as $\mathbf{x}^{k+1}$, increase $k \leftarrow k+1$, and repeat this step. ■

7.4.2 Implementation

Algorithm 7A is a very general concept requiring detail on at least two issues. First, the procedure commences at Step 0 with a feasible solution $\mathbf{x}^1 \in F$. How can a good start be obtained when one is not readily available?

One approach is to begin with another form of nonexact algorithm. Greedy methods often serve. Local improvement then starts with the output of the greedy procedure.

When no feasible solution at all is known, we may adopt the familiar two phase approach. Beginning from an infeasible solution, we first use local improvement to minimize the infeasibility of the solution. Later, if this phase I local improvement succeeds in driving the total infeasibility to 0, we shift to an optimizing phase II that begins at the final point of phase I. The only necessary change in Algorithm 7A during phase I is to replace set (7-16) by

$$\{\mathbf{x} \in N(\mathbf{x}^k): d(\mathbf{x}) < d(\mathbf{x}^k)\}, \tag{7-17}$$

where function $d(\mathbf{x})$ measures the infeasibility of solution $\mathbf{x}$.

A more subtle issue left open in Algorithm 7A is the definition of a neighborhood $N(\mathbf{x})$. In continuous optimization it is standard to take $N(\mathbf{x})$ to be an open ball around $\mathbf{x}$. In discrete models, however, such balls may contain no new points of F because of the discontinuous nature of the solution space. On the other hand, if we include all points in $\mathbb{R}^n$ with components that differ from those of a given $\mathbf{x}$ by amounts 0 or 1, we might have to evaluate 2^n neighboring solutions.

Clearly, appropriate discrete neighborhoods must be large enough to include some discrete variants of the current solution and small enough to be surveyed within practical computation. In presenting some common neighborhoods we shall assume, for notational simplicity, that all variables are subject to $0-1$ constraints (i.e., all components of $\mathbf{x}$ in (7-15) belong to I). The usual extension to mixed integer cases involves the solution of a linear program to find the best values of continuous variables for each binary choice in the neighborhood.

- *Unit Neighborhood.* For a given solution $\mathbf{x}^k$, the unit neighborhood about $\mathbf{x}^k$ is the one formed by complementing components of $\mathbf{x}^k$ one at a time, i.e.,

$$N_1(\mathbf{x}^k) \triangleq \{\text{binary } \mathbf{x}: \Sigma_j |x_j - x_j^k| = 1\}. \tag{7-18}$$

- *t-Change Neighborhood.* The t-change neighborhood generalizes the unit neighborhood by allowing complementation of up to t solution components. Specifically,

$$N_t(\mathbf{x}^k) \triangleq \{\text{binary } \mathbf{x}: \Sigma_j |x_j - x_j^k| \leq t\}. \tag{7-19}$$

- *Pairwise Interchange.* Pairwise interchange neighborhoods change two binary components at a time, but in a complementary fashion.

$$N_{2p}(\mathbf{x}^k) \triangleq \{\text{binary } \mathbf{x}: \Sigma_j |x_j - x_j^k| = 2, \Sigma_j (x_j - x_j^k) = 0\} \tag{7-20}$$

- *t-Interchange Neighborhood.* A t-interchange neighborhood changes up to t values of the solution in the same complementary manner as pairwise interchange.

$$N_{tp}(\mathbf{x}^k) \triangleq \{\text{binary } \mathbf{x}\text{: } \Sigma_j |x_j - x_j^k| \geq t,\ \Sigma_j (x_j - x_j^k) = 0\}. \qquad (7\text{-}21)$$

Example 7.7. Two-Phase Local Improvement. To illustrate these concepts, consider the simple set covering problem.

$$\min \quad 20x_1 + 5x_2 + 4x_3 + 6x_4 + 2x_5$$

$$\text{s.t.} \quad \begin{pmatrix}1\\1\\1\\0\end{pmatrix} x_1 + \begin{pmatrix}0\\0\\1\\1\end{pmatrix} x_2 + \begin{pmatrix}1\\1\\0\\0\end{pmatrix} x_3 + \begin{pmatrix}1\\0\\0\\1\end{pmatrix} x_4 + \begin{pmatrix}0\\1\\1\\0\end{pmatrix} x_5 \geq \begin{pmatrix}1\\1\\1\\1\end{pmatrix}$$

$$\mathbf{x} \text{ binary}$$

It would be possible to start local improvement from the obviously feasible solution $\mathbf{x} = (1, 1, 1, 1, 1)$. We shall illustrate the two-phase concept, however, by beginning with the infeasible $\mathbf{x}^1 = (0, 0, 0, 0, 0)$ and applying unit neighborhoods in a phase I. The obvious measure $d(\mathbf{x})$ of infeasibility is the number of rows unsatisfied.

The unit neighborhood $N_1(\mathbf{0})$ about our $\mathbf{x}^1$ consists of the five unit vector solutions. Trying these in turn, we see that they reduce $d(\mathbf{x}^1)$ by 3, 2, 2, 2 and 2 units respectively. The greatest improvement defines $\mathbf{x}^2 \leftarrow (1, 0, 0, 0, 0)$.

This solution is not yet feasible. Thus, another phase I iteration is required. The unit neighborhood now consists of $(0, 0, 0, 0, 0)$, $(1, 1, 0, 0, 0)$, $(1, 0, 1, 0, 0)$, $(1, 0, 0, 1, 0)$ and $(1, 0, 0, 0, 1)$. Only the second and the fourth decrease infeasibility (by 1 unit each). Selection of the former terminates phase I with the feasible solution $\mathbf{x}^3 = (1, 1, 0, 0, 0)$ at cost 25.

We now begin phase II from $\mathbf{x}^3$. For variety, we switch to pairwise interchange neighborhoods in phase II. The pairwise neighborhood about $\mathbf{x}^3$ consists of the six solutions obtained by replacing either $x_1 = 1$ or $x_2 = 1$ in $\mathbf{x}^3$ by one of x_3, x_4, and x_5. All such solutions except $(0, 1, 1, 0, 0)$ and $(1, 0, 0, 1, 0)$ are infeasible, and the latter yields no cost saving. Thus, local improvement would set $\mathbf{x}^4 \leftarrow (0, 1, 1, 0, 0)$ with cost 9.

Review of the pairwise interchange neighborhood about this new solution will show that $\mathbf{x}^4$ cannot be locally improved. Algorithm 7A would terminate. It is easy to see, however, that the global optimum is to $\mathbf{x}^* = (0, 0, 0, 1, 1)$, cost 8. ■

The above example used the simplest of the neighborhoods we have defined. Many standard nonexact algorithms can be viewed as implementations of these or more complex forms. For example, Lin (1965)'s classic k-opt algorithm for the traveling salesman problem operates by swapping

k pairs of edges to reduce tour length. The neighborhood employed is exactly what we have termed t-interchange (with $t = 2k$).

7.4.3 *Performance of Local Improvement*

Local improvement algorithms pose performance questions of both time (i.e., search duration) and quality. In this section we shall examine both issues using the lucid conceptual tools of Tovey (1981, 1983, 1985).

To begin, we shall restrict our notion of local improvement in the interest of clarity. Throughout this section we take local improvement to be a minimizing search over vertices of the $0-1$ hypercube in $\mathbb{R}^n$. That is, we take the search to be one over n-component binary problem solutions, with any continuous variables being determined as part of objective function evaluation. Furthermore, we consider only the usual notion of adjacency in the $0-1$ hypercube. In terms of the above discussion that form of adjacency corresponds to neighborhood $N_1(\mathbf{x})$, the unit neighborhood about a current point.

Adjacency Trees

We can visualize the evolution of any local improvement algorithm over a finite set of solutions as a directed forest of *adjacency trees*. Vertices in the forest are the possible solutions. Arcs lead from a solution to the next (if any exists) to which local improvement would take us. Thus, solutions are adjacent in the trees of the forest if a single execution of Step 1 in Algorithm 7A would advance from one to the next.

Figure 7.2 illustrates the notion of adjacency trees for two cases. All solutions of the $n = 3$ case in part (a) belong to a single tree, so that local improvement will always yield an optimal solution. Starting search at $\mathbf{x}^1 = (0, 1, 1)$, for example, the tree indicates local search would proceed to $\mathbf{x}^2 = (0, 1, 0)$ and then to the optimal $\mathbf{x}^3 = (0, 0, 0)$.

Part (b) shows a case of local optima that are not globals for $n = 2$. Search from any of (1, 0), (1, 0) or (0, 0) leads to the (0, 0) solution. But (1, 1) is a separate local optimum.

Selection Rules

There are at least two rules that might be applied in selecting the adjacent member in the N_1 neighborhood to which a local search algorithm will next proceed. Under an *optimal adjacency* rule, search proceeds next to the unique minimum cost member of the neighborhood. If search proceeds to any improving point in the neighborhood, we are employing the weaker *better adjacency* rule. (Here we implicitly assume without loss

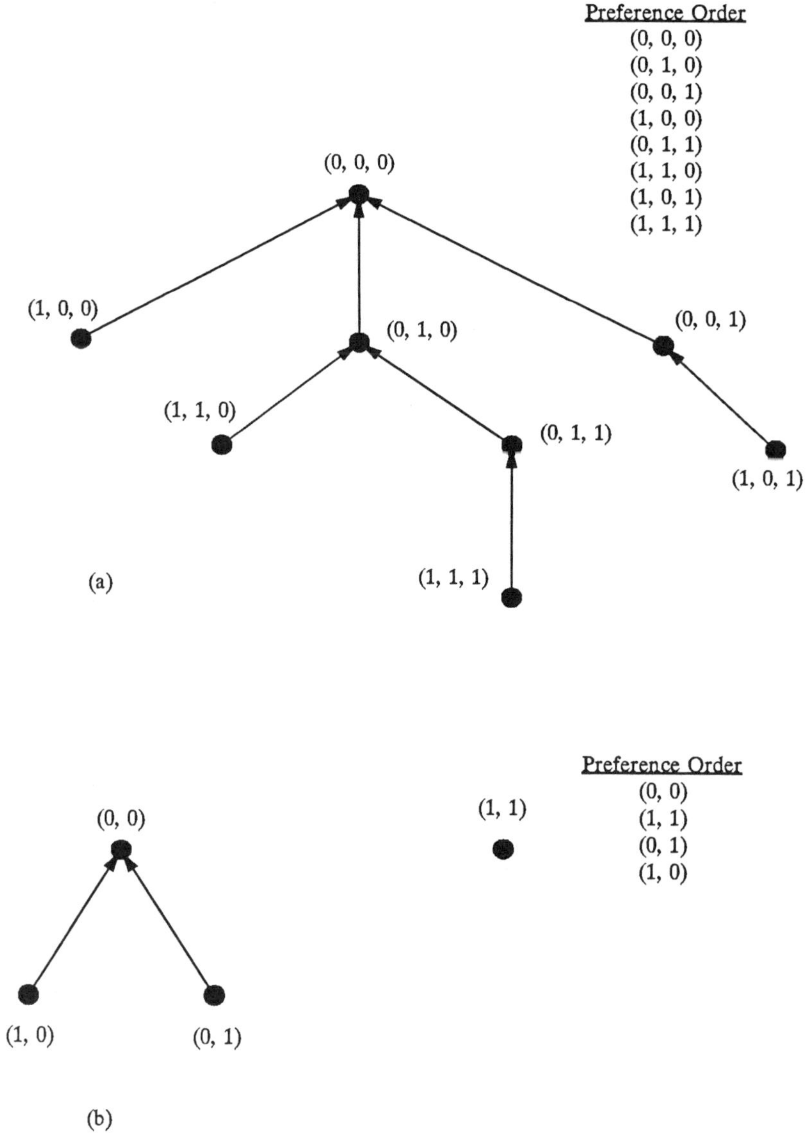

Fig. 7.2. Adjacency tree examples.

of generality that all points on the hypercube are totally ordered, e.g., by a lexicographic triple of infeasibility, cost and some tie breaker for equal cost solutions).

Worst Case Duration

It should be clear that the question of the duration of a local search in the context we have established resolves to the question "How long is the adjacency tree path from an initial solution?" One way to analyze the issue is to ask for the worst case, "What is the length of the longest path of an adjacency tree formed on the hypercube of order n?"

Although a precise answer to this question is not known, an exponential size lower bound on the length of the longest path can be established for even the demanding, optimal adjacency selection rule. We begin with a simple lemma.

Lemma 7.3. Neighbors in an Optimal Adjacency Path. *Let $\boldsymbol{x}^1$, $\boldsymbol{x}^2, \ldots, \boldsymbol{x}^r$ be the path of an optimal adjacency search algorithm from $\boldsymbol{x}^1$ to its adjacency tree root $\boldsymbol{x}^r$, and $P \triangleq \{\boldsymbol{x}^k\colon k = 1, 2, \ldots, r\}$. Then $N_1(\boldsymbol{x}^1) \cap P = \boldsymbol{x}^2$, $N_1(\boldsymbol{x}^r) \cap P = \boldsymbol{x}^{r-1}$, and $N_1(\boldsymbol{x}^k) \cap P = \{\boldsymbol{x}^{k-1}, \boldsymbol{x}^{k+1}\}$ for $1 < k < r$, where $N_1(\cdot)$ denotes the unit neighborhood defined above.*

Proof. We must show that the only unit neighbors of a point in the path P are the point's immediate path predecessor and successor (when they exist). This follows immediately because the definition of optimal adjacency search requires that $\mathbf{x}^1 \lesssim \mathbf{x}^2 \lesssim \ldots \lesssim \mathbf{x}^r$, and $\mathbf{x}^{k+1} \lesssim \mathbf{x}$ for all $\mathbf{x} \in N_1(\mathbf{x}^k)$, where $\lesssim$ denotes the preference order of vertices. If a member of $N_1(\mathbf{x}^k)$ occurred in the path after $\mathbf{x}^{k+1}$, it would have to be both less and more preferred than $\mathbf{x}^{k+1}$. ■

By adding a return arc from the tree root to the beginning vertex of an optimal adjacency tree path, we form a cycle that (by Lemma 7.3) has each member vertex adjacent in the n-hypercube to exactly the two vertices to which it is adjacent in the cycle. Such a cycle has been studied by Danzer and Klee (1967) as a *snake in the box.* Although the maximum cardinality of a snake in the n-hypercube has not been established, it has been shown to be at least

$$\frac{7(2^n)}{4(n-1)}.$$

We immediately have the following theorem.

Theorem 7.4. Worst Case Duration of Optimal Adjacency Search. *The maximum number of iterations in any optimal adjacency search of vertices on the n-hypercube is at least $O(2^n/n)$.*

Expected Duration

As with all worst case results, the above bound may be a very pathological phenomenon. To see, we would like to compute instead an average or expected number of search iterations. This, in term, requires our introducing some form of probability distribution over the possible adjacency trees for the n-hypercube.

It is equivalent to think of introducing probability distributions over orderings of the vertices of the n-hypercube. Given such an ordering, every vertex with no preferred unit neighbor is a local minimum. Those that have preferred unit neighbors may be directed toward any such in the adjacency tree of the search. For example, in Figure 7.2(b) an ordering that corresponds to the given adjacency forest is $(1, 0) \lesssim (0, 1) \lesssim (1, 1) \lesssim (0, 0)$, where again $\lesssim$ denotes preference ordering.

Following Tovey (1981), we shall consider three results for three possible distributions on orderings. The easiest to think about is simply that all orderings are *equally likely*.

Theorem 7.5. Expected Duration with Equally Likely Orderings. *If all preference orderings on the vertices of the n-hypercube are equally likely, then the expected number of iterations of any better adjacency local improvement algorithm is at most $O(n)$.*

Proof. Pick any random ordering of the vertices of the n-hypercube, and let $\mathbf{x}^i$ be the *ith* best vertex in that ordering. The probability that at most $k < i$ iterations are required to complete a search from vertex $\mathbf{x}^i$ is simply the probability that there is a k-subsequence of vertices higher in the ordering than i. That is (using the usual binomial coefficient notation), the probability for any particular path,

$$\frac{\binom{i-1}{k}}{\binom{2^n-1}{k}k!}$$

times the number of k-paths which is bounded by n^k. Using this maximum, evaluating the above binomial coefficients, and considering all possible values of i, we see that the probability for a path greater than k is

$$\frac{n^k}{k!}\sum_{i=k+1}^{2^n}\frac{(i-1)!\,k!\,(2^n-k-1)!}{k!\,(i-1-k)!\,(2^n-1)!}$$
$$=\frac{n^k(2^n-k-1)!}{k!\,(2^n-1)!}\sum_{j=0}^{2^n-k-1}\frac{(j+k)!}{j!}<\frac{2^n\,n^k}{(k+1)!}. \qquad (7\text{-}22)$$

Using Stirling's approximation, we can replace the last expression in (7-22) by $2^n(en)^k/k^k$, where e is the natural logarithmic constant. Evaluated at $k=(\frac{3}{2})en$, the latter is $2^n(\frac{2}{3})^{(3/2)en}$, which can be shown to be less than $(\frac{4}{9})^n$.

We are now in a position to bound the expected value of k. If a path of length $k \le (\frac{3}{2})en$ exists, we can incur no more than $(\frac{3}{2})en$ iterations. If no such path exists, we might incur 2^n iterations. Combining with the just computed probabilities shows

$$E[k]<((\tfrac{3}{2})en-1)(1-(\tfrac{4}{9})^n)+(2^n)((\tfrac{4}{9})^n)<(\tfrac{3}{2})en. \qquad (7\text{-}23)$$

This establishes the theorem. ■

The linear order of expected search duration implied by Theorem (7.5) is very comforting, especially since it requires only a better adjacency, as opposed to optimal adjacency selection rule. Still, it is rather difficult to discover any intuition for whether the equally likely probability assumption is realistic.

Tovey (1981) has proposed two different probability distributions that seem to have more intuitive appeal. To define these distributions, we must imagine constructing the preference order of n-hypercube vertices in top down or best to worst fashion. If we add the elements of the adjacency tree corresponding to any local minimum one at a time, each new tree vertex must be a unit neighbor of one of the higher ones in the ordering. We can then define the *boundary* probability distribution of vertex sequences as the one obtained when all unassigned vertices adjacent to at least one member of the current partial sequence are equally likely to be the next in the sequence. The *coboundary* distribution is the related one obtained when unassigned vertices adjacent to at least one member of the current partial sequence have a probability of being selected that is proportional to the number of already present vertices to which they are adjacent.

Under these distributions, we also obtain low-order polynomial bounds on the duration of local search. The next two theorems summarize Tovey's results. The reader is referred to Tovey (1981) for the rather technical proofs.

Theorem 7.6. Expected Duration Using Optimal Adjacency. *Suppose local improvement search of the vertices of the n-hypercube is operated according to the optimal adjacency decision rule. The expected duration of* *such a search is at worst $O(n^2)$ if corresponding adjacency trees arise according to the boundary distribution, and $O(n^2 \log n)$ if trees follow the co-boundary distribution.*

Theorem 7.7. Expected Duration Using Better Adjacency. *Suppose local improvement search of the vertices of the n-hypercube is operated according to the better adjacency decision rule. Then the expected duration of such a search is at worst $O(n^2)$ if corresponding adjacency trees arise according to the boundary distribution, and $O(n \log n)$ if trees follow the co-boundary distribution.*

Number of Local Minima

Besides the length of a local improvement search, of course, we must be concerned with its quality. One way to measure quality is to ask "How many adjacency trees could there be in the forest implied by a given ordering of the vertices of the n-hypercube?" That is, "How many distinct local minima might we expect?"

Recall that Theorem 7.5 showed the expected number of steps to complete local improvement was $O(n)$, if all vertex sequences were assumed equally likely. Our final result of this section demonstrates that this relatively short duration under the equally likely assumption is accompanied, sadly, by a very large expected number of local minima.

Theorem 7.8. Expected Number of Local Minima. *If all orderings of the vertices of the n-hypercube are taken to be equally likely, then the expected number of distinct adjacency trees in a corresponding better adjacency local search is at least $O(2^n/n)$.*

Proof. See Tovey (1981). ∎

7.5 Truncated Exponential Schemes

Chapters 5 and 6 presented generally worse-than-polynomial algorithms suitable for exact solution of discrete optimization problems. Such procedures, however, also give a source of nonexact methods. Instead of expending the generally exponential effort to find and prove an optimal solution, we can *truncate* the evolution of the algorithm in some fashion. The result is (hopefully) a reduced computation time at the cost of settling for a possibly suboptimal solution.

In principle, one could construct such a *truncated exponential* analog of any of the algorithms of Chapters 5 and 6. We shall limit our discussion, however, to *truncated enumeration,* i.e., abbreviated versions of Branch and Bound Algorithm 5A. The development is based on the work of Ibaraki (1976a).

7.5.1 Branch and Bound Environment

To provide an environment for discussing truncated enumeration algorithms, we briefly restate the algorithm descriptions of Sections 5.1 and 5.4. Branch and bound approaches a given discrete optimization problem (P) via a strategy of divide and conquer. More specifically, a given problem is subdivided into *candidate problems* (P_k) that correspond to distinct regions of the feasible space (note that here we abbreviate the $P(S_k)$ candidate notation of Chapter 5 as (P_k)). Active candidate problems are stored in a *candidate list* with associated lower bounds $\sigma(P_k)$.

Candidates are selected in turn until no active ones remain. If an optimal solution to the selected problem can be found, or if it can be proved that no solution to the candidate can improve upon a known *incumbent solution value,* v^*, the candidate is *fathomed* (deleted). Otherwise, it is *branched,* i.e., replaced in the candidate list by several more restricted versions that collectively span its part of the feasible space.

For a bounded problem to minimize $\mathbf{cx}$, a formal statement is as follows:

Algorithm 7B. Branch and Bound

STEP 0: *Initialization.* Establish the candidate list with (P) its sole entry, stored bound $\sigma(P) \leftarrow -\infty$, and incumbent value $v^* \leftarrow +\infty$.

STEP 1: *Selection.* Choose some member (P_k) from the current candidate list.

STEP 2: *Main Bounding.* Compute a lower bound $\beta(P_k) \geq \sigma(P_k)$ on the value $v(P_k)$ of an optimal solution to (P_k). If $\beta(P_k) \geq v^*$, go to Step 6 and fathom. Otherwise, go to Step 3.

STEP 3: *Optimal Candidate Solution.* If bound or other computations have produced an optimal solution $\mathbf{x}^k$ to (P_k) compare its value to v^*; when $\mathbf{cx}^k < v^*$, go to Step 4 with a new incumbent; when not, go to Step 6 to fathom. If an optimum was not obtained, proceed instead to Step 5.

STEP 4: *Incumbent Saving.* Update the incumbent solution via $\mathbf{x}^* \leftarrow \mathbf{x}^k$, $v^* \leftarrow \mathbf{cx}^k$. Then delete from the candidate list all members $(P_t) \neq (P_k)$ with $\sigma(P_t) \geq v^*$, and go to Step 6.

STEP 5: *Branching.* Replace (P_k) in the candidate list by one or more further restricted candidate problems $(P_{k1}), (P_{k2}), \ldots, (P_{kp})$, each stored with a lower bound $\sigma(P_{kt}) \geq \beta(P_k)$. Then return to Step 1.

STEP 6: *Fathoming.* Delete (P_k) from the candidate list. If the candidate list is now empty, stop; $\mathbf{x}^*$ is optimal unless $v^* = +\infty$ in which case (P) is infeasible. If candidates remain, return to Step 1. ■

Section 5.4 showed the importance of the sequences

$v^*(k) \triangleq$ the value of v^* at the *kth* execution of Step 1

$\sigma(k) \triangleq$ the value of $\sigma(P_k)$ at the *kth* execution of Step 1

$\sigma^*(k) \triangleq \min\{\sigma(P_t)$: (P_t) belongs to candidate list$\}$ at the *kth* execution of Step 1.

Discussion in Section 5.4 established the simple facts summarized in the next lemma.

Lemma 7.9. Global Upper and Lower Bound Sequences. *Sequence $\{v^*(k)\}$ as defined above is monotone nonincreasing with $v^*(k) \geq v(P)$ for all k. Sequence $\{\sigma^*(k)\}$ is monotone nondecreasing with $\sigma^*(k) \leq v(P)$ for all k.*

7.5.2 Bound Allowance Methods

The most common approach to developing a truncated, nonexact form of Algorithm 7B is to subtract a *bound allowance* $\epsilon(v^*)$ from the incumbent solution value in executing the bound tests of Steps 2 and 4. The tests become

$$\beta(P_k) \geq v^* - \epsilon(v^*) \text{ and } \sigma(P_t) \geq v^* - \epsilon(v^*), \tag{7-24}$$

respectively.

Typical choices for bound allowance functions are the constant one $\epsilon(v) \triangleq \epsilon$ and the proportional one $\epsilon(v) \triangleq p \times v$ for fixed $p \in (0, 1)$. We assume only

- *nonnegativity* ($\epsilon(v) \geq 0$ for all v), and
- *monotonicity* ($v_1 \leq v_2$ implies $v_1 - \epsilon(v_1) \leq v_2 - \epsilon(v_2)$).

Example 7.8. Bound Allowances. Let us first consider an example of the bound allowance approach. Figure 7.3 shows the branch and bound tree of a problem solved by Algorithm 7B without a bound allowance. As in Chapter 5, numbers in the vertices of the tree indicate the order in which candidate problems were selected; the upper and lower values to the right of each vertex indicate, respectively, the a priori stored bound $\sigma(P_k)$ above the full Step 2 bound $\beta(P_k)$.

It is easy to check that enumeration in Figure 7.3 implements Step 1 of Algorithm 7B *depth-first.* That is, the deepest active candidate problem in

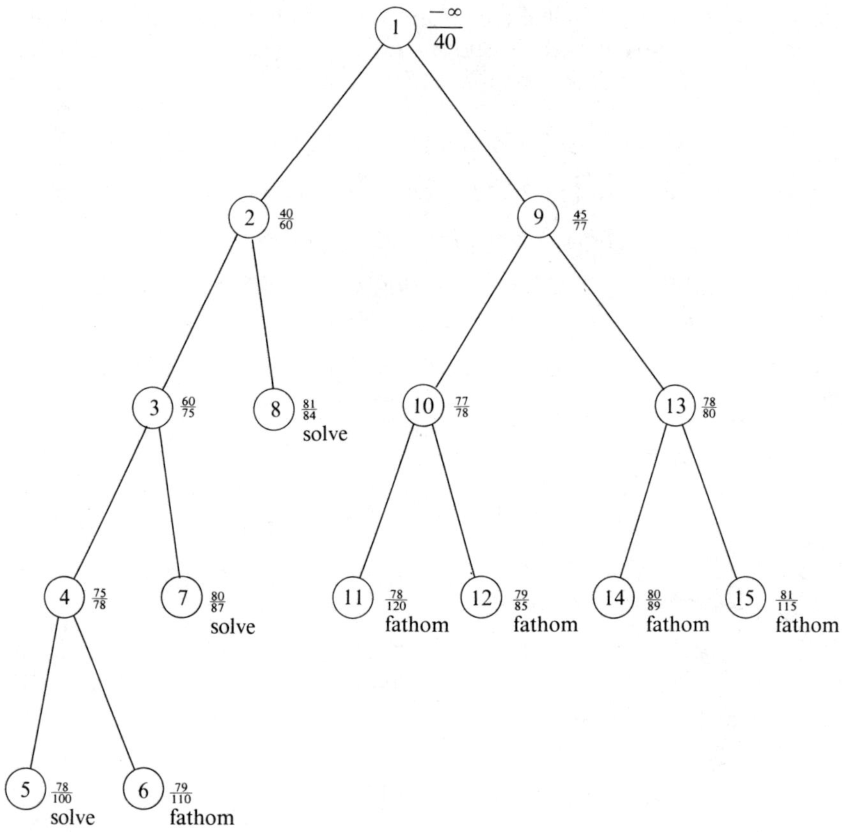

Fig. 7.3. Example 7.6 without bound allowance.

the tree is always the next selected. Ties are settled by choosing the candidate with least stored bound $\sigma(P_k)$.

The enumeration of Figure 7.3 requires a total of 15 candidate problem selections, i.e., 15 executions of the main bound Step 2. Figure 7.4 shows how this number can be reduced when a bound allowance is used. Part (a) has a constant $\epsilon_1 = 10$; part (b) employs the larger constant allowance $\epsilon_2 = 20$. Both parts of the figure preserve the candidate numbering of Figure 7.3.

To better understand the effects, consider ϵ_1 and the tree of part (a). Solution proceeds exactly as in Figure 7.3 through the first 7 candidate problems. What would have been candidate 8, however, is skipped because its stored bound $\sigma(P_k) = 81$ is already within $\epsilon_1 = 10$ of the incumbent solution value 87. That candidate is deleted from the list by Step 4 as soon as the 87 solution is discovered at candidate 7.

Processing thus passes immediately to candidate 9. There the full $\beta(P_k) = 77$. The problem is immediately fathomed and optimality concluded because that bound is within the allowed error of 10. Only a total of 8 candidates were required. The approximate optimal value produced is the current incumbent 87. This compares to the true optimum 84 computed in Figure 7.3.

The evolution of Figure 7.4(b), with $\epsilon_2 = 20$ is similar. Eleven candidate problems are required, and the approximate optimal solution value produced is 100. ■

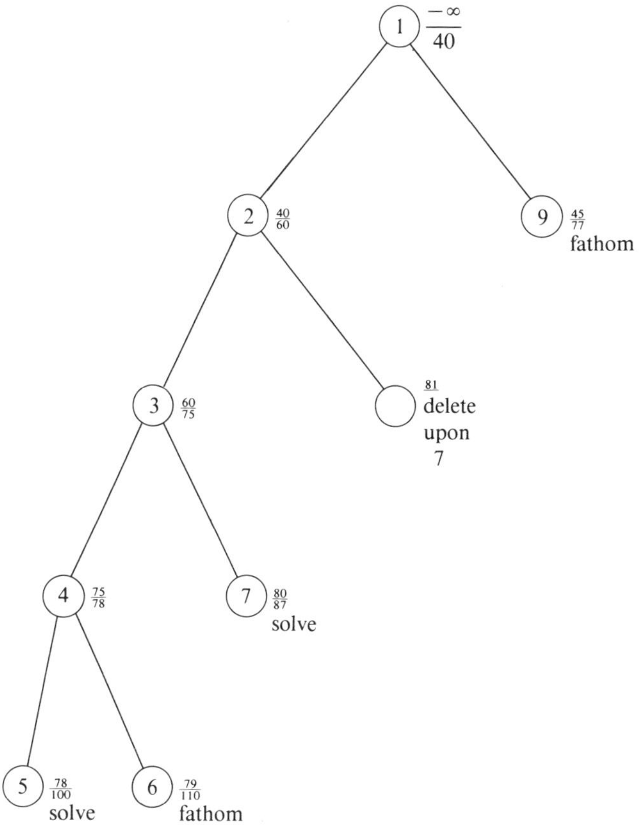

Fig. 7.4. Example 7.6 with bound allowance; (a) constant $\epsilon = 10$, (b) constant $\epsilon = 20$.

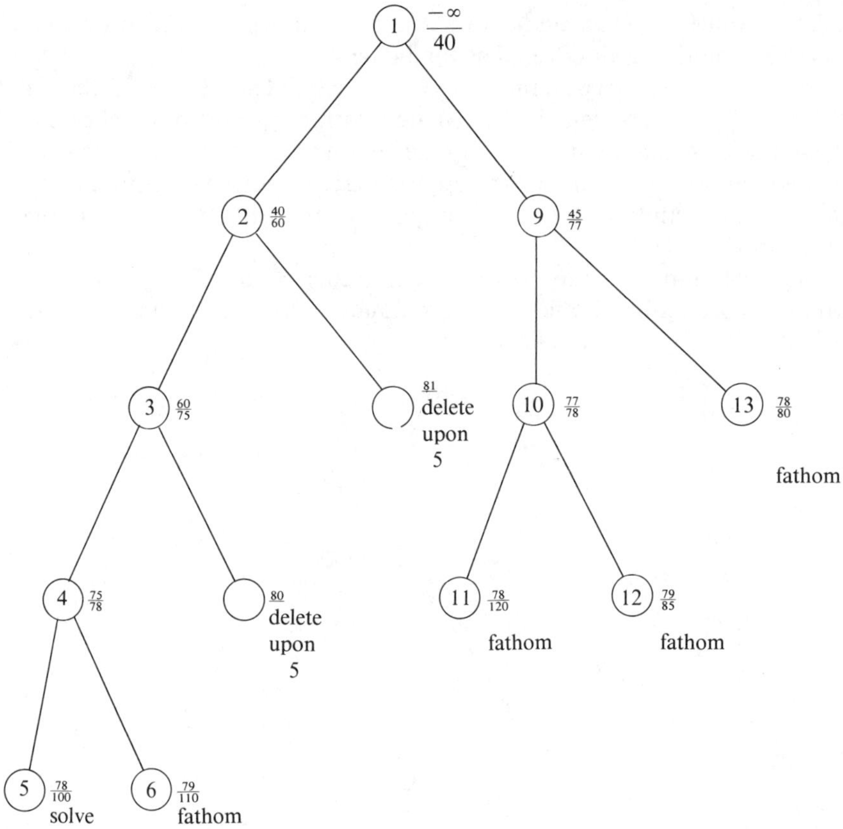

Fig. 7.4. *continued.*

Both bound allowance functions in Example 7.6 resulted in fewer candidate problems being solved than the exact case of Figure 7.3, and approximate solutions within the error allowance of optimal. We now proceed to show this will always be the case.

To do so, we require some mild restrictions on Algorithm 7B. We will term *nonadaptive* implementations of Steps 1 (candidate selection) and 5 (candidate creation) that are independent of the prior history of the enumeration. More specifically, the candidate creation process is taken to depend only on the definition of the candidate being branched and the results of bound calculations for that candidate. Selection is assumed to be effected by reference to a transitive total ordering of candidate problems that could (in principle) be specified before enumeration begins. Most well-known selection rules such as the depth-first approach employed in

Example 7.6, breadth-first and best-bound (least $\sigma(P_k)$) are nonadaptive. It is worth noting, however, that some common schemes, such as those based on pseudo-costs (estimates of solution values derived during the evolution of the enumeration), fail to be nonadaptive (see Section 5.4).

Lemma 7.10. Properties of Bound Allowance Sequences. *The sequence of candidate problems selected by an implementation of Algorithm 7B with a nonnegative, monotone bound allowance is a subsequence of the sequence of candidate problems selected by an exact implementation using the same nonadaptive candidate selection and branching rules, and the same bounds. Furthermore, at each selection k of the exact sequence,* $v^\epsilon(k) \geq v^*(k) \geq v^\epsilon(k) - \epsilon(v^\epsilon(k))$, *where* $v^\epsilon(k)$ *is the value of the allowance implementation incumbent solution after the most recent candidate of the subsequence it selected.*

Proof. For brevity let us refer to the bound allowance implementation as the ϵ-version and the exact one as the $*$-version. Both versions start by selecting the whole problem as the first candidate, and both start with the same incumbent solution value $v^\epsilon(1) = v^*(1) \geq v^\epsilon(1) - \epsilon(v^\epsilon(1))$ (the last by monotonicity of $\epsilon(\cdot)$). Thus the lemma holds for the first ϵ-selection.

Assume inductively that the lemma holds for the first i ϵ-selections. That is, it holds for ϵ-selected candidate subsequence $(P_{k[1]}), (P_{k[2]}), \ldots, (P_{k[i]})$ of $*$-selected sequence $(P_{k[1]}) \triangleq (P_1), (P_2), \ldots, (P_k) \triangleq (P_{k[i]})$. Define (P_n) as the final element of the $*$-sequence leading to the next candidate that will sometime be selected by the ϵ-algorithm. More precisely, $(P_n) \triangleq (P_k)$, if the next candidate selected by the $*$-algorithm is one that will ultimately be selected by the ϵ-procedure; otherwise, (P_n) is the last of the partial sequence $\{(P_l): l = k+1, k+2, \ldots, n\}$ of $*$-candidates that will never be considered by the ϵ-algorithm.

We first show that $v^\epsilon(l) \geq v^*(l)$ for all $k \leq l \leq n+1$. The inequality holds for $l = k$ by the above inductive hypothesis, and continues for $l = k+1$ unless at least the ϵ-sequence found a new incumbent in $v(P_k)$. But then, $v^\epsilon(k+1) \triangleq v(P_k) \geq v^*(k+1)$ because both sequences solved (P_k). When $k+1 < n+1$ some problems considered only in the $*$-sequence follow next. Through each of these the result must hold because $v^\epsilon(l)$ is unchanged and $v^*(l)$ is monotone nonincreasing.

We are now in a position to show that $v^*(l) \geq v^\epsilon(l) - \epsilon(v^\epsilon(l))$ for all $k \leq l \leq n+1$. When $v^\epsilon(l) = v^*(l)$, the inequality follows by nonnegativity of $\epsilon(\cdot)$. For any l with $v^\epsilon(l) > v^*(l)$, the $*$-incumbent is the value of some candidate problem (P_j) that will never be taken up by the ϵ-algorithm. This implies (P_j) or its ancestor (P_a) (problem from which (P_j) was derived by branching) must have satisfied one of the candidate deletion criteria in the

ϵ-algorithm. That is, using the fact that (P_j) is a restricted version of (P_a), $v^*(l) \triangleq v(P_j) \geq v(P_a) \geq v^\epsilon(q) - \epsilon(v^\epsilon(q))$, where q is the iteration at which deletion occurred. But then monotonicity of $v^\epsilon(\cdot)$ and $\epsilon(\cdot)$ yield $v^*(l) \geq v^\epsilon(q) - \epsilon(v^\epsilon(q)) \geq v^\epsilon(l) - \epsilon(v^\epsilon(l))$.

Summarizing, we have now proved that the lemma holds through $*$-selection n, and that its second, inequality part extends to $(n+1)$, if there is a (P_{n+1}). If $(P_{k[i]})$ was the last candidate considered by the ϵ-algorithm we are finished. If not, it remains only to prove that the next ϵ-candidate (call it $(P_{k[i+1]})$) is problem (P_{n+1}) in the $*$-sequence.

To see this, observe that all ancestors of $(P_{k[i+1]})$ were selected by both the ϵ- and the $*$-algorithms. This follows because $(P_{k[i+1]})$ would not be in the ϵ-candidate list if the ϵ-algorithm had not previously taken up all its ancestors, and ϵ-selections to date form a subsequence of $*$-selections. If $(P_{k[i+1]})$ is not in the $*$-candidate list after processing of (P_n), either its immediate ancestor (say (P_a)) was fathomed at Algorithm 7B Step 2 or the problem was deleted by an incumbent discovery Step 4. In the first case, using monotonicity of $v^*(\cdot)$, $\beta(P_a) \geq v^*(a) \geq v^*(n+1)$. In the second event, $\sigma(P_{k[i+1]}) \geq v^*(q) \geq v^*(n+1)$, where q is the iteration at which Step 4 deletion occurred. Either way, recalling σ(a candidate) $\geq \beta$(its ancestor), and using the already established inequality part of the lemma,

$$\sigma(P_{k[i+1]}) \geq v^*(n+1) \geq v^\epsilon(n+1) - \epsilon(v^\epsilon(n+1)) = v^\epsilon(k+1) - \epsilon(v^\epsilon(k+1)).$$

But this is a contradiction because the ϵ-algorithm will never select a problem with $\sigma(\cdot)$ already as large as its fathoming criterion level.

We can conclude that problem $(P_{k[i+1]})$ is in the $*$-candidate list at the beginning of iteration $n+1$. Under the assumption that the ϵ- and $*$-algorithms use the same nonadaptive candidate selection rule, $(P_{k[i+1]})$ must be preferred to any other $*$-candidate the ϵ-algorithm will eventually investigate because all have been in the $*$-list since completion of (P_k), and thus all also are members of the ϵ-list. It follows that $(P_n) = (P_{k[i+1]})$, and the proof is complete. ■

Theorem 7.11. Candidate Savings with Bound Allowance. *Any implementation of Algorithm 7B with a nonnegative, monotone bound allowance selects at Step 1 no more candidate problems than an exact implementation using the same nonadaptive candidate selection and branching rules, and the same bounds.*

Proof. Immediate from the fact established in Lemma 7.10 that candidates of the bound allowance version are a subsequence of those in the exact procedure. ■

Theorem 7.12. Maximum Error with Bound Allowance. *Any implementation of Algorithm 7B with a nonnegative, monotone bound allowance function $\epsilon(\cdot)$ produces a final incumbent solution v^ϵ within the error allowance of optimal, i.e, $v^\epsilon \geq v(P) \geq v^\epsilon - \epsilon(v^\epsilon)$.*

Proof. It is clear that $v^\epsilon \geq v(P)$ because v^ϵ is the value of some restricted version of (P). If $v^\epsilon = v(P)$, it follows from nonnegativity of $\epsilon(\cdot)$ that $v(P) \geq v^\epsilon - \epsilon(v^\epsilon)$. If, on the other hand, $v^\epsilon > v(P)$, each candidate problem (P_k) with $v(P_k) = v(P)$ or an ancestor (P_a) (less restricted) must have been deleted before solution by the ϵ-algorithm's reference to an incumbent of the moment, say $\hat{v}^\epsilon$. Then, using monotonicity of the sequence of ϵ-incumbent values and the $\epsilon(\cdot)$ function,

$$v(P) \triangleq v(P_k) \geq v(P_a) \geq \hat{v}^\epsilon - \epsilon(\hat{v}^\epsilon) \geq v^\epsilon - \epsilon(v^\epsilon). \qquad \blacksquare$$

Theorems 7.11 and 7.12 show that the performance of a bound allowance heuristic is predictable in comparison to its exact analog. An interesting anomaly is that there is no sense in which bound allowance algorithm performance is monotone with respect to the generosity of the allowance.

Example 7.8 illustrated for the number of candidate problems measure. Figure 7.4(a), with $\epsilon = 10$ required 8 candidates to obtain the approximate solution 87; part (b), with $\epsilon = 20$, required 11 problems to generate the solution 100. Greater allowable error did not result in fewer candidates because the larger allowance caused the incumbent-producing candidate 7 to be skipped.

For the numbers presented, it was true that larger error allowance produced a poorer approximate solution. It is easy to modify Figure 7.4, however, to see this will also not always hold. If candidate 12 solves at value 85, instead of merely producing a lower bound of 85, the $\epsilon = 20$ case would finish with a superior approximate solution to the $\epsilon = 10$ one.

7.6 Some Negative Results

We began this chapter by discussing some of the drawbacks one must consider when employing nonexact procedures. In this, the closing section, we return to issues raised earlier and examine, somewhat more substantively, their implications.

We showed previously that an instance of the traveling salesman problem could be constructed so that upon applying a given nonexact procedure (the nearest unvisited city heuristic), solutions were unavoidably generated that were arbitrarily distant from an optimal one. Of course, it is very easy to perform the sort of trickery we observed, since all that is required is an instance upon which the algorithm is made to ultimately encounter and make a terrible decision—one from which there is no recourse. Unless requirements are placed on the complexion of instances of various problems, such bad examples are typically quite easy to construct.

Now, let us suppose that in attempting to construct an instance of the traveling salesman problem for use in our demonstration earlier, we had been forced to draw from the family of instances having intercity distances that satisfy the triangle-inequality ($c_{ij} + c_{jk} \geq c_{ik}$). In this case, it would become evident very quickly that devising an example producing the stark effect we desire (arbitrarily poor solutions) would not be possible. This is not to suggest that when the triangle inequality holds, the nearest unvisited city heuristic performs well. It simply means that the extent to which we can trick the procedure is now limited. More precisely, we can now determine a worst case bound for the heuristic that is meaningful, (i.e., finite). This is important since such a bound is not presently known for any traveling salesman heuristic when applied to arbitrary instances. Furthermore, that this should change is no more likely than the advent of an efficient, exact algorithm for the problem.

7.6.1 Formal Limitations on Performance Bound Existence

The last statement above is somewhat disquieting, because taken at face value, it suggests that, for some problems, finding a procedure that is guaranteed to produce solutions even finitely close to an optimal solution is very difficult—just as difficult as finding an algorithm to solve the problem exactly. As we shall see, even in certain cases where finite performance bounds are known, it is possible to show that no improvement is possible unless the equivalence of *P* and *NP* is established.

Essentially, the mode of proof in establishing these sorts of so-called negative results is to demonstrate that should such finite performance bounds exist, the corresponding heuristic schemes could be used to solve a known, provably difficult problem (i.e., one which is *NP*-Complete). The following theorem provides a useful demonstration of this idea.

Theorem 7.13. Polynomial Time Approximation of the Knapsack Problem. *Let A be a polynomial time approximation scheme for the (optimization) knapsack problem. Then if $P \neq NP$, there can exist no value $K < \infty$ such that $|v_A(t) - v(t)| \leq K$ or every instance t.*

Proof. Assume that such an algorithm exists. Now, given an arbitrary instance of the knapsack problem, t, we can construct another instance, $\bar{t}$, which is identical to t but with objective function coefficients scaled up by

$(K + 1)$ times the respective coefficients in t. It is clear that the feasible solutions for $\bar{t}$ have values $K + 1$ times the corresponding solutions for t; hence, it must be true that $v_A(\bar{t}) - v(\bar{t}) = 0$, which follows since we can assume, without loss of generality, that K is an integer greater than zero. Thus, A can be used to solve optimally the knapsack problem; since the latter is *NP*-Hard, we contradict the assumption that $P \neq NP$. ∎

Could the result of Theorem 7.13 along with that alluded to relative to the traveling salesman problem be isolated phenomena? Unfortunately, the answer to this question is *no*, as the next, equally incriminating theorem demonstrates.

Theorem 7.14. Absolute Performance Bounds for Certain Integral-Valued Optimization Problem Heuristics. *Consider an optimization problem of the minimization form having strictly positive, integer solution values. Then for the NP-Hard problem of deciding whether or not $v(t) \leq K$ where K is a fixed, positive integer, if $P \neq NP$ there can exist no polynomial time approximation procedure A such that $v_A(t)/v(t) \leq \rho < 1 + (1/K)$ for arbitrary instance t.*

Proof. The proof is straightforward. Assume there is such an algorithm A with $\rho < K + 1/K$. Then, obviously, A could be used to solve the stated optimization problem since if $v(t) \leq K$, we must have $v_A(t) = v(t)$, because K is an integer. Similarly, if $v(t) > K$, then we have $v_A(t) \geq v(t) > K$. Since the optimization problem is *NP*-Hard, we again deny the assumption that P and NP are different and the proof is complete. ∎

Of course, Theorem 7.14 is useful in terms of some specific and rather well-known problems. The next two, easily proved, corollaries demonstrate this.

Corollary 7.15. Limitation on the Performance Bound for Bin-Packing Heuristics. *Unless $P = NP$, no polynomial time approximation procedure for bin-packing can have absolute performances bound, ρ, strictly less than $\frac{3}{2}$.*

Corollary 7.16. Limitation on the Performance Bound for Vertex-Coloring Heuristics. *Unless $P = NP$, no polynomial time approximation procedure for vertex-coloring can have absolute performance bound, ρ, strictly less than $\frac{4}{3}$.*

While the results of the two corollaries and in a broader sense, of Theorem 7.14 deal with absolute performance bounds, it is reasonable to ask if similarly discouraging results exist for asymptotic bounds. Indeed, while no absolute bound less than $\frac{3}{2}$ is likely for any bin-packing heuristic, we know that the first-fit, decreasing weight procedure (e.g., see Johnson (1973)) discussed earlier has asymptotic performance bound of $\frac{11}{9}$. The absolute bound in this case is $\frac{11}{9} + 4/v(t)$ but for large instances (large values of $v(t)$) the second term becomes negligible. Interestingly, in the case of vertex-coloring, the bound of Corollary 7.16 holds even for the asymptotic case (e.g., see Garey and Johnson (1974)).

It should also be noted that neither corollary is, in any way, suggesting that polynomial time approximation procedures exist that have bound values close to those in the corollary statements. In fact, Johnson (1974) has even shown that the best absolute performance bound currently known for any polynomial time approximation algorithm for vertex-coloring has value $cp/\log p$ where p is the order of the graph and c is a positive constant.

7.6.2 *Formal Limitations on Finite-Valued, Realizable Performance Bounds*

We just observed that if $P \neq NP$, then no polynomial time approximation procedure for vertex-coloring can have absolute performance bound less than $\frac{4}{3}$. Such an algorithm, if it existed, could be used to solve the *NP*-Complete problem of deciding if a graph is 3-colorable. In any event, the issue is somewhat academic at the present time since from Johnson's result, no such performance bound with finite value is known, let alone one approaching the $\frac{4}{3}$ value.

On the other hand, numerous hard problems do possess approximation procedures having finite performance bounds that are not known to be limited in nontrivial ways vis-à-vis Corollaries 7.15 and 7.16. In this section, we employ an example to demonstrate a phenomenon lying somewhere between these two extremes. In particular, we examine an approximation procedure for a problem having a finite-valued performance (absolute) bound that is both realizable and unimprovable (unless $P = NP$).

Our problem can be stated in the following manner: Given a graph $G(V, E)$ with edge weights $c_{ij} \geq 0$ for every $(i, j) \in E$, find a Hamiltonian cycle in G whose maximum weight edge is minimized. While this problem, known as the *minimax* or *bottleneck traveling salesman problem (BTSP)*, differs from its more celebrated relative (the minisum version) in objective, it is no different in difficulty. In fact, it is easy to show that solving this bottleneck version is equivalent to the problem of deciding if an arbitrary graph is Hamiltonian. We leave this as an exercise.

Let us assume, without loss of generality, that $G(V, E)$ is complete. Now, it is trivial (but helpful) to observe that any Hamiltonian cycle in G is a

2-connected subgraph of G and although the converse is not true, we can use this notion as the first step of our approximation procedure. Specifically, let us construct a bottleneck-optimal, 2-connected spanning subgraph of G and call this subgraph $\bar{G}$. Clearly, if $\bar{G}$ is Hamiltonian, then such a cycle (in G) would solve our problem. Thus, the maximum weight edge in G is either the value of an optimal bottleneck solution for the given instance or it is at least a valid lower bound on the optimal value. Regardless, the issue borders on the irrelevant since deciding the hamiltonicity of $\bar{G}$ is difficult. Suppose, however, we augment the edge set of $\bar{G}$ in such a way that the resultant graph is automatically made Hamiltonian.

Denoting the edge set of $\bar{G}$ by $\bar{E} \subset E$, suppose we augment $\bar{E}$ by the set $\{(i, j)\colon (i, k), (k, j) \in \bar{E}\}$. That is, we shall add edges to $\bar{G}$ that connect vertices i and j whenever there is a path (in $\bar{G}$) connecting i and j via vertex k (unless the stated edge is already present in $\bar{G}$). This new graph, say $\hat{G}$, is referred to as the *square* of $\bar{G}$. Other *powers* of a graph are defined in the obvious way. Having formed $\hat{G}$, we can capitalize on a result due to Fleischner (1974) that states that the square of every 2-connected graph is Hamiltonian. We thus have the second and final step of our procedure: Given $\bar{G}$, we construct its square, $\hat{G}$ after which we find a Hamiltonian cycle in $\hat{G}$. This cycle yields our (approximate) solution to the particular instance of *BTSP*.

The overall notion is easily demonstrated in Figure 7.5. The graph with solid edges represents the bottleneck-optimal, 2-connected subgraph $\bar{G}$. The dotted edges form a Hamiltonian cycle in its square, $\hat{G}$ (which is not shown explicitly). There may be, of course, numerous Hamiltonian cycles in a given graph, $\hat{G}$—here, we simply generate one, the maximum weight edge of which becomes our heuristic value.

Naturally, the procedure just described will be of value only if it can be efficiently implemented. Happily, this is the case. Step 1 can be accomplished by a greedy-like process in conjunction with a scheme for checking 2-connectivity. Both activities can be performed in polynomial time. Step 2 is substantially more complicated; however, it also can be implemented in polynomial time by rather lengthy algorithms detailed in Rardin and Parker (1982) and Lau (1980).

How well will this approximation algorithm perform? In order to answer this question in a meaningful way, we must confine our interest to instances of the *BTSP* satisfying the triangle-inequality. Otherwise, we experience the same phenomenon present in the classic (minisum) traveling salesman problem. To make this clear, we have

Lemma 7.17. Importance of the Triangle-Inequality in the BTSP. *If $P \neq NP$, then there can exist no polynomial time approximation algorithm*

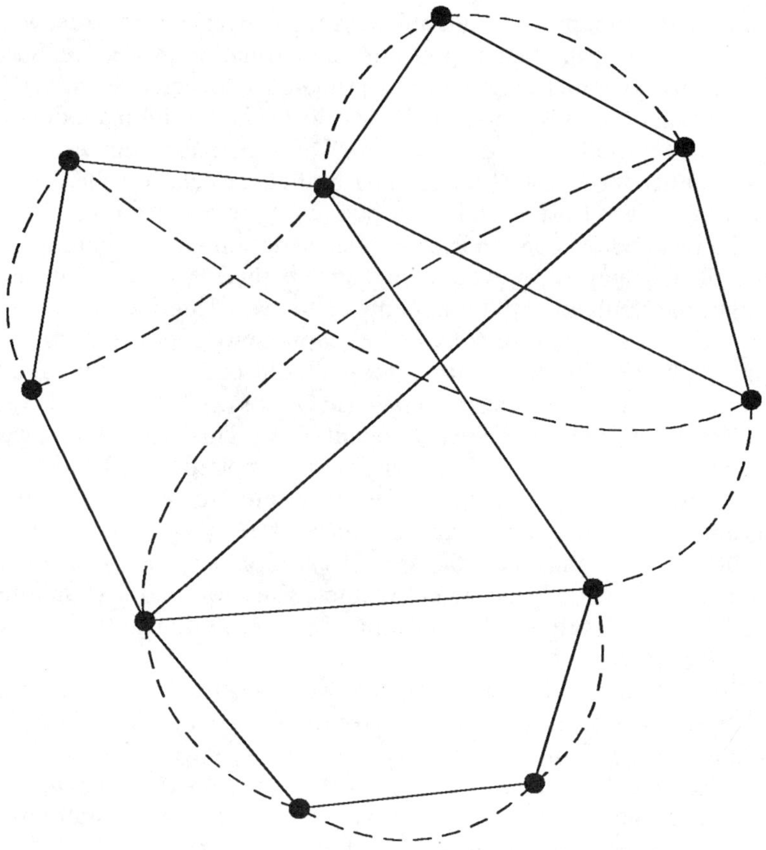

Fig. 7.5. $\overline{G}$ and a Hamiltonian cycle in its square $\hat{G}$.

for the arbitrary bottleneck traveling salesman problem having finite absolute performance bound.

Proof. Assume such an approximation algorithm did exist having worst case performance bound $\rho = K < \infty$. Now, consider the problem of deciding the hamiltonicity of an arbitrary graph and let an instance of this problem be given by graph $G(V, E)$. Let us now construct, from $G(V, E)$, an instance of *BTSP* on a complete graph with edge set E' and edge weights defined as follows:

$$c_{ij} = \begin{cases} 1 \text{ if } (i, j) \in E \\ K + 1 \text{ if } (i, j) \notin E \end{cases}$$

If G is Hamiltonian, then clearly an optimal *BTSP* solution must have value 1 and our theoretical heuristic can have value no greater than K (by hypothesis). Moreover, since edges in E' have weight 1 or $K+1$, our approximate solution must be optimal. Conversely, if G is not Hamiltonian, then the *BTSP* optimum (and thus the approximate solution) has value $K+1$. Hence, our hypothesized heuristic would solve the *NP*-Complete problem of deciding which graphs are Hamiltonian and the assumption that $P \neq NP$ is contradicted. This completes the proof. ■

Within the context of the triangle-inequality, we can look for a finite performance bound on our two-step *BTSP* approximation procedure. Accordingly, we can state the following result.

Theorem 7.18. Performance Bound of the BTSP Approximation Algorithm (with the Triangle-Inequality). *For an arbitrary instance of the BTSP where the triangle-inequality holds, the stated approximation algorithm will produce a solution having value no more than twice an optimal value.*

Proof. For any instance of the *BTSP*, let the maximum weight edge in a bottleneck-minimum 2-connected spanning subgraph $\bar{G}$ of G have value $\bar{c}$ and denote an optimal *BTSP* solution value by v. Clearly, $v \geq \bar{c}$. Furthermore, no edge in $\hat{G}$ can have weight more than twice $\bar{c}$ by the triangle-inequality. Thus, letting the solution value produced by the stated approximation algorithm be $\hat{v}$, we have $\hat{v} \leq 2\bar{c} \leq 2v$ and the theorem is established. ■

We have then that in the worst case, our proposed heuristic could miss the optimal solution by 100%. Our final theorem both shows no better bound is likely and (in its proof) provides a class of examples where the value of 2 will occur.

Theorem 7.19. Formal Limitation on the Performance Bound for any BTSP Approximation Procedure. *If $P \neq NP$, no polynomial time approximation algorithm for the BTSP can have absolute performance bound less than 2, even if the triangle inequality holds.*

Proof. The proof follows that given for Lemma 7.17. For an arbitrary graph, $G(V, E)$, an instance of the *BTSP* is constructed on a complete graph G' having edge weights equal to 1 if the corresponding edge is in G and 2 otherwise. Clearly, this assignment of edge weights satisfies the triangle-inequality. Moreover, any hypothesized heuristic would be optimal for the stated instance which implies that it could be used to decide the hamiltonicity of arbitrary graphs. As before, this denies that P and NP are distinct, and we have the desired contradiction. ■

EXERCISES

7-1. The bin packing problem is to arrange n given items of size $s_i \in [0, 1]$, $i = 1, \ldots, n$ into the minimum number of unit capacity bins. The *First Fit* algorithm for this problem imagines we are presented with an ordered supply of empty bins. Taking items in subscript order, each item is placed in the first bin with enough capacity to accommodate it.

(a) Show that the number of bins used by the First Fit heuristic is at most $\lceil 2 \sum_{i=1}^{n} s_i \rceil$.
(b) Use (a) to establish that First Fit has guaranteed performance ratio less than or equal to 2.

7-2. Given a graph $G(V, E)$ the minimum vertex cover problem seeks a minimum cardinality $S \subseteq V$ such that every edge $e \in E$ meets at least one $v \in S$. Consider a heuristic for this problem that (i) constructs a maximal matching of G by adding edges to a set M until no more can be added without having two members meeting a common vertex, then (ii) takes as S the vertices touched by members of M.

(a) Show that any S constructed in this way is a vertex cover of G.
(b) Show that this algorithm has guaranteed performance ratio at least 2.

7-3. Suppose we are interested in empirically evaluating nonexact methods for a discrete optimization problem in the form

$$\min \quad \mathbf{cx} \tag{1}$$

$$\text{s.t.} \quad \mathbf{Ax} \geq \mathbf{b} \tag{2}$$

$$\mathbf{x} \text{ integer} \tag{3}$$

and that a system of valid inequalities

$$\mathbf{Gx} \geq \mathbf{h} \tag{4}$$

is also known for the problem. Further suppose that a random generation is available to produce positive instances $(\mathbf{A},\mathbf{b})$, a random sample of corresponding inequalities $(\mathbf{G},\mathbf{h})$ and an integer vector $\mathbf{x}^*$ satisfying $\mathbf{Ax}^* \geq \mathbf{b}$. Pilcher and Rardin (1987) propose using such generators to produce test instances on which heuristic algorithms can be conveniently evaluated by choosing $\mathbf{c}$ so that $\mathbf{x}^*$ is a provable optimal solution to (1)-(3).

(a) Show that if such an $\mathbf{x}^*$ solves the linear program (1)-(2), (4), it solves the discrete problem (1)-(3).
(b) State linear programming optimality conditions for (1)-(2), (4).
(c) Devise a construction of dual multipliers in the conditions of (b) and of cost row $\mathbf{c}$, so that $\mathbf{x}^*$ will satisfy the optimality conditions for (1)-(2), (4) with the chosen dual solution.
(d) Establish that for the $\mathbf{c}$ of (c), $\mathbf{x}^*$ is optimal in (1)-(3).
(e) Produce an $\mathbb{R}^2$ numerical example illustrating this construction and showing that although $\mathbf{x}^*$ solves (1)-(2), (4) it need not solve the usual linear programming relaxation (1)-(2).
(f) Explain why it is desirable for this construction that constraints in the sample $(\mathbf{G}, \mathbf{h})$ are tight for $\mathbf{x}^*$, i.e., $\mathbf{Gx}^* = \mathbf{h}$.

7-4. Consider the symmetric traveling salesman problem (*STSP*) on graph $G(V, E)$ in the form

$$\begin{aligned} \min \quad & \sum_i \sum_{j>i} c_{ij} x_{ij} \\ \text{s.t.} \quad & \sum_{i<k} x_{ik} + \sum_{j>k} x_{kj} = 2 \text{ for every } k \in V \\ & \sum_{i,j \in S} x_{ij} \leq |S| - 1 \text{ for every } S \subset V \\ & x_{ij} \geq 0 \text{ and integer for every } i, j > i \end{aligned}$$

Apply the Pilcher and Rardin (1987) approach of **7-3** to generate $\{c_{ij}\}$ for which a chosen $\{x^*_{ij}\}$ is optimal. Specifically,

(a) State linear programming optimality conditions for the relaxation when "x_{ij} integer" is dropped.
(b) For any tour solution $\mathbf{x}^*$ show that the collection of sets S such that $\Sigma_{i,j \in S}\, x^*_{ij} = |S| - 1$ is the collection of subintervals from the $\mathbf{x}^*$ tour.
(c) Use results of (a) and (b) to specify a construction of c_{ij} such that $\mathbf{x}^*$ solves the corresponding *STSP*.

7-5. Consider a discrete optimization problem in the form

$$\begin{aligned} \min \quad & \mathbf{cx} \\ \text{s.t.} \quad & \mathbf{x} \in F \end{aligned}$$

and suppose F is bounded, $\mathbf{cx} > 0$ for every $\mathbf{x} \in F$, and in addition assume every $\mathbf{x} \in F$ satisfies the valid equality $\mathbf{px} = q$ with $q > 0$.

(a) Show that for any multiplier λ and any $\mathbf{x}^1, \mathbf{x}^2 \in F$, $\mathbf{cx}^1 < \mathbf{cx}^2$ if and only if $\mathbf{c}^\lambda \mathbf{x}^1 < \mathbf{c}^\lambda \mathbf{x}^2$ where $c^\lambda_i = c_j + \lambda p_i$, i.e., that feasible solution rankings are identical under $\mathbf{c}$ and any $\mathbf{c}^\lambda$.

(b) For a fixed instance with optimal solution $\mathbf{x}^*$ and nonexact solution $\mathbf{x}^H$ under costs $\mathbf{c}$, the performance ratio (7-1) is $\mathbf{cx}^H/\mathbf{cx}^*$. Show that for any ratio $\rho \in (1, \mathbf{cx}^H/\mathbf{cx}^*)$ there exists a λ such that the same nonexact algorithm will achieve performance ratio ρ by using costs $\mathbf{c}^\lambda$ of (a), i.e., that the ratio can be made arbitrarily small.

(c) Illustrate (b) with the valid equality $\Sigma_i \Sigma_{i>j} x_{ij} = |V|$ for the traveling salesman problem formulation of **7-4.**

★7-6. Consider the symmetric traveling salesman problem in the $\mathbb{R}^2$ plane where costs are the Euclidean distance between points. Karp's (1977) probabilistic analysis approaches this problem with a nonexact procedure that proceeds by dividing the plane into subrectangles containing a fixed number of cities, solving the *TSP* optimally in each of these rectangles, combining the resulting tours to obtain a spanning Eulerian (even degree) subgraph, and shortcutting around duplicated vertices via the triangle inequality to yield a tour. Rectangles are subdivided by a line parallel to the shortest side and through the median city as ordered in the direction of the longer side.

(a) Show that this splitting decision requires only polynomial time.

(b) Show that the union of tours for rectangles formed in this way is a spanning Eulerian subgraph of the original graph.

(c) Show that the Eulerian subgraph of (b) can be reduced to a *TSP* form of no greater length in polynomial time via shortcutting around repeated vertices.

(d) Show that if rectangles are all divided each round, $\lceil \log_2(n/t) \rceil \triangleq k$ rounds of partitioning are required to assure no rectangle has more than t cities.

(e) For any particular rectangle Y, let $T(Y)$ be the optimal tour of cities in Y and $T^*(Y)$ the portions of an optimal tour for all cities that intersect Y. Show that (length $T(Y)$) $\leq$ (length $T^*(Y)$) + $\frac{3}{2}$ (perimeter (Y)).

(f) After k rounds of rectangle splitting of an a by b rectangle the perimeter of any particular rectangle can be shown to be at most

$$F_k(a, b) \triangleq \min_{0 \leq q \leq k} \{2(2^q a + 2^{k-q} b)\}$$

Apply (c) to show that after k rounds the length of the walk formed by continuing optimal tours for each rectangle exceeds the optimal tour for the whole rectangle by at most $\frac{3}{2} F_k(a, b)$.

(g) Combine (d) and (f) to conclude the error of Karp's heuristic algorithm is at most $O(\sqrt{n/t})$.

(h) Beardwood, J., J. H. Halton and J. M. Hammersley (1959) show that the length of a tour through n random points in the plane tends to

$O(\sqrt{n})$ growth. Combine this result with (g) to conclude the performance ratio of Karp's heuristic tends to $1 + O(t^{-1/2})$.

7-7. Return to the nearest unvisited city heuristic and the traveling salesman problem instance of Section 7.1.2. Show how to modify costs (1, 3), (1, 4) and (4, 1) so that the algorithm produces arbitrarily poor solutions.

7-8. Return to the change making problem of Example 7.3. Show that the greedy algorithm will be optimal for any n coins valued at integers $v_1 < v_2 < v_3 < \cdots < v_n$ such that $v_1 = 1$ and $\Sigma_{i=1}^{k} v_i < v_{k+1}$ for $k = 1, 2, \ldots, n-1$.

7-9. Dobson (1982) proposes a greedy algorithm for the general form

$$\begin{aligned} \min \quad & \mathbf{cx} \\ \text{s.t.} \quad & \mathbf{Ax} \geq \mathbf{b} \\ & \mathbf{x} \geq \mathbf{0} \\ & \mathbf{x} \text{ integer} \end{aligned}$$

with $\mathbf{c} > \mathbf{0}$, $\mathbf{b} > \mathbf{0}$, $\mathbf{A} \geq \mathbf{0}$ integer data. The algorithm begins with solution $\bar{\mathbf{x}} \leftarrow \mathbf{0}$. Then while $\mathbf{b} \neq \mathbf{0}$, it successively sharpens $a_{ij} \leftarrow \min\{a_{ij}, b_i\}$ for all i and j, picks the k with minimum $(c_k/\Sigma_{i=1}^{m} a_{ij})$ and updates $\bar{x}_k \leftarrow \bar{x}_k + 1$, $b_i \leftarrow b_i - a_{ik}$ for all i. Dobson proves the algorithm has guaranteed performance ratio $\Sigma_{l=1}^{s} l^{-1}$, where s is the maximum column sum of the original matrix $\mathbf{A}$.

(a) Apply this greedy heuristic to compute an approximate solution to the instance

$$\begin{aligned} \min \quad & 10x_1 + 8x_2 + 3x_3 + 11x_4 \\ \text{s.t.} \quad & x_1 \phantom{{}+ 5x_2} + x_3 + 7x_4 \geq 10 \\ & 3x_1 + 5x_2 + x_3 + x_4 \geq 5 \\ & \phantom{3x_1 +{}} 2x_2 + x_3 \phantom{{}+ x_4} \geq 2 \\ & x_1, x_2, x_3, x_4 \geq 0 \text{ and integer.} \end{aligned}$$

(b) Compute Dobson's performance ratio for the problem.

(c) Determine an optimal solution for the instance of (a) by enumeration and establish that the solution of (a) is indeed within the guarantee of (b).

7-10. Interpret the change making Example 7.3 as a special case of the Dobson algorithm **7-9**, and evaluate Dobson's guaranteed performance ratio for the instance with the 5 U.S. coins.

7-11. The famous *column-generating* truncated exponential scheme of Gilmore and Gomory (1961) seeks to cut quantities $b_1, b_2, \ldots, b_m$ of pieces length $l_1, l_2, \ldots, l_m$ using the minimum number of stock pieces of length L.

(a) Formulate this optimization as an integer program of the Dobson **7-9** form with columns of **A** consisting of feasible ways to divide L into different mixes of lengths, l_i.

(b) Explain why an optimal solution to the linear programming relaxation of the form in (a) could easily be rounded to an approximately optimal integer solution.

(c) Explain why solving even the linear programming relaxation of the formulation in (a) could require exponential effort for the best available linear programming algorithms.

(d) Suppose that only a subset J of columns are actually at hand, that the corresponding restricted linear programming relaxation is solved by the simplex algorithm, and that the optimal dual solution resulting is $\bar{u}_1, \bar{u}_2, \ldots, \bar{u}_m$. Show that if any column $j \notin J$ can improve the current linear relaxation solution one is the column with coefficients optimal in the knapsack problem

$$\begin{aligned} \max \quad & \sum_{i=1}^{m} \bar{u}_i a_i \\ \text{s.t.} \quad & \sum_{i=1}^{m} l_i a_i \le L \\ & a_i \ge 0 \text{ and integer for all } i. \end{aligned}$$

(e) Use the result of (d) to structure an algorithm that solves such restricted linear programs and generates new columns as needed in an ultimately exponential quest for a solution to the full linear programming relaxation that can be rounded to an approximate integer optimum.

***7-12.** The guaranteed performance ratio for the pure greedy algorithm of Theorem 7.1 may equal 1, equal some other constant ρ, or vary with problem instance. For each of the independence systems detailed in Exercise **3-19** establish which of these three outcomes derives.

7-13. Consider the binary knapsack problem to

$$\begin{aligned} \max \quad & \sum_{j=1}^{n} c_j y_j \\ \text{s.t.} \quad & \sum_{j=1}^{n} a_j y_j \le b \\ & y_j = 0 \text{ or } 1 \qquad j = 1, 2, \ldots, n, \end{aligned}$$

where b and all c_j and a_j are positive integers.

(a) Establish that feasible sets for this model may be viewed as independent sets of an independence system defined over $N \triangleq \{1, 2, \ldots, n\}$.

(b) Devise and show correctness of procedures for computing upper and lower rank functions ur(N) and lr(N) for the whole set N.

(c) Show that for this model an $S \subseteq N$ establishing the performance guarantee of Theorem 7.1 is always $S = N$.

(d) Extend vertex packing Example 7.6 to show that the knapsack property of (c) does not hold for all independence systems.

*(e) Define a structural condition for independence systems that would assure property (c).

7-14. Example 7.1 describes a sequential algorithm for nonnegative integer knapsack problems that can be described as greedy yet differs from the pure greedy of Theorem 7.1.

(a) Show that problems (*IKP*) of Example 7.1 can be expressed in the form of Exercise **7-13** by substituting $x_j \leftarrow \Sigma_{l=1}^{\lfloor b/a_j \rfloor} y_j^l$, where all y_j^l are binary.

(b) Produce a numerical example illustrating that the $\max\{c_j/a_j\}$ heuristic of Example 7.1 gives different results than the pure max $\{c_j\}$ method of Theorem 7.1 applied to the restated version of (a).

(c) Compare the guaranteed performance ratios of Example 7.1 and Theorem 7.1 that result from your example of (b). Discuss their relative merits.

7-15. The classic *2-Opt* procedure for the symmetric (undirected) traveling salesman problem on complete graphs begins with any tour of vertices $\{1, 2, \ldots, p\}$ and applies a version of Algorithm 7A that searches over the pair interchange neighborhood where tour edges (i, j) and (k, l) are replaced by non-tour edges (i, k) and (j, l) (here i, j, k, l are distinct vertices occurring in that order along the current tour).

(a) Show that such interchanges always yield feasible tours.

(b) Describe the meaning of the optimal adjacency and better adjacency notions of Section 7.4.3 in the 2-Opt context.

(c) Apply the optimal adjacency approach on the instance with edge costs

$$\begin{array}{c|ccccc} & 1 & 2 & 3 & 4 & 5 \\ \hline 1 & \infty & 2 & 3 & 5 & 4 \\ 2 & 2 & \infty & 2 & 4 & 2 \\ 3 & 3 & 2 & \infty & 6 & 2 \\ 4 & 5 & 4 & 6 & \infty & 4 \\ 5 & 4 & 2 & 2 & 4 & \infty \end{array}$$

starting from the tour 1-2-3-4-5-1.

(d) Establish that with either better or optimal adjacency, 2-Opt neighborhood searches to find an improvement on a current tour or prove none exists can always be completed in polynomial time.
(e) Construct a 2-Opt adjacency forest like Figure 7.2 for the 4-city instance of Section 7.1.2.
★(f) Produce a family of examples where 2-Opt under better adjacency requires an exponentially growing number of local improvement steps to reach a local optimum.

7-16. Return to the instance of Example 7.7 under the unit neighborhood N_1.

(a) Construct an adjacency forest for this example like the ones of Figure 7.2 using the true objective function (Phase II).
(b) Determine from (a) how many local optima exist.
(c) Determine from (a) the longest better adjacency search that might occur for this problem.
(d) Determine from (a) the mean better adjacency duration of local improvement on this example using Tovey's boundary and coboundary distributions of Section 7.4.3.

7-17. Johnson, Papadimitriou and Yannakakis (1986) have proposed a complexity theory for local search centered on a class *PLS* of problem-neighborhood pairings (Q, N), where $(Q, N) \in PLS$ if (i) a feasible solution to Q can be found in polynomial time and (ii) given $\mathbf{x}$ feasible for Q we can determine in polynomial time whether $\mathbf{x}$ is a local minimum over neighborhood $N(\mathbf{x})$.

(a) Show that all four of the neighborhoods (7-18)–(7-21) yield local searches satisfying (ii) (assuming the dimension of $\mathbf{x}$ measures problem size).
(b) Show that the 2-Opt algorithm for traveling salesman problems described in **7-15** yields a member of *PLS*.

7-18. The graph partitioning problem is posed on an undirected graph with $2p$ vertices and given edge weights w_e. The vertices must be arranged in 2 subsets of size p so that the total weight of edges joining vertices in different sets is minimized. The famous Kernighan-Lin (1970) procedure for this problem begins from an arbitrary partition and performs local improvement by swapping a vertex in one set for a vertex in the other. Show that graph partitioning under this concept of neighborhood belongs to class *PLS* of **7-17**.

7-19. Suppose a minimization-form discrete optimization problem with an always-positive objective function can be enumerated by Algorithm 7B in such a way that a known nonexact algorithm will achieve constant guaranteed-performance ratio ρ on all candidate problems that might be encountered. Further suppose that nonexact algorithm is applied during any occurrence of Step 3 of Algorithm 7B for which an optimal $\mathbf{x}^k$ is not found in Step 2 in order to produce possible incumbent solutions for Step 4.

(a) Develop a nonnegative and monotone bound allowance function $\epsilon(v)$ for this enumeration that assumes exactly one candidate problem will be processed by Step 2 of our modified Algorithm 7B.
(b) For any polynomial function $p(n)$ of the problem size n, extend the concept of (a) to derive a nonnegative and monotone bound allowance function for which at most $p(n)$ candidates will be processed by Step 2 in enumeration by the modified Algorithm 7B.

7-20. Balas and Martin (1980) developed a *Pivot and Complement* algorithm for binary optimization problems in the form

$$\begin{array}{lll} \max & \mathbf{cx} \\ \text{s.t.} & \mathbf{Ax} \le \mathbf{b} \\ & \mathbf{0} \le \mathbf{x} \le \mathbf{1} \\ & \mathbf{x} \text{ integer} \end{array} \qquad \text{or equivalently} \qquad \begin{array}{lll} \max & \mathbf{cx} \\ \text{s.t.} & \mathbf{Ax} + \mathbf{y} = \mathbf{b} \\ & \mathbf{0} \le \mathbf{x} \le \mathbf{1} \\ & \mathbf{x} \text{ integer} \\ & \mathbf{y} \ge \mathbf{0}, \end{array}$$

where y_i is the slack in constraint i. The pivot phase of their algorithm solves the linear programming relaxation of the second form above by the simplex algorithm and then seeks to pivot (generally making costs worse) so that the linear programming solution remains feasible and all y_i become basic.

(a) Show that if such a pivot sequence exists the $\mathbf{x}$ solution derived will be a feasible solution for the original discrete problem.

The complement phase of the algorithm takes nonbasic x_j at value 0 or 1 and considers moving them to 1 or 0 respectively. Either one, two or three variables may be complemented in this way at any particular step.

(b) Describe the local improvement neighborhood implicit in this complement phase.
(c) Establish that finding an improved solution or establishing that the present one is a local optimum under the specified complementations requires polynomial-time effort.

7-21. Refer to the enumeration of Figure 7.3.

(a) Deduce a 0–1 integer program that realizes the σ and β bounds shown with β's the linear programming relaxation values and σ's including post-optimality penalties as in Section 5.3.2.
(b) Develop conditions on σ and β values in such an enumeration tree that render them realizable as linear programming-based branch and bound sequences like (a).

7-22. Prove Corollaries 7.15 and 7.16.

7-23. Given a complete graph G with edge weights w_e and a tree T, the *tree imbedding problem* (*TIP*) is to find a minimum weight subgraph of G isomorphic to T. The related problem of deciding whether a given incomplete graph H has a subgraph isomorphic to T is *NP*-Complete. Use this fact to show that unless $P = NP$ there can be no nonexact algorithm for *(TIP)* with data independent guaranteed performance ratio r.

7-24. The *NP*-Hard *k-center problem* (*k-CP*) on a graph $G(V,E)$ with edge weights $\{w_{ij}: (i,j) \in E\}$ is to find a subset $S \subseteq V$ with $|S| < k$ such that every vertex of V either belongs to S or is adjacent to a member of S and such that $\max_{i \in V \setminus S} \{\min_{j \in S}\{w_{ij}\}\}$ is as small as possible. Assume (i) G is complete (ii) $\{w_{ij}\}$ satisfy the triangle inequality and (iii) for any subgraph $\tilde{G}$ of G there exists a polynomial-time algorithm (Hochbaum and Shmoys (1985)) that either establishes $\tilde{G}$ contains no feasible k-center or exhibits one in the square $\tilde{G}^2$. Then parallel the bottleneck traveling salesman problem results of Section 7.6.2 to

(a) Show that the algorithm of assumption (iii) can be used to derive a nonexact procedure for (*k-CP*) of the assumed form with guaranteed performance ratio 2.
(b) Show that no algorithm can improve on this ratio unless $P = NP$.

7-25. Exercise **7-2** offers a straightforward guaranteed performance heuristic for minimum vertex covering and Section 2.3.2 detailed a simple reduction from maximum clique to minimum vertex cover. Can the heuristic for vertex cover be used to obtain one for clique? Discuss.

Appendix A

Vectors, Matrices and Convex Sets

Many results of this book draw freely on standard notions from linear algebra and convex analysis. In this appendix we briefly survey the most important.

A.1 Vectors

A *scalar* is an element of $\mathbb{R}^1$, the set of real numbers. For integer n, an *n-vector* is a finite array of n scalars, i.e., an element of $\mathbb{R}^n$, the n-fold Cartesean product of $\mathbb{R}^1$ termed *Euclidean n-space.* Often we are interested only in the *nonnegative orthant* $\mathbb{R}^n_+ \triangleq \{\mathbf{x} \in \mathbb{R}^n : \mathbf{x} \geq \mathbf{0}\}$.

In this book, scalars are generally denoted by lower case Greek letters (e.g., α, λ), or by lower case Latin letters in italics (e.g., n, x_i, a_{ij}). Vectors are indicated by bold lower case letters, with their scalar *components* indicated by subscripts. Thus, the vector $\mathbf{x} = (-3, \frac{2}{3}, \sqrt{5})$ is an element of $\mathbb{R}^3$ with component $x_2 = \frac{2}{3}$.

Sometimes it is convenient to array the components of a vector horizontally and other times, vertically. We take these as interchangeable, so that

$$(2, -\tfrac{1}{4}, 0) = \begin{pmatrix} 2 \\ -\frac{1}{4} \\ 0 \end{pmatrix}$$

Vectors of equal length may be added or subtracted componentwise. Thus, $\mathbf{x} \pm \mathbf{y} \triangleq (x_1 \pm y_1, x_2 \pm y_2, \ldots, x_n \pm y_n)$ for n-vectors $\mathbf{x}$ and $\mathbf{y}$. Similarly the product of a scalar and a vector is the vector of products, i.e., $\lambda\mathbf{x} \triangleq (\lambda x_1, \lambda x_2, \ldots, \lambda x_n)$.

Vectors of equal length may be multiplied in the *dot* or *inner product* to produce a scalar that is the sum of one vector's components weighted by those of the others. Throughout this book when two vectors are written next to each other, this dot product is indicated (except where explicitly stated otherwise). Thus for n-vectors $\mathbf{x}$ and $\mathbf{y}$, $\mathbf{xy} \triangleq \Sigma_{j=1}^{n} x_j y_j$.

A *norm* is a real-valued measure of the size or length of a vector. The most familiar is the *Euclidean norm* denoted by $\|\cdot\|$. For n-vector $\mathbf{x}$. $\|\mathbf{x}\| \triangleq [\Sigma_{j=1}^{n} (x_j)^2]^{1/2}$. Other norms measure lengths differently. For example, the l_1 or *rectilinear norm* is $\Sigma_{j=1}^{n} |x_j|$ and the l_∞ or *sup norm* is $\max \{|x_j|, j = 1, \ldots, n\}$.

It is easy to derive from the definition of the Euclidean norm some frequently used identities and inequalities.

$$\|\mathbf{x}\| \geq 0$$

$$\|\mathbf{x}\|^2 = \mathbf{x}\,\mathbf{x}$$

$$\|\mathbf{x} \pm \mathbf{y}\|^2 = \|\mathbf{x}\|^2 + \|\mathbf{y}\|^2 \pm 2\mathbf{xy}$$

$$\|\mathbf{x} + \mathbf{y}\| \leq \|\mathbf{x}\| + \|\mathbf{y}\|$$

$$|\mathbf{xy}| \leq \|\mathbf{x}\|\,\|\mathbf{y}\|$$

The last of these is called the *Cauchy-Swartz inequality.*

Vectors of equal length (say n) are partially ordered by *inequalities* $\mathbf{x} \leq \mathbf{y}$ signifying $x_1 \leq y_1, x_2 \leq y_2, \ldots, x_n \leq y_n$. Similarly, $\mathbf{x} < \mathbf{y}$ implies $x_1 < y_1, x_2 < y_2, \ldots, x_n < y_n$. When a total ordering is required *lexicographic order* is usually employed. Denoted $\mathbf{x} \overset{L}{\leq} \mathbf{y}$, an n-vector $\mathbf{x}$ is lexicographically less than or equal to $\mathbf{y}$ if the first nonzero component of $\mathbf{x} - \mathbf{y}$ is negative. Thus for example, $\mathbf{x} = (1, -2, 100, 17)$ and $\mathbf{y} = (1, 4, -100, -\sqrt{2})$ cannot be compared by normal inequalities, but $\mathbf{x} \overset{L}{\leq} \mathbf{y}$.

The zero, one and unit vectors in $\mathbb{R}^n$ play a special role in many computations. The *zero vector* (denoted $\mathbf{0}$) is an n-vector of zeroes. The *one vector* (denoted $\mathbf{1}$) is an n-vector of 1's. A *unit vector* (usually denoted $\mathbf{e}^j$) is a vector with all components zero except the *jth*, which is a 1. That is, $\mathbf{e}^j \triangleq (0, 0, 0, 1, 0, \ldots, 0)$ with the 1 in the *jth* position.

A collection of vectors $\{\mathbf{x}^k: k = 1, 2, \ldots, t\}$ is said to be *linearly dependent* if there exists any set of scalars $\{\lambda_k: k = 1, 2, \ldots, t\}$ such that $\Sigma_{k=1}^{t} \lambda_k\, \mathbf{x}^k = 0$. That is, vectors are linearly dependent if they can be combined in a *linear combination* to produce the zero vector. Equivalently, they are linearly dependent if one can be expressed as a linear combination of the others. Collections of vectors that are not linearly dependent are *linearly independent*. Examples of a linearly independent set in $\mathbb{R}^3$ are $\mathbf{x}^1 = (1, 0, 0)$, $\mathbf{x}^2 = (3, -\frac{2}{3}, 0)$ and $\mathbf{x}^3 = (-4, \sqrt{2}, \frac{8}{9})$. The vectors $\mathbf{y}^1 = (2, 1, 0)$ $\mathbf{y}^2 = (0, 1, 2)$ and $\mathbf{y}^3 = (1, 1, 1)$ are linearly dependent in $\mathbb{R}^3$ because $\frac{1}{2}\mathbf{y}^1 + \frac{1}{2}\mathbf{y}^2 - \mathbf{y}^3 = \mathbf{0}$.

A closely related concept is *affine* independence. A collection of vectors $\{\mathbf{x}^k\colon k = 1, 2, \ldots, t\}$ is said to be *affinely independent* if there exists no set of scalars $\{\lambda_k\colon k = 1, 2, \ldots, t\}$ such that $\Sigma_{k=1}^t \lambda_k \mathbf{x}^k = 0$ and $\Sigma_{k=1}^t \lambda_k = 1$. Clearly, linearly independent vectors are affinely independent. The reverse need not be true as the following affinely independent, but linearly dependent vectors illustrate.

$$\mathbf{x}^1 = \begin{pmatrix} 1 \\ 2 \end{pmatrix}, \quad \mathbf{x}^2 = \begin{pmatrix} 2 \\ 1 \end{pmatrix}, \quad \mathbf{x}^3 = \begin{pmatrix} 5 \\ 3 \end{pmatrix}.$$

A.2 Matrices

For integers m and n, a *matrix* is a rectangular array of elements of $\mathbb{R}^1$ with m rows and n columns. In this book, matrices are denoted by bold upper case Latin letters with their component entries shown lower case in italics, with row and column subscripts. An example of a 2 by 5 matrix $\mathbf{A}$ is

$$\mathbf{A} = \begin{bmatrix} 1 & 0 & 0 & -1 & 4 \\ 3 & \sqrt{2} & 9 & 3 & \frac{4}{3} \end{bmatrix}.$$

Entry $a_{2,3} = 9$ in this example. As with vectors, $\mathbf{0}$ denotes a matrix consisting entirely of zeros, and $\mathbf{1}$ a matrix entirely of ones.

Matrices of equal row by column size are added or subtracted componentwise. Thus for m by n matrices $\mathbf{A}$ and $\mathbf{B}$,

$$\mathbf{A} \pm \mathbf{B} \triangleq \begin{bmatrix} a_{11} \pm b_{11}, & a_{12} \pm b_{12}, & \ldots, & a_{1n} \pm b_{1n} \\ a_{21} \pm b_{21}, & a_{22} \pm b_{22}, & \ldots, & a_{2n} \pm b_{2n} \\ \vdots & \vdots & & \vdots \\ a_{m1} \pm b_{m1}, & a_{m2} \pm b_{m2}, & \ldots, & a_{mn} \pm b_{mn} \end{bmatrix}.$$

Scalar multiples of matrices are defined similarly with the i, j entry of $\lambda\mathbf{A}$ being λa_{ij}.

Matrix multiplication by vectors or other matrices is defined by thinking of a matrix as a collection of vectors forming its columns, say $\mathbf{A} \triangleq [\mathbf{a}^1, \mathbf{a}^2, \ldots, \mathbf{a}^n]$, or its rows

$$\mathbf{A} \triangleq \begin{bmatrix} \mathbf{r}^1 \\ \mathbf{r}^2 \\ \cdot \\ \cdot \\ \cdot \\ \mathbf{r}^m \end{bmatrix}.$$

A matrix may be *premultiplied by a vector* with as many components as the matrix has rows. This operation is denoted by simply writing the vector before and next to the matrix. The result is a weighted sum of the row vectors. In particular, for m by n matrix $\mathbf{A}$ with row vectors $\mathbf{r}^i$ and m-vector $\mathbf{u}$, $\mathbf{uA} \triangleq \sum_{i=1}^m u_i\mathbf{r}^i$. In like manner a matrix may be *postmultiplied by a vector* (denoted by writing the vector next to and after the matrix) if the vector has as many components as the matrix has columns. The result is a weighted sum of columns. Specifically, if $\mathbf{x}$ is an n-vector and $\mathbf{A}$ is as above, $\mathbf{Ax} \triangleq \sum_{j=1}^n \mathbf{a}^j x_j$.

These notions are illustrated by $\mathbf{A} = \begin{bmatrix} 2 & 0 & 5 \\ 3 & 1 & -9 \end{bmatrix}$ that has column vectors $\mathbf{a}^1 = (2, 3)$, $\mathbf{a}^2 = (0, 1)$, $\mathbf{a}^3 = (5, -9)$ and row vectors $\mathbf{r}^1 = (2, 0, 5)$, $\mathbf{r}^2 = (3, 1, -9)$. Premultiplication by $\mathbf{u} = (1, -1)$ gives $\mathbf{uA} = (-1, -1, 14)$. Postmultiplication by $\mathbf{x} = (1, 2, 1)$ yields $\mathbf{Ax} = (7, -4)$.

These notions are easily extended to products of matrices $\mathbf{C} = \mathbf{AB}$, where the number of columns of $\mathbf{A}$ is the same as the number of rows of $\mathbf{B}$. The i, j entry of the result is the product of the *ith* row of $\mathbf{A}$ with the *jth* column of $\mathbf{B}$. Thus if $\mathbf{A}$ is m by k and $\mathbf{B}$ is k by n, $\mathbf{C} = \mathbf{AB}$ is m by n with entries

$$c_{ij} \triangleq \sum_{t=1}^{k} a_{it}\, b_{tj}.$$

On occasions when matrices are multiplied, it is sometimes first necessary to convert a k by n matrix to an n by k matrix by exchanging columns and rows. This operation is called *transposing* the matrix and denoted $\mathbf{A}^T$ Thus for the $\mathbf{A}$ above

$$\mathbf{A}^T = \begin{bmatrix} 2 & 3 \\ 0 & 1 \\ 5 & -9 \end{bmatrix}$$

The *identity matrix* is a special n by n (square) matrix formed by the n unit vectors of $\mathbb{R}^n$. That is

$$\mathbf{I} = \begin{bmatrix} 1 & 0 & \dots & & 0 \\ 0 & 1 & \dots & & 0 \\ \cdot & & & & \cdot \\ \cdot & & & & \cdot \\ \cdot & & & & \cdot \\ 0 & & \dots & 1 & 0 \\ 0 & & \dots & & 1 \end{bmatrix}.$$

It is easy to check that for any n-vectors $\mathbf{u}$, $\mathbf{uI} = \mathbf{Iu} = \mathbf{u}$. Similarly for m by n matrix $\mathbf{A}$, $\mathbf{IA} = \mathbf{A} = \mathbf{AI}$ although the first identity matrix is m by m and the second n by n.

The *rank* of an m by n matrix $\mathbf{A}$ (denoted rank($\mathbf{A}$)) is the maximum number of linearly independent row vectors in $\mathbf{A}$, or equivalently the maximum number of linearly independent columns. An m by n matrix $\mathbf{A}$ matrix with rank $= \min\{m, n\}$ is said to be *full rank*. For example, $\mathbf{A} = \begin{bmatrix} 1 & -2 & 0 \\ 0 & 0 & 3 \end{bmatrix}$ is full rank (rank($\mathbf{A}$) = 2) because column $\mathbf{a}^1 = (1, 0)$ and $\mathbf{a}^3 = (0, 3)$ are linearly independent.

For square matrices $\mathbf{B}$, the *determinant* of $\mathbf{B}$ (denoted det($\mathbf{B}$)) is a real number defined recursively as follows

(i) For a 1 by 1 matrix $\mathbf{B}$

$$\det(\mathbf{B}) \triangleq \mathbf{B},$$

(ii) For an m by m matrix $\mathbf{B}$, $m \geq 2$,

$$\det(\mathbf{B}) \triangleq \sum_{i=1}^{m} (-1)^{i-1}\, b_{ij} \det(\mathbf{B}_i),$$

where $1 \leq j \leq m$ is any column number, and $\mathbf{B}_i$ is the $(m-1)$ by $(m-1)$ *minor* matrix derived from $\mathbf{B}$ by deleting row i and column j.

For example, using $j = 1$

$$\det\begin{bmatrix} 3 & -1 & 7 \\ 1 & 0 & 0 \\ 2 & 1 & 4 \end{bmatrix} = 3 \det\begin{bmatrix} 0 & 0 \\ 1 & 4 \end{bmatrix} - 1 \det\begin{bmatrix} -1 & 7 \\ 1 & 4 \end{bmatrix} + 2 \det\begin{bmatrix} -1 & 7 \\ 0 & 0 \end{bmatrix} = 11.$$

It is easy to check that for m by m matrices $\mathbf{A}$ and $\mathbf{B}$, $\det(\mathbf{AB}) = \det(\mathbf{A}) \cdot \det(\mathbf{B})$.

An m by m square matrix $\mathbf{B}$ is said to be *singular* if $\det(\mathbf{B}) = 0$, or equivalently if $\text{rank}(\mathbf{B}) < m$. If $\det(\mathbf{B}) \neq 0$ ($\text{rank}(\mathbf{B}) = m$), $\mathbf{B}$ is *nonsingular.*

One sufficient condition for a matrix $\mathbf{B}$ to be nonsingular is if the matrix is *positive definite.* A matrix is positive definite if $\mathbf{xBx} > 0$ for all $\mathbf{x} \in \mathbb{R}^m$ or equivalently if $\det(\mathbf{B}_k) > 0$ for all *principal minor* matrices $\mathbf{B}_k$ formed from $\mathbf{B}$ by deleting columns and rows with index $> k$.

A nonsingular matrix $\mathbf{B}$ has *inverse* matrix (denoted $\mathbf{B}^{-1}$) of the same size. $\mathbf{B}^{-1}$ is the unique matrix satisfying

$$\mathbf{B}\mathbf{B}^{-1} = \mathbf{B}^{-1}\mathbf{B} = \mathbf{I}.$$

For example, $\begin{bmatrix} 1 & 0 \\ 3 & 2 \end{bmatrix}^{-1} = \begin{bmatrix} 1 & 0 \\ -\frac{3}{2} & \frac{1}{2} \end{bmatrix}$.

One use of the inverse of a nonsingular m by m matrix $\mathbf{B}$ is in solving the system of linear equations of the form $\mathbf{Bx} = \mathbf{d}$, where $\mathbf{d}$ is a given m-vector and $\mathbf{x}$ is sought. The unique solution is $\bar{\mathbf{x}} \leftarrow \mathbf{B}^{-1}\mathbf{d}$.

Another (equivalent) method for solving systems $\mathbf{Bx} = \mathbf{d}$ with $\mathbf{B}$ nonsingular is *Cramer's rule.* Components of the solution vector $\bar{\mathbf{x}}$ are constructed as $\bar{x}_j \leftarrow \det(\mathbf{B}^j)/\det(\mathbf{B})$, where $\mathbf{B}^j$ is derived from $\mathbf{B}$ by substituting $\mathbf{d}$ for column j.

A.3 Convex Sets

A *line segment* in $\mathbb{R}^n$ is the set of points along the line between two vectors $\mathbf{x}^1, \mathbf{x}^2 \in \mathbb{R}^n$. Such a set can be described algebraically as $\{\lambda\mathbf{x}^1 + (1 - \lambda)\,\mathbf{x}^2 \colon 0 \le \lambda \le 1\}$.

A set $C \subseteq \mathbb{R}^n$ is said to be *convex* if the entire line segment between any two vectors in C is contained in C. That is, C is convex if and only if $\{\lambda\mathbf{x}^1 + (1 - \lambda)\,\mathbf{x}^2 \colon 0 \le \lambda \le 1\} \subseteq C$ for every $\mathbf{x}^1, \mathbf{x}^2 \in C$. The following depicts some convex and nonconvex sets in $\mathbb{R}^2$.

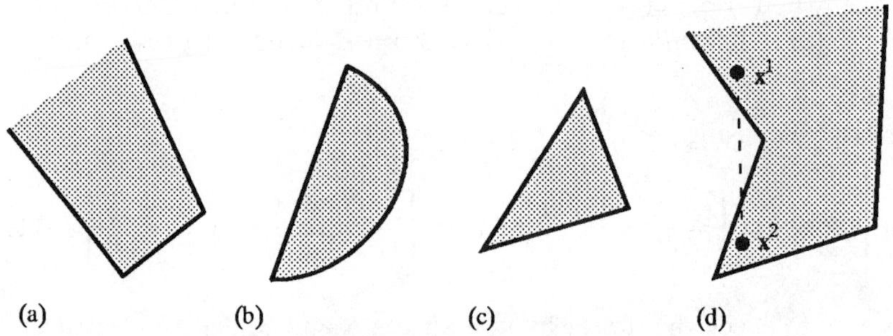

Cases (a)–(c) are convex, but (d) is not (points that fail the definition are indicated $\mathbf{x}^1, \mathbf{x}^2$).

One important property of convex sets is that their intersection is also convex. That is $C \triangleq \bigcap_{i=1}^{p} C_i$ is convex if $C_1, C_2, \ldots, C_p$ are.

A set C is *bounded* if there exists a constant δ_C such that $\|\mathbf{x}\| \leq \delta_C$ for every $\mathbf{x} \in C$. In the above, (b) and (c) are bounded; (a) and (d) are not.

A number of special forms of convex sets in $\mathbb{R}^n$ recur throughout this book.

- A *hyperplane* is a set C of the form $\{\mathbf{x}\colon \mathbf{ax} = a_0\}$ where $\mathbf{a}$ is a given n-vector and a_0 is a given scalar. For example $\{(x_1, x_2)\colon 3x_1 - 4x_2 = 7\}$ is a hyperplane of $\mathbb{R}^2$.
- A *half-space* is a set C of the form $\{\mathbf{x}\colon \mathbf{ax} \geq a_0\}$ for given n-vector $\mathbf{a}$ and scalar a_0. For example, $\{\mathbf{x}\colon 4x_1 - \frac{1}{2}x_2 + 7x_3 \geq 14\}$ is a half-space of $\mathbb{R}^3$.
- A *polyhedron* is a set C defined as the intersection of finitely many half-spaces in $\mathbb{R}^n$. That is C is a polyhedron if $C = \bigcap_{i=1}^{m} \{\mathbf{x}\colon \mathbf{r}^i\mathbf{x} \geq r_i\}$

  or, writing $\mathbf{R} \triangleq \begin{bmatrix} \mathbf{r}^1 \\ \mathbf{r}^2 \\ \cdot \\ \cdot \\ \cdot \\ \mathbf{r}^m \end{bmatrix}$, $C = \{\mathbf{x}\colon \mathbf{Rx} \geq \mathbf{r}\}$.

- A *polytope* is a polyhedron that is bounded. For example $\{\mathbf{x} \in \mathbb{R}^n\colon 0 \leq x_j \leq 1 \text{ for all } j = 1, 2, \ldots, n\}$ is a polytope.
- An *open Euclidean ball* (denoted $\mathbb{B}(\mathbf{x}, \rho)$) of $\mathbb{R}^n$ is the set of points of $\mathbb{R}^n$ within distance ρ of vector $\mathbf{x}$. That is $\mathbb{B}(\mathbf{x}, \rho) \triangleq \{\mathbf{y} \in \mathbb{R}^n\colon \|\mathbf{x} - \mathbf{y}\| < \rho\}$.
- The *convex hull* of any set $T \subseteq \mathbb{R}^n$, denoted $[T]$ is the intersection of all convex sets containing T (see Section 6.1 for cases).
- The *simplex* of $\mathbb{R}^n$ (denoted $\mathbb{S}$) is the convex hull of the set of unit vectors in $\mathbb{R}^n$. Specifically,

$$\mathbb{S} \triangleq \left\{ \sum_{j=1}^{n} \lambda_j \mathbf{e}^j\colon \sum_{j=1}^{n} \lambda_j = 1 \text{ and } \begin{matrix} \lambda_j \geq 0 \text{ for all} \\ j = 1, 2, \ldots, n \end{matrix} \right\}.$$

- The *polar* of a convex set C (denoted C^*) is $C^* = \{\mathbf{p}\colon \mathbf{px} \leq 1 \text{ for all } \mathbf{x} \in C\}$. It is easy to show $(C^*)^* = C$ and that $C_1 \subseteq C_2$ implies $C_1^* \supseteq C_2^*$.

A *convex combination* of vectors in $\mathbb{R}^n$ is the set of vectors obtained by nonnegative weights summing to 1. That is $\mathbf{x} = \sum_{i=1}^{p} \lambda_i \mathbf{x}^i$ is a convex combination of vectors $\{\mathbf{x}^i\colon i = 1, 2, \ldots, p\}$, if $\sum_{i=1}^{p} \lambda_i = 1$ and $\lambda_i \geq 0$ for $i = 1, 2, \ldots, p$. It is easy to see that convex combinations of $\mathbf{x}^i$ in a convex set C must all belong to C.

A distinguished member $\mathbf{p}$ of a convex set C such that $\mathbf{p}$ cannot be expressed as positive a convex combination of distinct members of C is called an *extreme point.* That is, p is an extreme point of C if $\mathbf{p} = \lambda \mathbf{x}^1 + (1-\lambda)\mathbf{x}^2$; $0 < \lambda < 1$; $\mathbf{x}^1, \mathbf{x}^2 \in C$ implies $\mathbf{p} = \mathbf{x}^1 = \mathbf{x}^2$. In the $\mathbb{R}^2$ convex sets below, all extreme points are indicated.

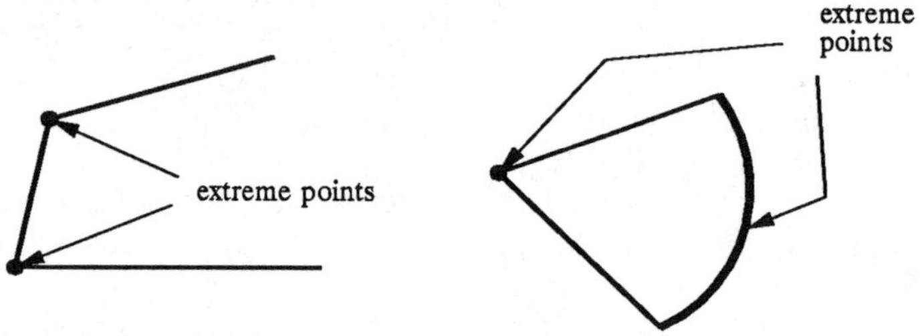

A *direction*, $\mathbf{d}$, of a set $S \subseteq \mathbb{R}^n$ is an n-vector such that $\mathbf{x} + \lambda \mathbf{d} \in S$ for every $\mathbf{x} \in S$ and $\lambda \geq 0$. The direction is said to be an *extreme direction* if it cannot be written as a positive linear combination of distinct directions. That is, $\mathbf{d}$ is an extreme direction of S if

$$\mathbf{d} = \mu_1 \mathbf{d}^1 + \mu_2 \mathbf{d}^2;\ \mu_1, \mu_2 > 0;\ \mathbf{d}^1, \mathbf{d}^2 \text{ directions of } S$$

imply there exist α such that $\mathbf{d}^2 = \alpha \mathbf{d}^1$.

Obviously, a bounded set (e.g., (a) below) has no directions. The $\mathbb{R}^2$ figure (b) illustrates both extreme and nonextreme directions.

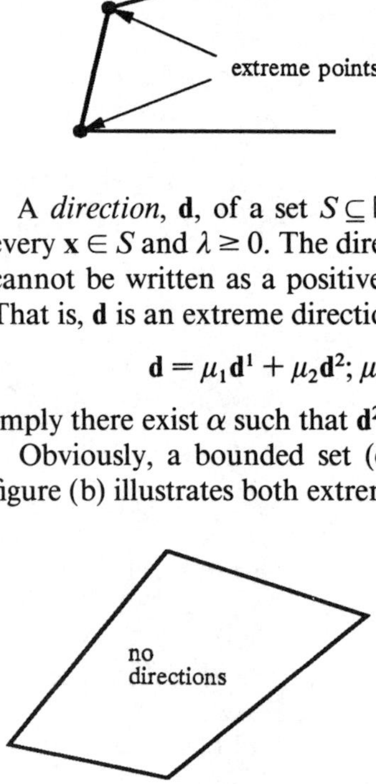

(a)

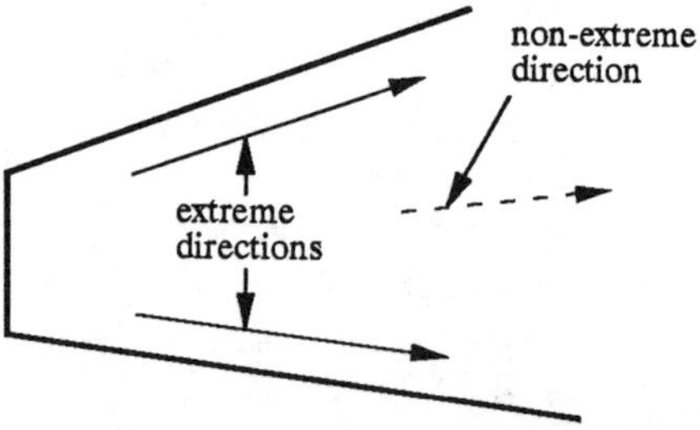

(b)

Convex sets are characterized by their extreme points and extreme directions. Every point of a convex set C can be expressed as a convex combination of extreme points of C plus a nonnegative combination of extreme directions.

A fundamental fact about a polyhedron is that it has only finitely many extreme points and extreme directions (none of the latter if it is also a polytope). Thus, every polyhedron C can be characterized

$$C = \left\{ \sum_{j=1}^{|P|} \lambda_j \mathbf{p}^j + \sum_{k=1}^{|D|} \mu_k \mathbf{d}^k : \sum_{j=1}^{|P|} \lambda_j = 1,\ \lambda_j \geq 0 \text{ for all } j = 1, 2, \ldots, |P|, \right.$$
$$\left. \mu_k \geq 0 \text{ for all } k = 1, 2, \ldots, |D| \right\},$$

where $\{\mathbf{p}^j: j \in P\}$ is a finite list of extreme points and $\{\mathbf{d}^k: k \in D\}$ a finite set of extreme directions.

The dimension of a convex set C (denoted dim(C)) is the maximum number of linearly independent difference vectors $\{(\mathbf{x}^1 - \mathbf{x}^0), (\mathbf{x}^2 - \mathbf{x}^0), \ldots\}$ such that $\mathbf{x}^0, \mathbf{x}^1, \mathbf{x}^2, \ldots, \in C$. (See also Section 6.1). A set $C \subseteq \mathbb{R}^n$ is said to be *full dimension* if $\dim(C) = n$. For $\mathbb{R}^2$ examples below, (a) is full dimension, (b) is not.

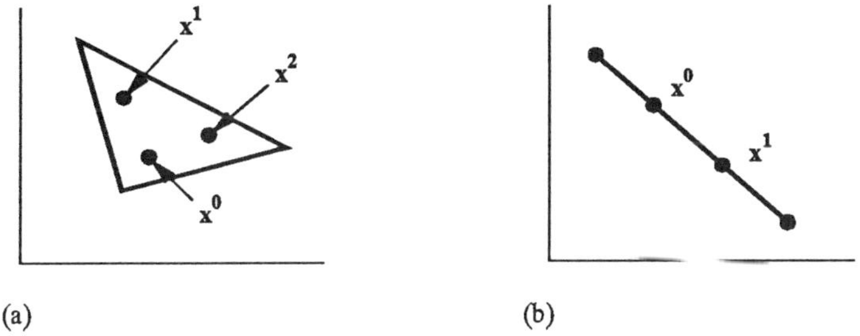

A *face* of a polyhedron C is the lower-dimensional set defined as the intersection of a half-space with the boundary of C. More precisely F is a face of C if there exists $(\mathbf{a}, a_0)$ such that

$$\varnothing \neq F = \{\mathbf{x} \in C: \mathbf{ax} = a_0\} = \{\mathbf{x} \in C: \mathbf{ax} \geq a_0\}.$$

Faces of dimension zero are called *vertices* of the polyhedron (or equivalently extreme points); faces of highest dimension are termed *facets.*

The $\mathbb{R}^3$ polytope below has faces of dimension 0, 1 and 2. Those of dimension 0 are its vertices $\{\mathbf{v}^1, \mathbf{v}^2, \mathbf{v}^3, \mathbf{v}^4\}$; those of dimension 2 are its facets (for example, the one shaded). Line segments between vertices are faces of dimension 1.

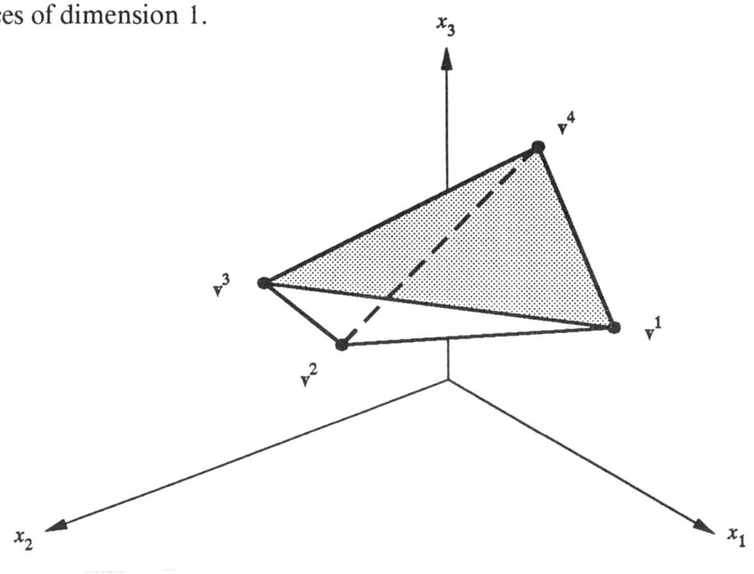

A closely related notion to that of a convex set is a convex function. A *function* θ: $\mathbb{R}^n \rightarrow \mathbb{R}^1$ is said to be *convex* if $\{(\mathbf{x}, x_0) \in \mathbb{R}^{n+1};\ \theta(\mathbf{x}) \leq x_0\}$ is a convex set. That is θ is convex on C if $\lambda\theta(\mathbf{x}^1) + (1-\lambda)\theta(\mathbf{x}^2) \geq \theta(\lambda\mathbf{x}^1 + (1-\lambda)\mathbf{x}^2)$ for every $\mathbf{x}^1$, $\mathbf{x}^2 \in C$, and $0 \leq \lambda \leq 1$. It is easy to check that the maximum or nonnegative sum of finitely many convex functions on the same domain is again convex.

The negative of a convex function is *concave.* Thus if θ: $\mathbb{R}^n \rightarrow \mathbb{R}^1$ is convex, α: $\mathbb{R}^n \rightarrow \mathbb{R}^1$ defined by $\alpha(\mathbf{x}) \triangleq -\theta(\mathbf{x})$ is concave. The finite minimum or nonnegative sum of concave functions is concave.

A *linear function* θ: $\mathbb{R}^n \rightarrow \mathbb{R}^1$ is one defined by an n-vector $\mathbf{a}$ as $\theta(\mathbf{x}) \triangleq \mathbf{ax}$. An *affine function* is the related idea where a constant a_0 is included. For example, $\theta(\mathbf{x}) \triangleq \mathbf{ax} + a_0$ is affine. Affine and linear functions are both convex and concave.

Appendix B

Graph Theory Fundamentals

In this appendix we present a very cursory treatise of basic definitions and concepts from the theory of graphs. Both the depth and the breadth of the ensuing coverage have been dictated entirely by the expected level of graph-theoretic sophistication exhibited in the chapters and exercises. Readers interested in full-fledged treatments of graph results are directed to any of a number of well-known works, some of which are listed in the bibliography. We particularly recommend Bondy and Murty (1976).

B.1 Fundamental Definitions

A graph is a pair (V, E), where V is a set of *vertices* and E a set of two-element (not necessarily distinct) subsets of V referred to as *edges*. Denoted by $G(V, E)$, the usual representation of graphs is a pictorial one with points associated with vertices and lines with edges. We say that a pair of distinct vertices are *adjacent* if they define an edge, and that an edge is said to be *incident* to or to *meet* its defining vertices. The *degree* of a vertex is the number of edges incident to the vertex.

An edge defined by nondistinct vertices is called a *loop,* and *multiple edges* result when a pair of vertices are connected by more than one edge. A graph is *simple* if it possesses no loops or multiple edges. Graph (a) below is simple, while that depicted by (b) is not. Unless otherwise noted, graphs are usually assumed to be simple. In this case edges may be denoted by simply listing the vertex pair in increasing order (e.g., $e_1 \triangleq (1, 2)$ in (a) below).

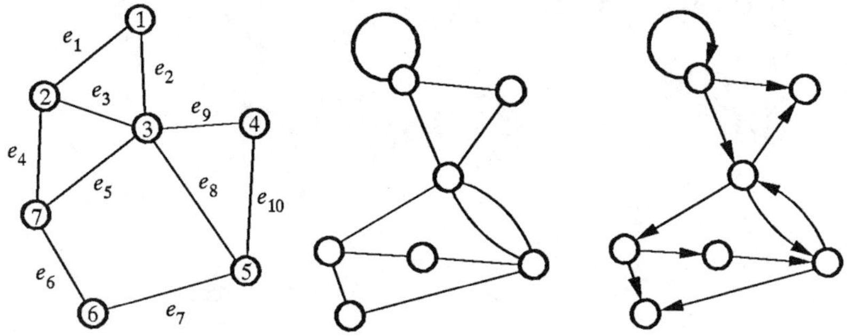

(a) (b) (c)

If the defining vertices for edges are ordered, a graph is said to be a *directed graph* or *digraph* and is denoted as $G(V, A)$ where A is a set of directed edges or *arcs.* Pictorially, the first vertex in the ordered pair has an arc directed from it, while the second has the arc directed into it. The underlying undirected graph associated with a directed one is the undirected graph obtained when order is ignored in arc vertex pairs. The graph in (c) above is directed; graph (b) is its underlying undirected graph. The degree of a vertex in a directed graph is generally specified in terms of the number of arcs directed into and out of the vertex (termed the *indegree* and *outdegree*).

Given two graphs $G_1(V_1, E_1)$ and $G_2(V_2, E_2)$ we call G_1 a *subgraph* of G_2 if $V_1 \subseteq V_2$, $E_1 \subseteq E_2$ and V_1 contains the set of vertices met by members of E_1. G_1 is a *spanning subgraph* of G_2 if $V_1 = V_2$. Edges $e_1, e_3, e_4, e_6, e_8, e_9, e_{10}$, in graph (a) above form a spanning subgraph. For some nonempty subset $\hat{V} \subseteq V$, the subgraph of $G(V, E)$ *induced* by $\hat{V}$ is the subgraph having vertex set $\hat{V}$ and all edges of E that are incident only to vertices in $\hat{V}$. An *edge-induced* subgraph is defined in a similar manner relative to an edge subset $\hat{E} \subseteq E$. Both notions are demonstrated below.

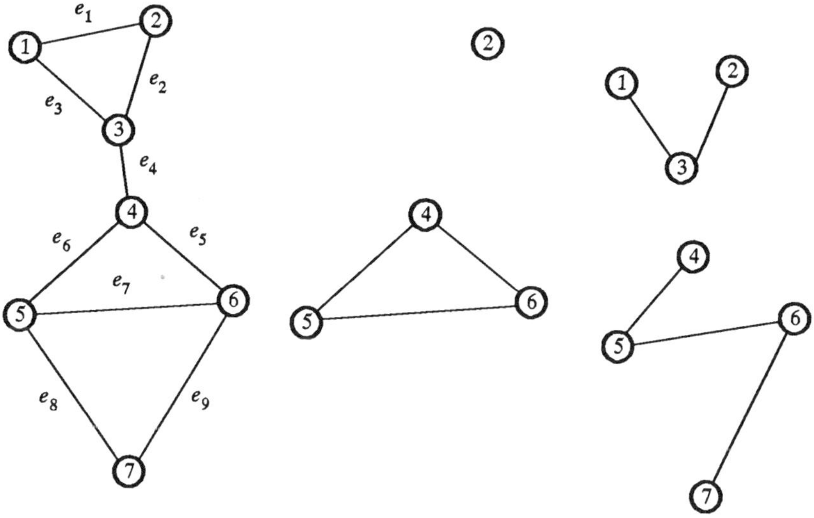

G $\qquad$ $\hat{V} = \{2, 4, 5, 6\}$ $\qquad$ $\hat{E} = \{e_2, e_3, e_6, e_7, e_9\}$

A *walk* in an undirected graph is an alternating sequence of vertices and incident edges beginning at some vertex v and ending at another u. If all vertices are distinct, the walk is called a *path*. If only the first and last are the same, it is a *cycle*. These notions are illustrated.

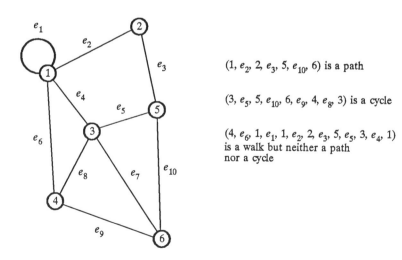

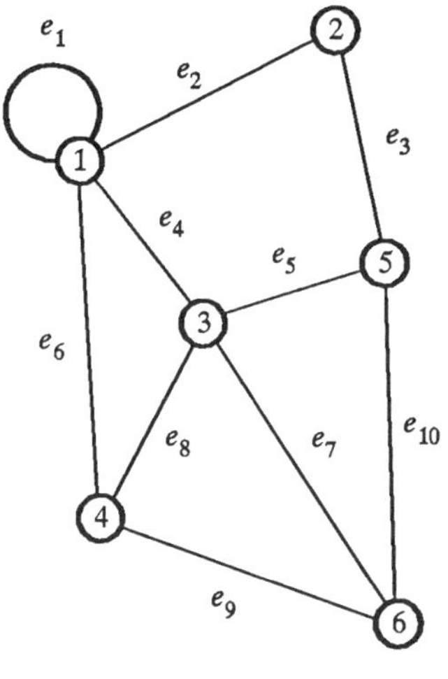

$(1, e_2, 2, e_3, 5, e_{10}, 6)$ is a path

$(3, e_5, 5, e_{10}, 6, e_9, 4, e_8, 3)$ is a cycle

$(4, e_6, 1, e_1, 1, e_2, 2, e_3, 5, e_5, 3, e_4, 1)$
is a walk but neither a path
nor a cycle

When a graph is directed, a path must conform to direction; the corresponding notion ignoring direction is a *chain.* A cycle need not conform to direction; the corresponding directed notion is a *circuit.* A graph with no circuits is (somewhat confusingly) termed *acyclic* and is illustrated in the following.

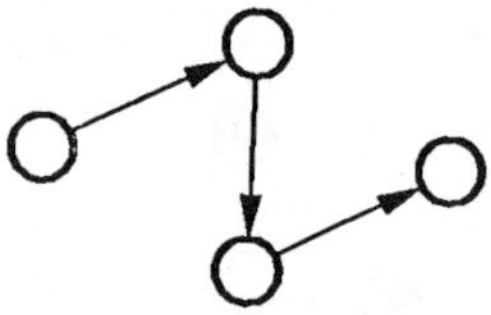

(directed) path

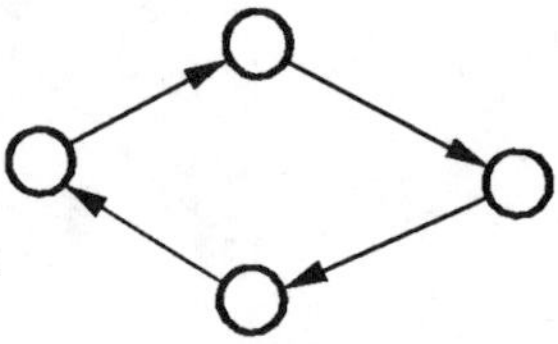

circuit

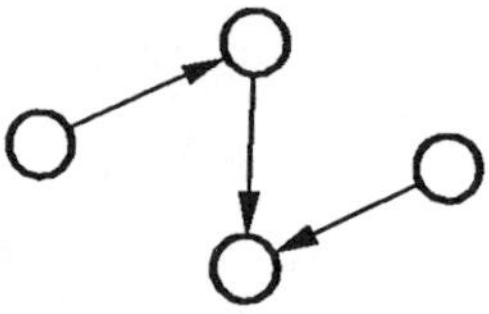

chain but not a path

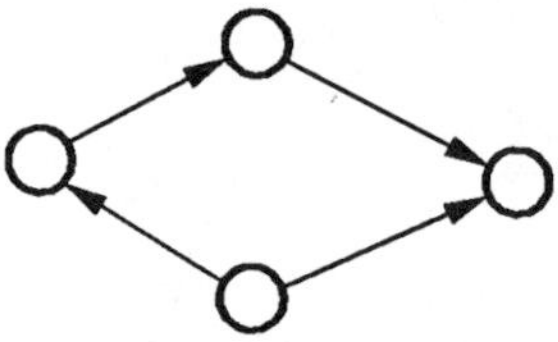

cycle but not a circuit

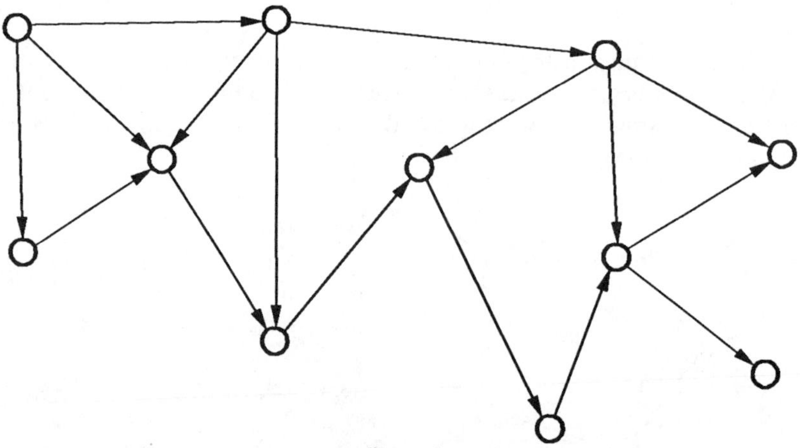

acyclic graph

A graph is *connected* if there exists a path between every pair of vertices in the graph. A directed graph is connected if its underlying undirected graph is connected (or equivalently if every pair of vertices is joined by a chain). The *vertex connectivity* off a graph is the minimum number of vertices whose removal leaves a disconnected or *trivial* graph (a graph consisting of a single vertex and no edges). *Edge connectivity* is defined in the same fashion relative to edge removal.

If the vertex set of a graph can be partitioned into two subsets V_1 and V_2 such that every edge in the graph is incident to a vertex in one subset and a vertex in the other, the graph is said to be *bipartite*. Equivalently, a graph is bipartite if and only if it is free of odd (edge cardinality) cycles.

A connected graph having the same degree r at every vertex is said to be *regular* in degree r or simply *r-regular*, and a simple graph where every pair of vertices are adjacent is called a *complete* graph. Similarly, a *complete bipartite* graph possesses edges connecting every vertex in V_1 with every one in V_2. We denote the complete graph on n vertices as K_n and by $K_{m,n}$ the complete bipartite graph having bipartition components with m and n vertices respectively. The first graph below is the complete K_5 and not bipartite. The second is not complete, but is a bipartite subgraph of $K_{4,4}$.

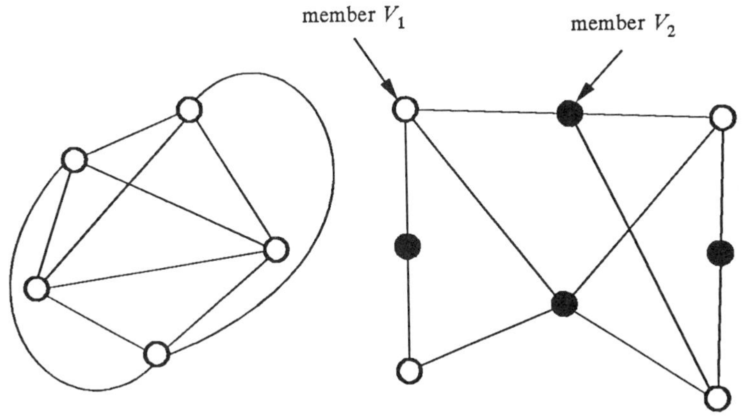

Given a pair of graphs G_1 and G_2, we say that G_1 and G_2 are *isomorphic* if there exists a one-to-one mapping between their vertex sets, which preserves adjacency. The graphs (a) and (b) on page 422 are isomorphic but neither is isomorphic to (c).

G_1 and G_2 are said to be *homeomorphic* if both can be obtained from the same graph by a sequence of edge subdivisions (i.e., adding degree-2 vertices). Equivalently, two graphs are homeomorphic if they are isomorphic to within vertices of degree two. All four graphs above are homeomorphic.

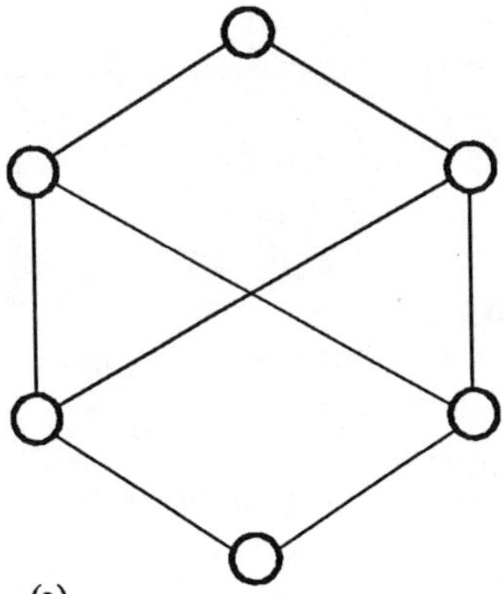
(a)

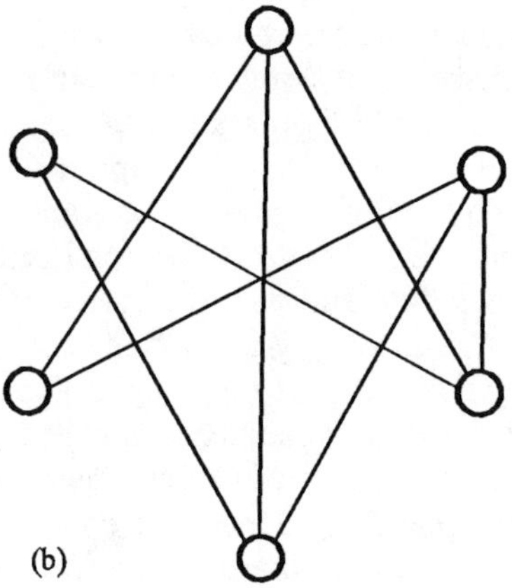
(b)

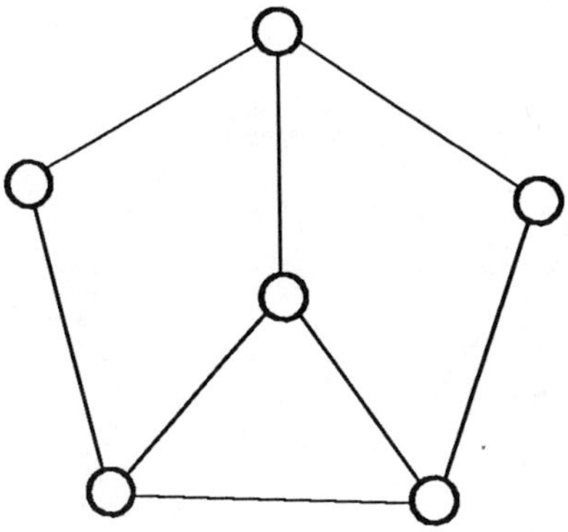
(c)

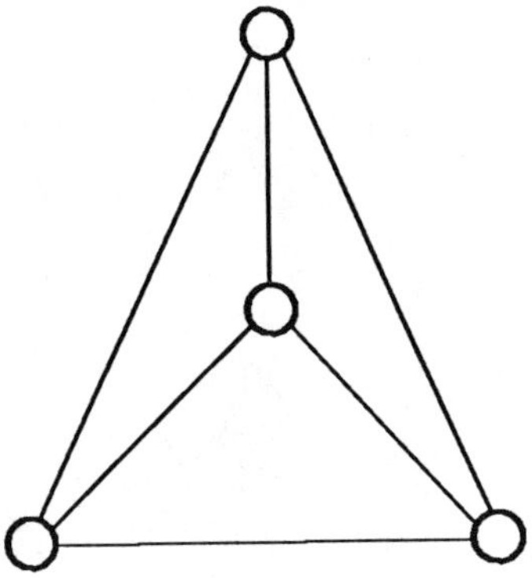
(d)

B.2 Planarity

A graph is *planar* if it can be drawn in the plane without edge crossings. Such a drawing of a planar graph is called a *plane* graph. The graph (a) below is planar and (b) is an *imbedding* of it as a plane graph. Deciding if a graph is planar can be accomplished in polynomial time (Hopcroft and Tarjan (1974)) and from one of the more important results in graph theory, planar graphs are characterized as those free of subgraphs homeomorphic to either K_5 or $K_{3,3}$ (Kuratowski (1930)). The third graph below is not planar having a K_5 homeomorph depicted by the bold edges.

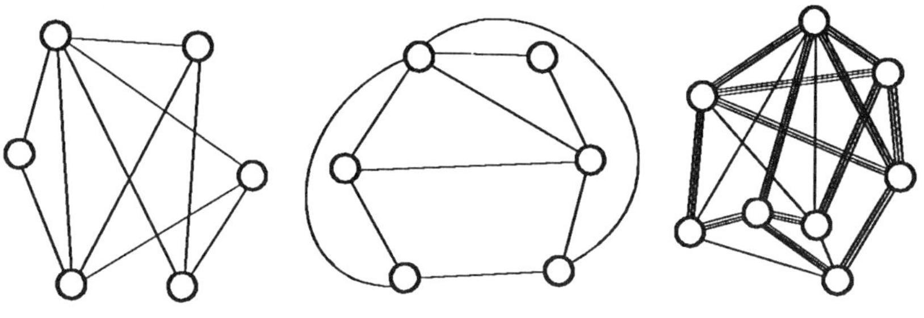

(a) (b) (c)

Every plane graph partitions the plane into connected regions that are called *faces*. The face that is unbounded (there will be only one) is the *exterior* or *infinite* face. Given a plane graph, G, we can construct another (plane) graph, say G^*, in the following way: for every face in G, define a corresponding vertex for G^* and connect every pair of vertices in G^* if the corresponding faces in G share a boundary (are separated by an edge). G^* is called the *dual* of G. A plane graph and its dual are shown in (a) below. Interestingly, isomorphic graphs need not have isomorphic duals as the graphs in (b) and (c) demonstrate.

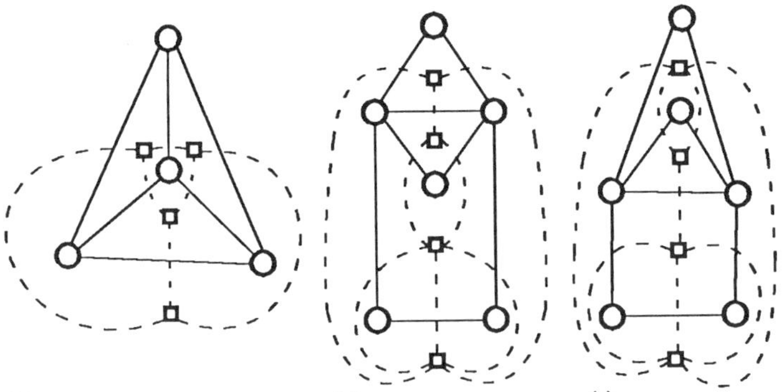

(a) (b) (c)

Euler showed that for a connected plane graph, the number of faces (f), edges (e), and vertices (v) were related as $f = e - v + 2$. Using this fact, it is easy to show that the number of edges in any maximal (simple) planar graph cannot exceed $3v - 6$. This number reduces to $2v - 4$ for maximal (simple) bipartite planar graphs.

B.3 Trees, Forests, Arborescences, and Branchings

A *forest* is a graph with no cycles. If a forest is connected, it is called a *tree*. A *spanning tree* (resp. forest) of a connected graph, $G(V, E)$, is a tree (forest) that has the same vertex set as G. The problem of producing a minimum (maximum) weight spanning tree (forest) subgraph in a graph is a well-solved problem (e.g., Boruvka (1926), Kruskal (1956)).

A *branching* is a directed graph with no cycles and no vertex having more than one arc directed into it. If a branching is connected, it is an *arborescence.* Finding optimum weight subgraphs that are branchings (resp. arborescences) is efficiently solved by an algorithm of Edmonds (1967a).

In graphs (a) and (b) below, the marked edges (resp. arcs) form a spanning tree (arborescence). Removing any edge (arc) of the tree (arborescence) leaves a forest (branching) that is not a tree (arborescence). Adding any edge (arc) creates a cycle and thus precludes the resulting subgraph from being a forest (branching).

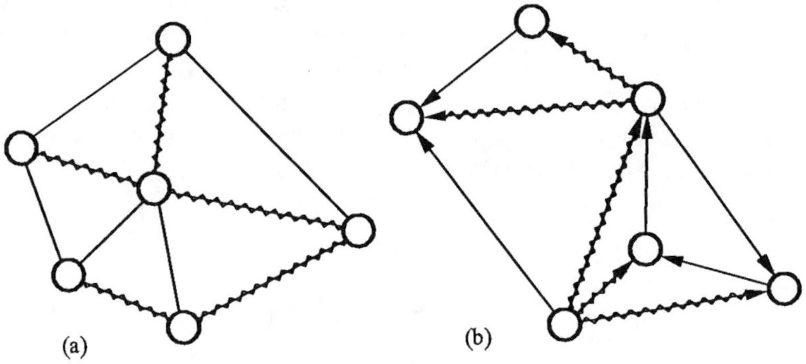

B.4 Traversals

Given a graph, $G(V, E)$, a *Hamiltonian cycle* or a *tour* is a cycle that includes every vertex exactly once. Any graph admitting such a cycle is said to be *Hamiltonian.* Although a host of sufficient conditions are available, there is no known (efficient) method for deciding which graphs are Hamiltonian. In fact, the problem remains *NP*-Complete on bipartite and planar, 3-regular graphs.

Alternatively, the problem of determining if a graph possesses a closed walk that includes every edge exactly once is well-solved. Referred to as an *Eulerian traversal,* graphs possessing such a traversal are said to be *Eulerian.* We have the following characterization (Euler (1736)): a connected graph is Eulerian if and only if the degree of every vertex is even. The corresponding issue for directed graphs is similarly resolved requiring symmetry at every vertex (i.e., in-degree and out-degree the same). In either case, producing existing traversals is easily accomplished.

The graph below is both Hamiltonian and Eulerian. Corresponding traversals are indicated at right.

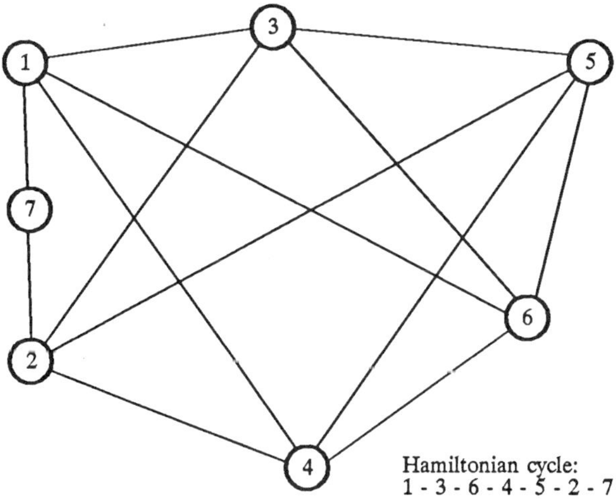

Hamiltonian cycle:
1 - 3 - 6 - 4 - 5 - 2 - 7 - 1

Eulerian traversal:
1 - 3 - 5 - 4 - 6 - 1 - 4 - 2 - 3 - 6 - 5 - 2 - 7 - 1

B.5 Covering, Matching, and Independence

A subset of vertices in a graph is *independent* if no two vertices in the subset are adjacent. Finding a maximum cardinality independent set in arbitrary graphs is difficult but solvable in polynomial time on bipartite graphs as well as on other special classes.

The corresponding problem of finding a subset $\hat{E}$, no two elements of which are incident to the same vertex, is called *matching*. The problem of finding maximum matchings is well-solved for arbitrary graphs (Edmonds (1965c)).

Given a graph $G(V, E)$, any subset $\hat{V} \subseteq V$ having the property that every edge in E is incident to at least one vertex in $\hat{V}$ is called a *cover* of G. In general, the problem of determining minimum cardinality covers is difficult but can be solved in polynomial time for various special classes of graphs among which is the class of bipartite graphs. The analogous problem of finding minimum cardinality *edge covers* (subset of E having the property that every vertex in V is incident to at least one edge in the subset) is well-solved for arbitrary graphs.

An important relationship between vertex covers and independent sets holds that the respective minimum and maximum cardinalities of each always sum to the number of vertices in the graph. That is, if $\hat{V}$ is a minimum vertex cover, then $V \backslash \hat{V}$ is a maximum independent set.

A *clique* is a subset of vertices that induces a complete graph. If $G(V, E)$ is a simple graph, then clearly any independent set in G is also a clique in the edge complement of G, i.e., the graph having the same vertex set as G but only edges connecting vertices that are not connected in G. It follows from the relationship between covers and independent sets that determination of cliques would also yield the former two subsets. In particular, ability to resolve any of the three would resolve the issue relative to the other two.

The following graphs are useful for illustrating covers, matchings, independent sets, and cliques.

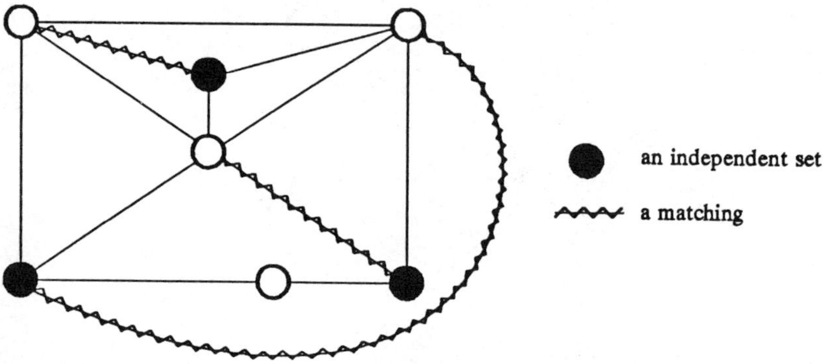

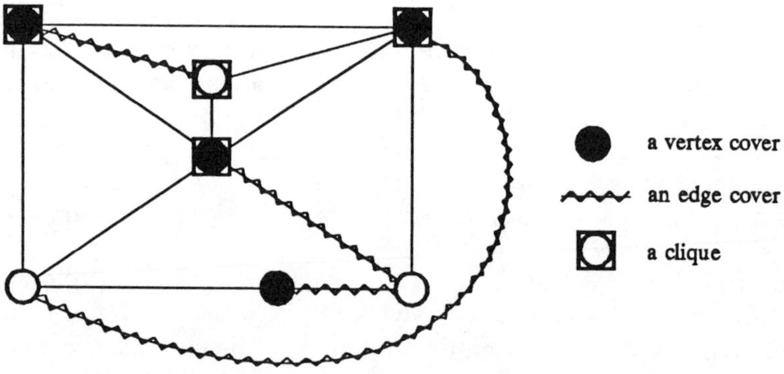

B.6 Network Flows

Let $G(V, A)$ be a directed graph and suppose we associate with each arc (i, j), parameters l_{ij} and u_{ij}, which will denote lower and upper bounds on *flow* through the arc. Then if c_{ij} measures the unit shipping cost along (i, j)

and if we require flow to be conserved at every vertex in V, the *minimum cost network flow problem* seeks a feasible *circulation* of flow in $G(V, A)$ of minimum total cost. Letting x_{ij} be the amount of flow in arc (i, j), the problem can be formulated by (NF).

$$\min \sum_{(i,j)\in A} c_{ij}x_{ij}$$

$$(NF) \quad \text{s.t.} \quad \sum_{\{(j,k):\,(j,k)\in A\}} x_{jk} - \sum_{\{(i,j):\,(i,j)\in A\}} x_{ij} = 0 \text{ for } j = 1, 2, \ldots, |V|.$$

$$l_{ij} \leq x_{ij} \leq u_{ij} \text{ for all } (i, j) \in A$$

The flow conservation constraints in (NF) yield, in general, a highly structured coefficient matrix. Letting this matrix be $\mathbf{A}$, then rows of $\mathbf{A}$ correspond to vertices in V and columns to arcs in A. Further, an element of $\mathbf{A}$, say a_{ik} will be $+1$ if arc e_k is directed out of vertex i, -1 if it is directed into i and 0 if e_k is not incident to i. Accordingly, $\mathbf{A}$ is called the *vertex-arc incidence matrix* of $G(V, A)$ and its construction is demonstrated for the graph.

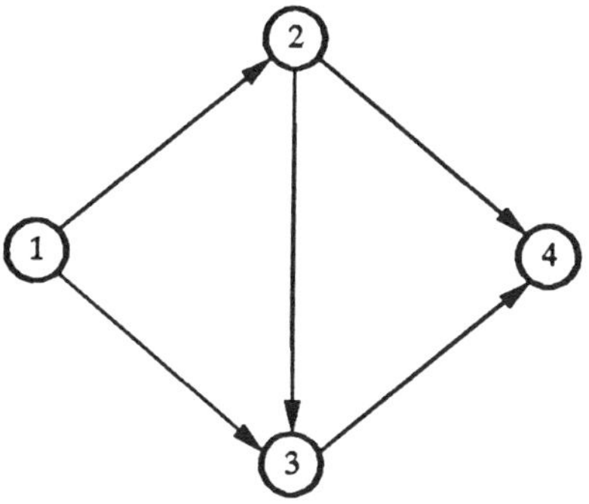

$$\mathbf{A} = \begin{array}{c} \\ 1 \\ 2 \\ 3 \\ 4 \end{array} \begin{array}{c} \begin{array}{ccccc} (1,2) & (1,3) & (2,3) & (2,4) & (3,4) \end{array} \\ \left[\begin{array}{ccccc} 1 & 1 & 0 & 0 & 0 \\ -1 & 0 & 1 & 1 & 0 \\ 0 & -1 & -1 & 0 & 1 \\ 0 & 0 & 0 & -1 & -1 \end{array}\right] \end{array}$$

A popular variant of the problem of determining a minimum cost flow in a capacitated network is the *maximal flow problem.* In the latter, we assume the lower bound on an arc's flow is zero and seek a feasible flow pattern in $G(V, A)$ of maximum total flow value where flow emanates at some prespecified *source* and concludes at a prespecified *sink.*

Letting the value of flow be f and assuming vertices s and t represent the source and sink respectively, the maximal flow problem can be written as:

$$\max \quad f$$

$$(MF): \qquad \text{s.t.} \qquad -\sum_i x_{ij} + \sum_k x_{jk} = \begin{cases} f & \text{if } j = s \\ 0 & \text{if } j \neq s, t \\ -f & \text{if } j = t \end{cases}$$

$$0 \leq x_{ij} \leq u_{ij}, \text{ for all } 1 \leq i, j \leq n$$

For a graph $G(V, A)$ and two distinct vertices say s and t, let X be any subset of vertices that contains s but not t and let $\bar{X}$ be its complement (i.e., $\bar{X} = V \backslash X$). Then a *cut-set* in $G(V, A)$ is a set of arcs $\{(i, j): i \in X, j \in \bar{X}, (i, j) \in A\}$. Clearly, removal of the arcs in a cut-set will result in a subgraph of G that will not allow flow from s to t.

If $(X, \bar{X})$ is a cut-set then the *capacity* of $(X, \bar{X})$ is simply the sum of capacities of arcs in the cut-set. For the structure below, where capacities are specified on the arcs, a cut-set separating flow between vertices 1 and 4 is given by $X = \{1, 2\}$ and $\bar{X} = \{3, 4\}$ having capacity 14.

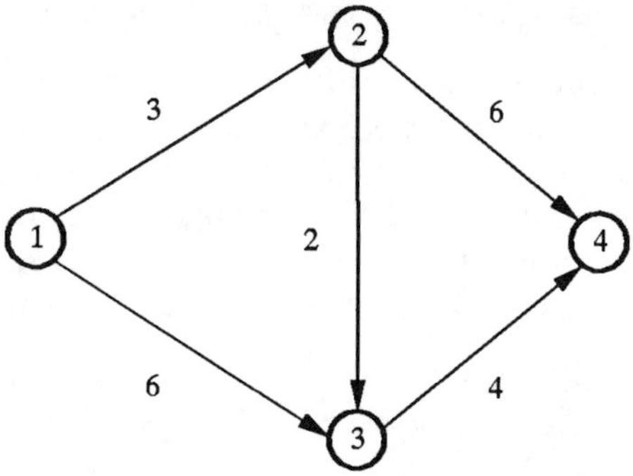

In a classic result of Ford and Fulkerson (1956), cut-sets and flows are shown to be related in a particularly powerful way. Following from the *max-flow, min-cut theorem,* it is the case that the maximum amount of flow in a capacitated network is equal to the value of a minimum capacity cut-set in the network. Thus, the value of maximum flow in the example above is seven following from the capacity of $(X, \bar{X})$ where $X = \{1, 3\}$. A suitable flow pattern is 3 units of flow along the directed path 1-2-4 and 4 along 1-3-4.

Appendix C

Linear Programming Fundamentals

Following, we present a concise overview of the fundamentals of linear programming, the purpose of which is to provide a suitable background for the subject as it interfaces with coverage of this book. Our treatment here is not intended to be pedagogic. For those desiring such a presentation, a wealth of books exist including Charnes, Cooper, and Henderson (1953), Hadley (1962), Dantzig (1963), Simmonard (1966), Zionts (1974), Murty (1976), Bazaraa and Jarvis (1977), Gass (1975), Papadimitriou and Steiglitz (1982), Chvátal (1983).

C.1 Duality and Optimality

Every linear program can be rearranged to have the matrix form (called *primal*)

$$(P) \qquad \begin{aligned} \min \quad & \mathbf{c}^1\mathbf{x}^1 + \mathbf{c}^2\mathbf{x}^2 \\ \text{s.t.} \quad & \mathbf{A}_{11}\mathbf{x}^1 + \mathbf{A}_{12}\mathbf{x}^2 \geq \mathbf{b}^1 \\ & \mathbf{A}_{21}\mathbf{x}^1 + \mathbf{A}_{22}\mathbf{x}^2 = \mathbf{b}^2 \\ & \quad \mathbf{x}^1 \geq \mathbf{0}, \ \mathbf{x}^2 \text{ unrestricted.} \end{aligned}$$

Each such problem also has a *dual* linear program given as

$$(D) \qquad \begin{aligned} \max \quad & \mathbf{u}^1\mathbf{b}^1 + \mathbf{u}^2\mathbf{b}^2 \\ \text{s.t.} \quad & \mathbf{u}^1\mathbf{A}_{11} + \mathbf{u}^2\mathbf{A}_{21} \le \mathbf{c}^1 \\ & \mathbf{u}^1\mathbf{A}_{12} + \mathbf{u}^2\mathbf{A}_{22} = \mathbf{c}^2 \\ & \mathbf{u}^1 \ge \mathbf{0},\ \mathbf{u}^2 \text{ unrestricted.} \end{aligned}$$

The dual of a minimization is a maximization over variables identified with primal constraints. Here $\mathbf{u}^1$ is the vector of dual multipliers for the first rows of the primal, and $\mathbf{u}^2$ applies to the second. Since the dual of the dual is the primal, we may also take $\mathbf{x}^1$ to be the vector of variables assigned to the first dual constaints and $\mathbf{x}^2$ those associated with the second. Notice that equality constraints in one problem are associated with unrestricted variables in the other; inequalities yield nonnegative variables in the other.

Every feasible $(\mathbf{u}^1, \mathbf{u}^2)$ in (D) provides a lower bound on the optimal value, $v(P)$, of the primal, i.e.,

$$\mathbf{u}^1\mathbf{b}^1 + \mathbf{u}^2\mathbf{b}^2 \le v(P)$$

for all $(\mathbf{u}^1, \mathbf{u}^2)$ feasible in (D). Similarly, every feasible $(\mathbf{x}^1, \mathbf{x}^2)$ in the primal (P) provides an upper bound on the optimal value, $v(D)$, of the dual, i.e.,

$$\mathbf{c}^1\mathbf{x}^1 + \mathbf{c}^2\mathbf{x}^2 \ge v(D)$$

for all $(\mathbf{x}^1, \mathbf{x}^2)$ feasible in the primal (P). We thus have that $v(P) \ge v(D)$.

Of course, it is possible that both the primal and dual problems have no feasible solutions, i.e., $v(P) = +\infty$, $v(D) = -\infty$. If at least one of (P) and (D) is feasible, however, then $v(D) = v(P)$. Specifically,

(P) feasible, (D) infeasible $\Leftrightarrow v(P) = v(D) = -\infty$, i.e., (P) unbounded

(D) feasible, (P) infeasible $\Leftrightarrow v(P) = v(D) = +\infty$, i.e., (D) unbounded

(P) feasible, (D) feasible $\Leftrightarrow -\infty < v(D) = v(P) < +\infty$, i.e., both have finite optima.

When both the primal and dual are feasible, we can state mutual conditions sufficient for optimality. Specifically, vectors $(\bar{\mathbf{x}}^1, \bar{\mathbf{x}}^2)$ and $(\bar{\mathbf{u}}^1, \bar{\mathbf{u}}^2)$ are optimal in (P) and (D) respectively if they satisfy

$$\left.\begin{aligned} \mathbf{A}_{11}\bar{\mathbf{x}}^1 + \mathbf{A}_{12}\bar{\mathbf{x}}^2 &\ge \mathbf{b}^1 \\ \mathbf{A}_{21}\bar{\mathbf{x}}^1 + \mathbf{A}_{22}\bar{\mathbf{x}}^2 &= \mathbf{b}^2 \\ \bar{\mathbf{x}}^1 &\ge \mathbf{0} \end{aligned}\right\} \quad \text{Primal Feasibility}$$

$$\left.\begin{aligned}\bar{\mathbf{u}}^1\mathbf{A}_{11} + \bar{\mathbf{u}}^2\mathbf{A}_{21} &\le \mathbf{c}^1\\ \bar{\mathbf{u}}^1\mathbf{A}_{12} + \bar{\mathbf{u}}^2\mathbf{A}_{22} &= \mathbf{c}^2\\ \bar{\mathbf{u}}^1 &\ge \mathbf{0}\end{aligned}\right\} \quad \text{Dual Feasibility}$$

$$\left.\begin{aligned}\bar{\mathbf{u}}^1(\mathbf{b}^1 - \mathbf{A}_{11}\bar{\mathbf{x}}^1 - \mathbf{A}_{12}\bar{\mathbf{x}}^2) &= 0\\ \bar{\mathbf{x}}^1(\mathbf{c}^1 - \bar{\mathbf{u}}^1\mathbf{A}_{11} - \bar{\mathbf{u}}^2\mathbf{A}_{21}) &= 0\end{aligned}\right\} \quad \begin{array}{l}\text{Complementary Slackness}\\ \text{(on Inequalities).}\end{array}$$

The above conditions are also necessary in that $(\bar{\mathbf{x}}^1, \bar{\mathbf{x}}^2)$ optimal in (P) implies the existence of a corresponding $(\bar{\mathbf{u}}^1, \bar{\mathbf{u}}^2)$ optimal in (D) and vice versa.

C.2 Representation and Bases

By (i) adding (nonnegative) *slack* or *surplus* variables to convert any inequalities to equalities, (ii) replacing any unrestricted variables by differences of nonnegative variables, (iii) deleting any redundant rows, and (iv) taking the negative of a maximize objection function, a linear program can be written in the *simplex standard form*

$$\begin{aligned}\min \quad & \mathbf{cx}\\ \text{s.t.} \quad & \mathbf{Ax} = \mathbf{b}\\ & \mathbf{x} \ge \mathbf{0},\end{aligned}$$

where $\mathbf{A}$ is full rank.

The feasible set F of a linear program is a polyhedron and thus a convex set. (See Appendix A). Extreme points and directions of that set play a central role. In particular, a feasible linear program has a finite optimal (min) solution if and only if its cost row $\mathbf{c}$ satisfies

$$\mathbf{cd} \ge 0 \text{ for all extreme directions } \mathbf{d} \text{ of } F.$$

Moreover, if a linear program has a finite optimal solution, it has one at an extreme point of F (although there may also be nonextreme point optima).

A basis submatrix $\mathbf{B}$ of $\mathbf{A}$ is a maximal column submatrix having linearly independent columns. For $\mathbf{A}$ full rank we may solve for basic variables, say $\mathbf{x}^B$, corresponding to a basic matrix $\mathbf{B}$ in terms of the other, *nonbasic* variables $\mathbf{x}^N$. Specifically if $\mathbf{A} \triangleq (\mathbf{B}, \mathbf{N})$ we have

$$\mathbf{x}^B = \mathbf{B}^{-1}\mathbf{b} - \mathbf{B}^{-1}\mathbf{N}\mathbf{x}^N.$$

We refer to a solution $(\mathbf{x}^B, \mathbf{x}^N)$ as *basic* if it is formed by $\mathbf{x}^N = \mathbf{0}$ and $\mathbf{x}^B = \mathbf{B}^{-1}\mathbf{b}$ for some basis matrix $\mathbf{B}$. A basic solution is said to be *basic feasible* if in addition, $\mathbf{x}^B = \mathbf{B}^{-1}\mathbf{b} \ge \mathbf{0}$.

Extreme points of F are precisely the basic feasible solutions. If an extreme point, however, has basic variables with zero values (i.e., is *degenerate*) several basic matrices $\mathbf{B}$ may compute the same extreme point solution.

Extreme directions of F are also obtained from feasible bases, $\mathbf{B}$. They arise from columns $\mathbf{B}^{-1}\mathbf{N}^j \leq \mathbf{0}$ for any nonbasic component x_j^N. The full direction has $+1$ on x_j^N, $-\mathbf{B}^{-1}\mathbf{N}^j$ on basic variables, $\mathbf{x}^B$, and $\mathbf{0}$ elsewhere.

C.3 Simplex Algorithms

Using the equation $\mathbf{x}^B = \mathbf{B}^{-1}\mathbf{b} - \mathbf{B}^{-1}\mathbf{N}\mathbf{x}^N$ the problem in the form

$$\begin{aligned} \min \quad & \mathbf{cx} \\ \text{s.t.} \quad & \mathbf{Ax} = \mathbf{b} \\ & \mathbf{x} \geq \mathbf{0} \end{aligned}$$

can be rewritten as

$$\begin{aligned} \min \quad & \sum_{j \in N} \bar{c}_j x_j + \bar{v} \\ \text{s.t.} \quad & x_k = \bar{x}_k - \sum_{j \in N} \bar{a}_{kj} x_j \qquad \text{for all basic } k \notin N \\ & \mathbf{x} \geq \mathbf{0}, \end{aligned}$$

where $\bar{c}_j \triangleq c_j - \sum_{k \notin N} c_k \bar{a}_{kj}$

$\bar{a}_{kj} \triangleq$ the column j entry of $\mathbf{B}^{-1}\mathbf{N}$ corresponding to basic variable x_k.

$\bar{x}_k \triangleq (\mathbf{B}^{-1}\mathbf{b})_k$

$\bar{v} \triangleq$ current objective function value

or $\sum_k c_k \bar{x}_k$

$N \triangleq$ {nonbasic variable subscripts}.

The *primal simplex algorithm* for linear programming begins with a basic feasible solution $\mathbf{x}^1$, $t = 1$, and corresponding representation

$$\begin{aligned} \min \quad & \sum_{j \in N_t} \bar{c}_j^t x_j + \bar{v}^t \\ \text{s.t.} \quad & x_k = \bar{x}_k^t - \sum_{j \in N_t} \bar{a}_{kj}^t x_j \qquad k \notin N_t \\ & \mathbf{x} \geq \mathbf{0} \end{aligned}$$

STEP 1: If $\bar{c}_j^t \geq 0$ for all $j \in N_t$, stop; $\mathbf{x}^t$ is optimal. Otherwise, pick any p with $\bar{c}_p^t < 0$, and consider increasing nonbasic x_p.

STEP 2: If $\bar{a}_{kp}^t \leq 0$ for all basic $k \notin N_t$, stop; the linear program is unbounded. (The extreme direction constructed from the $\{\bar{a}_{kp}^t\}$ permits unlimited improvement in the solution). If some $\bar{a}_{kp}^t > 0$, an increase in x_p decreases some basic variables x_k. To maintain feasibility, select r so that

$$\frac{\bar{x}_r^t}{\bar{a}_{rp}^t} = \min\left\{\frac{\bar{x}_k^t}{\bar{a}_{kp}^t} : \bar{a}_{kp}^t > 0\right\}.$$

STEP 3: Replace x_r in the basis by x_p. This is accomplished by row operations or *pivots* as follows:

$$\begin{aligned}
&N_{t+1} \longleftarrow N_t \backslash \{x_p\} \cup \{x_r\} \\
&\bar{x}_p^{t+1} \longleftarrow \bar{x}_r^t / \bar{a}_{rp}^t \\
&\bar{a}_{pj}^{t+1} \longleftarrow \bar{a}_{rj}^t / \bar{a}_{rp}^t && \text{for } j \in N_t \cap N_{t+1} \\
&\bar{a}_{pr}^{t+1} \longleftarrow 1/\bar{a}_{rp}^t \\
&\bar{x}_k^{t+1} \longleftarrow \bar{x}_k^t - \bar{a}_{kp}^t(\bar{x}_r^t/\bar{a}_{rp}^t) && \text{for all } k \notin N \cup N_{t+1} \\
&\bar{a}_{kj}^{t+1} \longleftarrow \bar{a}_{kj}^t - \bar{a}_{kp}^t(\bar{a}_{rj}^t/\bar{a}_{rp}^t) && \text{for all } k \notin N \cup N_{t+1} \\
&\bar{a}_{kr}^{t+1} \longleftarrow -\bar{a}_{kp}^t/\bar{a}_{rp}^t && \text{for all } k \notin N_t \cap N_{t+1} \\
&\bar{c}_j^{t+1} \longleftarrow \bar{c}_j^t - \bar{c}_p^t(\bar{a}_{rj}^t/\bar{a}_{rp}^t) && \text{for all } j \in N_t \cap N_{t+1} \\
&\bar{c}_r^{t+1} \longleftarrow -\bar{c}_p^t/\bar{a}_{rp}^t \\
&\bar{v}^{t+1} \longleftarrow \bar{v}_t^t + \bar{c}_p^t \bar{x}_p^{t+1}.
\end{aligned}$$

Set $t \leftarrow t+1$ and return to Step 1.

A primal simplex solution $\bar{\mathbf{x}} = (\bar{\mathbf{x}}^B, \bar{\mathbf{x}}^N)$ corresponding to basis matrix $\mathbf{B}$ always satisfies primal feasibility. It is easy to see that the corresponding dual solution $\mathbf{u} = \mathbf{c}^B\mathbf{B}^{-1}$, where $\mathbf{c}^B$ is the vector of $\mathbf{x}^B$ costs, satisfies the complementarity conditions given earlier. Thus the stated procedure can be viewed as one of searching for dual feasibility while maintaining primal feasibility and complementary slackness. Stopping is precisely checking if dual feasibility has been obtained.

The *dual simplex algorithm* works in similar fashion to the primal, but maintains dual feasibility and complementary slackness, while seeking primal feasibility. Thus, it too moves through bases by pivoting, but not feasible bases. We always have $\bar{c}_j^t \geq 0$ and seek $\bar{x}_k^t \geq 0$. The algorithm begins with a dual feasible basic solution $\mathbf{x}^1$ (i.e. $\mathbf{u} = \mathbf{c}^B\mathbf{B}^{-1}$ is dual feasible) and $t = 1$.

STEP 1: If $\bar{x}_k^t \geq 0$ for all $k \notin N_t$, stop; $\mathbf{x}^t$ is optimal. If not, pick any r with $\bar{x}_r^t < 0$ and consider removing x_r from the basis.

STEP 2: If $\bar{a}_{rj}^t \geq 0$ for all nonbasic $j \in N_t$, no nonbasic variable can move x_r toward zero by increasing. Stop; the problem is infeasible since x_r cannot be rendered nonnegative, while the presently nonbasic variables remain so. If any $\bar{a}_{rj}^t < 0$, an increase of corresponding nonbasic x_j will drive $\bar{c}_j^t$ toward zero. To prevent a loss of dual feasibility select p so that

$$\frac{\bar{c}_p^t}{\bar{a}_{rp}^t} = \max\left\{\frac{\bar{c}_j^t}{\bar{a}_{rj}^t} : \bar{a}_{rj}^t < 0\right\}$$

STEP 3: Replace x_r in the basis by x_p as in Step 3 for the primal simplex, set $t \leftarrow t+1$ and return to Step (i).

The dual simplex is particularly useful in integer programming since we often wish to add a new constraint, say $\mathbf{gx} \geq g_0$, to an existing optimal linear programming relaxation. We can construct a situation where the dual simplex may be directly applied by

(i) Picking as the basic variable of the new row, surplus variable

$$x_s = g_0 - \mathbf{gx}.$$

(ii) Constructing simplex data by substituting for any basic variables in expression $\mathbf{gx}$ via their equivalent sum of nonbasics, i.e.,

$$\bar{x}_s^t \longleftarrow g_0 - \sum_{k \notin N} g_k \bar{x}_k^t$$

$$\bar{a}_{sj}^t \longleftarrow g_j - \sum_{k \notin N} g_k \bar{a}_{kj}^t \qquad \text{for all } j \in N.$$

Observe that the above $\bar{x}_s^t < 0$ if the current basic solution $\bar{\mathbf{x}}^t$ fails to satisfy $\mathbf{gx}^t \geq g_0$. Since the nonbasic set is the same however, $\bar{c}_j^t \geq 0$ still holds and dual simplex can continue.

There are obviously only finitely many bases of a constraint matrix $\mathbf{A}$, so that finite convergence of both simplex methods is assured if bases do not repeat. In the absence of degeneracy the sequence $\{\bar{v}^t\}$ is strictly decreasing for the primal simplex and strictly increasing for the dual. Thus the algorithms converge in a finite number of steps.

When degeneracy is present (some basic $\bar{x}_k^t = 0$ for the primal or some nonbasic $\bar{c}_j^t = 0$ for the dual) several iterations may have the same solution value, $\bar{v}^t$. In such cases *cycling* is possible, i.e., bases can repeat.

Fortunately, however, there are a variety of minor modifications in the algorithms that restore finite convergence by preventing basis repeats. The most familiar depend on *lexicographic* ordering of vectors. (See Appendix A.) Suitable modification of the min (respectively max) rule in Step 2 of the above primal (respectively dual) simplex algorithm can assure the sequence of adjusted cost vectors $\{(\bar{v}^t, \bar{\mathbf{c}}^t)\}$ (respectively primal solution vectors $\{\bar{v}^t, \bar{\mathbf{x}}^t\}$) is lexicographically monotone. Thus, although first, optimal solution value components may repeat, bases may not.

References

Adler, I. (1983), "The Expected Number of Pivots Needed to Solve Parametric Linear Programs and the Efficiency of the Self-Dual Simplex Method," manuscript, Dept. of Industrial Engineering and Operations Research, University of California, Berkeley, CA.

Adler, I., R. Karp, and R. Shamir (1983a), "A Family of Simplex Variants Solving an $m \times d$ Linear Program in Expected Number of Pivot Steps Depending On d Only," Report No. UCB/CSD 83/157, Computer Science Division, University of California, Berkeley, CA.

Adler, I., R. M. Karp, and R. Shamir (1983b), "A Simplex Variant Solving An $m \times d$ Linear Program in $0(\min(m^2,d^2))$ Expected Number of Pivot Steps," Report UCB/CSD 83/158, Computer Science Division, University of California, Berkeley, CA.

Adler, I. and N. Megiddo (1984), "A Simplex Algorithm Whose Average Number of Steps is Bounded between Two Quadratic Functions of Smaller Dimension," *Journal of the Association for Computing Machinery,* **32**, 871–895.

Aho, A., J. E. Hopcroft, and J. D. Ullman (1974), *The Design and Analysis of Computer Algorithms,* Addison-Wesley, Reading, MA.

Aráoz, J. (1973), *Polyhedral Neopolarities,* Doctoral Thesis, University of Waterloo, Waterloo, Ontario.

Aráoz, J. (1979), "Blocking and Anti-blocking Extensions," *Operations Research Verfahren/ Methods of Operations Research,* **32** (III Symposium on O.R., Mannheim, 1978; W. Oettli and F. Steffens, eds.) Athenäum/Hain/Scriptor/Hanstein, Königstein/Ts., 5–18.

Aráoz, J., J. Edmonds, and V. J. Griffin (1983), "Polarities Given by Systems of Bilinear Inequalities," *Mathematics of Operations Research,* **8**, 34–41.

Aráoz, J. and E. L. Johnson (1981), "Some Results on Polyhedra of Semigroup Problems," *SIAM Journal on Algebraic and Discrete Methods,* **2**, 244–258.

Aspvall, B. and R. E. Stone (1980), "Khachiyan's Linear Programming Algorithm," *J. Algorithms,* **1**, no. 1.

Balas, E. (1963), "Programare Liniara cu Variabile Bivalente," (Rumanian), *Proceedings Third Scientific Session on Statistics,* Bucharest.

Balas, E. (1965), "An Additive Algorithm for Solving Linear Programs with Zero-One Variables," *Operations Research,* **13**, 517–546.

Balas, E. (1967), "Discrete Programming by the Filter Method," *Operations Research,* **15**, 915–957.

Balas, E. (1970a), "Minimax and Duality for Linear and Nonlinear Mixed-Integer Programming" in: *Integer and Nonlinear Programming,* (J. Abadie, ed.), North Holland, Amsterdam, 353–365.

Balas, E. (1970b), "Duality in Discrete Programming" in: *Proceedings of the Princeton Symposium on Mathematical Programming* (Princeton, 1967; H. W. Kuhn, ed.), Princeton University Press, Princeton, NJ, 179–197.

Balas, E. (1971), "Intersection Cuts—A New Type of Cutting Planes for Integer Programming," *Operations Research,* **19**, 19–39.

Balas, E. (1972), "Integer Programming and Convex Analysis: Intersection Cuts from Outer Polars," *Mathematical Programming,* **2**, 330–382.

Balas, E. (1973), "A Note on the Group Theoretic Approach to Integer Programming and the 0–1 Case," *Operations Research,* **21**, 321–322.

Balas, E. (1975a), "Facets of the Knapsack Polytope," *Mathematical Programming,* **8**, 146–164.

Balas, E. (1975b), "Disjunctive Programming: Cutting Planes from Logical Conditions," in: *Nonlinear Programming 2,* (Proc. Symp. Madison, Wisconsin, 1974; O. L. Mangasarian, R. R. Meyer, and S. M. Robinson, eds.) Academic Press, New York, 279–312.

Balas, E. (1977), "Some Valid Inequalities for the Set Partitioning Problem," in: *Studies in Integer Programming* (P. L. Hammet et al., eds.), *Annals of Discrete Mathematics* **1**, 13–47.

Balas, E. (1979), "Disjunctive Programming," *Annals of Discrete Mathematics,* **5**, 3–51.

Balas, E. (1980), "Cutting Planes from Conditional Bounds: A New Approach to Set Covering," *Mathematical Programming Study,* **12**, 19–36.

Balas, E. (1984), "A Sharp Bound on the Ratio Between Optimal Integer and Fractional Covers," *Mathematics of Operations Research,* **9**, 1–5.

Balas, E. and N. Christofides (1981), "A Restricted Lagrangean Approach to the Traveling Salesman Problem," *Mathematical Programming,* **21**, 19–46.

Balas, E. and A. Ho (1980), "Set Covering Algorithms Using Cutting Planes, Heuristics and Subgradient Optimization: A Computational Study," *Mathematical Programming Study,* **12**, 37–60.

Balas, E. and R. G. Jeroslow (1980), "Strengthening Cuts for Mixed Integer Programs," *European Journal of Operations Research,* **4**, 224–334.

Balas, E. and C. H. Martin (1980), "Pivot and Complement—A Heuristic for 0–1 Programming," *Management Science,* **26**, 86–96.

Balas, E. and M. W. Padberg (1972), "On the Set-Covering Problem," *Operations Research,* **20**, 1152–1161.

Balas, E. and M. W. Padberg (1975), "On the Set-Covering Problem, II: An Algorithm for Set Partitioning," *Operations Research,* **23**, 74–90.

Balas, E. and M. W. Padberg (1976), "Set Partitioning: A Survey," *SIAM Review,* **18**, 710–760.

Balas, E. and M. W. Padberg (1979), "Adjacent Vertices of the All 0–1 Programming Polytope," *Revue Francaise d'Automatique d'Informatique et de Recherche Operationelle,* **13**, 3–12.

Balas, E. and W. Pulleyblank (1983), "The Perfectly Matchable Subgraph Polytope of a Bipartite Graph," *Networks,* **13**, 495–516.

Balas, E. and E. Zemel (1976), "Solving Large Knapsack Problems," presented at the Joint National Meeting of ORSA/TIMS, Miami, FL.

Balas, E. and E. Zemel (1977), "Critical Cutsets of Graphs and Canonical Facets of Set-Packing Polytopes," *Mathematics of Operations Research,* **2**, 15–19.
Balas, E. and E. Zemel (1978), "Facets of the Knapsack Polytope from Minimal Covers," *SIAM Journal of Applied Mathematics,* **34**, 119–148.
Balas, E. and E. Zemel (1980), "An Algorithm for Large Zero-One Knapsack Problems," *Operations Research,* **28**, 1130–1154.
Balas, E. and E. Zemel (1984), "Lifting and Complementing Yields All the Facets of Positive Zero-One Programming Polytopes," in: *Mathematical Programming,* (R. W. Cottle, M. L. Kelmanson, and B. Korte, eds.), North Holland, Amsterdam, 13–24.
Balinski, M. L. (1961), "An Algorithm for Finding All Vertices of Convex Polyhedral Sets," *Journal of the Society for Industrial and Applied Mathematics,* **9**, 72–88.
Balinski, M. L. (1965), "Integer Programming: Methods, Uses, Computation," *Management Science,* **12(A)**, 253–313.
Balinski, M. L. (1967), "Some General Methods in Integer Programming," in: *Nonlinear Programming,* (J. Abadie, ed.), North-Holland, Amsterdam, 221–247.
Balinski, M. L. (1970), "On Recent Developments in Integer Programming," in: *Proceedings of the Princeton Symposium on Mathematical Programming,* (Princeton, 1967; A. W. Kuhn, ed.), Princeton University Press, Princeton, NJ, 267–302.
Balinski, M. L. and K. Spielberg (1969), "Methods for Integer Programming: Algebraic, Combinatorial and Enumerative," in: *Progress in Operations Research, Relationship Between Operations Research and the Computer, Vol. III,* (J. S. Aranofsky, ed.), John Wiley, New York, 195–292.
Barahona, F. and A. R. Mahjoub (1983), "On the Cut Polytope," Institut für Operations Research, Universität Bonn, WP 83271-OR.
Bazaraa, M. S. and J. J. Goode (1979), "Survey of Various Tactics for Generating Lagrangean Multipliers in the Context of Lagrangean Duality," *European Journal of Operations Research,* **3**, 322–338.
Bazaraa, M. S., J. J. Goode, and R. L. Rardin (1978), "An Algorithm for Finding the Shortest Element of a Polyhedral Set with Application to Lagrangean Duality," *Journal of Math, Analysis and Applications,* **65**, 278–288.
Bazaraa, M. S. and J. J. Jarvis (1977), *Linear Programming and Network Flows*, John Wiley, New York.
Bazaraa, M. S. and C. M. Shetty (1979), *Nonlinear Programming: Theory and Algorithms,* John Wiley, New York.
Beale, E. M. L. and R. E. Small (1965), "Mixed Integer Programming by a Branch and Bound Technique," *Proceedings of the IFIP Congress,* Vol. 2, Spartan Press, Washington, D.C.
Beardwood, J., J. H. Halton, and J. M. Hammersly (1959), "The Shortest Path Through Many Points," *Proc. Cambridge Philos. Soc.,* **55**, 299–327.
Benders, J. F. (1962), "Partitioning Procedures for Solving Mixed-Variables Programming Problems," *Numerische Mathematik,* **4**, 238–252.
Berge, C. (1971), *Principles of Combinatorics,* Academic Press, New York.
Berge, C. (1973), *Graphs and Hypergraphs,* North Holland, Amsterdam.
Bixby, R. E. (1976), "A Strengthened Form of Tutte's Characterization of Regular Matroids," *Journal of Combinatorial Theory (B),* **20**, 216–221.
Bixby, R. E. (1977), "Kuratowski's and Wagner's Theorems for Matroids," *Journal of Combinatorial Theory (B),* **22**, 31–53.
Bixby, R. E. (1981), "Matroids and Operations Research," in: *Advanced Techniques in the Practice of Operations Research,* (H. J. Greenberg, F. H. Murphy, and S. H. Shaw, eds.), North Holland, New York, 333–458.
Bixby, R. E. and C. R. Coullard (1986), "On Chains of 3-Connected Matroids," *Discrete Applied Mathematics,* **15**, 155–166.

Bixby, R. E. and W. H. Cunningham (1980), "Converting Linear Programs to Network Problems," *Mathematics of Operations Research,* **5**, 321–357.

Bixby, R. E. and W. H. Cunningham (1984), "Short Cocircuits in Binary Matroids," Report no. 84344-OR, Institute für Okonometrie und Operations Research, Universität Bonn.

Bixby, R. E. and D. K. Wagner (1988), "An Almost Linear-Time Algorithm for Graph Realization," *Mathematics of Operations Research,* **13**, 99–123.

Blair, C. (1978), "Minimal Inequalities for Mixed-Integer Programs," *Discrete Mathematics,* **24**, 147–151.

Blair, C. (1980), "Facial Disjunctive Programs and Sequences of Cutting-Planes," *Discrete Applied Mathematics,* **2**, 173–179.

Blair, C. (1983), "Random Linear Programs with Many Variables and Few Constraints," Faculty Working Paper No. 946, College of Commerce and Business Administration, University of Illinois at Urbana-Champaign, IL.

Blair, C. and R. G. Jeroslow (1977), "The Value Function of a Mixed Integer Program: I," *Discrete Mathematics,* **19**, 121–138.

Blair, C. and R. G. Jeroslow (1979), "The Value Function of a Mixed Integer Program: II," *Discrete Mathematics,* **25**, 7–19.

Blair, C. and R. G. Jeroslow (1982), "The Value Function of An Integer Program," *Mathematical Programming,* **23**, 237–273.

Blair, C. and R. G. Jeroslow (1984), "Constructive Characterizations of the Value Function of a Mixed-Integer Program: I," *Discrete Applied Mathematics,* **9**, 217–233.

Blair, C. and R. G. Jeroslow (1985), "Constructive Characterizations of the Value Function of a Mixed-Integer Program: II," *Discrete Applied Mathematics,* **10**, 227–240.

Bland, R. G. (1977a), "New Finite Pivoting Rules for the Simplex Method," *Mathematics of Operations Research,* **2**, 103–107.

Bland, R. G. (1977b), "A Combinatorial Abstraction of Linear Programming," *Journal of Combinatorial Theory (B),* **23**, 33–57.

Bland, R. G. (1978), "Elementary Vectors and Two Polyhedral Relaxations," *Mathematical Programming Study,* **8**, 159–166.

Bland, R. G., D. Goldfarb, and M. J. Todd (1981), "The Ellipsoid Method: A Survey," *Operations Research,* **29**, 1039–1091.

Bollobás, B. (1979), *Graph Theory,* Springer-Verlag, New York.

Bondy, J. A. and U. S. R. Murty (1976), *Graph Theory with Applications,* American Elsevier, New York.

Borgwardt, K.-H. (1982a), "Some Distribution-Independent Results About the Asymptotic Order of the Average Number of Pivot Steps of the Simplex Method," *Mathematics of Operations Research,* **7**, 441–462.

Borgwardt, K.-H. (1982b), "The Average Number of Pivot Steps Required by the Simplex-Method is Polynomial," *Zeitschrift für O.R.,* **26**, 157–177.

Boruvka, O. (1926), "On a Minimal Problem" *Prace Morawske Predovedecke Spolecrosti,* **3**.

Bowman, V. J. (1972a), "Sensitivity Analysis in Linear Integer Programming," *AIIE Transactions,* **4**, 284–289.

Bowman, V. J. (1972b), "A Structural Comparison of Gomory's Fractional Cutting Planes and Hermitian Basic Solution," *SIAM Journal on Applied Mathematics,* **23**, 460–462.

Bowman, V. J. (1974), "The Structure of Integer Programs Under the Hermitian Normal Form," *Operations Research,* **22**, 1067–1080.

Bowman, V. J. and G. L. Nemhauser (1970), "A Finiteness Proof for Modified Dantzig Cuts in Integer Programming," *Naval Research Logistics Quarterly,* **17**, 309–313.

Bowman, V. J. and G. L. Nemhauser (1971), "Deep Cuts in Integer Programming," *Opsearch,* **8**, 89–111.

Bradley, G. H. (1970), "Equivalent Integer Programs and Canonical Problems," *Management Science,* **17**, 354–366.

Bradley, G. H. (1971), "Transformation of Integer Programs to Knapsack Problems," *Discrete Mathematics,* **1**, 29–45.
Bradley, G. H. and P. N. Wahi (1973), "An Algorithm for Integer Linear Programming: A Combined Algebraic and Enumeration Approach," *Operations Research,* **21**, 45–60.
Burdet, C.-A. (1974), "Generating All the Faces of a Polyhedron," *SIAM Journal on Applied Mathematics,* **26**, 479–489.
Burdet, C.-A. and E. L. Johnson (1974), "A Subadditive Approach to the Group Problem of Integer Programming," *Mathematical Programming Study* **2**, 51–71.
Burdet, C.-A. and E. L. Johnson (1977), "A Subadditive Approach to Solve Linear Integer Programs," in: *Studies in Integer Programming,* (P. L. Hammer et al., eds.), *Annals of Discrete Mathematics,* **1**, 117–143.
Camion, P. (1963), "Caractérisation des Matrices Unimodulaires," *Cahiers du Centre d'Etudes de Recherche Operationelle,* **5**, 181–190.
Camion, P. (1965), "Characterizations of Totally Unimodular Matrices," *Proceedings of the American Mathematical Society,* **16**, 1068–1073.
Camion, P. (1968), "Modules Unimodulaires," *Journal of Combinatorial Theory,* **4**, 301–362.
Chandrasekaran, R. (1969), "Total Unimodularity of Matrices," *SIAM Journal on Applied Mathematics,* **17**, 1032–1034.
Chandrasekaran, R. (1981), "Polynomial Algorithms for Totally Dual Integral Systems and Extension," in: *Studies on Graphs and Discrete Programming* (P. Hansen, ed.), *Annals of Discrete Mathematics,* **11**, 39–51.
Chandrasekaran, R., S. N. Kabadi, and K. G. Murty (1981), "Some NP-Complete Problems in Linear Programming," *Operations Research Letters,* **1**, 101–104.
Chandrasekaran, R. and S. Shirali (1984), "Total Weak Unimodularity: Testing and Applications," *Discrete Mathematics,* **51**, 137–145.
Chandru, V. (1984), "Notes on Karmarkar's New Algorithm for Linear Programming," unpublished notes.
Chandru, V. and B. S. Kochar (1987). "A Class of Algorithms for Linear Programming," Technical Report, School of Industrial Engineering, Purdue University.
Chandru, V. and M. A. Trick (1987), "On the Complexity of Lagrange Multiplier Search," Report CC-87-15, School of Industrial Engineering, Purdue University.
Charnes, A., W. W. Cooper, and A. Henderson (1953), *An Introduction to Linear Programming,* John Wiley and Sons, New York.
Charnes, A., D. Granot, and F. Granot (1977), "On Intersection Cuts in Interval Integer Linear Programming," *Operations Research,* **25**, 352–355.
Christofides, N. (1975), *Graph Theory: An Algorithmic Approach,* Academic Press, London.
Christofides, N. (1976), "Worst-Case Analysis of a New Heuristic for the Traveling Salesman Problem," Technical Report, Graduate School of Industrial Administration, Carnegie-Mellon University, Pittsburgh, PA.
Christofides, N., A. Mingozi, P. Toth, and C. Sandi (1979), *Combinatorial Optimization,* John Wiley, Chichester.
Chvátal, V. (1973), "Edmonds Polytopes and a Hierarchy of Combinatorial Problems," *Discrete Mathematics,* **4**, 305–337.
Chvátal, V. (1975a), "On Certain Polytopes Associated with Graphs," *Journal of Combinatorial Theory (B),* **18**, 138–154.
Chvátal, V. (1975b), "Some Linear Programming Aspects of Combinatorics," in: *Proceedings of the Conference on Algebraic Aspects of Combinatorics* (Toronto, 1975; D. Corneil and E. Mendelsohn, eds.) Congressus Numerantium XIII, Utilitas, Winnipeg, 2–30.
Chvátal, V. (1979), "A Greedy Heuristic for the Set-Covering Problem," *Mathematics of Operations Research,* **4**, 233–235.
Chvátal, V. (1980), "Hard Knapsack Problems," *Operations Research,* **28**, 1402–1411.

Chvátal, V. (1983), *Linear Programming,* W. H. Freeman Co., San Francisco, CA.
Chvátal, V. (1984), "Cutting-Plane Proofs and the Stability Number of a Graph," Report No. 84326, Institut für Okonometrie and Operations Research, Rheinische Friedrich-Wilhelms-Universität, Bonn.
Chvátal, V. (1985), "Cutting Planes in Combinatorics," *European Journal of Combinatorics,* **6**, 217–226.
Chvátal, V. and P. L. Hammer (1977), "Aggregation of Inequalities in Integer Programming," in: *Studies in Integer Programming* (P. L. Hammer et al., eds.), *Annals of Discrete Mathematics,* **1**, 145–162.
Cobham, A. (1965), "The Intrinsic Computational Difficulty of Functions," in: *Logic, Methodology and Philosophy of Science,* (Proc. Intern. Congress 1964; Y. Bar-Hillel, ed.), North Holland, Amsterdam, 24–30.
Cook, S. A. (1971), "The Complexity of Theorem-Proving Procedures," *Proceedings 3rd Annual ACM Symposium on Theory of Computing,* Association for Computing Machinery, New York, 151–158.
Cook, W. (1983), "Operations That Preserve Total Dual Integrality," *Operations Research Letters,* **2**, 31–35.
Cook, W. (1986), "On Box Totally Dual Integral Polyhedra," *Mathematical Programming,* **34**, 48–61.
Cook, W., C. R. Coullard, and G. Turán (1987), "On the Complexity of Cutting–Plane Proofs," *Discrete Applied Mathematics,* **18,** 25–38.
Cook, W., J. Fonlupt, and A. Schrijver (1986), "An Integer Analogue of Carathéodory's Theorem," *Journal of Combinatorial Theory (B),* **40**, 63–70.
Cook, W., A. M. H. Gerards, A. Schrijver, and E. Tardos (1986), "Sensitivity Theorems in Integer Linear Programming," *Mathematical Programming,* **34**, 251–264.
Cook, W., L. Lovász, and A. Schrijver (1984), "A Polynomial-Time Test for Total Dual Integrality in Fixed Dimension," *Mathematical Programming Study,* **22**, 64–69.
Cottle, R. W. and G. B. Dantzig (1968), "Complementary Pivot Theory of Mathematical Programming," *Linear Algebra and Its Applications,* **1**, 103–125.
Crowder, H. P. and E. L. Johnson (1973), "Use of Cyclic Group Methods in Branch-and-Bound," in: *Mathematical Programming* (T. C. Hu and S. M. Robinson, eds), Academic Press, New York, 213–226.
Crowder, H., E. L. Johnson, and M. W. Padberg (1983), "Solving Large-Scale Zero–One Linear Programming Problems," *Operations Research,* **31**, 803–834.
Crowder, H. and M. W. Padberg (1980), "Solving Large-Scale Symmetric Travelling Salesman Problems to Optimality," *Management Science,* **26**, 495–509.
Cullen, F. H., J. J. Jarvis, and H. D. Ratliff (1981), "Set Partitioning Based Heuristics for Interactive Routing," *Networks,* **11**, 125–143.
Cunningham, W. H. (1982), "Separating Cocircuits in Binary Matroids," *Linear Algebra and Its Applications,* **43**, 69–86.
Cunningham, W. H. (1987), personal communication.
Cunningham, W. H. and J. Edmonds (1980), "A Combinatorial Decomposition Theory," *Canadian Journal of Mathematics,* **32**, 734–765.
Cunningham, W. H. and H. B. Marsh, III (1978). "A Primal Algorithm for Optimum Matching," *Mathematical Programing Study* **8**, 50–72.
Dahlhaus, E., D. S. Johnson, C. H. Papadimitriou, P. Seymour, and M. Yannakakis (1984), "The Complexity of Multiway Cuts," AT&T Bell Laboratories, Murray Hill, New Jersey.
Dakin, R. J. (1965), "A Tree-Search Algorithm for Mixed Integer Programming Problems," *The Computer Journal,* **8**, 250–255.
Dannenbring, D. G. (1977), "Procedures for Estimating Optimal Solution Values for Large Combinatorial Problems," *Management Science,* **23**, 1273–1283.

Dantzig, G. B. (1951), "Maximization of a Linear Function of Variables Subject to Linear Inequalities," in: *Activity Analysis of Production and Allocation* (T. C. Koopmans, ed.), John Wiley & Sons, New York, 339–347.
Dantzig, G. B. (1957), "Discrete-Variable Extremum Problems," *Operations Research,* **5**, 266–277.
Dantzig, G. B. (1959), "Note on Solving Linear Programs in Integers," *Naval Research Logistics Quarterly,* **6**, 75–76.
Dantzig, G. B. (1960), "On the Significance of Solving Linear Programming Problems With Some Integer Variables," *Econometrica,* **28**, 30–44.
Dantzig, G. B. (1963), *Linear Programming and Extensions,* Princeton University Press, Princeton, NJ.
Dantzig, G. B. and B. C. Eaves (1973), "Fourier-Motzkin Elimination and Its Dual," *Journal of Combinatorial Theory (A),* **14**, 288–297.
Dantzig, G. B., R. Fulkerson, and S. Johnson (1954), "Solution of a Large-Scale Traveling Salesman Problem," *Operations Research,* **2**, 393–410.
Dantzig, G. B., D. R. Fulkerson, and S. M. Johnson (1959), "On A Linear-Programming Combinatorial Approach to the Traveling-Salesman Problem," *Operations Research,* **7**, 58–66.
Danzer, L. and V. Klee (1967), "Lengths of Snakes in Boxes," *Journal of Combinatorial Theory,* **2**, 258–265.
Dilworth, R. P. (1944), "Dependence Relations in a Semi-Modular Lattice," *Duke Mathematical Journal,* **11**, 575–587.
Dilworth, R. P. (1950), "A Decomposition Theorem for Partially Ordered Sets," *Annals of Mathematics (2),* **51**, 161–166.
Dobson, G. (1982), "Worst-Case Analysis of Greedy Heuristics for Integer Programming with Nonnegative Data," *Mathematics of Operations Research,* **7**, 515–531.
Driebeek, N. J. (1966), "An Algorithm for the Solution of Mixed Integer Programming Problems," *Management Science,* **12**, 576–587.
Dyer, M. E. (1980), "Calculation Surrogate Constraints," *Mathematical Programming,* **19**, 255–278.
Dyer, M. E. (1983), "The Complexity of Vertex Enumeration Methods," *Mathematics of Operations Research,* **8**, 381–402.
Dyer, M. E. and L. G. Proll (1977), "An Algorithm for Determining All Extreme Points of a Convex Polytope," *Mathematical Programming,* **12**, 81–96.
Edmonds, J. (1965a), "Paths, Trees, and Flowers," *Canadian Journal of Mathematics,* **17**, 449–467.
Edmonds, J. (1965b), "Minimum Partition of a Matroid Into Independent Subsets," *Journal of Research of the National Bureau of Standards (B),* **69**, 67–72.
Edmonds, J. (1965c), "Maximum Matching and a Polyhedron With 0,1-Vertices," *Journal of Research of the National Bureau of Standards (B),* **69**, 125–130.
Edmonds, J. (1967a), "Optimum Branchings," *Journal of Research of the National Bureau of Standards (B),* **71**, 233–240.
Edmonds, J. (1967b), "Systems of Distinct Representatives and Linear Algebra," *Journal of Research of the National Bureau of Standards (B),* **71**, 241–245.
Edmonds, J. (1970), "Submodular Functions, Matroids and Certain Polyhedra," in: *Combinatorial Structures and Their Applications, Proceedings of the Calgary International Conference,* (R. Guy, ed.), Gordon and Breach, New York, 69–87.
Edmonds, J. (1973), "Edge-Disjoint Branchings," in: *Combinatorial Algorithms* (Courant Computer Science Symposium, Monterey, Cal., 1972; R. Rustin, ed.), Academic Press, New York, 91–96.
Edmonds, J. and R. Giles (1977), "A Min-Max Relation for Submodular Functions on

Graphs," in: *Studies in Integer Programming* (P. L. Hammer et al., eds.), *Annals of Discrete Mathematics,* **1**, 184–204.

Edmonds, J. and R. Giles (1984), "Total Dual Integrality of Linear Inequality Systems," in: *Progress in Combinatorial Optimization* (Jubilee Conference, University of Waterloo, Waterloo, Ontario, 1982; W. R. Pulleyblank, ed.), Academic Press, Toronto, 117–129.

Edmonds, J. and E. L. Johnson (1973), "Matching, Euler Tours and the Chinese Postman," *Mathematical Programming,* **5**, 88–124.

Edmonds, J., L. Lovász, and W. R. Pulleyblank (1982), "Brick Decompositions and the Matching Rank of Graphs," *Combinatorica,* **2**, 247–274.

Egerváry, E. (1931), "Matrixok kombinatorius tulajdonságairol," (Hungarian) ["On Combinatorial Properties of Matrices"], *Matematikai és Fizikai Lapok,* **38**, 16–28.

Ellwein, L. B. (1974), "A Flexible Enumeration Scheme for Zero-One Programming," *Operations Research,* **22**, 144–150.

Euler, L. (1736), "Solutio Problematis ad Geometriam Situs Pertinentis," *Commentarii Academiae Petropolitanae* **8**, 128–140.

Fisher, M. L. (1981), "The Lagrangian Method for Solving Integer Programming Problems," *Management Science,* **27**, 1–18.

Fisher, M. L., W. D. Northup, and J. F. Shapiro (1975), "Using Duality to Solve Discrete Optimization Problems: Theory and Computational Experience," *Mathematical Programming Study,* **3**, 56–94.

Fisher, M. L., and J. F. Shapiro (1974), "Constructive Duality in Integer Programming," *SIAM Journal on Applied Mathematics,* **27**, 31–52.

Fleischmann, B. (1967), "Computational Experience with the Algorithm of Balas," *Operations Research,* **15**, 153–155.

Fleischner, H. (1974), "On Spanning Subgraphs of a Connected Bridgeless Graph and Their Application to DT-Graphs," *Journal of Combinatorial Theory (B),* **16**, 17–28.

Ford, L. R. Jr., and D. R. Fulkerson (1956), "Maximal Flow Through a Network," *Canadian Journal of Mathematics,* **8**, 399–404.

Ford, L. R. Jr., and D. R. Fulkerson (1962), *Flows in Networks,* Princeton University Press, Princeton, NJ.

Forrest, J. J. H, J. P. H. Hirst, and J. A. Tomlin (1974), "Practical Solution of Large Mixed Integer Programming Problems with UMPIRE," *Management Science,* **20,** 736–773.

Frank, A. and E. Tardos (1984), "Matroids From Crossing Families," in: *Finite and Infinite Sets (Vol. I)* (Proceedings Sixth Hungarian Combinatorial Colloquium, Eger, 1981; A. Hajnal, L. Lovász, and V. T. Sós, eds.), North Holland, Amsterdam, 295–304.

Frank, A. and E. Tardos (1985), "An Application of Simultaneous Approximation in Combinatorial Optimization," in: *26th Annual Symposium on Foundations of Computer Science, IEEE,* New York, 459–463.

Fulkerson, D. R. (1968), "Networks, Frames, Blocking Systems," in: *Mathematics of the Decision Sciences, Part 1,* (G. B. Dantzig and A. F. Veinott, Jr., eds.), Lectures in Applied Mathematics Vol. 11, American Mathematical Society, Providence, RI, 303–334.

Fulkerson, D. R. (1970a), "Blocking Polyhedra," in: *Graph Theory and Its Applications* (B. Haris, ed.), Academic Press, New York, 93–112.

Fulkerson, D. R. (1970b), "The Perfect Graph Conjecture and Pluperfect Graph Theorem," in: *Proceedings of the Second Chapel Hill Conference on Combinatorial Mathematics and Its Applications* (1970; R. C. Bose et al., eds.), University of North Carolina, Chapel Hill, NC, 171–175.

Fulkerson, D. R. (1971), "Blocking and Anti-blocking Pairs of Polyhedra," *Mathematical Programming,* **1**, 168–194.

Fulkerson, D. R. (1972), "Anti-blocking Polyhedra," *Journal of Combinatorial Theory (B),* **12**, 50–71.

Fulkerson, D. R. (1973), "On the Perfect Graph Theorem," in: *Mathematical Programming* (T. C. Hu and S. M. Robinson, eds.), Academic Press, New York, 69–76.

Fulkerson, D. R., A. J. Hoffman, and R. Oppenheim (1974), "On Balanced Matrices," *Mathematical Programming Study,* **1**, 120–132.

Gács, P. and Lovász (1981), "Khachiyan's Algorithm for Linear Programming," *Mathematical Programming Study,* **14**, 61–68.

Garey, M. R. and D. S. Johnson (1976), "The Complexity of Near-Optimal Graph Coloring," *J. Assoc. Comput. Mach.,* **23**, 43–49.

Garey, M. R. and D. S. Johnson (1979), *Computers and Intractability: A Guide to the Theory of NP-Completeness,* W. H. Freeman, San Francisco.

Garfinkel, R. S. (1979), "Branch and Bound Methods for Integer Programming," in: *Combinatorial Optimization* (N. Christofides, A. Mingozzi, P. Toth, and C. Sandi, eds.), Wiley, Chichester, 1–20.

Garfinkel, R. S. and G. L. Nemhauser (1972), *Integer Programming,* Wiley, New York.

Garfinkel, R. S. and G. L. Nemhauser (1973), "A Survey of Integer Programming Emphasizing Computation and Relations Among Models," in: *Mathematical Programming* (Proceedings Advanced Seminar, Madison, Wis., 1972; T. C. Hu and S. M. Robinson, eds.), Academic Press, New York, 77–155.

Gass, S. I. (1975), *Linear Programming Methods and Applications,* McGraw Hill, New York.

Gavish, B. (1978), "On Obtaining the 'Best' Multipliers for a Lagrangean Relaxation for Integer Programming," *Computers and Operations Research,* **5**, 55–71.

Geoffrion, A. M. (1967), "Integer Programming by Implicit Enumeration and Balas' Method," *SIAM Review,* **9**, 178–190.

Geoffrion, A. M. (1969), "An Improved Implicit Enumeration Approach for Integer Programming," *Operations Research,* **17**, 437–454.

Geoffrion, A. M. (1972), "Generalized Benders Decomposition," *Journal of Optimization Theory and Applications,* **10**, 237–260.

Geoffrion, A. M. (1974), "Lagrangean Relaxation for Integer Programming," *Mathematical Programming Study,* **2**, 82–114.

Geoffrion, A. M. (1976), "A Guided Tour of Recent Practical Advances in Integer Linear Programming," *Omega,* **4**, 49–57.

Geoffrion, A. M. and R. E. Marsten (1971-2), "Integer Programming Algorithms: A Framework and State-of-the Art Survey," *Management Science,* **18**, 465–491.

Geoffrion, A. M. and R. Nauss (1976-77), "Parametric and Postoptimality Analysis in Integer Linear Programming," *Management Science,* **23**, 453–466.

Gerards, A. M. H. and A. Schrijver (1985), "Matrices with the Edmonds-Johnson Property," Report No. 85363-OR, Institut für Okonometrie und Operations Research, Universität Bonn.

Giles, R. and J. B. Orlin (1981), "Verifying Total Dual Integrality," manuscript.

Giles, F. R. and W. R. Pulleyblank (1979), "Total Dual Integrality and Integer Polyhedra," *Linear Algebra and Its Applications,* **25**, 191–196.

Giles, R. and L. E. Trotter, Jr. (1981), "On Stable Set Polyhedra for $K_{1,3}$-Free Graphs," *Journal of Combinatorial Theory (B),* **31**, 313–326.

Gill, P. E., W. Murray, and M. H. Wright (1981), *Practical Optimization,* Academic Press, London.

Gilmore, P. C. (1979), "Cutting Stock, Linear Programming, Knapsacking, Dynamic Programming and Integer Programming, Some Interconnections," in: *Discrete Optimization I* (P. L. Hammer, E. L. Johnson, and B. H. Korte, eds.) *Annals of Discrete Mathematics,* **4**, 217–235.

Gilmore, P. C. and R. E. Gomory (1961), "A Linear Programming Approach to the Cutting Stock Problem," *Operations Research,* **9**, 849–859.

Gilmore, P. C. and R. E. Gomory (1963), "A Linear Programming Approach to the Cutting Stock Problem: Part II," *Operations Research,* **11**, 863–88.

Gilmore, P. C. and R. E. Gomory (1965), "Multistage Cutting Stock Problems of Two and More Dimensions," *Operations Research,* **13**, 94–120.

Gilmore, P. C. and R. E. Gomory (1966), "The Theory and Computation of Knapsack Functions," *Operations Research,* **14**, 1045–1074 [erratum: **15**(1967)366].

Glover, F. (1965), "A Multiphase-Dual Algorithm for the Zero–One Integer Programming Problem," *Operations Research,* **13**, 879–919.

Glover, F. (1966), "Generalized Cuts in Diophantine Programming," *Management Science,* **13**, 254–268.

Glover, F. (1968a), "A New Foundation for a Simplified Primal Integer Programming Algorithm," *Operations Research,* **16**, 727–740.

Glover, F. (1968b), "A Note on Linear Programming and Integer Feasibility," *Operations Research,* **16**, 1212–1216.

Glover, F. (1969), "Integer Programming Over a Finite Additive Group," *SIAM Journal on Control,* **7**, 213–231.

Glover, F. (1970), "Faces of the Gomory Polyhedron," in: *Integer and Nonlinear Programming* (J. Abadie, ed.), North Holland, Amsterdam, 367–379.

Glover, F. (1972), "Cut Search Methods in Integer Programming," *Mathematical Programming,* **3**, 86–100.

Glover, F. (1973), "Convexity Cuts and Cut Search," *Operations Research,* **21**, 123–134.

Glover, F. (1975a), "New Results on Equivalent Integer Programming Formulations," *Mathematical Programming,* **8**, 84–90.

Glover, F. (1975b), "Polyhedral Annexation in Mixed Integer and Combinatorial Programming," *Mathematical Programming,* **9**, 161–188.

Glover, F. and R. E. Woolsey (1972), "Aggregating Diophantine Constraints," *Zeitschrift für Operations Research,* **16**, 1–10.

Glover, F. and S. Zionts (1965), "A Note on the Additive Algorithm of Balas," *Operations Research,* **13**, 546–549.

Gomory, R. E. (1958), "Outline of An Algorithm for Integer Solutions to Linear Programs," *Bulletin of the American Mathematical Society,* **64**, 275–278.

Gomory, R. E. (1960a), "Solving Linear Programming Problems in Integers," in: *Combinatorial Analysis* (R. Bellman and M. Hall, Jr., eds.), Proceedings of Symposia in Applied Mathematics X, AMS, Providence, RI, 211–215.

Gomory, R. E. (1960b), "An Algorithm for the Mixed Integer Problem," RM-2597, RAND Corp.

Gomory, R. E. (1963a), "An Algorithm for Integer Solutions to Linear Programs," in: *Recent Advances in Mathematical Programming* (R. L. Graves and P. Wolfe, eds.), McGraw-Hill, New York, 269–302.

Gomory, R. E. (1963b), "An All-Integer Programming Algorithm," in: *Industrial Scheduling* (J. F. Muth and G. L. Thompson, eds.), Prentice-Hall, Englewood Cliffs, NJ, 193–206.

Gomory, R. E. (1965), "On the Relation Between Integer and Noninteger Solutions to Linear Programs," *Proceedings of the National Academy of Sciences of the United States of America,* **53**, 260–265.

Gomory, R. E. (1967), "Faces of an Integer Polyhedron," *Proceedings of the National Academy of Sciences of the United of America,* **57**, 16–18.

Gomory, R. E. (1969), "Some Polyhedra Related to Combinatorial Problems," *Linear Algebra and Its Applications,* **2**, 451–458.

Gomory, R. E. (1970), "Properties of a Class of Integer Polyhedra," in: *Integer and Nonlinear Programming* (J. Abadie, ed.), North Holland, Amsterdam, 353–365.

Gomory, R. E. and W. J. Baumol (1960), "Integer Programming and Pricing," *Econometrica*, **28**, 521–550.
Gomory, R. E. and A. J. Hoffman (1963), "On the Convergence of an Integer-Programming Process," *Naval Research Logistics Quarterly*, **10**, 121–123.
Gomory, R. E. and T. C. Hu (1961), "Multi-Terminal Network Flows," *SIAM Journal of Applied Mathematics*, **9**, 551–570.
Gomory, R. E. and E. L. Johnson (1972), "Some Continuous Functions Related to Corner Polyhedra," *Mathematical Programming*, **3**, 23–85.
Gomory, R. E. and E. L. Johnson (1973a), "Some Continuous Functions Related to Corner Polyhedra II," *Mathematical Programming*, **3**, 359–389.
Gomory, R. E. and E. L. Johnson (1973b), "The Group Problem and Subadditive Functions," in: *Mathematical Programming* (T. C. Hu and S. M. Robinson, eds.), Academic Press, New York, 157–184.
Gondran, M. and M. Minoux (1984), *Graphs and Algorithms*, John Wiley & Sons, Inc., New York.
Graham, R. L. (1966), "Bounds for Certain Multiprocessing Anomalies," *Bell System Technical Journal*, **45**, 1563–1581.
Graham, R. L. (1969), "Bounds on Multiprocessing Timing Anomalies," *SIAM Journal of Applied Mathmatics*, **17**, 416–429.
Graham, R. L. (1976), "Bounds on the Performance of Scheduling Algorithms," in: *Computer and Job Shop Scheduling Theory* (E. G. Coffman ed.), John Wiley and Sons, New York, 165–227.
Greenberg, H. (1969), "A Dynamic Programming Solution to Integer Linear Programs," *Journal of Mathematical Analysis and Applications*, **26**, 454–459.
Greenberg, H. (1971), *Integer Programming*, Academic Press, New York.
Greenberg, H. (1975), "An Algorithm for Determining Redundant Inequalities and All Solutions to Convex Polyhedra," *Numerische Mathematik*, **24**, 19–26.
Greenberg, H. (1980), "An Algorithm for a Linear Diophantine Equation and a Problem of Frobenius," *Numerische Mathematik*, **34**, 349–352.
Greenberg, H. and R. L. Hegerich (1969-70), "A Branch Search Algorithm for the Knapsack Problem," *Management Science*, **16**, 327–332.
Greenberg, H. J. and W. P. Pierskalla (1970), "Surrogate Mathematical Programming," *Operations Research*, **18**, 924–999.
Griffin, V. (1977), *Polyhedral Polarity*, Doctoral Thesis, University of Waterloo, Waterloo, Ontario.
Griffin, V., J. Aráoz, and J. Edmonds (1982), "Polyhedral Polarity Defined by a General Bilinear Inequality," *Mathematical Programming*, **23**, 117–137.
Grötschel, M. (1980), "On the Symmetric Travelling Salesman Problem: Solution of a 120-City Problem," *Mathematical Programming Study*, **12**, 61–77.
Grötschel, M., L. Lovász, and A. Schrijver (1981), "The Ellipsoid Method and Its Consequences in Combinatorial Optimization," *Combinatorica*, **1**, 169–197 [corrigendum: **4**, 291–295].
Grötschel, M., L. Lovász, and A. Schrijver (1984), "Geometric Methods in Combinatorial Optimization," in: *Progress in Combinatorial Optimization* (Jubilee Conference, University of Waterloo, Waterloo, Ontario, 1982; W. R. Pulleyblank, ed.), Academic Press, Toronto, 167–183.
Grötschel, M., L. Lovász, and A. Schrijver (1986), *The Ellipsoid Method and Combinatorial Optimization*, Springer, Heidelberg, to appear.
Grötschel, M. and M. W. Padberg (1979a), "On the Symmetric Travelling Salesman Problem I: Inequalities," *Mathematical Programming*, **16**, 265–280.

Grötschel, M. and M. W. Padberg (1979b), "On the Symmetric Travelling Salesman Problem II: Lifting Theorems and Facets," *Mathematical Programming,* **16**, 281–302.

Grötschel, M. and M. W. Padberg (1985), "Polyhedral Theory," in: *The Traveling Salesman Problem: A Guided Tour of Combinatorial Optimization* (E. L. Lawler et al., eds.), Wiley, Chichester, 251–305.

Grünbaum, B. (1967), *Convex Polytopes,* Interscience–Wiley, London.

Guignard-Spielberg, M. (1984), "Lagrangean Decomposition: An Improvement over Lagrangean and Surrogate Duals," Wharton School, University of Pennsylvania, November.

Guignard, M. M. and K. Speilberg (1972). "Mixed-Integer Algorithms for the (0,1) Knapsack Problem," *IBM Journal of Research and Development,* **16**, 424–430.

Hadley, G. (1962), *Linear Programming,* Addison-Wesley, Reading, MA.

Haimovich, M. (1983), "The Simplex Method is Very Good!—On the Expected Number of Pivot Steps and Related Properties of Random Linear Programs," preprint.

Haimovich, M. (1984a), "On the Expected Behavior of Variable Dimension Simplex Algorithms," preprint.

Haimovich, M. (1984b), "A Short Proof of Results on the Expected Number of Steps in Dantzig's Self Dual Algorithm," preprint.

Haldi, J. and L. M. Isaacson (1965), "A Computer Code for Integer Solutions to Linear Problems," *Operations Research,* **13**, 946–959.

Hammer, P. L., E. L. Johnson, and U. N. Peled (1975), "Facets of Regular 0–1 Polytopes," *Mathematical Programming,* **8**, 179–206.

Harary, F. (1969), *Graph Theory,* Addison-Wesley, Reading, MA.

Hausmann, D. (ed.) (1978), *Integer Programming and Related Areas, A Classified Bibliography, 1976–1978,* Lecture Notes in Economics and Math. Systems 160, Springer, Berlin.

Hausmann, D., B. Korte, and T. A. Jenkyns (1980), "Worst Case Analysis of Greedy Type Algorithms for Independence Systems," *Mathematical Programming Study,* **12**, 120–131.

Hayes, A. C. and D. G. Larman (1983), "The Vertices of the Knapsack Polytope," *Discrete Applied Mathematics,* **6**, 135–138.

Held, M. and R. M. Karp (1970), "The Traveling Salesman Problem and Minimum Spanning Trees," *Operations Research,* **18**, 1138–1162.

Held, M. and R. M. Karp (1971), "The Traveling Salesman Problem and Minimum Spanning Trees: Part II," *Mathematical Programming,* **1**, 6–25.

Held, M., P. Wolfe, and H. P. Crowder (1974), "Validation of Subgradient Optimization," *Mathematical Programming,* **6**, 62–88.

Heller, I. (1957), "On Linear Systems with Integral Valued Solutions," *Pacific Journal of Mathematics,* **7**, 1351–1364.

Heller, I. (1963), "On Unimodular Sets of Vectors," in: *Recent Advances in Mathematical Programming,* (R. L. Graves and P. Wolfe, eds.), McGraw-Hill, New York, 39–53.

Heller, I. and A. J. Hoffman (1962), "On Unimodiular Matrices," *Pacific Journal of Mathematics,* **12**, 1321–1327.

Hochbaum, D. S. (1982), "Approximation Algorithms for the Set Covering and Vertex Cover Problems," *SIAM Journal on Computing,* **11**, 555–556.

Hochbaum, D. S. and D. B. Shmoys (1985), "A Best Possible Heuristic for the *k*-Center Problem" *Mathematics of Operations Research,* **10**, 180–184.

Hoffman, A. J. (1976), "Total Unimodularity and Combinatorial Theorems," *Linear Algebra and Its Applications,* **13**, 103–108.

Hoffman, A. J. (1979a), "Linear Programming and Combinatorics," in: *Relations Between Combinatorics and Other Parts of Mathematics* (Proceedings of Symposia in Pure Mathematics Vol. XXXIV; D. K. Ray-Chaudhuri, ed.), American Mathematical Society, Providence, RI, 245–253.

Hoffman, A. J. (1979b), "The Role of Unimodularity in Applying Linear Inequalities to Combinatorial Theorems," in: *Discrete Optimization* (P. L. Hammer, E. L. Johnson and B. H. Korte, eds.), *Annals of Discrete Mathematics,* **4**, 73–84.

Hoffman, A. J. and J. B. Kruskal (1956), "Integral Boundary Points of Convex Polyhedra," in: *Linear Inequalities and Related Systems* (H. W. Kuhn and A. W. Tucker, eds.), Princeton University Press, Princeton, NJ, 223–246.

Hoffman, A. J. and R. Oppenheim (1978), "Local Unimodularity in the Matching Polytope," in: *Algorithmic Aspects of Combinatorics* (D. Alspach, P. Hell, and D. J. Miller, eds.), *Annals of Discrete Mathematics* **2**, 201–209.

Hopcroft, J. E. and R. E. Tarjan (1974), "Efficient Planarity Testing," *Journal of the Association of Computing Machinery,* **21**, 549–558.

Hu, T. C. (1969), *Integer Programming and Network Flows,* Addison-Wesley, Reading, MA.

Hu, T. C. (1970a), "On the Asymptotic Integer Algorithm," in: *Integer and Nonlinear Programming* (J. Abadie, ed.), North Holland, Amsterdam, 381–383.

Hu, T. C. (1970b), "On the Asymptotic Integer Algorithm, " *Linear Algebra and Its Applications,* **3**, 279–294.

Hu, T. C. (1982), *Combinatorial Algorithms,* Addison-Wesley, Reading.

Ibaraki, T. (1976a), "Computational Efficiency of Approximate Branch-and-Bound Algorithms," *Mathematics of Operations Research,* **1**, 287–298.

Ibaraki, T. (1976b), "Integer Programming Formulation of Combinatorial Optimization Problems," *Discrete Mathematics,* **16**, 39–52.

Ibarra, O. H. and C. E. Kim (1975), "Fast Approximation Algorithms for the Knapsack and Sum of Subset Problems," *Journal of the Association for Computing Machinery,* **22**, 463–468.

Ingargiola, G. P. and J. F. Korsh (1977), "A General Algorithm for One-Dimensional Knapsack Problems," *Operations Research,* **25**, 752–759.

Jeroslow, R. G. (1971), "Comments on Integer Hulls of Two Linear Constraints," *Operations Research,* **19**, 1061–1069.

Jeroslow, R. G. (1973a), "There Cannot be Any Algorithm for Integer Programming with Quadratic Constraints," *Operations Research,* **21**, 221–224.

Jeroslow, R. G. (1973b), "The Simplex Algorithm With the Pivot Rule of Maximizing Criterion Improvement," *Discrete Mathematics,* **4**, 367–377.

Jeroslow, R. G. (1974), "Trivial Integer Programs Unsolvable by Branch-and-Bound," *Mathematical Programming,* **6**, 105–109.

Jeroslow, R. G. (1975), "A Generalization of a Theorem of Chvátal and Gomory," in: *Nonlinear Programming,* **2**(0. L. Mangasarian, R. R. Meyer, and S. M. Robinson, eds.), Academic Press, New York, 313–331.

Jeroslow, R. G. (1977), "Cutting-Plane Theory: Disjunctive Methods," *Annals of Discrete Mathematics,* **1**, 293–330.

Jeroslow, R. G. (1978a), "Cutting-Plane Theory: Algebraic Methods," *Discrete Mathematics,* **23**, 121–150.

Jeroslow, R. G. (1978b), "Some Basis Theorems for Integral Monoids," *Mathematics of Operations Research,* **3**, 145–154.

Jeroslow, R. G. (1979a), "The Theory of Cutting-Planes," in: *Combinatorial Optimization* (N. Christofides, A. Mingozzi, P. Toth, and C. Sandi, eds.), John Wiley, Chichester, 21–72.

Jeroslow, R. G. (1979b), "An Introduction to the Theory of Cutting-Planes," *Annals of Discrete Mathematics,* **5**, 71–95.

Jeroslow, R. G. (1979c), "Minimal Inequalities," *Mathematical Programming,* **17**, 1–15.

Jeroslow, R. G. (1979d), "Some Relaxation Methods for Linear Inequalities," *Cahiers du Centre d'Etudes de Recherche Opérationnelle,* **21**, 43–53.

Jeroslow, R. G. (1986), "Ten Lectures on Mixed Integer Model Formulation for Logic-Based Decision Support," Presented to the 1st Annual Advanced Research Institute of Discrete Applied Mathematics, Rutgers University, May.

Jeroslow, R. G. and K. O. Kortanek (1971), "On An Algorithm of Gomory," *SIAM Journal of Applied Mathematics,* **21**, 55–60.

Johnson, D. S. (1973), "Near Optimal Bin Packing Algorithms," Ph.D. Thesis, Department of Mathematics, Massachusetts Institute of Technology, Cambridge, MA.

Johnson, D. S. (1974), "Worst Case Behavior of Graph Coloring Algorithms," *Proc. 5th Southeastern Conference on Combinatorics, Graph Theory, and Computing,* Utilitas Mathematica Publishing, Winnipeg, 513–527.

Johnson, D. S., C. H. Papadimitriou, and M. Yannakakis (1986), "How Easy is Local Search?" AT&T Bell Laboratories report, Murry Hill, New Jersey.

Johnson, E. L. (1973), "Cyclic Groups, Cutting Planes, Shortest Paths," in: *Mathematical Programming* (T. C. Hu and S. M. Robinson, eds.), Academic Press, New York, 185–211.

Johnson, E. L. (1974), "On the Group Problem for Mixed Integer Programming," *Mathematical Programming Study,* **2**, 137–179.

Johnson, E. L. (1978), "Support Functions, Blocking Pairs, and Anti-blocking Pairs," *Mathematical Programming Study,* **8**, 167–196.

Johnson, E. L. (1979), "On the Group Problem and a Subaddative Approach to Integer Programming," in: *Discrete Optimization II* (P. L. Hammer, E. L. Johnson, and B. H. Korte, eds.), *Annals of Discrete Mathematics,* **5**, 97–112.

Johnson, E. L. (1980a), *Integer Programming—Facets, Subadditivity, and Duality for Group and Semi-group Problems,* [CBMS-NSF Regional Conference Series in Applied Mathematics **32**] Society for Industrial and Applied Mathematics, Philadelphia, PA.

Johnson, E. L. (1980b), "Subaddative Lifting Methods for Partitioning and Knapsack Problems," *Journal of Algorithms,* **1**, 75–96.

Johnson, E. L. (1981a), "On the Generality of the Subaddative Characterization of Facets," *Mathematics of Operations Research,* **6**, 101–112.

Johnson, E. L. (1981b), "Characterization of Facets for Multiple Right-Hand Choice Linear Programs," *Mathematical Programming Study,* **14**, 112–142.

Johnson, E. L. and U. H. Suhl (1980), "Experiments in Integer Programming," *Discrete Applied Mathematics,* **2**, 39–55.

Kannan, R. (1983a), "Polynomial-Time Aggregation of Integer Programming Problems," *Journal of the Association for Computing Machinery,* **30**, 133–145.

Kannan, R. (1983b), "Improved Algorithms for Integer Programming and Related Lattice Problems," in: *Proceedings of the Fifteenth Annual ACM Symposium on Theory of Computing,* The Association of Computing Machinery, New York, 193–206.

Karmarkar, N. (1984), "A New Polynomial-Time Algorithm for Linear Programming," *Proceedings of the 16th Annual ACM Symposium on Theory of Computing,* 302–311.

Karp, R. M. (1972), "Reducibility Among Combinatorial Problems," in: *Complexity of Computer Computations,* (R. E. Miller and J. W. Thatcher, eds.), Plenum Press, New York, 85–103.

Karp, R. M. (1975), "On the Computational Complexity of Combinatorial Problems," *Networks,* **5**, 45–68.

Karp, R. M. (1976), "The Probabilistic Analysis of Some Combinatorial Search Algorithms," in: J. F. Traub, ed. *Algorithms and Complexity: New Directions and Recent Results* (Academic Press, New York) 1–19.

Karp, R. (1977), "Probabilistic Analysis of Partitioning Algorithms for the Traveling Salesman Problem in the Plane," *Mathematics of Operations Research,* **2**, 209–224.

Karp, R. M. and C. H. Papadimitriou (1982), "On Linear Characterizations of Combinatorial Optimization Problems," *SIAM Journal on Computing,* **11**, 620–632.

Karwan, M. H. and R. L. Rardin (1979), "Some Relationships Between Lagrangean and Surrogate Duality in Integer Programming," *Mathematical Programming*, **17**, 320–334.
Karwan, M. H. and R. L. Rardin (1980), "Searchability of the Composite and Multiple Surrogate Dual Functions," *Operations Research*, **28**, 1251–1257.
Karwan, M. H. R. L. Rardin and S. Sarin (1987), "A New Surrogate Dual Multiplier Search Procedure," *Naval Research Logistics*, **34**, 431–450.
Kendall, K. E. and S. Zionts (1977), "Solving Integer Programming Problems by Aggregating Constraints," *Operations Research*, **25**, 346–351.
Kennington, J. L. and R. V. Helgason (1980), *Algorithms for Network Programming*, John Wiley and Sons, New York.
Kernighan, B. W. and S. Lin (1970), "An Efficient Heuristic Procedure for Partitioning Graphs," *BSTJ*, **49**, 291–307.
Khachian, L. G. (1979), "A Polynomial Algorithm for Linear Programming," *Doklady Akad. Nauk USSR*, **244**, No. 5, 1093–1096. Translated in *Soviet Mathematics Doklady*, **20**, 191–194.
Klee, V. and G. T. Minty (1972), "How Good is the Simplex Algorithm?" in: *Inequalities III* (O. Shisha, ed.), Academic Press, New York, 159–175.
Kolesar, P. J. (1967), "A Branch-and-Bound Algorithm for the Knapsack Problem," *Management Science*, **13**, 723–735.
König, D. (1931), "Graphs and Matrices," (Hungarian), *Mat. Fiz. Lapok*, **38**, 116–119.
Korte, B. and D. Hausmann (1978), "An Analysis of the Greedy Heuristic for Independence Systems," *Annals of Discrete Mathematics*, **2**, 65–74.
Kruskal, J. B. (1956), "On the Shortest Spanning Subtree of a Graph and the Traveling Salesman Problem," *Proceedings of the American Mathematical Society*, **17**, 48–50.
Kuratowski, K. (1930), "Sur le probleme des Courbes Gauches en Topologie," *Fundamentals of Mathematics*, **15**, 271–283.
Land, A. and S. Powell (1979), "Computer Codes for Problems of Integer Programming," in: *Discrete Optimization II* (P. L. Hammer, E. J. Johnson and B. H. Korte, eds.), *Annals of Discrete Mathematics* **5**, 221–269.
Lau, H. (1980), "Finding a Hamiltonian Cycle in the Square of a Block," Ph.D. Dissertation, McGill University, Montreal, Quebec.
Lawler, E. L. (1975), "Matroid Intersection Algorithms," *Mathematical Programming*, **9**, 31–56.
Lawler E. L. (1976), *Combinatorial Optimization: Networks and Matroids*, Holt, Reinhart and Winston, New York.
Lawler, E. L. (1977), "Fast Approximation Algorithms for Knapsack Problems," in: *18th Annual Symposium on Foundation of Computer Science*, (Providence, R.I.,) IEEE, New York, 206–213.
Lawler, E. L. (1979), "Fast Approximation Algorithms for Knapsack Problems," *Mathematics of Operations Research*, **4**, 339–356.
Lawler, E. L. and M. D. Bell (1966), "A Method for Solving Discrete Optimization Problems," *Operations Research*, **14**, 1098–1112 [errata: **15**, 578].
Lawler, E. L., J. K. Lenstra, A. H. G. Rinnooy Kan, and D. Shmoys (1985), *The Traveling Salesman Problem: A Guided Tour of Combinatorial Optimization*, John Wiley, Chichester.
Lawler, E. L. and D. E. Wood (1966), "Branch-and-Bound Methods: A Survey," *Operations Research*, **14**, 699–719.
Lemke, C. E. and K. Spielberg (1967), "Direct Search Algorithms for Zero-One and Mixed Integer Programming," *Operations Research*, **15**, 892–914.
Lenstra, A. K., H. W. Lenstra, Jr., and L. Lovász (1982), "Factoring Polynomials With Rational Coefficients," *Mathematische Annalen*, **261**, 515–534.

Lenstra, Jr., H. W. (1983), "Integer Programming with a Fixed Number of Variables," *Mathematics of Operations Research,* **8**, 538–548.

Lin, B. W. and R. Rardin (1977), "Development of a Parametric Generating Procedure for Integer Programming Test Problems," *Journal of the Association for Computing Machinery,* **24**, 465–472.

Lin, B. W. and R. L. Rardin (1979), "Controlled Experimental Design for Comparison of Integer Programming Algorithms," *Management Science,* **25**, 1258–1271.

Lin, S. (1965), "Computer Solutions of the Traveling Salesman Problem," *Bell System Tech. Journal,* **44**, 2245–2269.

Lovász, L. (1977), "Certain Duality Principles in Integer Programming," in: *Studies in Integer Programming* (P. L. Hammer et al., eds.), *Annals of Discrete Mathematics,* **1**, 363–374.

Lovász, L. (1978), "The Matroid Matching Problem," *Proceedings of Conference on Algebraic Graph Theory,* Szeged, Hungary.

Lovász, L. (1979a), "Graph Theory and Integer Programming," in: *Discrete Optimization I* (P. L. Hammer, E. L. Johnson, and B. H. Korte, eds.), *Annals of Discrete Mathematics* **4**, 141–158.

Lovász, L. (1979b), *Combinatorial Problems and Exercises,* Akadémiai Kiadó, Budapest and North Holland, Amsterdam.

Lovász, L. (1981), "The Matroid Matching Problem," in: *Algebraic Methods in Graph Theory* (L. Lovász and V. T. Sós, eds.), North Holland, Amsterdam, 495–517.

Luenberger, D. G. (1984), *Linear and Nonlinear Programming,* Addison-Wesley, Reading.

Magazine, M., G. Nemhauser, and L. E. Trotter, Jr. (1975), "When the Greedy Solution Solves a Class of Knapsack Problems," *Operations Research,* **23**, 207–217.

Magnanti, T. G. and R. T. Wong (1984), "Network Design and Transportation Planning: Models and Algorithms," *Transportation Science,* **18**, 1–55.

Marsten, R. E. and T. L. Morin (1977), "A Hybrid Approach to Discrete Mathematical Programming," *Mathematical Programming,* **14**, 21–40.

Martello, S. and P. Toth (1977), "An Upper Bound for the Zero–One Knapsack Problem and a Branch-and-Bound Algorithm," *European Journal of Operational Research,* **1**, 169–175.

Martello, S. and P. Toth (1978), "Algorithm 37, Algorithm for the Solution of the 0–1 Single Knapsack Problem," *Computing,* **21**, 81–86.

Martello, S. and P. Toth (1979), "The 0–1 Knapsack Problem," in: *Combinatorial Optimization,* (N. Christofides, A. Mingozzi, P. Toth, and C. Sandi, eds.), Wiley, Chichester, 237–279.

Martin, R. K. (1987), "Generating Alternative Mixed-Integer Linear Programming Models Using Variable Redefinition," *Operations Research,* forthcoming.

Martin, R. K., R. L. Rardin, and B. Campbell (1987), "Polyhedral Characterization of Discrete Dynamic Programming," Report CC-87-24 School of Industrial Engineering Report Series, Purdue University.

Megiddo, N. (1982), "Linear-time Algorithms for Linear Programming in $\mathbb{R}^3$ and Related Problems," in: *23rd Annual Symposium on Foundations of Computer Science,* IEEE, New York, 329–338 [also: *SIAM Journal on Computing,* **12** (1983), 759–779].

Megiddo, N. (1983), "Towards a Genuinely Polynomial Algorithm for Linear Programming" *SIAM Journal on Computing,* **12**, 347–353.

Megiddo, N. (1984), "Linear Programming in Linear Time when the Dimension is Fixed," *Journal of the Association for Computing Machinery,* **31**, 114–127.

Megiddo, N. (1986), "Improved Asymptotic Analysis of the Average Number of Steps Performed by the Self-dual Simplex Algorithm," *Mathematical Programming,* **35**, 140–172.

Meyer, R. R. (1974), "On the Existence of Optimal Solutions to Integer and Mixed Integer Programming Problems," *Mathematical Programming,* **7**, 223–235.

Minieka, E. (1978). *Optimization Algorithms for Networks and Graphs,* Marcel Dekker, New York.

Murty, K. G. (1976), *Linear and Combinatorial Programming,* John Wiley and Sons, New York.

Murty, K. G. (1983), *Linear Programming,* John Wiley and Sons, New York.

Nash-Williams, C. St. J. A. (1964), "Decomposition of Finite Graphs into Forests," *J. London Math. Soc.,* **39**, 12.

Nauss, R. M. (1976), "An Efficient Algorithm for the 0–1 Knapsack Problem," *Management Science,* **23**, 27–31.

Nemhauser, G. L. and Z. Ullman (1968), "A Note on the Generalized Lagrange Multiplier Solution to an Integer Programming Problem," *Operations Research,* **16**, 450–453.

Orlin, J. B. (1982), "A Polynomial Algorithm for Integer Programming Covering Problems Satisfying the Integer Round-up Property," *Mathematical Programming,* **22**, 231–235.

Orlin, J. and J. Vande Vate (1986), "On the Non-Simple Parity Problem," ISyE Report Series J-86-4, Georgia Institute of Technology.

Orlin, J., J. Vande Vate, E. Gugenheim, and J. Hammond (1983), "Linear Matroid Parity Made Almost Easy," presented at the June 1983 Conference of the Society of Industrial and Applied Mathematics, Boston.

Padberg, M. W. (1972), "Equivalent Knapsack-Type Formulations of Bounded Integer Linear Programs: An Alternative Approach," *Naval Research Logistics Quarterly,* **19**, 699–708.

Padberg, M. W. (1973), "On the Facial Structure of Set Packing Polyhedra," *Mathematical Programming,* **5**, 199–215.

Padberg, M. W. (1974), "Perfect Zero–One Matrices," *Mathematical Programming,* **6**, 180–196.

Padberg, M. W. (1975a), "Characterizations of Totally Unimodular, Balanced and Perfect Matrices," in: *Combinatorial Programming: Methods and Applications* (B. Roy, ed.), Reidel, Dordrecht, Holland, 275–284.

Padberg, M. W. (1975b), "A Note on Zero-One Programming," *Operations Research,* **23**, 833–837.

Padberg, M. W. (1976), "A Note on the Total Unimodularity of Matrices," *Discrete Mathematics,* **14**, 273–278.

Padberg, M. W. (1977), "On the Complexity of Set Packing Polyhedra," in: *Studies in Integer Programming* (P. L. Hammer et al., eds.), *Annals of Discrete Mathematics,* **1**, 421–434.

Padberg, M. W. (1979), "Covering, Packing, and Knapsack Problems," in: *Discrete Optimization I* (P. L. Hammer et al., eds.), *Annals of Discrete Mathematics,* **4**, 265–287.

Padberg, M. W. and M. Grötschel (1985), "Polyhedral Computations," in: *The Traveling Salesman Problem, A Guided Tour of Combinatorial Optimization,* (E. L. Lawler et al., eds.), John Wiley, Chichester, 307–360.

Padberg, M. W. and S. Hong (1980), "On the Symmetric Traveling Salesman Problem: A Computational Study," *Mathematical Programming Study,* **12**, 78–107.

Padberg, M. W. and M. R. Rao (1980), "The Russian Method and Integer Programming," GBA Working Paper, New York University, New York.

Pabderg, M. W. and M. R. Rao (1982), "Odd Minimum Cut-Sets and *b*-Matchings," *Mathematics of Operations Research,* **7**, 67–80.

Papadimitriou, C. H. (1981), "On the Complexity of Integer Programming," *Journal of the Association for Computing Machinery,* **28**, 765–768.

Papadimitriou, C. H. and K. Steiglitz (1982), *Combinatorial Optimization; Algorithms and Complexity,* Prentice-Hall, Englewood Cliffs, NJ.

Papadimitriou, C. H. and M. Yannakakis (1982), "The Complexity of Facets (and Some Facets of Complexity)," in: *Proceedings of the Fourteenth Annual ACM Symposium on Theory of Computing* (San Francisco, CA., 1982), The Association for Computing Machinery, New York, 255–260 [also: *Journal of Computer and System Science,* **28**(1984)244–259].

Parker, R. G. and R. L. Rardin (1982a), "An Overview of Complexity Theory in Discrete Optimization: Part I. Concepts," *IIE Transactions,* **14**, 3–10.

Parker, R. G. and R. L. Rardin (1982b), "An Overview of Complexity Theory in Discrete Optimization: Part II. Results and Implications," *IIE Transactions,* **14**, 83–89.

Parker, R. G. and R. L. Rardin (1983), "The Traveling Salesman Problem: An Update of Research," *Naval Research Logistics Quarterly,* **30**, 69–96.

Parker, R. G. and R. Rardin (1984), "Guaranteed Performance Heuristics for the Bottleneck Traveling Salesman Problem," *Operations Research Letters,* **2**, 269–272.

Pearl, J. (1984), *Heuristics: Intelligent Search Strategies for Computer Problem Solving,* Addison-Wesley, Reading, MA.

Picard, J.-C. and M. Queyranne (1977), "On the Integer-Valued Variables in the Linear Vertex Packing Problem," *Mathematical Programming,* **12**, 97–101.

Pilcher, M. G. and R. L. Rardin (1987a), "Invariant Problem Statistics and Generated Data Validation: Symmetric Traveling Salesman Problems," Report CC-87-16, School of Industrial Engineering, Purdue University.

Pilcher, M. G. and R. L. Rardin (1987b), "A Random Cut Generator for Symmetric Traveling Salesman Problems with Known Optimal Solutions," Report CC-87-4, School of Industrial Engineering, Purdue University.

Polyak, B. T. (1967), "A General Method of Solving Extremum Problems," *Soviet Mathematics Doklady,* **8**, 593–597.

Prodon, A., T. M. Lieblin, and H. Gröflin (1985), "Steiner's Problem on Two-Trees," Report R0850315, Départment de Mathématiques, EPF Lausanne, March.

Pulleyblank, W. and J. Edmonds (1974), "Facets of 1-Matching Polyhedra," in: *Hypergraph Seminar,* (Ohio State University, Columbus, Ohio, 1972; C. Berge and D. Ray-Chaudhuri, eds.), Lecture Notes in Mathematics 411, Springer, Berlin, 214–242.

Rardin, R. L. and M. H. Karwan (1984), "Surrogate Dual Multiplier Search Procedures in Integer Programming," *Operations Research,* **32**, 52–69.

Rardin, R. and R. G. Parker (1982), "An Efficient Algorithm for Producing a Hamiltonian Cycle in the Square of a Biconnected Graph," ISyE Report Series, J-82, Georgia Institute of Technology, Atlanta, GA.

Rardin, R. L., C. A. Tovey, and M. G. Pilcher (1987), "Traveling Salesman Problems of Intermediate Complexity," Technical Report, School of Industrial Engineering, Purdue University.

Rardin, R. L. and V. E. Unger (1976a), "Surrogate Constraints and the Strength of Bounds in 0–1 Mixed Integer Programming," *Operations Research,* **24**, 1169–1175.

Rardin, R. and V. E. Unger (1976b), "Solving Fixed Charge Network Problems with Group Theory-Based Penalties," *Naval Research Logistics Quarterly,* **23**, 67–84.

Richey, M. B. and R. G. Parker (1986), "On Multiple Steiner Subgraph Problems," *Networks,* **16**, 423–438.

Richey, M. B. and R. G. Parker (1988), "On Minimum-Maximal Matching in Series-Parallel Graphs," *European Journal of Operations Research,* **33,** 98–105.

Rockafellar, R. T. (1970), *Convex Analysis,* Princeton University Press, Princeton, NJ.

Rosenberg, I. G. (1974), "Aggregation of Equations in Integer Programming," *Discrete Mathematics,* **10**, 325–341.

Rosenberg, I. G. (1975), "On Chvátal's Cutting Planes in Integer Linear Programming," *Mathematische Operationsforschung und Statistik,* **6**, 511–522.

Rubin, D. S. (1970), "On the Unlimited Number of Faces in Integer Hulls of Linear Programs With a Single Constraint," *Operations Research,* **18**, 940–946.

Rubin, D. S. (1971–72), "Redundant Constraints and Extraneous Variables in Integer Programs," *Management Science,* **18**, 423–427.

Rubin, D. S. (1984–85), "Polynomial Algorithms for $m \times (m+1)$ Integer Programs and $m \times (m+k)$ Diophantine Systems," *Operations Research Letters,* **3**, 289–291.

Rubin, D. S. and R. L. Graves (1972), "Strengthened Dantzig Cuts for Integer Programming," *Operations Research,* **20**, 178–182.

Sahni, S. (1975), "Approximate Algorithms for the 0/1 Knapsack Problem," *Journal of the Association of Computing Machinery,* **22**, 115–124.

Sahni, S. and T. Gonzalez (1976), "*P*-Complete Approximation Problems," *Journal of the Association of Computing Machinery,* **23**, 555–565.

Saigal, R. (1983), "On Some Average Results for Linear Complementarity Problems," manuscript, Dept. of Industrial Engineering and Management Sciences, Northwestern University, Evanston, IL.

Sakarovitch, M. (1975), "Quasi-balanced Matrices," *Mathematical Programming,* **8**, 382–386.

Sakarovitch, M. (1976), "Quasi-balanced Matrices—An Addendum," *Mathematical Programming,* **10**, 405–407.

Salkin, H. M. (1971), "A Note on Gomory Fractional Cuts," *Operations Research,* **19**, 1538–1541.

Salkin, H. M. (1973), "A Brief Survey of Algorithms and Recent Results in Integer Programming," *Opsearch,* **10**, 81–123.

Salkin, H. M. (1975), *Integer Programming,* Addison-Wesley, Reading, MA.

Salkin, H. M. and C. A. de Kluyver (1975), "The Knapsack Problem: A Survey," *Naval Research Logistics Quarterly,* **22**, 127–144.

Salkin, H. M. and R. D. Koncal (1973), "Set Covering by an All Integer Algorithm: Computational Experience," *Journal of the Association for Computing Machinery,* **20**, 189–193.

Savage, J. E. (1976), *The Complexity of Computing,* John Wiley & Sons, New York.

Schrader, R. (1982), "Ellipsoid Methods," in: *Modern Applied Mathematics—Optimization and Operations Research* (B. Korte, ed.), North Holland, Amsterdam, 265–311.

Schrader, R. (1983), "The Ellipsoid Method and Its Implication," *OR Spektrum,* **5**, 1–13.

Schrage, L. and L. Wolsey (1985), "Sensitivity Analysis for Branch and Bound Integer Programming," *Operations Research,* **33**, 1008–1023.

Schrijver, A. (1980), "On Cutting Planes," *Annals of Discrete Mathematics,* **9**, 291–296.

Schrijver, A. (1981), "On Total Dual Integrality," *Linear Algebra and Its Applications,* **38**, 27–32.

Schrijver, A. (1982), "Submodular Functions," working paper, Institute voor Actuariaat en Econometri, Universiteit van Amsterdam.

Schrijver, A. (1986), *Theory of Linear and Integer Programming,* John Wiley & Sons, New York.

Schrijver, A. and P. D. Seymour (1977), "A Proof of Total Dual Integrality of Matching Polyhedra," Mathematical Centre report ZN 79/77, Mathematical Centre, Amsterdam.

Seymour, P. D. (1980), "Decomposition of Regular Matroids," *Journal of Combinatorial Theory (B),* **28**, 305–359.

Seymour, P. D. (1981a), "Recognizing Graphic Matroids," *Combinatorica,* **1**, 75–78.

Seymour, P. D. (1981b), "Applications of the Regular Matroid Decomposition," in: *Matroid Theory* (Proceedings Matroid Theory Colloquium, Szeged, 1982; L. Lovász and A. Recski, eds.), North Holland, Amsterdam, 345–357.

Shapiro, J. F. (1968a), "Dynamic Programming Algorithms for the Integer Programming

Problem-I: The Integer Programming Problem Viewed as a Knapsack Type Problem," *Operations Research,* **16**, 103–121.

Shapiro, J. F. (1968b), "Group Theoretic Algorithms for the Integer Programming Problem II: Extension to a General Algorithm," *Operations Research,* **16**, 928–947.

Shapiro, J. F. (1971), "Generalized Lagrange Multipliers in Integer Programming," *Operations Research,* **19**, 68–76.

Shapiro, J. F. (1977), "Sensitivity Analysis in Integer Programming," in: *Studies in Integer Programming* (P. L. Hammer et al., eds.), *Annals of Discrete Mathematics,* **1**, 467–477.

Shapiro, J. F. (1979), "A Survey of Lagrangean Techniques for Discrete Optimization," in: *Discrete Optimization II* (P. L. Hammer, E. L. Johnson, and B. H. Korte, eds.), *Annals of Discrete Mathematics,* **5**, 113–138.

Shapiro, J. F. (1979), *Mathematical Programming: Structures and Algorithms,* John Wiley and Sons, New York.

Sherali, H. D. and C. M. Shetty (1980), *Optimization with Disjunctive Constraints,* Lecture Notes in Economics and Mathematical Systems, No. 181, Springer-Verlag, New York.

Shor, N. Z. (1970a), "Utilization of the Operation of Space Dilatation in the Minimization of Convex Functions," (in Russian), *Kibernetika* (Kiev), **1**, 6–12 [English translation: *Cybernetics,* **6** (1970), 7–15].

Shor, N. Z. (1970b), "Convergence Rate of the Gradient Descent Method with Dilatation of the Space," (in Russian), *Kibernetika* (Kiev), **2**, 80–85 [English translation: *Cybernetics,* **6** (1970), 102–108].

Shor, N. Z. (1977), "Cut-off Method with Space Extension in Convex Programming Problems," (in Russian), *Kibernetika* (Kiev), **1**, 94–95. [English translation: *Cybernetics,* **13** (1977), 94–96].

Simonnard, M. A. (1966), *Linear Programming,* Prentice-Hall, Englewood Cliffs, NJ.

Smale, S. (1983a), "The Problem of the Average Speed of the Simplex Method," in: *Mathematical Programming, The State of the Art—Bonn 1982* (A. Bachem, M. Grötschel, and B. Korte, eds.), Springer, Berlin, 530–539.

Smale, S. (1983b), "On the Average Number of Steps in the Simplex Method of Linear Programming," *Mathematical Programming,* **27**, 241–262.

Spielberg, K. (1979), "Enumerative Methods in Integer Programming," in: *Discrete Optimization II* (P. L. Hammer, E. L. Johnson, and B. H. Korte, eds.), *Annals of Discrete Mathematics,* **5**, 139–183.

Srinivasan, A. V. (1965), "An Investigation of Some Computational Aspects of Integer Programming," *Journal of the Association for Computing Machinery,* **12**, 525–535.

Stone, R. (1985), "Karmarkar's Algorithm," unpublished notes.

Suhl, U. (1977), "An Algorithm and Efficient Data Structures for the Binary Knapsack Problem," Arbeitspapier Nr. **20**/1976, Fachbereich Wirtschaftswissenschaft, Freie Universität Berlin.

Swart, G. (1985), "Finding the Convex Hull Facet by Facet," *Journal of Algorithms,* **6**, 17–48.

Syslo, M. M., N. Deo and J. S. Kowalik (1983), *Discrete Optimization Algorithms with PASCAL Programs,* Prentice-Hall, Englewood Cliffs.

Taha, H. A. (1975), *Integer Programming, Theory, Applications, and Computations,* Academic Press, New York.

Tamin, A. (1976), "On Totally Unimodular Matrices," *Networks,* **6**, 373–382.

Tind, J. (1974), "Blocking and Antiblocking Sets," *Mathematical Programming,* **6**, 157–166.

Tind, J. (1977), "On Antiblocking Sets and Polyhedra," in: *Studies in Integer Programming* (P. L. Hammer et al., eds.), *Annals of Discrete Mathematics,* **1**, 507–515.

Tind, J. (1979), "Blocking and Antiblocking Polyhedra," in: *Discrete Optimization I* (P. L.

Hammer, E. L. Johnson, and B. H. Korte, eds.), *Annals of Discrete Mathematics,* **4**, 159–174.

Tind, J. (1980), "Certain Kinds of Polar Sets and Their Relation to Mathematical Programming," *Mathematical Programming Study,* **12**, 206–213.

Tind, J. and L. A. Wolsey (1981), "An Elementary Survey of General Duality Theory in Mathematical Programming," *Mathematical Programming,* **21**, 241–261.

Tind, J. and L. A. Wolsey (1982),"On the Use of Penumbras in Blocking and Antiblocking Theory," *Mathematical Programming,* **22**, 71–81.

Todd, M. J. (1976), "A Combinatorial Generalization of Polytopes," *Journal of Combinatorial Theory (B),* **20**, 229–242.

Todd, M. J. (1977), "The Number of Necessary Constraints in an Integer Program: A New Proof of Scarf's Theorem," Technical Report 355, School of Operations Research and Industrial Engineering, Cornell University, Ithaca, NY.

Todd, M. J. (1979), "Some Remarks on the Relaxation Method for Linear Inequalities," Tech. Rep. 419, School of Operations Research and Industrial Engineering, Cornell University, Ithaca, NY.

Todd, M. J. (1980), "The Monotonic Bounded Hirsch Conjecture is False for Dimension at Least 4," *Mathematics of Operations Research,* **5**, 599–601.

Todd, M. J. (1982), "On Minimum Volume Ellipsoids Containing Part of a Given Ellipsoid," *Mathematics of Operations Research,* **7**, 253–261.

Todd, M. J. (1986), "Polynomial Expected Behavior of a Pivoting Algorithm for Linear Complementarity and Linear Programming Problems," *Mathematical Programming,* **35**, 173–192.

Tomlin, J. A. (1971), "An Improved Branch-and-Bound Method for Integer Programming," *Operations Research,* **19**, 1070–1075.

Toth, P. (1980), "Dynamic Programming Algorithms for the Zero–one Knapsack Problem," *Computing,* **25**, 29–45.

Tovey, C. A. (1981), "Polynomial Local Improvement Algorithms in Combinatorial Optimization," Ph.D. Dissertation, Stanford University, Palo Alto, CA.

Tovey, C. A. (1983), "On the Number of Iterations of Local Improvement Algorithms," *Operations Research Letters,* **2**, 231–238.

Tovey, C. A. (1985), "Hill Climbing with Multiple Local Optima," *SIAM J. Alg. Disc. Meth.,* **6**, 384–393.

Tovey, C. A. (1986), "Low Order Polynomial Bounds on the Expected Performance of Local Improvement Algorithms," *Mathematical Programming,* **35**, 193–224.

Traub, J. F. and H. Wozniakowski (1981), "Complexity of Linear Programming," *Operations Research Letters,* **1**, 59–62.

Trauth, Jr., C. A. and R. E. Woolsey (1968), "Integer Linear Programming: A Study in Computational Efficiency," *Management Science,* **15**, 481–493.

Trotter, Jr., L. E. (1975), "A Class of Facet Producing Graphs for Vertex Packing Polyhedra," *Discrete Mathematics,* **12**, 373–388.

Trotter, Jr., L. E. and C. M. Shetty (1974), "An Algorithm for the Bounded Variable Integer Programming Problem," *Journal of the Association for Computing Machinery,* **21**, 505–513.

Truemper, K. (1977), "Unimodular Matrices of Flow Problems with Additional Constraints," *Networks,* *7*, 343–358.

Truemper, K. (1978a), "Algebraic Characterizations of Unimodular Matrices," *SIAM Journal on Applied Mathematics,* **35**, 328–332.

Truemper, K. (1978b), "On Balanced Matrices and Tutte's Characterization of Regular Matroids," preprint.

Truemper, K. (1980), "Complement Total Unimodularity," *Linear Algebra and Its Applications,* **30**, 77–92.

Truemper, K. (1982), "On the Efficiency of Representability Tests for Matroids," *European Journal of Combinatorics,* **3**, 275–291.

Truemper, K. and Y. Soun (1979), "Minimal Forbidden Subgraphs of Unimodular Multicommodity Networks," *Mathematics of Operations Research,* **4**, 379–389.

Tutte, W. T. (1958), "A Homotopy Theorem for Matroids I, II," *Transactions of the American Mathematical Society,* **88**, 144–174.

Tutte, W. T. (1959), "Matroids and Graphs," *Transactions of the American Mathematical Society,* **90**, 527–552.

Tutte, W. T. (1965), "Lectures on Matroids," *Journal of Research of the National Bureau of Standards,* (B) **69**, 1–48.

Vaidya, P. (1987), "An Algorithm for L. P. Which Requires $O((m+n)n+(m+n)^{1.5}nL)$ Operations," *Proceedings of ACM STOC.*

Vande Vate, J. (1984), "Linear Matroid Parity," Ph.D. Dissertation, Massachusetts Institute of Technology, Cambridge, MA.

Veinott, A. F. Jr., and G. B. Dantzig (1968), "Integral Extreme Points," *SIAM Review,* **10**, 371–372.

Welsh, D. J. A. (1976), *Matroid Theory,* Academic Press, London.

de Werra, D. (1981), "On Some Characterisations of Totally Unimodular Matrices," *Mathematical Programming,* **20**, 14–21.

Whitney, H. (1935), "On the Abstract Properties of Linear Dependence," *American Journal of Mathematics,* **57**, 509–533.

Williams, H. P. (1976), "Fourier-Motzkin Elimination Extension to Integer Programming Problems," *Journal of Combinatorial Theory (A),* **21**, 118–123.

Williams H. P. (1983), "A Characterisation of all Feasible Solutions to an Integer Program," *Discrete Applied Mathematics,* **5**, 147–155.

Williams, H. P. (1984), "A Duality Theorem for Linear Congruences," *Discrete Applied Mathematics,* **7**, 93–103.

Wolfe, P. (1980), "A Bibliography for the Ellipsoid Algorithm," IBM Research Center Report (July 7).

Wolsey, L. A. (1971a), "Extensions of the Group-Theoretic Approach in Integer Programming," *Management Science,* **18**, 74–83.

Wolsey, L. A. (1971b), "Group-Theoretic Results in Mixed Integer Programming," *Operations Research,* **19**, 1691–1697.

Wolsey, L. A. (1973), "Generalized Dynamic Programming Methods in Integer Programming," *Mathematical Programming,* **4**, 222–232.

Wolsey, L. A. (1974a), "A View of Shortest Route Methods in Integer Programming," *Cahiers du Centre D'Etudes de Recherche Opèrationelle,* **16**, 317–335.

Wolsey, L. A. (1974b), "A Number Theoretic Reformulation and Decomposition Method for Integer Programming," *Discrete Mathematics,* **7**, 393–403.

Wolsey, L. A. (1974c), "Groups, Bounds, and Cuts for Integer Programming Problems," in: *Mathematical Programs for Activity Analysis* (P. van Moeseke, ed.), North Holland, Amsterdam, 73–78.

Wolsey, L. A. (1975), "Faces for a Linear Inequality in 0–1 Variables," *Mathematical Programming,* **8**, 165–178.

Wolsey, L. A. (1976a), "Facets and Strong Valid Inequalities for Integer Programs," *Operations Research,* **24**, 367–372.

Wolsey, L. A. (1976b), "Further Facet Generating Procedures for Vertex Packing Polytopes," *Mathematical Programming,* **11**, 158–163.

Wolsey, L. A. (1977), "Valid Inequalities and Superadditivity for 0–1 Integer Programs," *Mathematics of Operations Research,* **2**, 66–77.

Wolsey, L. A. (1979), "Cutting Plane Methods," in: *Operations Research Support Methodology* (A. G. Holzman, ed.), Marcel Dekker, New York, 441–466.

Wolsey, L. A. (1980), "Heuristic Analysis, Linear Programming and Branch-and-Bound," *Mathematical Programming Study,* **13**, 121–134.

Wolsey, L. A. (1981a), "Integer Programming Duality: Price Functions and Sensitivity Analysis," *Mathematical Programming,* **20**, 173–195.

Wolsey, L. A. (1981b), "The *b*-Hull of an Integer Program," *Discrete Applied Mathematics,* **3**, 193–201.

Yannakakis, M. (1980), "On a Class of Totally Unimodular Matrices," in: *21st Annual Symposium on Foundations of Computer Science,* IEEE, New York, 10–16.

Yannakakis, M. (1985), "On a Class of Totally Unimodular Matrices," *Mathematics of Operations Research,* **10**, 280–304.

Yao, A. C.-C. (1981), "A Lower Bound to Finding Convex Hulls," *Journal of the Association for Computing Machinery,* **28**, 780–787.

Young, R. D. (1965), "A Primal (All-Integer) Integer Programming Algorithm," *Journal of Research of the National Bureau of Standards (B),* **69**, 213–250.

Young, R. D. (1968), "A Simplified Primal (All-Integer) Integer Programming Algorithm," *Operations Research,* **16**, 750–782.

Yudin, D. B. and A. S. Nemirovskii (1976a), "Evaluation of the Informational Complexity of Mathematical Programming Problems," (in Russian), *Ekonomika i Matematicheskie Metody,* **12**, 128–142 [English translation: *Matekon,* **13**(2) (1976–77), 3–25].

Yudin, D. B. and A. S. Nemirovskii (1976b), "Informational Complexity and Efficient Methods for the Solution of Convex Extremal Problems," (in Russian), *Ekonomika i Matematicheskie Metody,* **12**, 357–369 [English translation: *Matekon,* **13**(3) (1977), 25–45].

Zadeh, N. (1973), "A Bad Network Problem for the Simplex Method and Other Minimum Cost Flow Algorithms," *Mathematical Programming,* **5**, 255–266.

Zemel, E. (1978), "Lifting the Facets of Zero–one Polytopes," *Mathematical Programming,* **15**, 268–277.

Zionts, S. C. (1974), *Linear and Integer Programming,* Prentice-Hall, Englewood Cliffs, NJ.

Zoltners, A. A. (1975), "A Direct Descent Binary Knapsack Algorithm," Working Paper 75–31, School of Business Administration, University of Massachusetts, Amherst, MA.

Zoltners, A. A. (1978), "A Direct Descent Binary Knapsack Algorithm," *Journal of the Association for Computing Machinery,* **25**, 304–311.

Index

C

D

M

N

O

P

S